Palaeoproterozoic of India

IUGS/GSL publishing agreement

This volume is published under an agreement between the International Union of Geological Sciences and the Geological Society of London and arises from IGCP 509 on Palaeoproterozoic Supercontinents and Global Evolution.

GSL is the publisher of choice for books related to IUGS activities, and the IUGS receives a royalty for all books published under this agreement.

Books published under this agreement are subject to the Society's standard rigorous proposal and manuscript review procedures.

It is recommended that reference to all or part of this book should be made in one of the following ways:

Mazumder, R. & Saha, D. (eds) 2012. *Palaeoproterozoic of India*. Geological Society, London, Special Publications, **365**.

Chattopadhyay, N., Mukhopadhyay, D. & Sengupta, P. 2012. Reactivation of basement: example from the Anasagar Granite Gneiss Complex, Rajasthan, western India. *In*: Mazumder, R. & Saha, D. (eds) *Palaeoproterozoic of India*. Geological Society, London, Special Publications, **365**, 219–245, http://dx.doi.org/10.1144/SP365.12.

GEOLOGICAL SOCIETY SPECIAL PUBLICATION NO. 365

Palaeoproterozoic of India

EDITED BY

R. MAZUMDER and D. SAHA
Geological Studies Unit, Indian Statistical Institute, Kolkata, India

2012
Published by
The Geological Society
London

THE GEOLOGICAL SOCIETY

The Geological Society of London (GSL) was founded in 1807. It is the oldest national geological society in the world and the largest in Europe. It was incorporated under Royal Charter in 1825 and is Registered Charity 210161.

The Society is the UK national learned and professional society for geology with a worldwide Fellowship (FGS) of over 10 000. The Society has the power to confer Chartered status on suitably qualified Fellows, and about 2000 of the Fellowship carry the title (CGeol). Chartered Geologists may also obtain the equivalent European title, European Geologist (EurGeol). One fifth of the Society's fellowship resides outside the UK. To find out more about the Society, log on to www.geolsoc.org.uk.

The Geological Society Publishing House (Bath, UK) produces the Society's international journals and books, and acts as European distributor for selected publications of the American Association of Petroleum Geologists (AAPG), the Indonesian Petroleum Association (IPA), the Geological Society of America (GSA), the Society for Sedimentary Geology (SEPM) and the Geologists' Association (GA). Joint marketing agreements ensure that GSL Fellows may purchase these societies' publications at a discount. The Society's online bookshop (accessible from www.geolsoc.org.uk) offers secure book purchasing with your credit or debit card.

To find out about joining the Society and benefiting from substantial discounts on publications of GSL and other societies worldwide, consult www.geolsoc.org.uk, or contact the Fellowship Department at: The Geological Society, Burlington House, Piccadilly, London W1J 0BG: Tel. +44 (0)20 7434 9944; Fax +44 (0)20 7439 8975; E-mail: enquiries@geolsoc.org.uk.

For information about the Society's meetings, consult *Events* on www.geolsoc.org.uk. To find out more about the Society's Corporate Affiliates Scheme, write to enquiries@geolsoc.org.uk.

Published by The Geological Society from:
The Geological Society Publishing House, Unit 7, Brassmill Enterprise Centre, Brassmill Lane, Bath BA1 3JN, UK

The Lyell Collection: www.lyellcollection.org
Online bookshop: www.geolsoc.org.uk/bookshop
Orders: Tel. +44 (0)1225 445046, Fax +44 (0)1225 442836

British Library Cataloguing in Publication Data

A catalogue record for this book is available from the British Library.
ISBN 978-1-86239-345-5
ISSN 0305-8719

Distributors
For details of international agents and distributors see:
www.geolsoc.org.uk/agentsdistributors

Typeset by Techset Composition Ltd, Salisbury, UK
Printed by CPI Antony Rowe, Chippenham, Wilts, UK

Contents

MAZUMDER, R. & SAHA, D. Palaeoproterozoic of India: an introduction 1

SAHA, D. & MAZUMDER, R. An overview of the Palaeoproterozoic geology of Peninsular India, and key stratigraphic and tectonic issues 5

MAZUMDER, R., VAN LOON, A. J., MALLIK, L., REDDY, S. M., ARIMA, M., ALTERMANN, W., ERIKSSON, P. G. & DE, S. Mesoarchaean–Palaeoproterozoic stratigraphic record of the Singhbhum crustal province, eastern India: a synthesis 31

MAZUMDER, R., ERIKSSON, P. G., DE, S., BUMBY, A. J. & LENHARDT, N. Palaeoproterozoic sedimentation on the Singhbhum Craton: global context and comparison with Kaapvaal 51

ABBOTT, D. H., MAZUMDER, R. & BREGER, D. Native iron in the Palaeoproterozoic Chaibasa Formation: primary or secondary? 77

SENGUPTA, N. Stability of chloritoid + biotite-bearing assemblages in some metapelites from the Palaeoproterozoic Singhbhum Shear Zone, eastern India and their implications 91

SANYAL, S. & SENGUPTA, P. Metamorphic evolution of the Chotanagpur Granite Gneiss Complex of the East Indian Shield: current status 117

BHATTACHARYA, S., CHAUDHARY, A. K. & BASEI, M. Original nature and source of khondalites in the Eastern Ghats Province, India 147

SAHA, D. & TRIPATHY, V. Palaeoproterozoic sedimentation in the Cuddapah Basin, south India and regional tectonics: a review 161

CHALAPATHI RAO, N. V., WU, F.-Y. & SRINIVAS, M. Mesoproterozoic emplacement and enriched mantle derivation of the Racherla alkali syenite, Palaeo-Mesoproterozoic Cuddapah Basin, southern India: insights from *in situ* Sr–Nd isotopic analysis on apatite 185

JOY, S., JELSMA, H. A., PRESTON, R. F. & KOTA, S. Geology and diamond provenance of the Proterozoic Banganapalle conglomerates, Kurnool Group, India 197

CHATTOPADHYAY, N., MUKHOPADHYAY, D. & SENGUPTA, P. Reactivation of basement: example from the Anasagar Granite Gneiss Complex, Rajasthan, western India 219

SRIVASTAVA, P. Morphodiversity, complexity and macroevolution: revealed by the megascopic life of the Palaeo-Neoproterozoic Vindhyan Supergroup, India 247

SANTOSH, M. India's Palaeoproterozoic legacy 263

Index 289

Palaeoproterozoic of India: an introduction

RAJAT MAZUMDER* & DILIP SAHA

Geological Studies Unit, Indian Statistical Institute, 203 B.T. Road, Kolkata 700108, India

**Corresponding author (e-mail: mazumder.rajat@gmail.com)*

A relatively rapid change in the Earth's surface processes has been anticipated across the Archaean–Palaeoproterozoic boundary as a consequence of changes in the crust–mantle system and tectonic regime (Condie 1989, 1997; Eriksson *et al.* 2004; Reddy & Evans 2009). The Palaeoproterozoic era (2500–1600 Ma: Plumb 1991) represented perhaps the first supercontinent cycle, from the amalgamation and dispersal of a Neoarchaean supercontinent to the formation of the 1.9–1.8 Ga supercontinent Nuna (Reddy & Evans 2009), and encompasses one or more global tectonic event that coincides with fundamental changes in the integrated system of core, mantle, lithosphere, hydrosphere, atmosphere and biosphere (i.e. an integrated Earth System). An integration of seemingly disparate geoscience disciplines is therefore an essential prerequisite to understand these changes (Reddy & Evans 2009); and that was the aim of the UNESCO-IGCP 509 project (2005–2009) on Palaeoproterozoic Supercontinents and Global Evolution.

The fifth and final conference and post-conference field workshop (in the Singhbhum Craton) related to the UNESCO-IGCP 509 project was organized by the Indian Statistical Institute (ISI), Kolkata with financial support from the UNESCO, the ISI, and the Council of Scientific and Industrial Research, Government of India during the period 26 October–3 November 2009. An entire session was devoted to the Palaeoproterozoic geology of India, as the Indian Shield represents a vast repository of the Palaeoproterozoic geological record. While most of the papers presented in this session were essentially on Palaeoproterozoic geology of the Indian Shield, a few were on the geological record during the Archaean–Palaeoproterozoic and Palaeo-Mesoproterozoic transitions. Some clues to the tectonic regime changes during these transitions also add to our understanding of the Palaeoproterozoic Earth processes, and how they connect with earlier or later periods of Earth's evolutionary history. A thematic set of 13 papers on various regional as well as broader issues of the Indian Palaeoproterozoic geology and their global context constitute the present Special Publication. Built over the four major amalgamated Archaean nucleii, the major and minor Palaeoproterozoic sedimentary basins and supracrustal sequences in India contain potential information comparable only in scales to those of North America, Africa, Australia and Brazil. The deformation of these supracrustal sequences, attendant metamorphism and emplacement of plutonic bodies hold important clues to their connection with major orogenies, and facilitate investigations into global correlations. The latter has a direct bearing on refining models of Palaeoproterozoic supercontinent assembly and break-up.

This volume begins with an overview of the Palaeoproterozoic geological records of the Indian Shield by **Saha & Mazumder**, wherein the key stratigraphic and tectonic issues are highlighted. Major Palaeoproterozoic sedimentary systems, according to these authors, developed in an intracratonic to marginal-marine setting. In significant contrast to the presence of banded iron formations (BIFs) in the Archaean of Singhbhum and Dharwar, the Indian Palaeoproterozoic supracrustals are devoid of BIFs. This article also provides a brief overview of the fold belts that straddle the suture between major Archaean nuclei, and the attendant tectonic processes. A comprehensive tectonic model for the amalgamation of the major Archaean nuclei of Dharwar, Bastar, Singhbhum and Aravalli–Bundelkhand, according to these authors, is yet to emerge and is of paramount importance to further our understanding of Palaeoproterozoic supercontinent development. An updated critical analysis of the Mesoarchaean–Palaeoproterozoic stratigraphic record of the Singhbhum Craton and a synthesis thereof has been undertaken by **Mazumder *et al.*** According to these authors, the stratigraphic record of the Singhbhum Craton across the Archaean–Palaeoproterozoic boundary implies a changing tectonic scenario. They have pointed out problematic and controversial regional stratigraphic issues that need to be resolved through further geochronological studies in order to understand the relationship between Singhbhum and other cratonic provinces of India, Australia and South Africa. In a complementary paper, **Mazumder *et al.*** have presented a critical analysis of the Singhbhum Palaeoproterozoic supracrustals, and compared the northern and southern Singhbhum successions, and their possible relationship with global-scale events and those identified on Kaapvaal. **Abbott *et al.*** have reported native Fe, native Si and (Fe, Mn)S

From: Mazumder, R. & Saha, D. (eds) 2012. *Palaeoproterozoic of India*. Geological Society, London, Special Publications, **365**, 1–3. http://dx.doi.org/10.1144/SP365.1

in the shale facies of the Chaibasa Formation, eastern India. The sample has a high magnetic susceptibility. The native Fe, according to the authors, is primary and they have argued for an impact origin of the native Fe. **Sengupta** has discussed the stability of chloritoid+ biotite-bearing assemblages in some metapelites from the Palaeoproterozoic Singhbhum Shear Zone, eastern India and their implications. **Sanyal & Sengupta** have critically reviewed the geological information on the Chotanagpur Granite Gneiss Complex, eastern India. These authors have identified four metamorphic events and correlated them with globally extensive orogenic processes. The earliest metamorphic events, that is recorded in granulite enclaves in the regionally extensive felsic gneisses, according to these authors, culminated in ultrahigh-temperature metamorphism ($T > 900\,^{\circ}\mathrm{C}$ at *c*. 5–8 kbar pressure (P)) at $\geq$1.87 Ga. The proposed metamorphic events have been correlated with the supercontinent cycles in Proterozoic time. The garnet–sillimanite–K-feldspar gneisses (khondalites) form a significant component in the regional granulite terrain of the Eastern Ghats, India. **Bhattacharya *et al*.** have made an effort to infer the source of these khondalites. Their Nd-model dates are indicative of a Palaeoproterozoic provenance, which could be the Eastern Dharwar Craton and/or the Napier Complex of east Antarctica.

The Cuddapah Basin, occurring along the eastern margin of the Dharwar Craton, is the largest of the Proterozoic intracratonic basins of south India. **Saha & Tripathy** have discussed the Palaeoproterozoic sedimentation history of the Cuddapah Basin and its regional tectonic implications. The late Palaeoproterozoic shallow-marine sedimentary sequence of the Nallamalai Fold Belt, according to Saha & Tripathy, was thrust over the Papaghni successions and the Kurnool Group in the western part of the Cuddapah Basin. They have discussed the inversion of the Papaghni sub-basin and the development of regional erosional unconformities within the Cuddapah succession in the context of Palaeoproterozoic and early Mesoproterozoic orogenic events at the SE margin of India. **Chalapathi Rao *et al*.** reported an *in situ* Sm–Nd isochron age of 1326 ± 73 Ma, determined by LA-MC-ICP-MS (laser ablation-multiple collector-inductively coupled plasma-mass spectrometry), on crystalline apatite grains from the Racherla alkali syenite occurring within the Nallamalai Fold Belt at the eastern part of the Palaeo-Mesoproterozoic Cuddapah Basin, southern India. The Banganapalle Quartzite Formation of the Cuddapah Basin is characterized by a basal diamond-bearing conglomerate horizon. De Beers India's exploration efforts have resulted in the discovery of a number of dykes within the basin, with petrographical and geochemical similarities to lamproites. **Joy *et al*.** proposed that the emplacement of lamproites occurred as dyke–sill complexes at 1.4–1.3 Ga and that the lamproites represent the source of the diamonds in the Banganapalle conglomerates. The above two papers have bearing on crust–mantle interactions during the Palaeoproterozoic–Mesoproterozoic transitions.

Chattopadhyay *et al*. have analysed the Anasagar Granite Gneiss of the Aravalli Craton, NW India and the associated supracrustals. Detailed structural analyses coupled with U–Pb dating of zircon undertaken by these authors reveal that the protolith of Anasagar Granite Gneiss was emplaced during the first phase of folding at 1.85 Ga. As suggested by Chattopadhyay *et al*., Anasagar Granite Gneiss and its enveloping supracrustals might represent the basement of the Delhi Supergroup, which was folded, thrust and domed up during the South Delhi Orogeny. According to these authors, the approximately 1.85 Ga event is probably synchronous with the Aravalli Orogeny. **Srivastava** has critically analysed the megascopic life of the Palaeo- to Neoproterozoic Vindhyan Supergroup, northern India. According to Srivastava, these megascopic life forms in the Vindhyan Supergroup document possibly two macroevolutionary modes; while the Palaeo-Mesoproterozoic transitional period (the Semri and Rewa groups) exhibits the domination of diversified prokaryotic cyanobacterial community and moderately diversified megascopic life, the Neoproterozoics (Bhander groups, uppermost Vindhyans) are characterized by diversified and morphologically complex megascopic eukaryotes, the emergence of multicellularity among plant and animal clades, viz. structures resembling bryophytes, sporophyte and Ediacara fauna. In the final paper, **Santosh** provides a critical synthesis of information from recent geological and geophysical studies across the major Palaeoproterozoic belts of Peninsular India in order to identify the various stages of an associated Wilson cycle and to compile the evidence for ocean closure through subduction–accretion–collision tectonics.

It is anticipated that a state-of-the-art exposition on the Indian Palaeoproterozoic may help to identify the key issues, gaps in the existing knowledge base and guide the interested researchers to understand better the Earth processes during the Palaeoproterozoic era.

We are grateful to A. Basu, A. R. Basu, R. Bhutani, T. Bhattacharyya, S. Bhowmik, I. Braun, N. V. Chalapathi Rao, N. Chatterjee, A. Collins, S. Dasgupta, A. Dubey, P. G. Eriksson, J. Ganguly, G. Ghosh, A. K. Gupta, S. Gupta, S. Kumar, A. Mohan, J. Mukhopadhyay, B. Runnegar, P. Sengupta, D. Shome, A. J. van Loon, V. Ravikant and U. Zimmermann for their critical review of the manuscripts. We are also grateful to several colleagues who have provided easy access to their published and/or unpublished works, and for many helpful discussions. The final draft of this Special Publication was completed with infrastructural support from the Indian Statistical Institute.

References

CONDIE, K. C. 1989. *Plate Tectonics and Crustal Evolution*, 3rd edn. Pergamon Press, Elmsford, NY.

CONDIE, K. C. 1997. *Plate Tectonics and Crustal Evolution*, 4th edn. Butterworth–Heinemann, Oxford.

ERIKSSON, P. G., ALTERMANN, W., NELSON, D. R., MUELLER, W. & CATENEAU, O. (eds) 2004. *Tempos and Events in Precambrian Time*. Developments in Precambrian Geology, **12**.

PLUMB, K. A. 1991. New Precambrian time scale. *Episodes*, **14**, 139–140.

REDDY, S. M. & EVANS, D. A. D. 2009. Palaeoproterozoic supercontinents and global evolution: correlations from core to atmosphere. *In*: REDDY, S. M., MAZUMDER, R., EVANS, D. A. D. & COLLINS, A. (eds) *Palaeoproterozoic Supercontinents and Global Evolution*. Geological Society, London, Special Publications, **323**, 1–26.

An overview of the Palaeoproterozoic geology of Peninsular India, and key stratigraphic and tectonic issues

DILIP SAHA* & RAJAT MAZUMDER

Geological Studies Unit and Fluvial Mechanics Laboratory, Indian Statistical Institute, 203 B.T. Road, Kolkata 700108, India

**Corresponding author (e-mail: sahad.geol@gmail.com)*

Abstract: The Archaean–Proterozoic rock successions in India have the potential to enrich the global database on Precambrian stratigraphic development, and to offer valuable clues to global tectonic reconstructions. Built over four distinct Archaean cratonic nuclei, the major Palaeoproterozoic supracrustal belts/cover sequences include the Dhanjori Group, the Singhbhum Group (including the Dalma volcanics and the Chandil Formation in eastern India), the Cuddapah Supergroup and the adjoining Nellore Schist Belt bordering the eastern Dharwar Craton, the Aravalli Supergroup in the Aravalli–Delhi Fold Belt in NW India, the lower Vindhyan (Semri Group) and the Mahakosal/Sausar/Betul belts close to the Central Indian Tectonic Zone (CITZ), and, possibly, the Dongargarh Supergroup in Bastar. A major erosional unconformity separates the Archaean tonalite–trondjhemite gneiss basement in these cratons from the overlying volcano-sedimentary successions. An overview of Palaeoproterozoic stratigraphic development in these cratonic blocks is presented to bring out the salient features for global comparison and to highlight issues requiring further attention. Multiple metamorphic, magmatic and deformation events are recorded in the fold belts at the join of the cratons or their margins, inviting application of the plate tectonic paradigm. However, a comprehensive tectonic model for the amalgamation of the Indian Archaean nuclei is yet to emerge, and is crucial for our understanding of Palaeoproterozoic supercontinent development.

Based primarily on the clustering of geochronological data constraining the timing of tectonothermal events on the Earth, the Archaean–Palaeoproterozoic transition (2600–2400 Ma) is dubbed as a period of global tectonic quiescence (Machado *et al.* 1996; Condie *et al.* 2011; Mazumder *et al.* 2012). The beginning of the Palaeoproterozoic also saw a significant change in the Earth's atmosphere that indirectly influenced sedimentation, including copious development of iron formations in the ancient seas (e.g. Cloud 1973; Klein 2005). Some authors also contemplate a thermal regime change across the transition, controlling the tectonics on Earth, the nature of orogens and the growth of supercontinents (e.g. Brown 2006, 2008). Keeping aside the recycling of older crust by subduction and the attendant bias of geological preservation, such generalizations represent reasonable and fairly robust first-order postulates. However, the database is far from complete, and the sampling is not uniform across the continents or over the geological time period. The Precambrian Indian Shield preserves large tracts of Archaean and Proterozoic rock successions. Its putative association with older supercontinents also makes these geological records important in the global reconstructions during the Precambrian, particularly the Palaeoproterozoic. In an attempt to bridge the above gap in the existing database, a review on the geology of Palaeoproterozoic India is presented here.

Four major Archaean nuclei – namely, the Aravalli–Bundelkhand Craton in northern India, the Singhbhum Craton in eastern India, the Bastar Craton in central India and the Dharwar (including western and eastern Dharwar) Craton in southern India – constitute the old foundation of Peninsular India (Fig. 1). Younger belts and terranes were built on or accreted to this foundation during the subsequent geological ages (Radhakrishna & Naqvi 1986; Ramakrishnan & Vaidyanadhan 2008; Meert *et al.* 2010). These nuclei, composed mainly of tonalite–trondjhemite gneisses (TTG), supracrustals rocks including greenstone belts and intrusive granitic plutons, were possibly amalgamated into a reasonably compact, large continental mass by the end of the Archaean or early Proterozoic. This is evident from the high-grade Proterozoic metamorphic belts along the join of adjacent cratons, such as the Karimnagar and Bhopalpatnam granulites at the NE margin of the Dharwar Craton and the SW margin of the Bastar Craton (e.g. Santosh *et al.* 2004), and the ultra-high-temperature (UHT) granulites along the Central Indian Tectonic Zone (CITZ: Bhowmik *et al.* 2005; Bhowmik 2006) suturing the Bundelkhand Craton in northern India to Bastar and the adjoining blocks of south India.

From: Mazumder, R. & Saha, D. (eds) 2012. *Palaeoproterozoic of India*. Geological Society, London, Special Publications, **365**, 5–29. http://dx.doi.org/10.1144/SP365.2

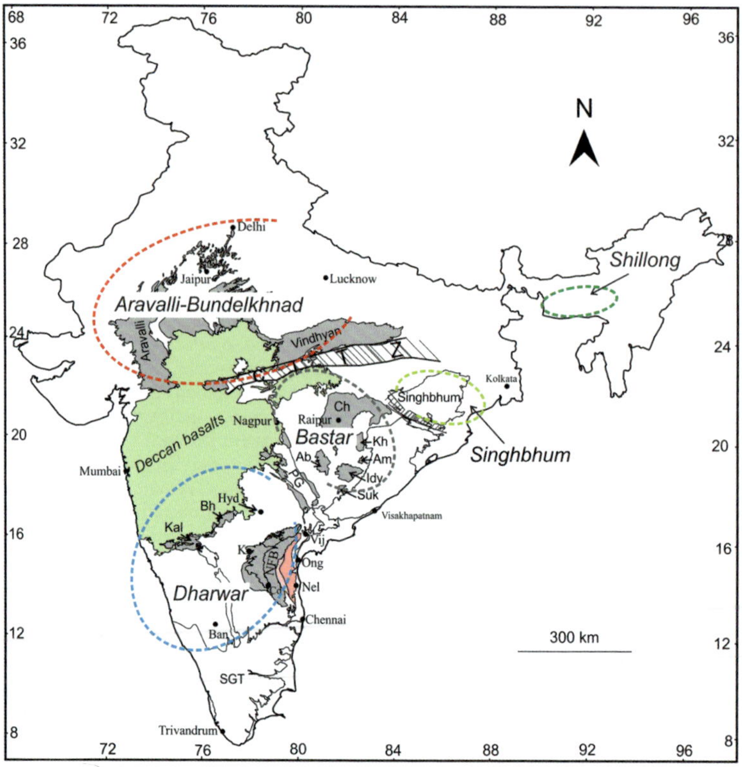

Fig. 1. Sketch map showing four Archaean nuclei in India. Note the Proterozoic sedimentary basins (grey patches), and the large Deccan basaltic province (green). SGT, Southern Granulite Terrain; NFB, Nallamalai Fold Belt; CITZ, Central Indian Tectonic Zone; PG, Pranhita–Godavari valley; Ab, Abujhmar Basin; Am, Ampani Basin; Bh, Bhima Basin; Ch, Chhattisgarh Basin; Idv, Indravati Basin; Kal, Kaladgi Basin; Suk, Sukma Basin; Ban, Bangalore; Cd, Cuddapah; Hyd, Hyderabad; K, Kurnool; Kh, Khariar; Nel, Nellore; Ong, Ongole; Vij, Vijaywada.

Widespread development of Proterozoic cratonic basins (Purana basins of Holland 1909), which host thick, generally undeformed and little metamorphosed sedimentary successions, marks the establishment of a stable continental mass that remained a coherent entity throughout the Phanerozoic. The preserved geological records from the cratonic nuclei and their envelopes suggest commonality, as well as significant variation in development of these blocks since the early Palaeoproterozoic. The Archaean–Proterozoic transition and subsequent development until the Mesoproterozoic is relatively better preserved in Singhbhum (Saha 1994; Mazumder 2005; Mazumder *et al.* 2012), while gaps in preserved record exist in other blocks. Based on published literature on Indian geology and our hitherto unpublished data, we examine here the geological records in these four cratonic blocks, focusing mainly on the Palaeoproterozoic stratigraphic development and correlation – or lack thereof – on a regional scale. On a larger canvas, the tectonics of the Indian subcontinent is largely anchored with Pangaea in the Palaeozoic and Mesozoic, with Gondwana during the late Neoproterozic–Cambrian interval (Collins *et al.* 2008), with Rodinia during the late Mesoproterozoic–Neoproterozoic and, possibly, with Columbia in the Palaeoproterozoic (Rogers & Santosh 2004; see also Reddy & Evans 2009). The broad aspects of tectonics and sedimentation in these cratonic blocks, and the adjoining fold belts, are examined in the light of available data that suggest a connection between India and other contemporary continental fragments, and the marginal seas and oceans (see also Eriksson *et al.* 1999).

In this contribution, we deal with the Aravalli–Bundelkhand Craton first, followed successively by the Bastar, Singhbhum and Dharwar cratons, and very briefly with the Shillong plateau and the adjoining lesser Himalaya. The overall geological framework in each craton is discussed with emphasis on those stratigraphic records and tectonothermal events that are regionally significant, and which provide clues to regional and tentative global correlation. Each of the cratonic blocks hosts Proterozoic sedimentary basins (Fig. 1) with several unconformity-bound stratigraphic successions with aggregate thickness up to 8 km (Radhakrishna 1987; Chaudhuri *et al.* 2002; Ramakrishnan & Vaidyanadhan 2008). The focus is on the Palaeoproterozoic records. However, the older Archaean basement over which these successions were deposited and the younger Proterozoic successions are also discussed to highlight the transitions and broader context. The fold belts adjoining the cratons are also important in the evolutionary history. These are discussed in the context of broad structures of the cratonic blocks, and probable tectonic convergence.

Aravalli–Bundelkhand Craton

Situated north of the CITZ (Sharma 2009), the Aravalli–Bundelkhand Craton is bordered on the west by the Aravalli–Delhi Fold Belt. The Son Valley Vindhyan sedimentary succession now separates the Archaean Banded Gneissic Complex (BGC), including the Berach Granite and the relatively younger Sandmata Complex of the Aravalli region, from the Bundelkhand Granite batholith of central India (Fig. 2a, b). Geophysical data indicate that the Aravalli–Bundelkhand Craton constitutes the basement beneath the Lesser Himalayan successions, and the Quaternary alluvium of the Indo-Gangetic plain (Kailasam 1976). The Palaeoproterozoic Aravalli Fold Belt is generally considered to be separated from the Mesoproterozoic Delhi Fold Belt occurring to the west by a tectonic contact (see the Geological Survey of India map, Gupta 2004). The Delhi Fold Belt is again subdivided into the North and South Delhi fold belts based on the age of the intrusive granites in these belts (Deb *et al.* 2001). A synoptic view of the stratigraphic and tectonic development, including the deformation and magmatism across the craton, is depicted in Figure 3. The main geochronological constraints, based on more recent age data, are shown in Table 1.

Rajasthan sector

West of the Great Boundary Fault (GBF) marking the NW boundary of the Vindhyan Basin, the Aravalli–Delhi Fold Belt hosts the bulk of the volcano-sedimentary successions with multiple deformation, metamorphism, migmatitic gneisses and intrusive granites. The Aravalli Supergroup, ranging in age from 2100 to 1800 Ma (Roy & Kröner 1996; Weidenbeck *et al.* 1996), overlies an Archaean basement (e.g. Mewar Gneissic Complex) that underwent peneplanation prior to the deposition of the basal Aravalli rocks (Roy 2006). A summary of the revised stratigraphy of the Aravalli Supergroup with a threefold subdivision has recently been suggested (table 2 in Roy 2006; see also Sinha-Roy *et al.* 1998; cf. Ramakrishnan & Vaidyanadhan 2008). The lower part of Aravalli Supergroup (Delwara Formation) has a preponderance of meta-lava (meta-basalt), meta-tuff, mixed with dolomites, having a broad similarity with the meta-basalts of the Dhanjori Group in Singhbhum. However, the development of extensive dolomitic limestone and stromatolitic phosphorite is unique to the Jhamarkotra Formation of Udaipur (Roy & Jakhar 2002), and is so far unreported from any other basins representing Palaeoproterozoic sedimentation. Lower–middle parts of the succession are known to represent shallow-shelf sedimentation, while the upper part (Jharol Group: Sinha-Roy *et al.* 1998), occurring to the west of a linear belt of serpentinite (around Rikhabdev), may represent deep-water sedimentation. The early deformation in the Aravalli Supergroup has been linked to the Aravalli Orogeny, culminating in the emplacement of syntectonic (with respect to F_1) granitic plutons such as Darwal Granite or Amet Granite (Sharma 2009). Similarly, the protolith of the Anasagar Granite Gneiss was emplaced within supracrustal metasedimentary rocks (Aravalli Supergroup) at around 1.85 Ga (Mukhopadhyay *et al.* 2000; Chattopadhyay *et al.* 2012).

The available LA-ICP-MS (laser ablation inductively coupled plasma mass spectrometry) zircon U–Pb dates indicate that the Sandmata Complex, occurring east of the Delhi Fold Belt, is comprised of charno-enderbite plutons. Therefore, the sheared granulite-facies metamorphic rocks and supracrustals within the Aravalli belt are distinct from the older basement gneisses, as well as the Mangalwar Complex, as they consist largely of felsic orthogneisses, amphibolites and supracrustal rocks, and the Hindoli Group (Deb *et al.* 2002; Buick *et al.* 2006).

In the North Delhi Fold Belt (NDFB), the Alwar Group (largely arenaceous) and the Ajabgarh Group (largely argillaceous) constitute the Delhi Supergroup, while the status of the Rialo Series (quartzites and marbles: Heron 1953) as an intervening rock unit between the Aravalli and Delhi supergroups has been debated. This rock unit, renamed as the Rayanhalla Group, is included within the Delhi Supergroup in the NDFB (Roy 2006). The intrusive granites within the Delhi Supergroup in NDFB could be as old as 1650 Ma (Bairat, Ajitgarh, Barodia and Harsoria granites; Rb–Sr age: Ramakrishnan & Vaidyanadhan 2008). In the South Delhi Fold Belt, a twofold classification into the Gogunda and Kumbhalgarh groups correspond, respectively, to the Alwar and Ajabgarh groups of

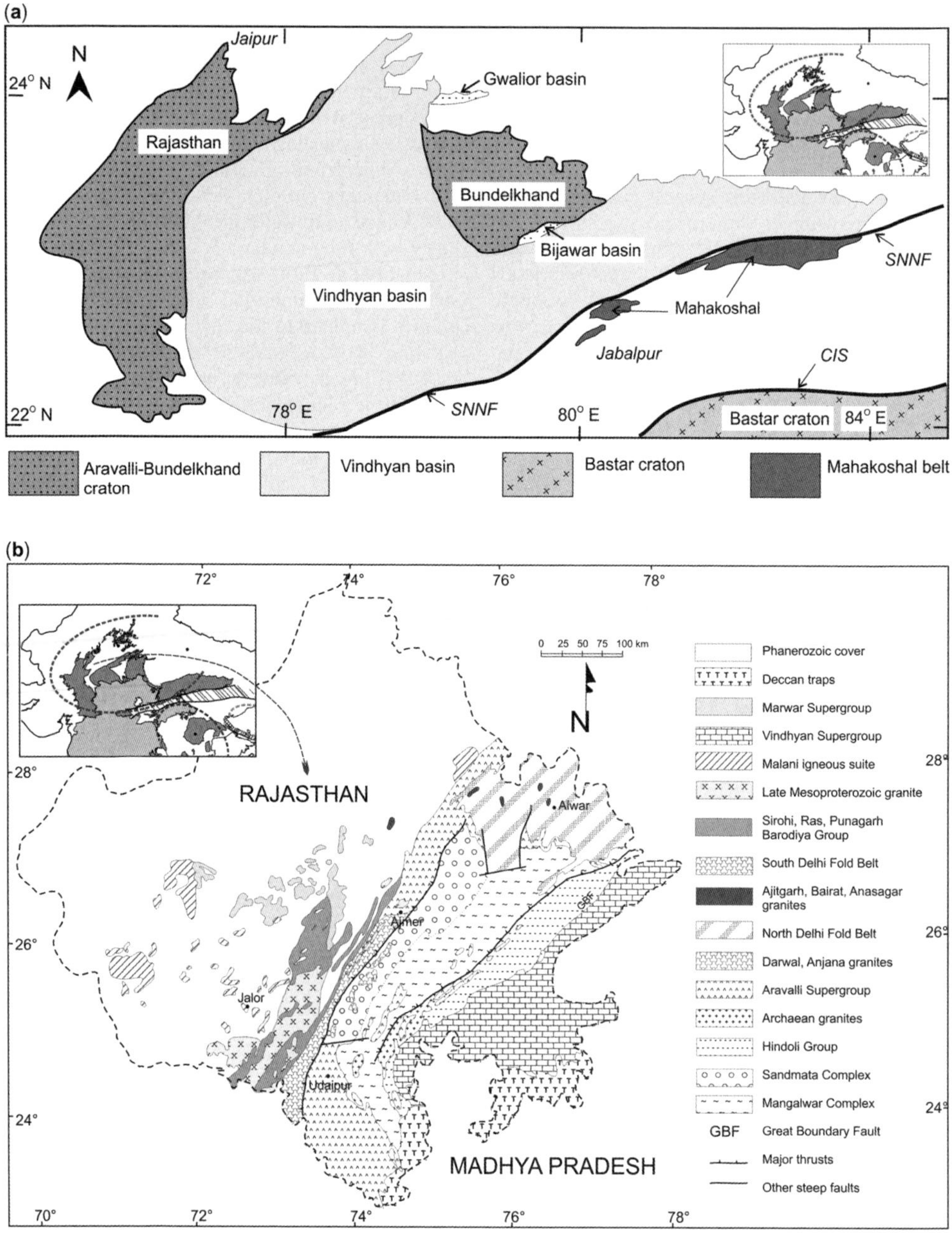

Fig. 2. Sketch map of the Aravalli–Bundelkhand Craton. (**a**) Outline map showing the disposition of the Aravalli–Bundelkhand (AB) Craton with the Vindhyan Basin separating the Rajasthan and Bundelkhand sectors. Son–Narmada North Fault separates the Mahakoshal belt from the AB Craton. The Bastar Craton lies to the south of the Central Indian Suture Zone (CIS). (**b**) Simplified geological map of Rajasthan sector. Incorporates data from the Geological Survey of India (GSI) maps (Gupta *et al.* 1980, 1997; Ramakrishnan & Vaidyanadhan 2008; Sharma 2009).

the NDFB. However, the above classification may represent only fault-bounded tectonostratigraphic units, with the exact stratigraphic relationship still unknown (Mukhopadhyay & Bhattacharyya 2000).

Granitic intrusions in the SDFB, varying in composition from granites to granodiorites/tonalites, show two age clusters. The Ranakpur Diorite in the Phulad Shear Zone (western margin of the SDFB)

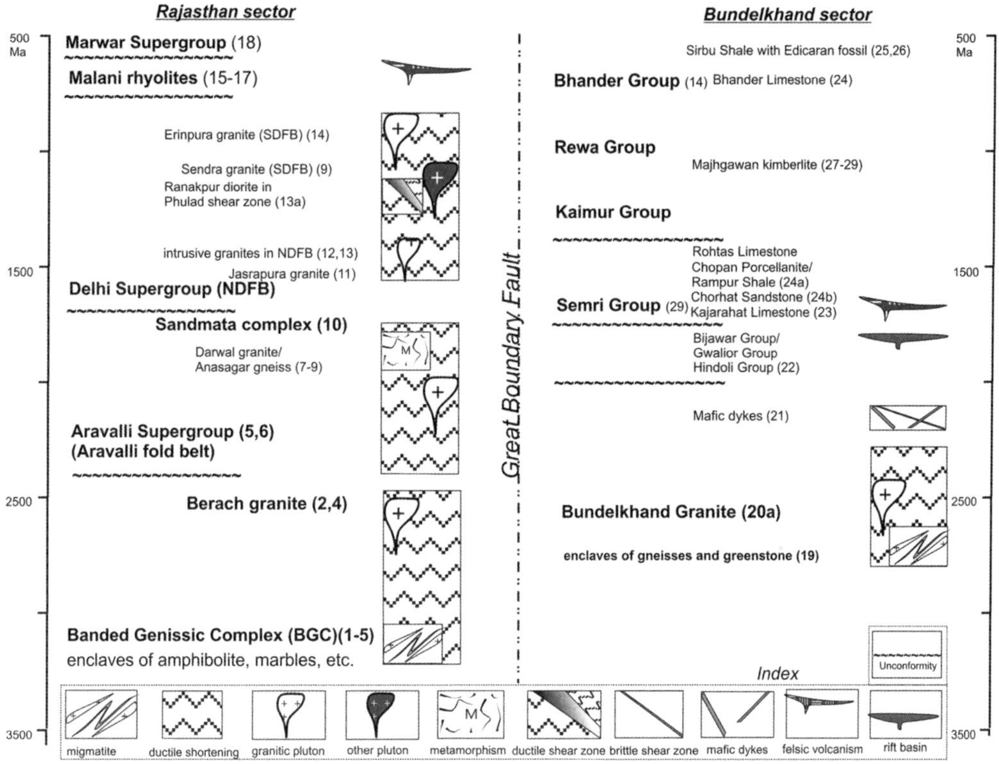

Fig. 3. Synoptic view of tectonostratigraphic development in the Aravalli–Bundelkhand Craton. The numbers refer to references dealing with key gechronological data and are cited in Table 1.

gives an age of 1012 ± 78 Ma (Sm–Nd: Shamim Khan *et al.* 2005), while the Sendra Granite gives an age of around 1000 Ma (Sm–Nd: Tobisch *et al.* 1994). The other suite, including Erinpura granitoids, are slightly younger (860–800 Ma: Malone *et al.* 2008). These granites probably represent the end of the Delhi Orogeny (Sharma 2009). Significant Neoproterozoic events west of the Delhi Fold Belt are represented by the Malani Rhyolites (800–700 Ma: Torsvik *et al.* 2001; Gregory *et al.* 2009) and the unconformably overlying late Precambrian–early Cambrian Marwar Supergroup (Pandit *et al.* 2001).

Bundelkhand sector

North of the Son–Narmada North Fault (SNNF) marking the northern limit of the CITZ, the cratonic block of Bundelkhand is bordered on the west and south by the sedimentary successions of the Vindhyan Basin. The northern part of the craton is covered by alluvium. Two smaller basins attached to the craton, namely the Bijawar Basin in the south and the Gwalior Basin in the north, host low-grade volcano-sedimentary successions older than the Vindhyan successions. The cratonic block consists of the Bundelkhand Granite batholith with associated NE–SW-trending quartz reefs and rare felsic volcanics, and very prominent doleritic dyke swarms with a dominant NW–SE trend (review in Ramakrishna & Vaidyanadhan 2008). The mafic dykes usually cut across the quartz reefs. The granitoid pluton hosts a number of enclaves with older supracrustals and TTG gneisses. The supracrustals consist of banded iron formations (BIF) with amphibolite, ultramafic rock, minor quartzite, metapelite and marble. A mafic enclave from Mahoba has yielded a zircon Pb/Pb age of 3249 ± 5 Ma (Gopalan *et al.* 1990). Four major components have been identified within the Bundelkhnad Granite batholith: (1) hornblende granodirite comparable in geochemical character to Archaean TTG; (2) porphyritic biotite granites; (3) coarse porphyritic granite; and (4) leucogranites, arranged in descending order of relative ages. Pb/Pb single zircon ages of these granitic phases range in age from 2550 to 2400 Ma (Mondal *et al.* 1998, 2002).

Sandwiched between the Bundelkhand Craton and the Vindhyan Group outcrops, the Bijawar Group has an unconformable contact with the

Table 1. *Abridged stratigraphy of the Aravalli–Bundelkhand Craton. Numbers refer to cited works for geochronological constraints**

Rajasthan sector		Bundelkhand sector	
Marwar Supergroup (635–515 Ma[18])			Sirbu Shale (Ediacara Discoids[25,26])
^^^^^^^^^			
Malani Rhyolites (800–700 Ma[15–17])	Upper Vindhyan	Bhander Group[14] (Bhander Limestone 750 Ma[24])	
^^^^^^^^^†		Rewa Group	
			Majhgawan Kimberlite (1075 Ma[27–29])
Erinpura Granite, etc. (860–800 Ma[14])		Kaimur Group	
Sendra Granite (1000–960 Ma[9])		^^^^^^^^^	
Ranakpur Diorite, Phulad Shear Zone, 1012 $\pm$ 78 Ma[13])			Rohtas Limestone
NDFB intrusive granites, 1780–1710 Ma[12,13])			Porcellanites/Rampur Shale (1631 Ma[24])
	Lower Vindhyan		Chorhat Sandstone, 1.6 Ga[24a])
Jasrapura Granite (1821 Ma[11])			Kajrahat Limestone (1720 $\pm$ 90 Ma[23])
Delhi Supergroup (NDFB)		Semri Group	
^^^^^^^^^		^^^^^^^^^	
Sandmata Complex 1720 Ma[10]) (granulite metamorphism)			Bijawar/Gwalior groups
Anasagar Gneiss (1849 $\pm$ 8 Ma[9])			Hindoli Group, 1854 $\pm$ 7 Ma[22])
Darwal Granite (1850 Ma[7,8])			^^^^^^^^^
		mafic dykes (2.15 Ga, 2.0 Ga[21])	
Aravalli Supergroup (2075–2150 Ma, Pb/Pb[5,6])		Bundelkhand Granite (2492 $\pm$ 10 Ma[20a])	
^^^^^^^^^			
Berach Granite (2530 Ma[2,4])			
Banded Gneissic Complex (BGC) (3.3–2.5 Ga[1–5])		Archaean enclaves of gneisses and greenstones[19,20]	

* [1]Gopalan *et al.* (1990); [2]Roy & Kröner (1996); [3]Weidenbeck *et al.* (1996); [4]Meert *et al.* (2010); [5]Deb (1999); [6]Deb & Thorpe (2004); [7]Choudhary *et al.* (1984); [8]Sharma (2009); [9]Tobisch *et al.* (1994); [10]Buick *et al.* (2006); [11]Kaur *et al.* (2009); [12]Kaur *et al.* (2007); [13]Biju-Shekhar *et al.* (2003); [13a]Shamim Khan *et al.* (2005); [14]Malone *et al.* (2008); [15]Torsvik *et al.* (2001); [16]Gregory *et al.* (2009); [17]Van Lente *et al.* (2009); [18]Pandit *et al.* (2001); [19]Sharma & Rahman (2000); [20]Mondal *et al.* (1998); [20a]Mondal *et al.* (2002); [21]Rao (2004); [22]Deb *et al.* (2002); [23]Sarangi *et al.* (2004); [24]Ray *et al.* (2003); [24a]Rasmussen *et al.* (2002); [25]De (2003); [26]De (2006); [27]Ray (2006); [27a]Gregory *et al.* (2006); [28]Gregory (2008); [29]Kumar *et al.* (1993).
†^^^^^^^^^ unconformity.

older granitoids and granite gneisses marked by a basal conglomerate, and an angular unconformity with the supracrustal enclaves within the cratonic block. The contact with the overlying Vindhyan Group is also unconformable (Kumar *et al.* 1990). The Bijawar Group, with an aggregate thickness not exceeding 1 km, consists of thick basalt above the basal conglomerate, dolomite, sandstone, chert-breccia and mafic sill (Pant & Banerjee 1990). Based on the Rb–Sr age of the Dargaon mafic sill and Kurat lava, the approximate time of deposition is constrained to be around 1700–1800 Ma.

The sediments of the Bijawar Group are interpreted to have been deposited in a shallow-water, high-energy near-shore environment within a rift basin (Pant & Banerjee 1990; Ramakrishnan & Vaidyanadhan 2008). The age is roughly similar to the mafic sills within the Papaghni/Chitravati groups and those in the Abujhmar Basin in the Bastar Craton.

In the northern periphery of the Bundelkhand Granite batholith, the east–west-trending outcrops of the Gwalior Group (<1 km thickness) have similar lithological associations to the Bijawar Group. However, the Gwalior Group consists of siliceous and ferruginous shale, chert (banded jasper), limestone and tuffaceous sandstone intercalated with tholeiitic basalt flows and gabbroic sills in the upper part. The Gwalior lavas have a Rb–Sr age similar to those obtained from the Bijawar Group basalts and one set of mafic dykes within the Bundelkhnad Granite (Ramakrishna & Vaidyanadhan 2008).

Vindhyan Basin

The startigraphy and sedimentation within the Vindhyan Basin has been summarized by Chakraborty (2006). Unconformably overlying the Bijawar Group and the Bundelkhand Granite Gneiss, two major sequences have been recognized within the Vindhyan succession with a cumulative thickness of 4.5 km (Bose *et al.* 2001). A major angular unconformity separates the Semri Group (lower Vindhyan) from the upper Vindhyan, consisting of the Kaimur, Rewa and Bhander groups (see a recent review by Chakraborti *et al.* 2010). Available geochronological data (summarized in Ray 2006; see also Gregory *et al.* 2006) suggest that the oldest sediments in the Vindhyan Basin are around 1721 ± 90 Ma (Sarangi *et al.* 2004), and that deposition of the youngest rock unit in the Semri Group was complete roughly around 1600 Ma (Ray *et al.* 2002; Sarangi *et al.* 2004; Chakraborti *et al.* 2007). A major hiatus in deposition followed, and the subsequent deposition was heralded by the emplacement of kimberlite pipes and related lamproites, indicating renewed crust–mantle interaction under the Bundelkhand Craton at around 1075 Ma (Gregory *et al.* 2006).

Deposition in the Semri Group started in a deltaic–fan-delta setting, dominated by siliclasts (Deoland Formation), grading upwards into a peritidal–subtidal carbonate platform (Kheinjua Formation). This carbonate sedimentation was interrupted by a phase of rhyolitic–rhyodacitic intrabasinal volcanism and volcaniclastic deposition (Srivastava *et al.* 2001; Chopan/Deoland Porcellanite: Mishra & Sen 2008). The middle part of the group consists of intercalated sandstone–shale (Koldaha Shale) representing inner–outer shelf deposition, and Chorhat Sandstone representing shallow-marine, shoreface–intertidal deposition (Banerjee & Jeevankumar 2003, 2005). The upper part of the Semri Group, again, shows carbonate sedimentation (Rohtas Limestone) representing a carbonate platform–distally steepened ramp. In the western part of the basin in the Satna district, carbonate deposition is, again, terminated by a thick (30 m) volcaniclastic deposit (review in Chakraborty 2006).

Bastar (Bhandara) Craton

Lying to south of the CITZ, the Bastar Craton (also referred to as the Bastar–Bhandara Craton) is flanked by two younger rifts hosting Upper Palaeozoic–Mesozoic successions, namely the Mahanadi graben in the NE and the Godavari graben in the SW (Fig. 4). While the Satpura mobile (orogenic) belt partly overlapping with the CITZ marks the northern part of the Bastar Craton, the Eastern Ghats Belt was accreted to the SE margin of the craton in the Neoproterozoic–early Palaeozoic period (Biswal *et al.* 2007; Vijaya Kumar & Leelanandam 2008). A synoptic view of the stratigraphic and tectonic development, including deformation and magmatism across the craton, is depicted in Figure 5. The main geochronological constraints, based on more recent age data, are shown in Table 2.

As in the Aravalli–Bundelkhand Craton, the oldest basement consists of 2600–2500 Ma TTG gneisses and greenstone belts (e.g. the Sonakhan belt). Still older crustal remnants are interpreted from tonalitic enclaves that yielded an age of 3651 ± 11 Ma (U–Pb upper intercept age: Ghosh 2004) and gneissic enclaves dated at $3509 + 14/-7$ Ma (Sarkar *et al.* 1993). The TTG gneisses enclose rafts of quartzite–carbonate–pelite and minor BIF. The major BIF-bearing supracrustals occur in the Bailadila Group outcropping in the Bailadila Hills, and are a major source of iron ore in this region. The Meso-Neoarchaean Bailadilla Group has a lithostratigraphy similar to the Iron

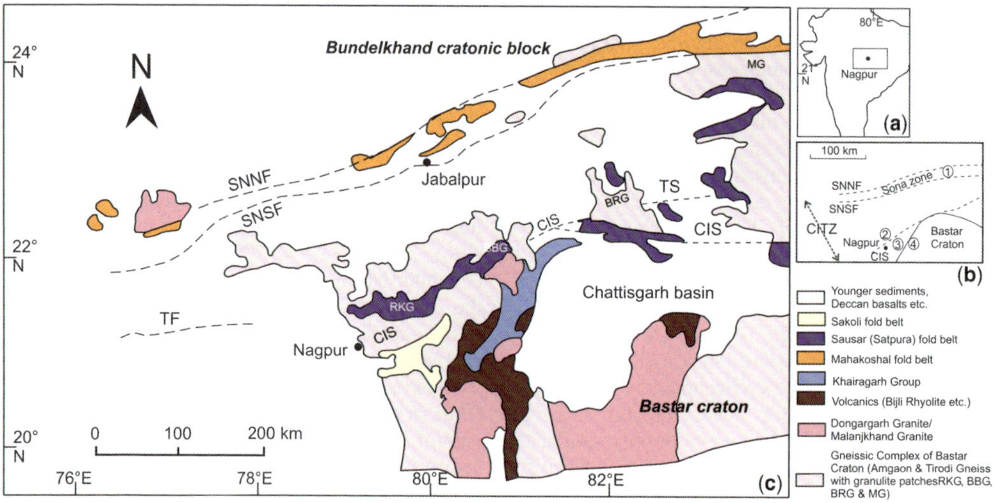

Fig. 4. Salient features of the Bastar Craton and CITZ. (**a**) Location map. (**b**) Index map showing different fold belts in the central Indian region. (1) Mahakoshal belt; (2) Satpura belt; (3) Sakoli belt; (4) Dongargarh belt; CITZ, Central Indian Tectonic Zone; SONA, Son–Narmada Lineament Zone; CIS, Central Indian Shear Zone; SNNF, Son–Narmada North Fault; SNSF, Son–Narmada South Fault. (**c**) Sketch geological map of central India (after Roy *et al.* 2002) showing basement gneisses, granite plutons and granulite patches of Ramakona–Katangi (RKG), Bhandara–Balaghat (BBG), Bilaspur–Raipur (BRG) and Makrohar (MG). TS, Tan Shear Zone; TF, Tapti Fault.

Ore Group in the Singhbhum Craton (see Mazumder *et al.* 2012). The Bastar Craton is marked by a number of Palaeoproterozoic supracrustal sequences that have undergone multiple deformation and metamorphism (some attaining UHT granulite facies: Bhowmik *et al.* 2005). The oldest of the supracrustal successions, the Dongargarh Supergroup, unconformably overlies the basement

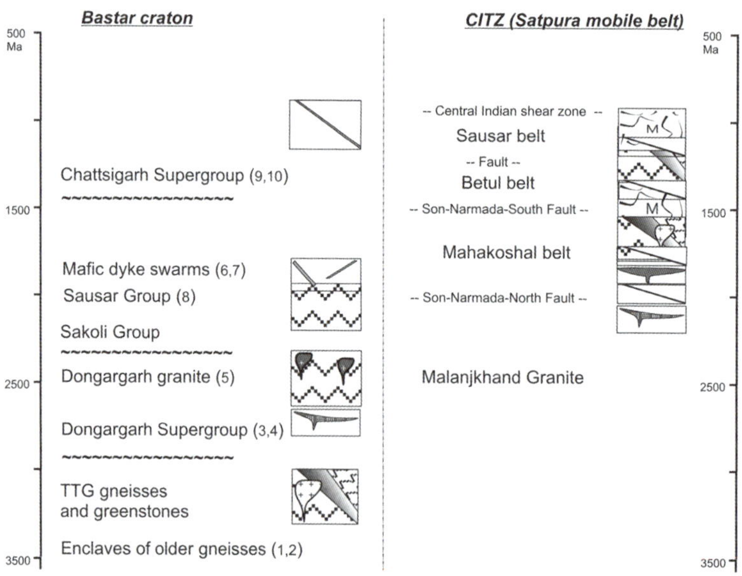

Fig. 5. Synoptic view of the tectonostratigraphic development in the Bastar Craton and the CITZ. The numbers refer to references dealing with key gechronological data and are cited in Table 2. Index as in Figure 3.

Table 2. *Abridged stratigraphy of the Bastar Craton. Numbers refer to cited works for geochronological constraints* *

Bastar Craton		CITZ/Satpura mobile belt
Chhattishgarh Supergoup		‡*Granulites, Ramakona–Katangi (RKG), Bhandara–Balaghat (BBG), Bilaspur–Raipur (BRG) and Makrohar (MG)*
(≤1400 Ma[9,10])		____________CIS________________
^^^^^^^^^^^^^^^^^^^^†		Sausar belt‡
mafic dyke swarms (1.89–183 Ga[6,7])		____________fault________________
		Betul belt‡
Sausar Group[8]		_____Narmada–Son South Fault_____
Sakoli Group		
		Mahakoshal belt
^^^^^^^^^^^^^^^^^^^^^		_____Narmada–Son North Fault_____
Dongargarh Granite/Malanjkhand Granite (2500 Ma[5])		
		Budelkhand Granite
	Khairagarh Group	
Dongargarh Supergroup		
	Bijli Rhyolite, Rb–Sr age 2180 ± 25 Ma[3]	
	2503 ± 35 Ma[4]	
	Nandgaon Group	
^^^^^^^^^^^^^^^^^^^^^		
TTG gneisses and greenstones (e.g. Sonakhan belt) enclaves of older gneisses with 3.5 Ga old zircon[1,2]		

* [1]Ghosh (2004); [2]Sarkar *et al.* (1993); [3]Sarkar *et al.* (1981); [4]Krishnamurthy *et al.* (1988, 1990); [5]Stein *et al.* (2004); [6]French *et al.* (2008); [7]Srivastava & Gautam (2008); [8]Srivastava & Singh (2003); Huin & Pal (2006); [9]Bickford *et al.* (2011); [10]Das *et al.* (2009).
†^^^^^^^^^ unconformity.

gneisses, and has been divided into volcano-sedimentary successions of the Nandgaon and Khairagarh groups. The Bijli Rhyolites belonging to the latter represent a major volcanic event, possibly at around 2108 Ma (Sarkar *et al.* 1981). A number of granitic plutons, namely the Dongargarh Granite and the Malanjkhand Granite, were emplaced during the stabilization of the Bastar nucleus. The Malanjkhand Granite has been dated at around 2500 Ma (Re–Os: Stein *et al.* 2004). The deformed and metamorphosed Sausar and Sakoli groups unconformably overlie the Donagargarh and other contemporary granitic plutons.

Taking into account field relationships, available emplacement ages, and petrological and geochemical character, Srivastava & Gautam (2008) recognized three distinct dyke swarms in the Bastar Craton: Meso-Neoarchaean subalkaline mafic dykes (BD1); Neoarchaean–Palaeoproterozoic boninite–norite dykes (BN); and Palaeoproterozoic subalkaline mafic dykes (BD2). Only the BD2 dykes from south Bastar have been precisely dated at 1891 ± 0.9 Ma (Baddeleyite, U–Pb: French *et al.* 2008). French *et al.* (2008) proposed that these NW-trending dykes in Bastar and the east–west dykes west of the Cuddapah Basin in the eastern Dharwar are part of a large igneous province of radiating dyke swarm related to a hypothetical approximately 1890 Ma mantle plume.

The Sakoli Group unconformably overlies the Amgaon Gneiss and consists of a volcano-sedimentary succession beginning with a conglomerate, followed upwards by coarse arenite, minor carbonaceous pelite and BIF, metabasalts, and mainly metapelites with minor felsic volcanics/tuffs in the upper part (Bandyopadhyay *et al.* 1995). The Sakoli succession is multiply folded and suffered low-pressure metamorphism, generally regarded as related to Palaeoproterozoic–Mesoproterozoic tectonothermal events, although the geochronological constraints are not unequivocal (Ramakrishnan & Vaidyanadhan 2008). The Sausar Group, rich in Mn ores, unconformably overlies the migmatitic Tirodi Gneiss and consists of conglomerate–quartzite at the bottom, biotite schist with marble, mica schist with gondites (Mn ore) in the middle, and marble and calc-silicate rocks in the upper part. The Tirodi Gneiss has a number of granulite patches (around Ramakona–Katangi, RKG; Balaghat–Bhandara, BBG; Bilaspur–Raipur, BRG; Makrohar, MG). These granulite patches have been interpreted as representing Mesoproterozoic reworking of Palaeoproterozoic UHT granulites (Bhowmik *et al.* 2005).

The Mahakoshal belt occurring at the join of the southern Indian block (Dharwar–Bastar–Singhbhum cratons) and the northern Indian block (Aravalli–Bundelkhand Craton) is important in understanding the amalagamation of the two blocks along the CITZ. The basal part of the Mahakoshal Group is dominated by subaqueous mafic–ultramafic lava overlain by quartzites with minor basalts, carbonates and BIF. A thick succession of argillites and turbidites constitute the upper part of the group (see Roy & Devarajan 2000 for a review of the stratigraphy of the belt). Outcrops of the deformed volcano-sedimentary assemblage of the Mahakoshal Group are generally restricted between the SNNF marking the southern boundary of the Bijawar Group in the Bundelkhnad Craton, and the Son–Narmada South Fault (SNSF) (Fig. 4). The Mahakoshal rocks have been intruded by syn- to post-tectonic granites (Tamkhan and Barambara) dated at 1800–1865 Ma (Sarkar *et al.* 1998).

Bhopalpatnam/Kondagaon granulites

The Pranhita–Godavari rift that hosts the Mesoproterozoic–Neoproterozoic succession of the Godavari Supergroup, and much younger Mesozoic Gondwana rocks, conceals the join between the Bastar Craton and the Dharwar Craton (Rogers 1986; Saha & Chaudhuri 2003). The Bhopalpatnam granulite on the eastern shoulder of the Godavari rift, which has been dated at 1.6–1.9 Ga, appears to be much younger than the 2.6 Ga Kondagaon granulite in the interior of the Bastar Craton. The latter is comparable in age to the Karimnagar granulite on the western shoulder of the Godavari rift (cf. Santosh *et al.* 2004).

Singhbhum (Singhbhum–North Orissa) Craton

The Singhbhum Craton (SC) – the easternmost cratonic block in Peninsular India – is separated from the Bastar Craton on the west by the Upper Palaeozoic–Mesozoic Son–Mahanadi Gondwana Basin, while the Damodar Valley Gondwana Basin occurs to the north of the Chotanagpur Granite Gneiss Complex (CGC, Fig. 6). The latter was accreted to the SC along a major shear zone, the North Singhbhum Shear Zone, alternatively named as the Tamar–Porapahar Shear Zone (Misra 2006). The latter is considered to be an extension of the Satpura mobile belt, which is an important tectonic element in the eastern continuation of the CITZ (Ramakrishna & Vaidyanadhan 2008; cf. Sharma 2009). The northernmost termination of the high-grade Eastern Ghats Belt abuts against southern Singhbhum, and the contact zone is referred to as the Sukinda Thrust (Saha 1994). A synoptic view of the stratigraphic and tectonic development, including deformation and magmatism, across the craton is depicted in Figure 7. The main geochronological constraints, based on more recent age data, are shown in Table 3.

The Mesoarchaean–Palaeoproterozoic development of the Singhbhum nucleus has been discussed in detail by Mazumder *et al.* (2012). Here we highlight only the major Palaeoproterozoic stratigraphic and tectonic issues. The composite Singhbhum Granite batholith, with enclaves of the approximately 3.44 Ga Older Metamorphic Tonalitic Gneiss and Older Metamorphic Group, constitutes the foundation (Fig. 6; Table 3). The overlying volcano-sedimentary successions with major BIF-bearing formations – namely the Iron Ore Group in three isolated basins of Jamda–Noamundi–Koira in the west, the Gorumahisani–Badampahar in the east and the Daitari in the extreme south – host the largest late Archaean iron ore deposits of eastern India. Mukhopadhyay *et al.* (2008) reported oldest zircon SHRIMP (Sensitive High Resolution Ion Microprobe) ages of 3.55 Ga from the Daitari Basin Iron Ore Group, representing the oldest shelf sedimentation in eastern India.

A number of smaller basins with intercalated metabasalts and fluvial–marginal marine sedimentary successions occur at the fringes of the Singhbhum Granite batholith. Although the age of these successions are yet to be firmly established, the Kolhan Group in the SW, the Jagannathpur–Malangtoli–Ongarbira mafic volcanics in the west (Saha 1994), and the Dhanjori Group and the Simlipal volcanics occurring in the eastern fringes constitute possible Palaeoproterozoic intracratonic development within the Singhbhum Craton. The Kolhan Group unconformably overlies the Singhbhum Granite in the Chaibasa–Noamundi area, and the Malangtoli and Jagannathpur lavas are overlain by the Kolhan succession of sandstone–conglomerate and limestone, which is locally stromatolitic (Saha 1994). While the Malangtoli lavas are tholeiitic basalts with uniform composition, the Jagannathpur lavas vary in composition from tholeiite to andesite. These lavas are interpreted to have erupted under a continental flood-basalt-like setting. In contrast, the Simplipal Basin hosts spilitic lavas and tuffs intercalated with quartzites, interpreted to represent explosive volcanism in a deep marine basin on a continental platform (Saha 1994). Similarly, the Dhanjori Group unconformably overlies the Singhbhum Granite, and consists of a basal arkosic sandstone–conglomerate horizon overlain by lavas and tuffs intercalated with quartzites and phyllites. The sedimentary succession of the Dhanjori Group is fluvial–fluvio-deltaic (Mazumder 2005). The Singhbhum Shear Zone passes through

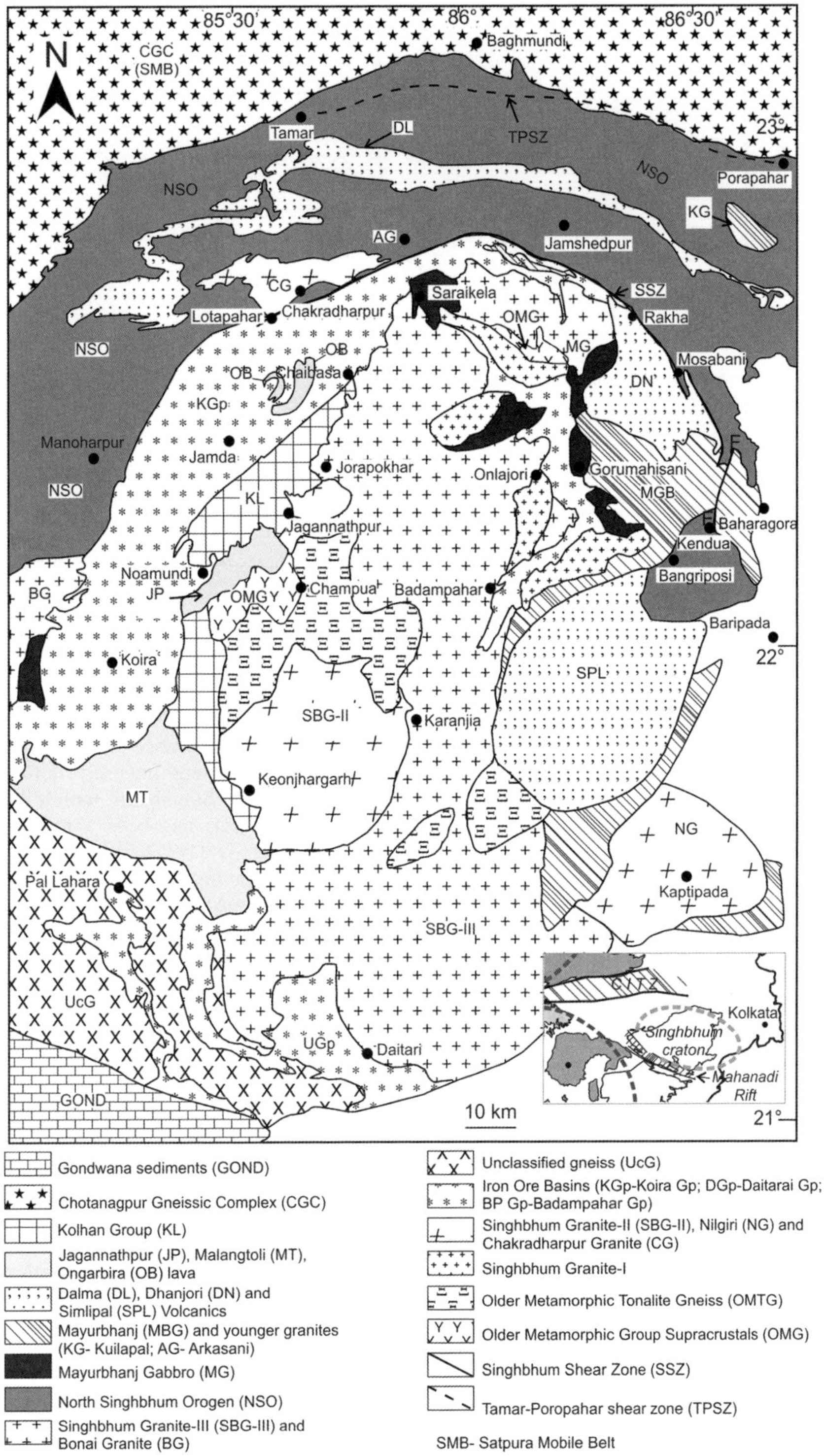

Fig. 6. Sketch geological map showing the salient features of the Singhbhum–North Orissa Craton. Incorporates data from Saha (1994) and Misra (2006).

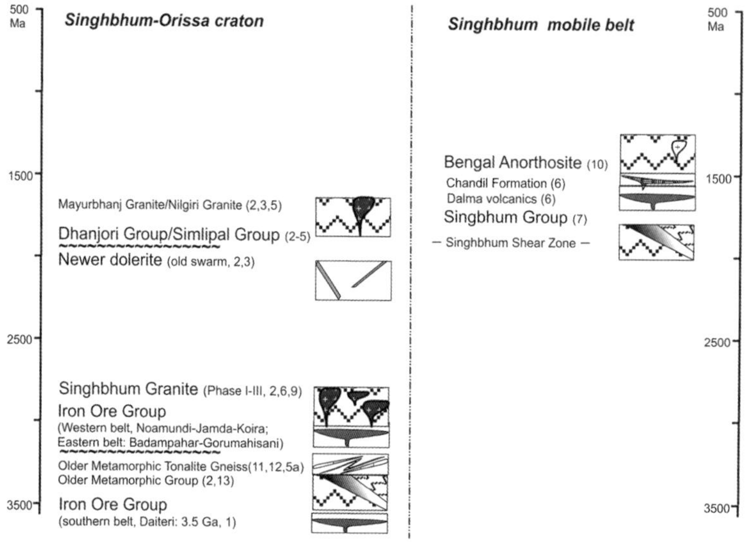

Fig. 7. Synoptic view of tectonostratigraphic development in the Singhbhum Craton. The numbers refer to references dealing with key gechronological data and are cited in Table 3. Index as in Figure 3.

the upper Dhanjori rocks, and the intensely deformed shallow-marine–shelf deposits of the Singhbhum Group of rocks occur to the north of the Singhbhum Shear Zone (Mukhopadhyay 1984; Joy & Saha 2000; Mazumder *et al.* 2012). The Singhbhum Group is divided into the Chaibasa Formation (lower part) and the Dhalbhum Formation (the upper part). The former consists of thick succession of quartzites–mica schists–amphibolites (mafic sills) and shows up to kyanite-grade metamorphism. Chloritic phyllites, carbon phyllites, thin psammites, acid tuffs and ferruginous cherts constitute the Dhalbhum Formation.

The extensive ultramafic–mafic Dalma volcanics constitute the middle part of an east–west belt sandwiched between the CGC and the Singhbhum Shear Zone. The North Singhbhum orogen includes this east–west belt of highly deformed and metamorphosed Singhbhum Group, the Dalma volcanics and the Chandil Formation located further north (e.g. Misra 2006). The Chandil Formation consists largely of phyllites, meta-mafics and rhyolitic tuffs. The Dalma volcanics and the Chandil Formation represent late Palaeoproterozoic development in this belt, as evident from the recent zircon SHRIMP ages (1629 $\pm$ 5 Ma) from the rhyolites of the Chandil Formation occurring north of the Dalma volcanic belt (Reddy *et al.* 2009). The domain north of the Dalma volcanics and south of the CGGC is also marked by the occurrence of abundant felsic tuffs, mafic bodies, nepheline syenite and intrusive granites; for example, the Rb–Sr age of Kuilapal Granite, is 1638 $\pm$ 38 Ma (Sengupta *et al.* 1994).

The tectonic significance of the Dalma volcanic belt and the adjoining intensely deformed rock successions in the Singhbhum mobile belt (North Singhbhum orogen) has been a matter of long-standing debate. The belt has been variously interpreted as representing island arc magmatism (Naha 1965), an ophiolitic assemblage at the suture arising out of the Singhbhum 'microplate' in the south subducting under the Chotanagpur continental block (now preserved as the CGC: Sarkar 1982) and a rift filling in a marginal basin or back-arc basin in a subduction-related environment (Bose & Chakraborty 1981; Bose 1994). Based on Sr–Nd systematics, Roy *et al.* (2002) suggested an approximately 1600 Ma plume-related activity in the Dalma belt as part of a global event, lead to continental rifting and the formation of an ocean basin that ultimately collapsed into the mobile belt. In the above-cited work, tectonic discriminations are based largely on geochemical signatures of the mafic–ultramafic lavas and associated intrusives, and the subduction polarity is not adequately constrained taking into account structural details, sediment characters and chronology of the related geological successions and their stratigraphic relationships. The nature of displacements along the major shear zones – namely the Singhbhum and Tamar–Porapahar shear zones – and the relative chronology of the tectonic movements have not been properly integrated into these models

Table 3. *Abridged stratigraphy of the Singhbhum Craton. Numbers refer to cited works for geochronological constraints**

Singhbhum Craton	Singhbhum mobile belt
	‡1.65–1.70 Ga Pb/Pb age of Cu mineralization[8]
	Chotanagpur Granite Gneiss Complex
	Bengal Anorthosite (1.5 Ga[10])
	_________Tamar–Porapahar Shear Zone__________
	Chandil Formation (1600 Ma[6])
	Dalma Formation (1750–1600 Ma[6])
	_____________?______________________
	Singhbhum Group
	Soda Granite (Pb/Pb, 2.2 Ga[7])/Chakradharpur Granite
Mayurbhanj Granite/Nilgiri Granite[2,3,5]	____Singhbhum Shear Zone‡/Copper belt thrust_____
Dhanjori Group/Simlipal Group (2.6–2.1 Ga[2–5])	
^^^^^^^^^^^^^^^^^^^^^^^^^^^^^^^^^^^†	
Newer dolerites[2,3] (older swarm ≥1960 Ma)	
Singhbhum Granite phases I–III (3.3–3.1 Ga[2,6,9])	
Iron Ore Group (western belt: Noamundi–Jamda–Koira; eastern belt: Badampahar–Gorumahisani)	
^^^^^^^^^^^^^^^^^^^^^^^^^^^^^^^^^^^^^	
Older Metamorphic Tonalite Gneiss (3.4Ga[11,12]; 3.5 Ga[5])	
Older Metamorphic Group (3.8 Ga[2]; 3.3 Ga[13])	
Iron Ore Group (southern belt around Daiteri, 3.5Ga[1])	

* [1]Mukhopadhyay *et al.* (2008); [2]Saha (1994); [3]Bose (2008); [4]Mazumder (2005); [5]Acharyya *et al.* (2010); [6]Reddy *et al.* (2009); [7]Sarkar *et al.* (1985); [8]Johnson *et al.* (1993); [9]Misra *et al.* (1999); [10]Chatterjee *et al.* (2008); [11]Moorbath & Taylor (1986); [12]Moorbath *et al.* (1988); [13]Sharma *et al.* (1994).
†^^^^^^^^^ unconformity.

based on our understanding of the modern plate tectonic set-up.

Dharwar (Karnataka) Craton

The Dharwar Craton is the largest of the Archaean nuclei in India and is bordered to the south by the Pan-African Pandyan Mobile Belt, also referred to as the Southern Granulite Terrain (Ramakrishnan & Vaidhyanadhan 2008). The Deccan basalts of end-Cretaceous age cover its northern fringes where two small Proterozoic basins (Bhima and Kaladgi) also occur (Fig. 8). The Pranhita–Godavari (PG) rift defines its NE limit, while the intracratonic Cuddapah Basin occurs in the eastern Dharwar Craton. The latter is amalgamated with the western Dharwar Craton by the large north–south-trending body of the Closepet Granite. South of the PG rift, the juxtaposition of the Eastern Ghats Belt, the Nellore Schist Belt (NSB) – with remnants of ocean plate stratigraphy – and the Nallamalai Fold Belt (NFB) along the eastern margin of the Dharwar Craton are important in constraining the Palaeoproterozoic–Mesoproterozoic tectonic development of the Indian Craton (Saha 2002, 2011; Saha & Chakraborty 2003; Saha *et al.* 2010). A synoptic view of the stratigraphic and tectonic development, including deformation and magmatism across the craton, is depicted in Figure 9. The main geochronological constraints, based on more recent age data, are shown in Table 4.

The bulk of the western Dharwar Craton consists of 3.3–3.0 Ga old TTG rocks of the Peninsular Gneiss that contain enclaves of a supposedly older greenstone belt including the Sargur Group of rocks, and a younger greenstone succession including the Bababudan and Chitradurga groups (Naha *et al.* 1991; Fig. 9). Approximately 2.6 Ga old granitoids (Chitradurga Granite, Chikmagalur Granite) intrude the above older successions (Chadwick *et al.* 2007). In the eastern Dharwar, the oldest preserved schist belts (Sandur, Ramgiri, Kolar–Kadiri) are relatively young compared to the greenstone

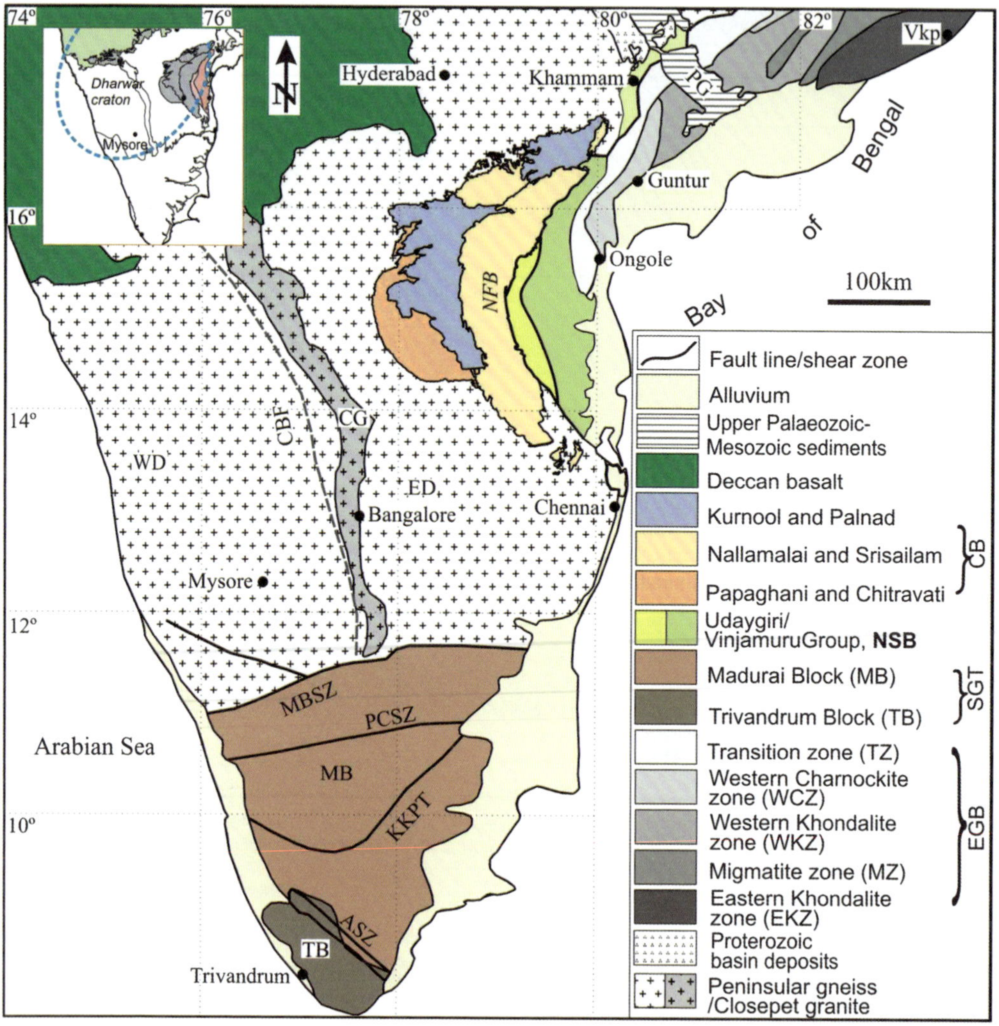

Fig. 8. Sketch geological map showing the salient features of the Dharwar Craton and the SE margin of India. ED, eastern Dharwar; WD, western Dharwar; CG, Closepet Granite; CBF, Chitradurga Boundary Fault; CB, Cuddapah Basin; NFB, Nallamalai Fold Belt; NSB, Nellore Schist Belt; EGB, Eastern Ghats Belt; SGT, Southern Granulite Terrain; MBSZ, Moyar–Bhavani Shear Zone; PCSZ, Palghat–Cauvery Shear Zone; ASZ, Achankovil Shear Zone; KKPT, Karur–Kambam–Painavu–Trissur Shear Zone. Incorporates data from Naqvi & Rogers (1987), French *et al.* (2008) and Saha *et al.* (2010).

belts of western Dharwar (e.g. Nutman *et al.* 1996; Balakrishnan *et al.* 1999). A Sm–Nd age of 2700 Ma is obtained from the Kadiri belt (Naqvi *et al.* 2002). Although older literature suggests continuity of Peninsular Gneiss in the eastern Dharwar, granites and gneisses east of the Closepet Granite outcrop with younger radiometric dates of between 2700 and 2500 Ma are renamed as the Dharwar batholith (Chadwick *et al.* 2000). Emplacement of the Closepet Granite represents amalgamation of the eastern and western Dharwar cratonic blocks at about 2500 Ma.

Of the Proterozoic intracratonic basins of the Dharwar Craton, the Cuddapah Basin is the largest and provides Palaeoproterozoic sedimentary records. The other smaller basins of Bhima and Kaladgi, as well as younger sub-basins around Cuddapah, suggest Mesoproterozoic–Neoproterozoic sedimentation including the development of carbonate platforms. The initiation of the Cuddapah Basin (Papaghni sub-basin) was preceded by an episode of dyke swarm emplacement at around 2200 Ma in the basement; for example, the Mahabubnagar swarm.

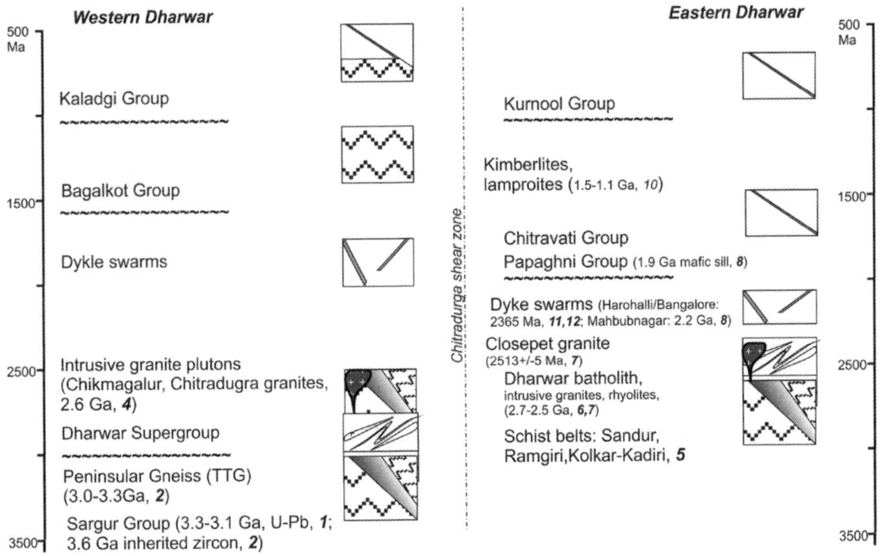

Fig. 9. Synoptic view of tectonostratigraphic development in the Dharwar Craton. The numbers refer to references dealing with key gechronological data and are cited in Table 4. Index as in Figure 3.

Table 4. *Abridged stratigraphy of the Dharwar Craton. Numbers refer to cited works for geochronological constraints**

Western Dharwar Craton	Eastern Dharwar Craton
	Kurnool Group ^^^^^^^^^^^^^^^^^^^^^^
	Kimberlites/lamproites (1500–1100 Ma[10])
Kaladgi Group ^^^^^^^^^^^^^^^^^^†[3]	Chitravati groups (1.9 Ga mafic dykes and sills[9])
Bagalkot Group ^^^^^^^^^^^^^^^^^^^^[3]	Papaghni Group ^^^^^^^^^^^^^^^^^^^^^^
	Dyke swarms, Harrohalli and Bangalore swarm 2365 Ma[11] Mahabubnagar swarm, 2.2 Ga[8]
Dyke swarms (undated[2])	Closepet Granite (2513 ± 5 Ma[7])
	Dharwar batholith (? Peninsular Gneiss) (2700–2500 Ma[6])
intrusive granite plutons (Chikmagalur Granite, 2605 ± 18, Pb/Pb) Chitradurga Granite, 2.6 Ga[4])	intrusive granites and rhyolites (2700–2500 Ma[7])
Dharwar Supergroup — Chitradurga Group ~~disconformity~~[3] Babbudan Group	Schist belts: Sandur, Ramgiri, Kolar–Kadiri (2700 Ma[5])
^^^^^^^^^^^^^^^^^^^^^^^^	
Peninsular Gneiss (TTG), 3.0–3.3 Ga[2]	
Sargur Group (older greenstones, 3.3–3.1 Ga[1,2])	

* [1]Jayananda *et al.* (2008); [2]Meert *et al.* (2010); [3]Ramakrishnan & Vaidyanadhan (2008); [4]Chadwick *et al.* (2007); [5]Vasudev *et al.* (2000); [6]Chadwick *et al.* (2000); [7]Friend & Nutman (1991); [8]French *et al.* (2004); [9]French *et al.* (2008); [10]Osborne *et al.* (2011); [11]Halls *et al.* (2007).
†^^^^^^^^^ unconformity.

Details of the stratigraphic development of the Cuddapah Basin have been dealt with in a companion paper (Saha & Tripathy 2012). The hypothesis of an approximately 1900 Ma old large igneous province (French *et al.* 2008) incorporating the Cuddapah dykes – that is, those occurring in the basement (Dharwar batholith) – is yet to be tested properly with data from other continental fragments.

Fold belts and accreted terranes

Saha & Tripathy (2012) propose an allochthonous origin for the Nallamalai Fold Belt that occurs at the eastern fringe of the eastern Dharwar Craton. Recent data from the Nellore Schist Belt (NSB) also suggest accretion of oceanic crust remnants to the cratonic margin in the Palaeoproterozoic (Saha *et al.* 2010; Vijaya Kumar *et al.* 2010; Saha 2011). Although some work indicates the presence of Neoarchaean protoliths in the NSB (Ravikant 2010), reports of supra-subduction zone ophiolites from the Kandra domain in the southern NSB (e.g. Saha 2011) and early Mesoproterozoic intrusive granites in the Vinjamuru domain of the NSB (e.g. Vinukonda Granite, 1589 Ma, Dobmeier *et al.* 2006) suggest major late Palaeoproterozoic–early Mesoproterozoic tectonic convergence in these domains. The presence of Mesoproterozoic ophiolitic mélange (the Kanigiri mélange: Dharma Rao *et al.* 2011), immediately east of the Vellikonda thrust front bordering the NFB, suggests multiple episodes of crustal extension and convergence, possibly affecting a loose aggregation of microcontinents/archipelago along the eastern fringe of the Dharwar Craton during the 1900–1100 Ma interval (Saha 2002, 2004; Vijaya Kumar & Leelanandam 2008).

The granulite terrain south of the Palghat–Cauvery Shear Zone (Fig. 1) is loosely described as the Southern Granulite Terrane (SGT), while Ramakrishna & Vaidyanadhan (2008) preferred to call it the Pandyan Mobile Belt. The SGT consists of a mosaic of blocks bound by steep shear zones (Chetty & Bhaskar Rao 2006). The main development of the SGT and its accretion to the Dharwar nucleus possibly took place during Edicaran–Cambrian time (e.g. Collins *et al.* 2007, 2008), and any detailed discussion is outside the scope of this article.

Pranhita–Godavari graben and the Somanpalli Group

Occurring at the join between the Bastar (Bhandara) Craton in the NE and the Dharwar Craton in the SW, the Pranhita–Godavari (PG) graben hosts on its two flanks largely Mesoproterozoic–Neoproterozoic sedimentary successions (Saha & Chaudhuri 2003). Recent Ar/Ar glauconite ages from the lower part of the Pakhal Group suggests that the oldest sedimentation in the western flank probably started at around 1686 ± 6 Ma and the Somanpalli Group occurring in the eastern flank may be as old as 1620 ± 6 Ma (Conrad *et al.* 2010). While the approximately 1900 Ma Bhopalpatnam granulites occur on the eastern shoulder of the Godavari graben, the Somanpalli Group is strongly deformed (Saha 1992; Saha & Ghosh 1998; Saha & Chaudhuri 2003), and the exact tectonic framework and timing of deformation of these intracratonic successions is yet to be solved. Younger, unconformity-bound fluvial–shallow-marine successions (Penganga and Albaka groups in the PG graben), however, have been found to be practically undeformed. Hence, the issue of repeated basin opening and closure in the same spatial domain during the upper Mesoproterozoic–Neoproterozoic requires further investigation into the tectonic underpinnings of stratigraphic development in the PG graben, and similar research is needed in the other Proterozoic basins.

Shillong plateau and Lesser Himalaya

A relatively small outcrop of Proterozoic basement gneisses and deformed metasedimentary rock groups occur in the Shillong plateau, which has been considered as part of the Indian Craton (Ramakrishnan & Vaidyanadhan 2008). As yet, there is no convincing evidence that these outliers of cratonic blocks contain any Archaean component, which is in contrast to the other cratonic blocks of Peninsular India. However, recently published geochronological dates suggest that the earliest phase of orthogneisses (migmatitic gneiss) within the Shillong–Meghalaya basement gneissic complex are at least as old as 1600 Ma (Yin *et al.* 2010*a*). The overlying Shillong Group, consisting of phyllites, mudstones, sandstones and cross-stratified quartz arenites, contains rare detrital zircons as old as 2500 Ma. Based on the age distribution of detrital zircon population, Yin *et al.* (2010*a*) concluded that the Shillong Group is possibly younger than 1100 Ma (see also Chatterjee *et al.* 2007; cf. Mitra & Mitra 2001). The Lesser Himalayan augen gneisses (Bomdila Group: Kumar 1997) in Arunachal Pradesh are dated at 1760–1745 Ma old (Yin *et al.* 2010*b*). As the gneisses are reported to be intrusive into a phyllite–quartzite succession in the Bomdila Group (Kumar 1997), Palaeoproterozoic sedimentation in the Lesser Himalaya may be broadly contemporaneous with the lower Vindhyan (Semri Group) sedimentation in the Aravalli–Bundelkhand Craton, or similar elements preserved in the Mahaloshal belt or the Chotanagpur Granite Gneiss terrain (see

Ramakrishnan & Vaidyanadhan 2008). On the whole, the available evidence suggests that the amalgamation of the Shillong plateau basement with the rest of the Indian Craton is a late Palaeoproterozic–Mesoproterozoic phenomenon.

Discussion

Epeiric seas and Proterozoic basins in India

The Proterozoic basins in India (Fig. 1) generally host extensive fluvial–shallow-marine successions (Table 5). Sedimentation in both the Vindhyan Basin in the Aravalli–Bundelkhand Craton and in the Cuddapah Basin of the Dharwar Craton of south India was initiated in the Palaeoproterozoic, at roughly 1800–1900 Ma. This is also the time of emplacement of a prominent phase of regional dyke swarms in Bundelkhand, Bastar and the eastern Dharwar Craton (French *et al.* 2008). Based on the mafic dyke swarm events in India, a plume-mediated large igneous province in India has been suggested to reflect the break-up of the Neoarchaean–Palaeoproterozoic supercontinent. The near-contemporaneous development of large Palaeoproterozoic sedimentary basins in India clearly calls for significant tectonothermal activity in the subcontinental lithospheric mantle beneath the Indian Craton where Archaean nuclei were largely stabilized by 2500 Ma. Extensive Palaeoproterozoic carbonate deposition in the Semri Group in the Vindhyan Basin, and in the Papaghni and Chitravati groups of the Cuddapah Basin, suggests flooding of the inland basins, possibly related to a spurt in global sea-level rise mediated by superplume activity and/or supercontinent break-up (Santosh *et al.* 2009; Santosh 2010). Deep-water sequences are rare, but tempestites are common in the Singhbhum Group and in the Cuddapah Basin, indicating open ocean connections for the host basins. In both of these large cratonic basins, sedimentation was terminated by the end of the Palaeoproterozoic, as evidenced by profound regional unconformities above the Semri Group in the Vindhyan Basin and above the Chitravati Group in the Cuddapah Basin, respectively (compare Figs 3 & 8) (see also Saha & Tripathy 2012). A second event of extensive Mesoproterozoic–Neoproterozoic intracratonic basin development occurred across the Indian Shield as evidenced by the upper Vindhyan sedimentary successions (Chakraborty 2006), the Proterozoic successions in the Pranhita–Godavari Valley (e.g. Saha & Ghosh 1998; Chaudhuri 2003), in the Chhattisgarh Basin (e.g. Patranbis-Deb & Chaudhuri 2008), in sedimentary successions in the smaller basins of Khariar, Indravati and Sukma (Chakraborti *et al.* 2010), and in the Kurnool Group of rocks in the Cuddapah Basin (Saha *et al.* 2006). It should be recalled here that extensive middle–upper Proterozoic successions are reported from the Shillong plateau in the NE of India, and from the Lesser Himalaya, which

Table 5. *Broad depositional environment of major Palaeoproterozoic successions in India*

Archaean cratonic nuclei	Sedimentary succession	Age	Key reference	Broad depositional environment
Aravalli–Bundelkhand	Aravalli Supergroup	2.2–1.85 Ga	Deb (1999); Deb & Thorpe (2004); Roy (2006)	Shallow to deep marine
Bundelkhand	Lower Vindihyan/Semri Group	1631 Ma	Chakraborty (2006); Rasmussen *et al.* (2002); Roy *et al.* (2002)	Coastal to inner shelf
CITZ	Mahakoshal Group	>1865 Ma	Sarkar *et al.* (1981); Sarkar *et al.* (1998); Roy & Devarajan (2000)	Shallow marine to deep shelf
Bastar	Dongargarh Supergroup	2180 ± 25 Ma	Sarkar *et al.* (1981); Divakara Rao *et al.* (2000)	Stable shelf, island arc
Singhbhum	Dhanjori Group, Singhbhum Group	2.1–1.6 Ga	Roy *et al.* (2002); Mazumder (2005); Reddy *et al.* (2009)	Fluvial/marginal marine to inner shelf
Eastern Dharwar	Lower Cuddapah: Papaghni and Chitravati groups	>1.8 Ga	Anand *et al.* (2003); French *et al.* (2008); Saha & Tripathy (2012)	Fan-delta, craton margin to inner shelf

were involved in the Himalayan Orogeny unlike the Proterozoic successions in Peninsular India (Kumar 1997). By the end of the Palaeoproterozoic, an extensive sea with connections to the intracratonic basins probably bordered the Indian Craton to the east and to the north (see Singh 1985; Chaudhuri *et al.* 2002).

Do the CITZ granulites represent major collisional events?

As mentioned earlier, a number of granulite patches occur in the CITZ and in the Sausar/Sakoli belts. Acharyya (2003) considered a south-directed subduction that led to the deformation of the Mahakosal rift succession, and to the emplacement of syn- to post-tectonic granites along the northern margin of the belt, and to fusion between the Bundelkhand and Bastar cratons and the intervening terrains. Roy & Hanuma Prasad (2003) proposed an alternative view endorsing northwards subduction and continent–continent collision at around 1500 Ma (see also Sharma 2009). However, there are reports of multiple deformation and metamorphism (e.g. Roy *et al.* 2001, 2006; Bhowmik *et al.* 2005). The timing and exact nature of a subduction–collision process in this belt is not adequately constrained. Multiple tectonothermal events including syn- to post-tectonic granite emplacement at different ages (2500 Ma for Malanjkhand Granite and 1800–1700 Ma for the Tamakhan–Barakhamba granites) and the later events in the Sausar belt (Roy & Hanuma Prasad 2003) in the linear domain straddling the CITZ also compound the problem. Understanding the process of basin opening and closure related to the volcano-sedimentary succession of different ages in the Mahkoshal, Sausar and Sakoli belts requires further work.

Initial amalgamation and tectonic rejuvenation of the cratonic nuclei and their joins

As discussed earlier, four main cratonic nuclei constitute the Indian Craton that possibly remained as one coherent continental entity since early Palaeoproterozoic times. One striking feature of the end-Archaean continental growth of the Indian Craton is the emplacement of large granite batholiths at around 2500 Ma, exemplified by the Closepet Granite in the Dharwar Craton, the Dongargarh Granite and its equivalents in the Bastar Craton, and the Berach Granite/Bundelkhand Granite in Aravalli–Bundelkhand Craton (Figs 3, 5 & 9). Large granite batholiths of this age are missing in the Singhbhum Craton. Although granulites of similar age are reported from the SW shoulder of the Godavari graben in Karimnagar granulites or from the interior of the Bastar Craton (Kondagaon granulite), it is still uncertain whether these represent products of large continental collision or are exhumed core complexes associated with crustal rifting (Lister & Davis 1989). Palaeoproterozoic UHT granulites are reported from the vicinity of the CITZ, and the clockwise pressure–temperature–time (P–T–t) path may represent a collisional event (e.g. Bhowmik & Spiering 2004) leading to the final amalgamation of the Aravalli–Bundelkhand and Bastar cratons. Similarly, the Sandmata Granulite Complex along the eastern margin of the Delhi Fold Belt may be a product of continental collision.

The Mesoproterozoic overprint of a second granulite metamorphism in the CITZ raises the issue of the nature of the tectonic setting for younger reworking. One can invoke multiple Wilson cycles affecting the CITZ in the interval 2500–1500 Ma to explain multiple deformations and the high-grade metamorphism in the Sakoli/Sausar belt. But the sporadic occurrence of the granulites or the paucity of recognizable Palaeoproterozoic oceanic crust remnants in this belt, renders testing of such hypothesis very difficult given the present state of knowledge. We do see more than one age cluster of Palaeoproterozoic dyke swarms in India (Ernst & Srivastava 2008), with a prominent development at around 1900 Ma at Bastar and, possibly, also at eastern Dharwar, but these are interpreted as expressions of plume-related large igneous provinces rather than continental rifts ultimately developing into oceanic basins. Alternatively, one may consider limited intracratonic subduction to explain the fold belts, such as the North Singhbhum orogen (Mukhopadhyay & Mukhopadhyay 2008) or a 'slab break off' model (Sharma 2009).

Concluding remarks

The following points seem to emerge from the above overview of the Palaeoproterozoic geology of India, and the salient discussion.

All of the four major Archaean nuclei were cratonized at around 2500 Ma. The small basement outlier in the Shillong plateau was possibly stabilized and amalgamated with the rest of the Indian Craton at a relatively late stage. The end-Archaean stabilization is coincident with the emplacement of major granitic plutons in Aravalli–Bundelkhand, Bastar and Dharwar. Banded iron formations (BIFs) most common in Dharwar, Bastar and Singhbhum in India are largely Archaean, and, unlike Australia and North America, Palaeoproterozoic peak development of BIF is missing in India.

The early Palaeoproterozoic (*c.* 2200 Ma) saw copious emplacement of mafic dyke swarms or mafic lavas in Bundelkhand, Bastar and Karnataka. This was possibly also a period of mafic lava eruption in the Dhanjori and Simlipal basins in Singhbhum. However, more precise geochronological data are required for any firm correlation across the cratonic blocks.

Sedimentation in the intracratonic basins of India probably did not start until 2150 Ma, as apparent from the Aravalli Supergroup. Smaller basins of Bijawar and Gwallior appeared on the fringes of the Bundelkhand Craton. Shallow-water marginal marine siliciclastic sedimentation was also in progress in the Singhbhum Craton. Mixed siliciclastic–carbonate platforms (Cuddapah Basin) were established in the eastern Dharwar Craton by 1900 Ma, heralding a major transgression of sea and the establishment of epicontinental seas. This event roughly coincides with rifting and ocean crust development in north China and in the eastern outboard of India, as evidenced by the Nellore Schist Belt and the adjoining terrain.

Development of circum-Indian pericratonic–intracratonic basins like the Vindhyan (Semri Group) and Cuddapah (Papaghni and Chitravati groups) and the Delhi Fold Belt is evident from thick platformal sequences including carbonates. The sedimentary successions in the Lesser Himalaya also attest to this relatively stable pericratonic sedimentation in the later part of the Palaeoproterozoic period.

Following the above crustal extension and widespread sedimentation, development of a number of fold belts in different parts of the Indian Craton, namely the Aravalli–Delhi fold belts, the CITZ/Satpura belt, and the Nallamalai Fold Belt, suggest crustal shortening. The latter affected not only the marginal parts but also the smaller basins at the join of the adjacent cratonic blocks, possibly as a result of far-field stresses. Although these deformations are locally associated with syn- to-late tectonic granites, clear-cut evidence of typical ocean crust subduction phenomenon is yet to emerge in these intracratonic fold belts, often referred to as mobile belts in the older literature (cf. Sharma 2009).

Throughout the Mesoproterozoic and through to the end of the Neoproterozoic, the amalgamated cratonic blocks remained a coherent entity, but sedimentation continued in the successor basins in eastern Dharwar (Bhima, Kurnool and Palnad sub-basins), in the Aravalli–Bundelkhand Craton (upper Vindhyan), and in the Marwar region west of the Delhi Fold Belt, in the PG Basin and in the Chhattisgarh Basin. Such widespread development of shallow epeiric seas engulfing large tracts of the Indian Craton is likely to have been connected to the break-up of Columbia, which could have occurred over a protracted time interval possibly starting at around 1500–1400 Ma. This roughly coincides with the earliest emplacement of lamproites and signalling renewed mantle activity beneath India.

We thankfully acknowledge facilities at Indian Statistical Institute. Our sincere thanks go to V. Tripathy and S. De, who helped in drafting some of the illustrations. An initial version of this contribution was presented in the IGCP-509 India conference. Review and editorial corrections by R. Law are gratefully acknowledged.

References

Acharyya, S. K. 2003. The nature of Mesoproterozoic Central Indian Tectonic Zone with exhumed and reworked older granulites. *Gondwana Research*, **6**, 197–214.

Anand, M., Gibson, S. A., Subbarao, K. V., Kelley, S. P. & Dickin, A. P. 2003. Early Proterozoic melt generation processes beneath that intra-cratonic Cuddapah Basin, southern India. *Journal of Petrology*, **44**, 2139–2171.

Balakrishnan, S., Hanson, G. N. & Rajamani, V. 1999. U–Pb ages for zircon and titanite from the Ramagiri area, southern India: evidence for accretionary origin of the Eastern Dharwar craton during the late Archean. *Journal of Geology*, **107**, 69–86.

Bandyopadhyay, B. K., Roy, A. & Huin, A. K. 1995. Structure and tectonics of a part of the central Indian shield. *In*: Sinha-Roy, S. & Gupta, K. R. (eds) *Continental Crust of Northwestern and Central India. Memoirs of the Geological Society of India*, **31**, 433–467.

Banerjee, S. & Jeevankumar, S. 2003. Facies motif and paleogeography of Kheinjua Formation, Vindhyan Supergroup, eastern Son valley. *In*: Roy, A. & Mohabey, D. M. (eds) *Advances of Precambrian of Central India. Gondwana Geological Magazine*, Special Volume, **7**, 363–370.

Banerjee, S. & Jeevankumar, S. 2005. Microbially originated wrinkle structures on sandstone and their stratigraphic context: Paleoproterozoic Koldaha Shale, central India. *Sedimentary Geology*, **176**, 211–224.

Bhowmik, S. K. 2006. Ultra high temperature metamorphism and its significance in the Central Indian tectonic Zone. *Lithos*, **92**, 485–505.

Bhowmik, S. K. & Spiering, B. 2004. Constraining the prograde and retrograde *P–T* paths of granulites using decomposition of initially zoned garnets: an example from the Central Indian Tectonic Zone. *Contributions to Mineralogy and Petrology*, **147**, 581–603.

Bhowmik, S. K., Basu Sarbadhikari, A., Speiring, B. & Raith, M. M. 2005. Mesoproterozic reworking of Palaeoproterozoic ultrahigh temperature granulites in the Central Indian Tectonic Zone. *Journal of Petrology*, **46**, 1085–1119.

Bickford, M. E., Basu, A., Patranabis-Deb, S., Dhang, P. & Scheiber, J. 2011. Depositional history of the Chhattisgarh Basin, Central India: Constraints from new SHRIMP zircon ages. *Journal of Geology*, **119**, 33–50.

BIJU-SHEKHAR, S., YOKOHAMA, K., PANDIT, M. K., OKUDAIRA, T. & SANTOSH, M. 2003. Paleoproterozoic magmatism in Delhi fold belt, NW India and its significance: evidence from EPMA chemical ages from zircons. *Journal of Asian Earth Sciences*, **22**, 189–207.

BISWAL, T. K., WAELE, B. & AHUJA, H. 2007. Timing and dynamics of the juxta- position of the Eastern Ghats mobile belt against the Bhandara craton, India: a structural and zircon U–Pb SHRIMP study of the fold-thrust belt and associated nepheline syenite plutons. *Tectonics*, **6**, 1–21.

BOSE, M. K. 1994. Sedimentation pattern and tectonic evolution of the Proterozoic Singhbhum basin in the eastern Indian shield. *Tectonophysics*, **231**, 325–346.

BOSE, M. K. 2008. Petrology and geochemistry of Proterozoic 'Newer Dolerite' and associated ultramafic dykes within Singhbhum granite pluton, eastern India. *In*: SRIVASTAVA, R. K. (ed.) *Indian Dykes: Geochemistry, Geophysics, and Geochronology* Narosa Publishing House Pvt. Ltd., New Delhi, India.

BOSE, M. K. & CHAKRABORTY, M. 1981. Fossil marginal basin from the Indian shield: a model for the evolution of the Singhbhum Precambrian belt, eastern India. *Geologische Rundschau*, **70**, 504–518.

BOSE, P. K., SARKAR, S., CHAKRABORTY, S. & BANERJEE, S. 2001. Overview of the Meso to Neoproterozoic evolution of the Vindhyan basin, central India (1.4–0.55 Ga). *Sedimentary Geology*, **141**, 395–419.

BROWN, M. 2006. A duality of thermal regimes is the distinctive characteristic of plate tectonics since the Neoarchean. *Geology*, **34**, 961–964.

BROWN, M. 2008. Geodynamic regimes and tectonic settings for metamorphism: relationship to the supercontinent cycle. *Indian Journal of Geology*, **80**, 3–21.

BUICK, I. S., ALLEN, C., PANDIT, M., RUBATTO, D. & HERMANN, J. 2006. The Proterozoic magmatic and metamorphic history of the banded gneissic complex, central Rajasthan, India: La-ICP-MS U/Pb zircon constraints. *Precambrian Research*, **151**, 119–142.

CHADWICK, B., VASUDEV, V. N. & HEGDE, G. V. 2000. The Dharwar craton, Southern India, interpreted as the result of late Archean oblique convergence. *Precambrian Research*, **99**, 91–111.

CHADWICK, B., VASUDEV, V. N., HEGDE, G. V. & NUTMAN, A. P. 2007. Structure and SHRIMP, U/Pb zircon ages of granites adjacent to the Chitradurga schist belt: implications for Neoarchaean convergence in the Dharwar craton, southern India. *Journal of the Geological Society of India*, **69**, 5–24.

CHAKRABORTI, C. C. 2006. Proterozoic intra-continental basin: the Vindhyan example. *Journal of Earth System Science*, **115**, 3–22.

CHAKRABORTI, P. P., DEY, S. & MOHANTY, S. 2010. Proterozoic platform sequences of Peninsular India: Implications towards basin evolution and supercontinent assembly. *Journal of Asian Earth Sciences*, **39**, 589–607, http://dx.doi.org/10.1016/j.jseaes.2010.04.030

CHAKRABORTI, R., BASU, A. R. & CHAKRABORTI, A. 2007. Trace element and Nd-isotopic evidence for sediment sources in the mid-Proterozoic Vindhyan basin, central India. *Precambrian Research*, **159**, 260–274.

CHATTERJEE, N., MAZUMADAR, A. C., BHATTACHARYA, A. & SAIKIA, R. R. 2007. Mesoproterozoic granulites of the Shillong–Meghalaya plateau: evidence of westward continuation of the Prydz Bay Pan-African suture into northeastern India. *Precambrian Research*, **152**, 1–26.

CHATTERJEE, N., CROWLEY, J. L. & GHOSE, N. C. 2008. Geochronology of the 1.55 Ga Bengal anorthosite and Grenvillian metamorphism in the Chotanagpur gneissic complex, eastern India. *Precambrian Research*, **161**, 303–316.

CHATTOPADHYAY, N., MUKHOPADHYAY, D. & SENGUPTA, P. 2012. Reactivation of basement: example from Anasagar Granite Gneiss Complex, Rajasthan, western India. *In*: MAZUMDER, R. & SAHA, D. (eds) *Palaeoproterozoic of India*. Geological Society, London, Special Publications, **365**, 217–242, http://dx.doi.org/10.1144/SP365.12.

CHAUDHURI, A. K. 2003. Stratigraphy and paleogeography of the Godavari Supergroup in the south-central Godavari valley, south India. *Journal of Asian Earth Sciences*, **21**, 595–611.

CHAUDHRI, A. K., SAHA, D., DEB, G. K., PATRANABIS-DEB, S., MUKHERJEE, M. K. & GHOSH, G. 2002. The Purana basins of southern cratonic of India – a case study for Mesoproterozoic fossil rifts. *Gondwana Research*, **5**, 23–33.

CHETTY, T. R. K. & BHASKAR RAO, Y. J. 2006. The Cauvery shear zone, Southern Granulite Terrain, India: a crustal scale flower structure. *Gondwana Research*, **10**, 77–85.

CHOUDHARY, A. K., GOPALAN, K. & SASTRY, C. A. 1984. Present status of the geochronology of the Precambrian rocks of Rajasthan. *Tectonophyics*, **105**, 131–140.

CLOUD, P. 1973. Paleoecological significance of the banded iron-formation. *Economic Geology*, **68**, 1135, doi: 10.2113/gsecongeo.68.7.1135.

COLLINS, A. S., CLARK, C., CHETTY, T. R. K. & SANTOSH, M. 2008. Ediacaran–Cambrian tectonic evolution of Southern India. *Indian Journal of Geology*, **80**, 23–40.

COLLINS, A. S., SANTOSH, M., BRAUN, I. & CLARK, C. 2007. Age and sedimentary provenance of the Southern Granulites, South India: U–Th–Pb SHRIMP secondary ion mass spectrometry. *Precambrian Research*, **155**, 125–138.

CONDIE, K. C., BICKFORD, M. E., ASTER, R. C., BELOUSAVA, E. & SCHOLL, D. W. 2011. Episodic zircon ages, Hf isotopic composition, and the preservation rate of continental crust. *Geological Society of America Bulletin*, **123**, 951–957.

CONRAD, J. E., HEIN, J. R., CHAUDHURI, A. K., PATRANABIS-DEB, S., MUKHOPADHYAY, J., DEB, G. K. & BEUKES, N. J. 2010. Constraints on the development of Proterozoic basins in central India from $^{40}Ar/^{39}Ar$ analysis of authigenic glauconitic minerals. *Geological Society of America Bulletin*, **123**, 158–167, doi: 10.1130/B30083.1.

DAS, K., YOKOYAMA, K., CHAKRABORTY, P. P. & SARKAR, A. 2009. Basal tuffs and contemporaneity of the Chattisgarh and Khariar basins based on new dates and geochemistry. *Journal of Geology*, **117**, 88–102.

DE, C. 2003. Possible organisms similar to Ediacaran forms of the Bhander Group, Vindhyan Supergroup, late Neoproterozoic of India. *Journal of Asian Earth Sciences*, **21**, 387–395.

De, C. 2006. Ediacara fossil assemblege in the upper Vindhyans of central India and its significance. *Journal of Asian Earth Sciences*, **27**, 660–683.

Deb, M. 1999. Metallic minerals deposits of Rajasthan. *In*: Kataria, P. (ed.) *Proceedings of Seminaron Geology of Rajasthan: Status and Perspective*. Geology Department, MLS University, Udaipur, India, 213–237.

Deb, M. & Thorpe, R. I. 2004. Geochronological constraints in the Precambrian geology of Rajasthan and their metallogenic implications. *In*: Deb, M. & Goodfellow, W. D. (eds) *Sediment Hosted Lead–Zinc Sulphide Deposits: Attributes and Models of Some Major Deposits in India, Australia and Canada*. Narosa, New Delhi, 246–263.

Deb, M., Thorpe, R. & Kristc, D. 2001. Hindoli Group of rocks in the eastern fringe of the Aravalli-Delhi orogenic belt – Archean secondary greenstone belt or Proterozoic supracrustals. *Gondwana Research*, **5**, 879–883.

Dharma Rao, C. V., Santosh, M. & Yuan-Bao, W. 2011. Mesoproterozoic ophiolitic melange from the SE periphery of the Indian plate: U–Pb zircon ages and tectonic implications. *Gondwana Research*, **19**, 384–401.

Divakara Rao, V., Narayana, B. L., Rama Rao, P., Murthy, N. N., Subba Rao, M. V., Mallikharjuna Rao, J. & Reddy, G. L. N. 2000. Precambrian acid volcanism in central India – geochemistry and origin. *Gondwana Research*, **3**, 215–226.

Dobmeier, C., Lütke, S., Hammerschmidt, H. & Mezger, K. 2006. Emplacement and deformation of the Vinukonda granite – implications for the geological evolution of peninsular India and for Rodinia reconstructions. *Precambrian Research*, **146**, 165–178.

Eriksson, P. G., Mazumder, R., Sarkar, S., Bose, P. K., Altermann, W. & Van Der Merwe, R. 1999. The 2.7–2.0 Ga volcano-sedimentary record of Africa, India and Australia: evidence for global and local changes in sea level and continental freeboard. *Precambrian Research*, **97**, 269–302.

Ernst, R. E. & Srivastava, R. K. 2008. India's place in the Proterozoic world: constraints from the large igneous province (LIP) record. *In*: Srivastava, R. K., Shivaji, Ch. & Chalapathi Rao, N. V. (eds) *Indian Dykes: Geochemistry, Geophysics and Geochronology*. Narosa, New Delhi, 41–56.

French, J. E., Heaman, L. M., Chacko, T. & Rivard, B. 2004. Global mafic magmatism and continentalbreakup at 2.2 Ga: evidence from the Dharwar Craton, India. *Geological Society America (Abstracts with Programs)*, **36**, 340.

French, J. E., Heaman, L. M., Chacko, T. & Rivard, B. 2008. 1891–1883 Ma southern Bastar craton-Cuddapah mafic igneous events, India: a newly recognized large igneous province. *Precambrian Research*, **160**, 308–322.

Friend, C. R. L. & Nutman, A. P. 1991. SHRIMP U–Pb, geochronology of the Closepet Granite and Peninsular Gneiss, Karnataka, South India. *Journal of the Geological Society of India*, **38**, 357–368.

Ghosh, J. G. 2004. 3.56 Ga tonalite in the central part of the Bastar craton, India: oldest Indian date. *Journal of Asian Earth Sciences*, **23**, 359–364.

Gopalan, K., McDougal, J. D., Roy, A. B. & Murali, A. K. 1990. Sm–Nd evidence for 3.3 Ga old rocks in Rajasthan, northwestern India. *Precambrian Research*, **48**, 287–297.

Gregory, L. 2008. *India in the Proterozoic: Two Key Spatial and Temporal Constraints*. MS thesis, University of Florida. World Wide Web Address: http://etd.fcla.edu/UF/UFE0022255/gregory_l.pdf.

Gregory, L. C., Meert, J. G., Pradhan, V., Pandit, M. K., Tamrat, E. & Malone, S. J., 2006. A paleomagnetic and geochronologic study of the Majhgawan Kimberlite, India: Implications for the age of the Vindhyan SuperGroup. *Precambrian Research*, **149**, 65–75.

Gregory, L. C., Meert, J. G., Bingen, B. A., Pandit, M. K. & Torsvik, T. H. 2009. Paleomagnetism and geochronology of the Malani igneous suite, northwest India: implications for the configuration of Rodinia and the assembly of Gondwana. *Precambrian Research*, **170**, 13–26.

Gupta, A. (ed.) 2004. *A Manual of The Geology of India, Vol. I : Precambrian, Part IV: Northern and Northwestern Part of the Peninsula*. Geological Survey of India, Special Publication **77**.

Gupta, S. N., Arora, Y. K., Mathur, R. K., Iqballuddin, Prasad, B., Sahai, T. N. & Sharma, S. B. 1980. *Lithostratigraphic Map of Aravalli Region*. Geological Survey of India, Hyderabad.

Gupta, S. N., Arora, Y. K., Mathur, R. K., Iqballuddin, Prasad, B., Sahai, T. N. & Sharma, S. B. 1997. The Precambrian geology of the Aravalli region, southern Rajasthan and Northeastern Gujarat. *Memoirs of the Geological Survey of India*, **123**, 1–262.

Halls, H. C., Kumar, A., Srinivasan, R. & Hamilton, M. A. 2007. Paleomagnetism and U–Pb geochronology of easterly trending dykes in the Dharwar craton, India: feldspar clouding, radiating dyke swarms and the position of India at 2.37 Ga. *Precambrian Research*, **155**, 47–68.

Heron, A. M. 1953. *Geology of Central Rajputana. Memoirs of the Geological Survey of India*, **79**.

Holland, T. H. 1909. Imperial Gazetteer of India, **Volume 1**. Clarendon Press, Oxford, 50–103.

Huin, A. K. & Pal, T. 2006. Rb–Sr and Sm–Nd dating of different metamorphic events from the Sausar Belt, central India: implications for Proterozoic crustal evolution. *Journal of Asian Earth Sciences*, **26**, 61–76.

Jayananda, M., Kano, T., Peucat, J. J. & Channabasappa, S. 2008. 3.35 ga komatiite volcanism in the western Dharwar craton, southern India: constraints from Nd isotopes and whole rock geochemistry. *Precambrian Research*, **162**, 160–179.

Joy, S. & Saha, D. 2000. Dynamically recrystallised quartz *c*-axis fabrics in greenschist facies quartzites, Singhbhum shear zone and its footwall, eastern India – influence of high fluid activity. *Journal of Structural Geology*, **22**, 777–793.

Kailasam, L. N. 1976. Geophysical studies of major sedimentary of basins of the Indian craton, their deep structural features and evolution. *Tectonophysics*, **36**, 225–245.

Kaur, P., Chaudhary, N., Raczek, I., Kroner, N. & Hoffman, A. W. 2007. Geochemistry, zircon ages and whole rock Nd isotopic systematic for

Paleoproterozoic A-type granitoids in the northern part of the Delhi belt, Rajasthan, NW India: implications for the late Paleoproterozoic crustal evolution of the Aravalli craton. *Geological Magazine*, **144**, 361–378.

KAUR, P., CHAUDHARY, N., RACZEK, I., KRONER, N. & HOFFMAN, A. W. 2009. Record of 1.82 Ga andean-type continental magmatism in NE Rajasthan, India: insights from zircon and Sm–Nd ages, combined with Nd–Sr isotope geochemistry. *Gondwana Research*, **16**, 56–71.

KLEIN, C. 2005. Some Precambrian banded iron-formations (BIFs) from around the world: their age, geologic setting, mineralogy, metamorphism, geochemistry, and origins. *American Mineralogist*, **90**, 1473–1499.

KRISHNAMURTHY, P., CHAKI, A., PANDEY, B. K., CHIMOTE, J. S. & SINGH, S. N. 1988. Geochronology of granite–rhyolite suites of the Dongargarh Supergroup, central India. *In*: *Proceedings of the Fourth National Symposium on Mass Spectrometry*, Hyderabad. Indian Society for Mass Spectrometry, Mumbai, EPS 2/1–EPS 2/3.

KRISHNAMURTHY, P., SINHA, D. K., RAI, A. K., SETH, D. K. & SINGH, S. N. 1990. Magmatic rocks of the Dongargarh Supergroup, central India: their petrological evolution and implications on metallogeny. *In*: *Precambrian of Central India*. Geological Survey of India, Special Publications, **28**, 309–319.

KUMAR, A., PADMAKUMARI, V. M., DAYAL, A. M., MURTHY, D. S. N. & GOPALAN, K. 1993. Rb–Sr ages of Proterozoic kimberlites of India: evidence for contemporaneous emplacement. *Precambrian Research*, **62**, 227–232.

KUMAR, B., SRIVASTAVA, R. K., JHA, D. K., PANT, N. C. & BHANDARI, B. K. 1990. A revised stratigraphy of the type area of the Bijawar Group in central India. *Indian Minerals*, **44**, 303–314.

KUMAR, G. 1997. *Geology of Arunachal Pradesh*. Geological Society of India, Bangalore.

LISTER, G. S. & DAVIS, G. A. 1989. The orogin of metamorphic core complexes and detachment faults formed during tertiary continental extension in the northern Colorado region, U.S.A. *Journal of Structural Geology*, **11**, 65–94.

MACHADO, N., SCHRANK, A., NOCE, M. & GAUTHIER, G. 1996. Ages of detrital zircon from Archean–Paleoproterozoic sequences: implications for Greenstone belt setting and evolution of a Transamazonian foreland basin in Quadrilátero Ferrífero, southeast Brazil. *Earth and Planetary Science Letters*, **141**, 259–276.

MALONE, S. J., MEERT, J. G. *ET AL.* 2008. Paleomagnetism and detrital zircon geochronology of the upper Vindhyan sequence of son Valley, Rajasathan, India: a *c.* 1000 Ma closure age for the Purana basins. *Precambrian Research*, **164**, 137–159.

MAZUMDER, R. 2005. Proterozoic sedimentation and volcanism in the Singhbhum crustal province, India and their implications. *Sedimentary Geology*, **176**, 167–193.

MAZUMDER, R., ERIKSSON, P. G., DE, S., BUMBY, A. J. & LENHARDT, N. 2012. Palaeoproterozoic sedimentation on the Singhbhum craton: global context and comparison with Kaapvaal. *In*: MAZUMDER, R. & SAHA, D. (eds) *Palaeoproterozoic of India*. Geological Society, London, Special Publications, **365**, 49–74, http://dx.doi.org/10.1144/SP365.4.

MEERT, J. G., PANDIT, M. K., PRADHAN, V. R., BANKS, J., STROUD, M., NEWSTEAD, B. & GIFFORD, J. 2010. Precambrian crustal evolution of Peninsular India: a 3.0 billion year odyssey. *Journal of Asian Earth Sciences*, **39**, 483–515.

MISHRA, M. & SEN, S. 2008. Geochemistry and origin of Proterozoic porcellanitic shales from Chopan, Vindhyan basin, India. *Indian Journal of Geology*, **80**, 157–171.

MISRA, S. 2006. Precambrian chronostratigraphic growth of Singhbhum–Orissa craton: an alternative model. *Journal of the Geological Society of India*, **67**, 356–378.

MISRA, S., DEOMURARI, M. P., WIDENBECK, M., GOSWAMI, J. N., RAY, S. & SAHA, A. K. 1999. $^{207}Pb/^{206}Pb$ zircon age ages and the evolution of the Singhbhum craton, eastern India: an ion microprobe study. *Precambrian Research*, **93**, 139–151.

MITRA, S. K. & MITRA, S. C. 2001. Tectonic setting of the Precambrian of Northeast India (Meghalaya plateau) and age of the Shillong Group of rocks. *In*: SAXENA, M. B. L. (ed.) *Recent Advances in the Field of Earth Sciences and Their Implications in National Development*. Geological Survey of India, Special Publications, **64**, 653–658.

MONDAL, M. E. A., GOSWAMI, J. N., DEOMURARI, M. P. & SHARMA, K. K. 2002. Ion microprobe $^{207}Pb/^{206}Pb$ ages of zircons from the Bundelkhand massif, northern India: implications for the crustal evolution of the Bundelkhand–Aravalli protocontinent. *Precambrian Research*, **117**, 85–100.

MONDAL, M. E. A., SHARMA, K. K., RAHMAN, A. & GOSWAMI, J. N. 1998. Ion microprobe Pb/Pb zircon ages for the gneiss-granitoid rocks from Bundelkhand massif: evidence for the Archean components. *Current Science*, **74**, 70–75.

MOORBATH, S. & TAYLOR, P. N. 1986. Geochronology and related isotope geochemistry of high grade metamorphic rocks from the lower continental crust. *In*: DAWSON, J. B., CARSWELL, D. A., HALL, J. & WEDEPOHL, K. H. (eds) *The Nature of the Lower Continental Crust*. Geological Society, London, Special Publications, **24**, 211–220.

MOORBATH, S., TAYLOR, P. N. & JONES, N. W. 1986. Dating the oldest terrestrial rocks – facts and fiction. *Chemical Geology*, **57**, 63–80.

MUKHOPADHYAY, D. 1984. The Singhbhum shear zone and its place in the evolution of the Precambrian mobile belt, north Singhbhum. *Indian Journal of Earth Sciences*, CEISM Volume, 205–212.

MUKHOPADHYAY, D. & BHATTACHARYYA, T. 2000. Tectono-stratigraphic framework of the South Delhi fold belt in the Ajmer–Beawar region, central Rajasthan, India: a critical review. *In*: DEB, M. (ed.) *Crustal Evolution and Metallogeny in the Northwestern Indian Shield*. Narosa, New Delhi, 126–137.

MUKHOPADHYAY, D. & MUKHOPADHYAY, C. 2008. Cycles of deformation and metamorphism during Mesoproterozoic orogenesis in North Singhbhum fold belt, eastern India. *In*: *International Conference on Tectonics of the Indian Subcontinent*. International Association for Gondwana Research (IAGR), Conference Series, **5** (Abstract Volume), 170–171.

Mukhopadhyay, D., Bhattacharyya, T., Chattopadhyay, N., Lopez, R. & Tobisch, O. T. 2000. Anasagar gneiss: a folded granitoid in the Proterozoic South Delhi fold belt, central Rajasthan. *Proceedings of the Indian Academy of Science (Earth and Planetary Science)*, **109**, 21–37.

Mukhopadhyay, J., Beukes, N. J., Armstrong, R. A., Zimmermann, U., Ghosh, G. & Medda, R. A. 2008. Dating the oldest greenstone in India: a 3.51-Ga precise U–Pb SHRIMP zircon age for dacitic lava of the Southern Iron Ore Group, Singhbhum craton. *Journal of Geology*, **116**, 449–461.

Naha, K. 1965. Metamorphism in relation to stratigraphy, structure and movements in parts of east Singhbhum, eastern India. *Quarterly Journal of the Geological, Mining & Metallurgical Society of India*, **37**, 41–95.

Naha, K., Srinivasan, R. & Jayaram, S. 1991. Sedimentological, structural and migmatitic history of the Archaean Dharwar tectonic province, southern India. *Proceedings of the Indian Academy of Science (Earth and Planetary Science)*, **100**, 413–433.

Naqvi, S. M. & Rogers, J. J. W. 1987. *Precambrian Geology of India*. Oxford University Press, Oxford.

Naqvi, S. M., Manikyamba, C., Gnaeshawar, R. T., Subba-Rao, D. V. & Ram Mohana, M. 2002. Geochemical and isotopic constraints of late Archaean fossil plume for the evolution of the volcanic rocks of the Sandur Greenstone belt. *Journal of the Geological Society of India*, **60**, 27–56.

Nutman, A. P., Chadwick, B., Krishna Rao, B. & Vasudev, V. N. 1996. SHRIMP U/Pb zircon ages of acid volcanic rocks in the Chitradurga and Sandur groups, and granites adjacent to the Sandur schist belt, Karnataka. *Journal of the Geological Society of India*, **47**, 153–164.

Osborne, I., Sherlock, S., Anand, M. & Argles, T. 2011. New Ar–Ar ages of southern Indian kimberlites and a lamproite and their geochemical evolution. *Precambrian Research*, **189**, 91–103, doi: 10.1016/j.precamres.2011.05.004.

Pandit, M. K., Sial, A. N., Jamrani, S. S. & Ferreira, V. P. 2001. Carbon isotope profiles across the Bilara Group rocks of trans-Aravalli Marwar Supergroup in western India: implications for Neoproterozoic–Cambrian transition. *Gondwana Research*, **4**, 387–394.

Pant, N. C. & Banerjee, D. M. 1990. Patterns of sedimentation in the type Bijawar Basin of central India. *In*: *Precambrian of Central India*. Geological Survey of India, Special Publications, **28**, 653–657.

Patranabis-Deb, S. & Chaudhuri, A. K. 2008. Straigraphic and tectonic evolution of the Mesoproterozoic Chattisgarh basin in central India. *Indian Journal of Geology*, **78**, 139–155.

Radhakrishna, B. P. (ed.) 1987. *Purana Basins of India. Memoirs of the Geological Society of India*, **6**.

Radhakrishna, B. P. & Naqvi, S. M. 1986. Precambrian continental crust of India and its evolution. *Journal of Geology*, **94**, 145–166.

Rao, J. M. 2004. The widespread 2 Ga dyke activity in the Indian shield – evidences from Bundelkhnad mafic dyke swarm, central India and their tectonic implications. *Gondwana Research*, **7**, 1219–1228.

Ramakrishnan, M. & Vaidyanadhan, R. 2008. *Geology of India*, Volume I. Geological Society of India, Bangalore.

Rasmussen, B., Bose, P. K., Sarkar, S., Banerjee, S., Fletcher, I. R. & McNaughton, N. J. 2002. 1.6 Ga U–Pb zircon age for the Chorhat Sandstone, lower Vindhyan, India: possible implications for early evolution of animals. *Geology*, **30**, 103–106.

Ravikant, V. 2010. Palaeoproterozoic (1.9 Ga) extension and breakup along the eastern margin of the Eastern Dharwar Craton, SE India: new Sm–Nd isochron age constraints from anorogenic mafic magmatism in the Neoarchean Nellore greenstone belt. *Journal of Asian Earth Sciences*, **37**, 67–81.

Ray, J. S. 2006. Age of the Vindhyan Supergroup: a review of recent findings. *Journal of Earth System Science*, **115**, 149–160.

Ray, J. S., Veizer, J. & Davis, W. J. 2003. C, O, Pb, Sr isotopic systematics of carbonate sequence of the Vindhyan Supergroup, India: age, diagenesis, correlations, and implications for global events. *Precambrian Research*, **121**, 103–140.

Ray, J. S., Martin, M. W., Veizer, J. & Bowring, S. A., 2002. U–Pb zircon dating and Sr isotope systematic of the Vindhyan SuperGroup, India. *Geology*, **30**, 131–134.

Reddy, S. M. & Evans, D. A. D. 2009. Palaeoproterozoic supercontinents and global evolution: correlations from core to atmosphere. *In*: Reddy, S. M., Mazumder, R., Evans, D. A. D. & Collins, A. (eds) *Palaeoproterozoic Supercontinents and Global Evolution.* Geological Society, London, Special Publications, **323**, 1–26.

Reddy, S., Clarke, C. & Mazumder, R. 2009. Temporal constraints on the evolution of the Singbhum Crustal Province from U–Pb SHRIMP data. *In*: Saha, D. & Mazumder, R. (eds) *Abstract Volume, International Conference on Paleoproterozoic Supercontinents and Global Evolution. Indian Statistical Institute, Kolkata.* International Association for Gondwana Research (IAGR), Conference Series, **9**, 17–18.

Rogers, J. J. W. 1986. The Dharwar craton and the assembly of peninsular India. *Journal of Geology*, **94**, 129–143.

Rogers, J. J. W. & Santosh, M. 2004. *Continents and Supercontinents*. Oxford University Press, Oxford.

Roy, A. & Devarajan, M. K. 2000. A reappraisal of the stratigraphy and tectonics of the Palaeo-Proterozoic Mahakoshal supracrustal belt, central India. *Geological Survey of India Special Publication*, **57**, 79–97.

Roy, A. & Hanuma Prasad, M. 2003. Tectonothermal events in Central Indian Tectonic Zone (CITZ) and its implications in Rodinia crustal assembly. *Journal of Asian Earth Sciences*, **22**, 115–129.

Roy, A., Hanuma Prashad, M. & Bhowmik, S. K. 2001. Recognition of Pre-Grenville and Grenville tectonothermal events in the Central Indian Tectonic Zone: implications on Rodinia crustal assembly. *Gondwana Research*, **4**, 755–757.

Roy, A., Kagami, H., Yoshida, M., Roy, A., Bandyopadhyay, B. K. & Chattopadhyay, A. 2006. Rb/Sr and Sm/Nd dating of different metamorphic events from the Sausar mobile belt, central India: implications for Proterozoic crustal evolution. *Journal of the Asian Earth Sciences*, **26**, 61–76.

Roy, A., Sarkar, A., Jayakumar, S., Aggrawal, S. K. & Ebihara, M. 2002. Mid-Proterozoic plume-related thermal event in eastern Indian craton: evidence from trace elements, REE geochemistry and Sr–Nd isotope systematics of basic-ultrabasic intrusives from Dalma volcanic belt. *Gondwana Research*, **5**, 133–146.

Roy, A. B. 2006. Proterozoic lithostratigraphy and geochronologic framework of the Aravalli mountains and adjoining areas in Rajasthan and neighbouring states, northwest India. *Indian Journal of Geology*, **78**, 5–18.

Roy, A. B. & Jakhar, S. R. 2002. *Geology of Rajasthan (Northwest India): Precambrian to Recent*. Scientific Publishers (India), Jodhpur.

Roy, A. B. & Kröner, A. 1996. Single zircon evaporation ages constraining the growth of the Archaean Aravalli craton, northwestern Indian shield. *Geological Magazine*, **133**, 333–342.

Saha, A. K. 1994. *Crustal Evolution of Singhbhum–North Orissa, Eastern India. Memoirs of the Geological Society of India*, **27**.

Saha, D. 1992. Geologic constraints on depths of tectonic mobility in a Proterozoic intra-cratonic basin. *Indian Minerals*, **46**, 339–355.

Saha, D. 2002. Multi-stage deformation in the Nallamalai fold belt, Cuddapah basin, South India – implications for Mesoproterozoic tectonism along southeastern margin of India. *Gondwana Research*, **5**, 701–719.

Saha, D. 2004. Structural asymmetry and plate tectonic set-up for a Proterozoic fold and thrust belt: Nallamalai Fold Belt and adjoining terrane, south India. *In*: *Proceedings of the Workshop on (IGCP-453): Uniformitarianism Revised Edition on Orogens of India*. Geological Survey of India, Special Publications, **84**, 101–119.

Saha, D. 2011. Dismembered ophiolites in Paleoproterozoic nappe complexes of Kandra and Gurramkonda, South India. *Journal of Asian Earth Sciences*, **42**, 158–175.

Saha, D. & Chakraborty, S. 2003. Deformation pattern in the Kurnool and Nallamalai Groups in the northeastern part (Palnad area) of the Cuddapah Basin, South India and its implication on Rodinia/Gondwana tectonics. *Gondwana Research*, **6**, 573–583.

Saha, D. & Chaudhuri, A. K. 2003. Deformation of Proterozoic successions in the Pranhita-Godavari basin, south India – regional perspectiva. *Journal of Asian Earth Sciences*, **21**, 557–565.

Saha, D. & Ghosh, G. 1998. Lithostratigraphy of deformed Proterozoic rocks from around the confluence of Godavari and Indravati rivers. *Indian Journal of Geology*, **70**, 217–230.

Saha, D. & Tripathy, V. 2012. Palaeoproterozoic sedimentation in the Cuddapah Basin, south India and regional tectonics: a review. *In*: Mazumder, R. & Saha, D. (eds) *Palaeoproterozoic of India*. Geological Society, London, Special Publications, **365**, 159–182, http://dx.doi.org/10.1144/SP365.9.

Saha, D., Chakraborti, S. & Tripathy, V. 2010. Intracontinental thrusts and inclined transpression along eastern margin of the East Dharwar Craton, India. *Journal of the Geological Society of India*, **75**, 323–337.

Saha, D., Ghosh, G., Chakraborty, A. K. & Chakraborti, S. 2006. Comparable Neoproterozoic sedimentary sequences in Palnad and Kurnool subbasins and their paleogeographic and tectonic implications. *Indian Journal of Geology*, **78**, 175–192.

Santosh, M. 2010. A synopsis of recent conceptual models on supercontinent tectonics in relation to mantle dynamics, life evolution and surface environment. *Journal of Geodynamics*, **50**, 116–133.

Santosh, M., Maruyama, S. & Yamamoto, S. 2009. The making and breaking of supercontinents: Some speculations based on superplumes, super downwelling and the role of tectosphere. *Gondwana Research*, **15**, 324–341.

Santosh, M., Yokoyama, K. & Acharyya, S. K. 2004. Geochronolgy and tectonic evolution of Karimnagar and Bhopalpatnam granulite belts, central India. *Gondwana Research*, **7**, 501–518.

Sarangi, S., Gopalan, K. & Kumar, S. 2004. Pb–Pb age of earliest mgascopic eucaryotic alga bearing Rhotas Formation, Vindhyan Supergroup, India: implications for Precambrian oxygen evolution. *Precambrian Research*, **132**, 107–121.

Sarkar, A., Bodas, M. S., Kindu, H. K., Mamgain, V. V. & Ravishankar. 1998. Geochronology and geochemistry of Mesoproterozoic intrusive plutonites from the eastern segment of the Mahakoshal greenstone belt, central India. (Abstract.) *In*: *Precambrian Crust in Eastern and Central India: Proceedings of the International Seminar, UNESCO–IUGS–IGCP-368, 29–30 October, 1998, Bhubaneswar, India*. Geological Survey of India, Special Publications, **57**, 82–86.

Sarkar, A. N. 1982. Precambrian tectonic evolution of eastern India: a model of converging microplates. *Tectonophysics*, **86**, 363–397.

Sarkar, G., Corfu, F., Paul, D. K., McNaughton, N. J., Gupta, S. N. & Bishui, P. K. 1993. Early Archaean crust in Bastar craton, central India: a geochemical and isotopic study. *Precambrian Research*, **62**, 127–137.

Sarkar, S. N., Gopalan, K. & Trivedy, J. 1981. New data on the geochronology of the Precambrians of Bhnadara-Durg, central India. *Indian Journal of Earth Sciences*, **8**, 131–151.

Sarker, S. N., Ghosh, D. K. & Lambert, R. J. St. 1985. Rubidium-strontium lead isotopic studies of the soda granites from Mosabani, Singhbhum Copper Belt, India. *Indian Journal of Earth Science*, **13**, 101–116.

Sengupta, S., Paul, D. K., Bishui, P. K., Gupta, S. N., Chakrabarti, R. & Sen, P. 1994. Geochemical and Rb–Sr isotopic study of Kuilapal granite and Arkasani granophyre from the eastern Indian craton. *Indian Minerals*, **48**, 77–88.

Shamim Khan, M., Smith, T. E., Raza, M. & Huang, J. 2005. Geology, geochemistry and tectonic significance of mafic–ultramafic rocks of the Mesoproterozoic Phulad Ophiolite Suite of south Delhi fold Belt, NW Indian shield. *Gondwana Research*, **8**, 553–566.

Sharma, K. K. & Rahman, A. 2000. The early Archaean–Palaeoproterozoic crustal growth of crustal growth of the Bundelkhand craton, northern Indian shield. *In*: Deb, M. & Goodfellow, W. D. (eds) *Sediment Hosted Lead–Zinc Sulphide Deposits: Attributes and Models of Some Major Deposits in India, Australia and Canada*. Narosa, New Delhi, 51–72.

Sharma, M., Basu, A. R. & Ray, S. L. 1994. Sm–Nd isotopic and geochemical study of the Archaean tonalite-amphibolite association from the eastern Indian craton. *Contributions to Mineralogy and Petrology*, **117**, 45–55.

SHARMA, R. S. 2009. *Cratons and Fold Belts of India*. Springer, Berlin.

SINGH, I. B. 1985. Paleogeography of the Vindhyan basin and its relationship with other Late Proterozoic basins of India. *Journal of the Paleontological Society of India*, **30**, 35–41.

SINHA-ROY, S., MALHOTRA, G. & MOHANTY, M. 1998. *Geology of Rajasthan*. Geological Society of India, Bangalore.

SRIVASTAVA, R. K. & GAUTAM, G. C. 2008. Precambrian mafic dyke swarms from the southern Bastar central Indian craton: present and future perspectives. *In*: SRIVASTAVA, R. K., SHIVAJI, C. H. & CHALAPATHI RAO, N. V. (eds) *Indian dykes: Geochemistry, Geophysics and Geochronology*. Narosa, New Delhi, 367–376.

SRIVASTAVA, R. K. & SINGH, R. K. 2003. The Paleoproterozoic dolerite dyke swarms of southern Bastar craton, Central East India: evidence for the Columbia supercontinent. *Memoirs of the Geological Society of India*, **52**, 163–177.

SRIVASTAVA, R. N., SRIVASTAVA, A. K., SINGH, K. N. & RADCLIFFE, R. P. 2001. Precambrian acid volcanism in the Lower Vindhyan Basin of the Sonbhadra District and adjoining areas of Uttar Pradesh: a review. *In*: *National Seminar Commemorating Dr M. S. Krishnan's Birth centenary (Proceedings Volume)*. Geological Survey of India, Special Publications, **55**, 205–218.

STEIN, H. J., HANNAH, J. L., ZIMMERMAN, A., MARKEY, R. J., SARKAR, S. C. & PAL, A. B. 2004. A 2.5 Ga porphyry Cu–Mo–Au deposit at Malanjkhand, central India: implication for late Archaean continental assembly. *Precambrian Research*, **134**, 189–226.

TOBISCH, O. T., COLLERSON, K. D., BHATTACHARYYA, T. & MUKHOPADHYAY, D. 1994. Structural relationship and Sr–Nd isotope systematics of polymetamorphic granitic gneisses and granitic rocks from central Rajasthan, India: implications for the evolution of the Aravalli craton. *Precambrian Research*, **65**, 319–339.

TORSVIK, T. H., ASHWAL, L. D., TUCKER, R. D. & EIDE, E. A. 2001. Neoproterozoic geochronology and Paleogeography of Seychelles microcontinent: the India link. *Precambrian Research*, **110**, 47–59.

VAN LENTE, B., ASHWAL, L. D., PANDIT, M. K., BOWRING, S. A. & TORSVIK, T. H. 2009. Neoproterozoic hydrothermally-altered basaltic rocks from Rajasthan, northwest India: implications or late Precambrian tectonic evolution of the Aravalli craton. *Precambrian Research*, **170**, 202–222.

VASUDEV, V. N., CHADWICK, B., NUTMAN, A. P. & HEGDE, G. V. 2000. Rapid development of the Hutti schist belt, northern Karnataka: Implications for new field data and SHRIMP U–PB ages. *Journal of the Geological Society of India*, **55**, 529–540.

VIJAYA KUMAR, K. & LEELANANDAM, C. 2008. Evolution of the Eastern Ghats Belt, India: a plate tectonic perspective. *Journal of the Geological Society of India*, **72**, 720–749.

VIJAYA KUMAR, K., ERNST, W. G., LEELANANDAM, C., WOODEN, J. L. & GROVE, M. J. 2010. First Paleoproterozoic ophiolite from Gondwana: geochronologic-geochemical documentation of ancient oceanic crust from Kandra, SE India. *Tectonophysics*, **487**, 22–32.

WEIDENBECK, M., GOSWAMI, J. N. & ROY, A. B. 1996. Stabilization of the Aravalli craton of northwestern India at 2.5 Ga: an ion microprobe zircon study. *Chemical Geology*, **129**, 325–340.

YIN, A., DUBEY, C. S., WEBB, A. A. G., KELTY, T. K., GROVE, M., GEHRELS, G. E. & BURGESS, W. P. 2010*a*. Geologic correlation of the Himalayan orogen and Indian craton: part 1. Structural geology, U–Pb zircon geochronology, and tectonic evolution of the Shillong Plateau and its neighboring regions in NE India. *Bulletin Geological Society of America*, **122**, 336–359, doi: 10.1130/B26460.1.

YIN, A., DUBEY, C. S., KELTY, T. K., WEBB, A. A. G., HARRISON, T. M., CHOU, C. Y. & CÉLÉRIER, J. 2010*b*. Geologic correlation of the Himalayan orogen and Indian craton: part 2. Structural geology, geochronology, and tectonic evolution of the Eastern Himalaya. *Bulletin Geological Society of America*, **122**, 360–395, doi: 10.1130/B26461.1.

Mesoarchaean–Palaeoproterozoic stratigraphic record of the Singhbhum crustal province, eastern India: a synthesis

R. MAZUMDER[1,4]*, A. J. VAN LOON[2], L. MALLIK[3], S. M. REDDY[4], M. ARIMA[5], W. ALTERMANN[6], P. G. ERIKSSON[6] & S. DE[1]

[1]*Geological Studies Unit and Fluvial Mechanics Laboratory, Indian Statistical Institute, 203 B.T. Road, Kolkata 700108, India*

[2]*Institute of Geology, Adam Mickiewicz University, Maków Polnych 16, 61-606 Poznan, Poland*

[3]*Department of Geological Sciences, Jadavpur University, Kolkata 700032, India*

[4]*Department of Applied Geology, The Institute of Geoscience Research, Curtin University of Technology, GPO Box U1987, Perth, WA 6845, Australia*

[5]*Department of Environment and Natural Sciences, Graduate School of Environment and Information Sciences, Yokohama National University, 79-7, Tokiwadai, Hodogaya Yokohama 240-8501, Japan*

[6]*Department of Geology, University of Pretoria, Pretoria 0002, South Africa*

**Corresponding author (e-mail: mazumder.rajat@gmail.com)*

Abstract: The Mesoarchaean–Palaeoproterozoic stratigraphic record of the Singhbhum crustal province, eastern India, implies sedimentation and volcanism in a changing tectonic scenario, and thus assumes immense geological significance. Although efforts have been made by many researchers in the past several decades to summarize various geological aspects of the Singhbhum crustal province, a critical synthesis of various stratigraphic issues was long overdue. The present contribution is an updated critical synthesis of the Mesoarchaean–Palaeoproterozoic stratigraphic record of the Singhbhum crustal province. We have pointed out the problematic stratigraphic issues of the Singhbhum crustal province that deserve careful scrutiny in order to gain better insights into the mode of stratigraphic sequence building.

A relatively rapid change in Earth's surface processes is anticipated across the Archaean–Proterozoic boundary as a consequence of changes in the crust–mantle system and tectonic regime (Condie 1989, 1997; Eriksson *et al.* 2004 and references therein). The Palaeoproterozoic era (2500–1600 Ma) was a time of profound change in Earth evolution and encompassed, perhaps, the first supercontinent cycle, from the amalgamation and dispersal of a possible Late Archaean supercontinent (Aspler & Chiarenzelli 1998) to the formation of the 1900–1800 Ma supercontinent Nuna (Reddy & Evans 2009). The crust was thickened and rigid enough to allow surface and internal geological processes comparable to those prevailing today. Archaean cratons were rifted and terrestrial/continental shelf assemblages were deposited along the truncated margins (Eriksson *et al.* 1999, 2004, 2006; Mazumder *et al.* 2000).

The Mesoarchaean–Palaeoproterozoic rocks in the Singhbhum crustal province encircling the Archaean Singhbhum Granitoid (SG) batholith (Fig. 1a, b) (Saha 1994; Mukhopadhyay 2001) record sedimentation and volcanism in a changing tectonic scenario, and thus assumes immense geological significance (Eriksson *et al.* 1999, 2006; Mazumder *et al.* 2000; Mazumder 2005). Although efforts have been made by many researchers in the last several decades to summarize various geological aspects of the Singhbhum crustal province (Dunn & Dey 1942; Sarkar & Saha 1962; Naha 1965; Mukhopadhyay 1976, 1988, 2001; Sarkar 1984; Bose 1994, 2009; Saha 1994; Gupta & Basu 2000; Mazumder *et al.* 2000; Sengupta & Mukhopadhyay 2000), a critical synthesis of various stratigraphic issues was long overdue (cf. Mazumder 2005). The present contribution presents an updated synthesis of the Mesoarchaean–Palaeoproterozoic stratigraphic record of the Singhbhum crustal province. An attempt has been made to highlight the problematic as well as the prospective stratigraphic issues that researchers may find useful.

From: MAZUMDER, R. & SAHA, D. (eds) 2012. *Palaeoproterozoic of India*. Geological Society, London, Special Publications, **365**, 31–49. http://dx.doi.org/10.1144/SP365.3

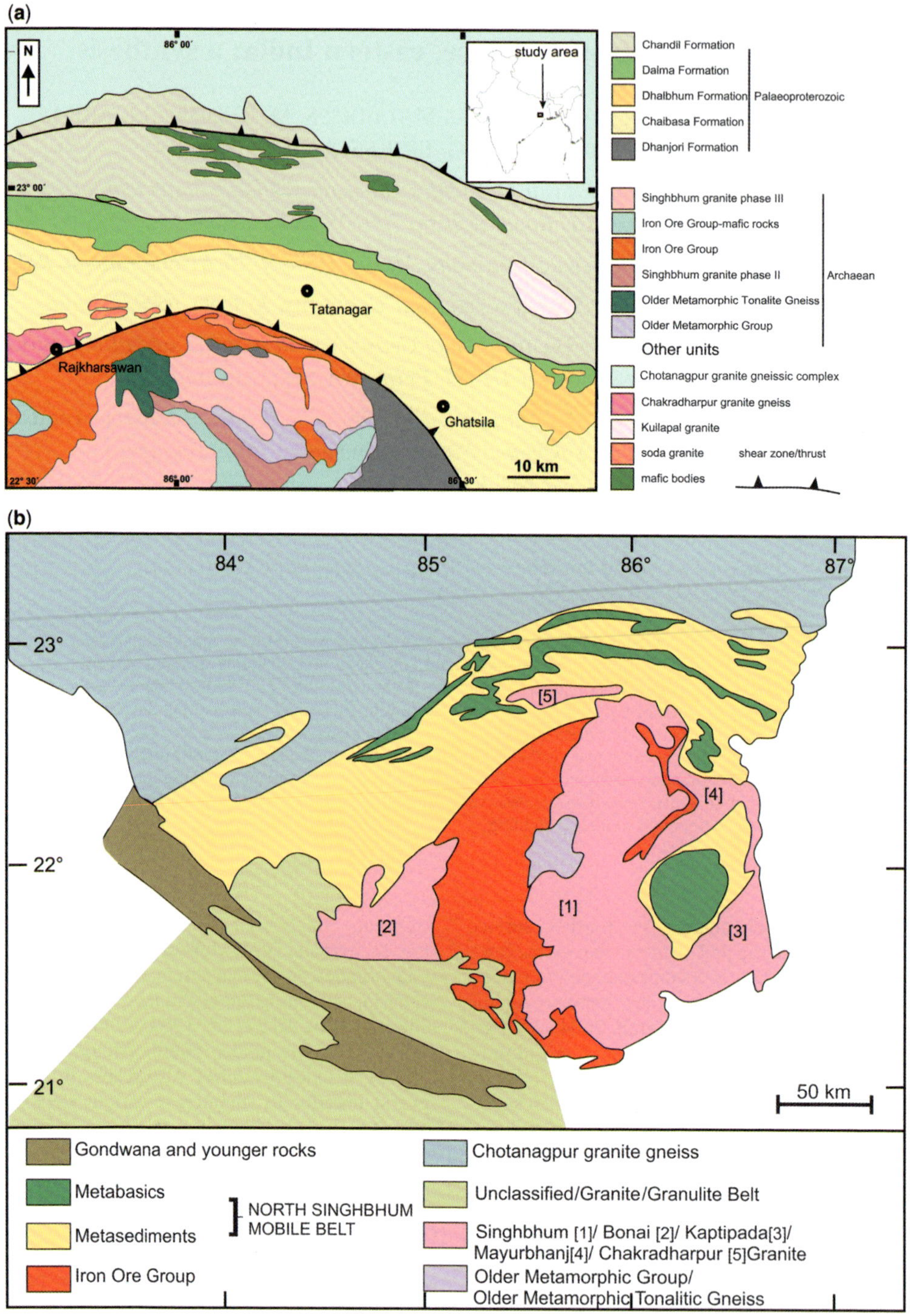

Fig. 1. (**a**) Geological map showing disposition of Archaean and Proterozoic units of the Singhbhum crustal province (after Saha 1994; Sengupta *et al.* 2000). The arcuate shear zone occurring to the north of the Singhbhum Granite and the IOG rocks is known as the Singhbhum Shear Zone (SSZ). (**b**) Simplified geological map (modified after Mukhopadhyay 2001) showing the disposition of the Older Metamorphic Group (OMG), the Older Metamorphic Tonalite Gneiss (OMTG), the Archaean granitoids and the Iron Ore Group (IOG). Note that the IOG occurs to the east, west and south of the Archaean Singhbhum Granitoid.

Mesoarchaean–Late Archaean stratigraphic record

The earliest supracrustal rocks in the Singhbhum crustal province, pristine sandstone and shale enclaves within the tonalitic gneissic basement, include pure orthoquartzite (Older Metamorphic Group (OMG): Fig. 1a, b). The $^{207}Pb/^{206}Pb$ ages of OMG zircons from the orthoquartzite cluster around 3.55, 3.4 and 3.2 Ga (see Goswami *et al.* 1995; Mishra *et al.* 1999). The approximately 3.55 Ga zircons are rounded–subrounded, indicating that these are detrital in origin. These detrital OMG zircons suggest an older limit of approximately 3.5 Ga for the age of sedimentation in the Singhbhum crustal province (Goswami *et al.* 1995; Mishra *et al.* 1999) (Table 1). Basu *et al.* (1996) analysed zircon grains from the pelitic schists of the Older Metamorphic Group and reported an upper intercept age of 3.35 Ga in the U–Pb Concordia diagram. These authors interpreted this age as the first Pb-loss event of older zircons (Basu *et al.* 1996). The OMG rocks are intruded by the Older Metamorphic Tonalite Gneiss (OMTG: Fig. 1a, b) (Sharma *et al.* 1994). Ion microprobe $^{207}Pb/^{206}Pb$ ages of zircons from the OMTG cluster around approximately 3.4 and 3.2 Ga (Mishra *et al.* 1999). Acharyya *et al.* (2010*a*) reported 3448 ± 19 and 3527 ± 17 Ma U–Pb zircon ages from the OMTG. The approximately 3.4 Ga age is believed to represent the time of formation of the OMTG (see also Acharyya *et al.* 2010*a*, *b*), and 3.2 Ga to be the time of a widespread metamorphic event that affected both the OMG and the OMTG (Mishra *et al.* 1999).

The Iron Ore Group rocks

The Mesoarchaean succession in the Singhbhum crustal province begins with the Iron Ore Group

Table 1. *Geochronological data from the Singhbhum crustal province and summary of Mesoarchaean–Palaeoproterozoic–Mesoproterozoic geodynamic events*

Event	Age and data source	Geochronological method
Thermal events (1500–1400 Ma)	1487 ± 34 Ma: Sengupta *et al.* (2000)	Rb–Sr age
	1547 ± 20 Ma: Saha (1994)	K–Ar age
Crystallization of Chandil Rhyolite and *c.* 1600 Ma events	1628.5 ± 4.3 Ma: Reddy *et al.* (2009)	U–Pb SHRIMP concordant age
	1619 ± 38 Ma: Roy *et al.* (2002*b*)	Rb–Sr WR age
	c. 1580 Ma: Krishna Rao *et al.* (1979)	Pb/Pb age
	1677 ± 11 Ma: Sarkar *et al.* (1986)	Rb–Sr, Pb/Pb ages
	1638 ± 38 Ma: Sengupta *et al.* (1994)	Rb–Sr age
Thermal events (2100–2000 Ma)	2072 ± 106 Ma: Roy *et al.* (2002*a*, *b*)	Sm–Nd age
	c. 2008 Ma: Iyenger *et al.* (1981)	Rb–Sr age
	c. 1960 Ma: Vohra *et al.* (1991)	Rb–Sr age
Thermal events *c.* 2800 Ma (emplacement of Temperkola Granite; emplacement of the main phase of Mayurbhanj Granite)	*c.* 2800 Ma: Bandyopadhyay *et al.* (2001)	$^{207}Pb/^{206}Pb$ age
	c. 2800 Ma: Acharyya et al. (2010*a*, *b*)	U–Pb zircon age
Metamorphic event	3241 ± 7 Ma: Mishra *et al.* (1999)	$^{207}Pb/^{206}Pb$ age
Emplacement of Singhbhum Granitoid phase III	3288 ± 8 Ma: Reddy *et al.* (2009)	U–Pb SHRIMP concordant age
Emplacement of Singhbhum Granitoid phase II*	3328 ± 7 Ma: Mishra *et al.* (1999)	$^{207}Pb/^{206}Pb$ age
	3290 ± 8.6 Ma: Tait *et al.* (2011)	U–Pb SHRIMP concordant age
Thermal Pb-loss event	*c.* 3350 Ma: Basu *et al.* (1996)	U–Pb age
Emplacement of OMTG–IOG sedimentation	3437 ± 9 Ma: Mishra *et al.* (1999)	$^{207}Pb/^{206}Pb$ age
	3448 ± 19 Ma: Basu *et al.* (2008), Acharyya *et al.* (2010*a*, *b*)	U–Pb zircon age
OMG sedimentation	*c.* 3500 Ma: Mishra *et al.* 1999	$^{207}Pb/^{206}Pb$ age
Dacitic volcanism	3506.8 ± 2.3 Ma: Mukhopadhyay *et al.* (2008) (oldest Indian greenstone succession)	U–Pb age
Crust formation including a significant granitoid component (detrital OMG zircons)	*c.* 3600 Ma: Mishra *et al.* (1999)	$^{207}Pb/^{206}Pb$ age

*No geochronological data are yet available for the Singhbhum Granite Phase I (Saha 1994).

(IOG) (Fig. 1a, b), characterized by banded iron formations (BIF), clastic sedimentary rocks and minor carbonates (Chakraborty & Majumder 1986; Saha 1994; Chakraborty 1996; Bhattacharya *et al.* 2007; Mukhopadhyay *et al.* 2008). Saha *et al.* (1988) and Saha (1994) suggested that phase III of the Singhbhum Granite (SG) is intrusive into the IOG rocks, while phases I and II of the SG are older than it. The IOG rocks flank the Singhbhum Granodiorite pluton on the eastern (Gorumahisani–Badampahar, the eastern belt), western (Noamundi–Jamda–Koira, the western belt) and southern flanks (Tomka–Daitari, the southern belt) (Fig. 1b). In all of these belts, the BIFs are closely associated with volcanic rocks and volcaniclastic deposits (Saha 1994; Mukhopadhyay 2001; Ghosh & Mukhopadhyay 2007; Bose 2009; Ghosh *et al.* 2010). In all three belts, the succession starts with a sandstone–conglomerate association at the base, followed upwards by ferruginous shales, tuffs, ultramafic–mafic lavas (locally felsic) and BIFs (Dunn & Dey 1942; Murthy & Acharya 1975; Saha 1994; Mukhopadhyay 2001; Mukhopadhyay *et al.* 2008). A podiform chromiferous ultramafic body (Augé *et al.* 2003; Mondal *et al.* 2006, 2007) overlies the BIF in the southern IOG belt with a thrust contact (Mukhopadhyay *et al.* 2008). On the basis of geochemical considerations, Dasgupta *et al.* (1999) suggested a freshwater environment for cherts associated with the BIFs. According to Rao & Dasgupta (1995), kaolinite-rich argillites and their inferred granitic provenance indicate that the argillites associated with the BIFs of the western IOG succession were probably river-borne sediments. According to some researchers, primary sedimentary structures preserved within the BIFs are indicative of a shallow-water environment with current action and occasional desiccation under subaerial conditions (Majumder & Chakraborty 1977; Rai *et al.* 1980; Majumder 1985). Clearly, the palaeogeographical setting of the IOG rocks is somewhat speculative and detailed sedimentary facies analysis will be of paramount importance.

Although carbonates are rare in the IOG succession, siliceous dolomites with stromatolites have been reported from a few places within the western IOG belt (Avasthy 1980; Grant *et al.* 1980; Sarkar 1989). Some spheroidal and filamentous structures in the BIF have been described as microfossils (Sarkar 1984). Stromatolites and associated microbiota are observed in rocks containing bands of chert, jasper and hematite, and also in nodular dolomite (Sarkar 1989). Sarkar (1989) has inferred an intertidal–subtidal palaeogeography for the BIF of the western IOG belt on the basis of the presence of columnar stromatolites. However, these studies are elementary, and detailed geobiological research of the inferred stromatolites, microfossils and associated sedimentary structures are urgently needed in order to trace the early history of life in these rocks (cf. Noffke 2010).

One of the major controversial issues of the Precambrian stratigraphy of the Singhbhum is whether or not the eastern, western and southern BIFs (Fig. 1b) are of the same age (Dunn 1940; Dunn & Dey 1942; Banerji 1977; Sarkar & Saha 1977, 1983; Iyenger & Murthy 1982; Acharya 1984; Saha 1994; Mukhopadhyay 2001 and references therein). Researchers have noted some differences in the mineralogy, geochemistry and lithological associations of the eastern, western and southern BIFs, particularly between the western and eastern BIFs (Acharya 1984; Chakraborty & Majumder 1986; Saha 1994; Mukhopadhyay 2001; Bose 2009). Metabasalts of the western IOG belt have a calc-alkaline trend, whereas the eastern IOG metavolcanic rocks show a tholeiitic trend (Sengupta *et al.* 1997). The western IOG succession comprises quartz arenites, overlain by BIF and tholeiites but, unlike other platformal greenstone belts, does not have komatiites (Bose 2009, p. 15). Komatiitic rocks are, however, reported from the eastern IOG belt (Sahu & Mukherjee 2001). Bose (2009) commented that the diagnostic trace element data in support of the komatiitic nature of these volcanics are lacking. An intensive geochemical and petrological study of these suspected komatiitic rocks may help us to gain valuable insights into the extant mantle processes and crust–mantle interactions. A group of researchers has speculated that the BIFs of all three belts are of the same age (Dunn 1940; Dunn & Dey 1942; Sarkar & Saha 1977, 1983; Basu *et al.* 2008). However, Banerji (1977), Iyenger & Murthy (1982), Acharya (1984) and Bose (2009) suggested that these BIFs have different ages. Acharya (1984) proposed that the BIF of the eastern belt is the oldest and is unconformably overlain by the BIF of the southern belt in the Malayagiri Hill area, whereas the BIF of the southern IOG belt is unconformably overlain by the BIF of the western belt. Thus, according to Acharya (1984), the BIF of the western belt is the youngest of the three BIFs. Subsequent field studies and structural analyses (see also Mukhopadhyay 2001, p. 314), however, did not confirm the existence of angular unconformities as suggested by Acharya (1984). Basu *et al.* (2008) reported a 3392 ± 29 Ma U–Pb concordant age of zircons collected from tuffs lying less than 1 m below the BIF of the western IOG belt. These authors have assigned this age to the immediately overlying BIF and suggested a similar age for all of the BIFs of the Singhbhum crustal province (Basu *et al.* 2008). Although Mukhopadhyay *et al.* (2008) reported a precise 3506.8 ± 2.3 Ma U–Pb SHRIMP (Sensitive High Resolution Ion

Microprobe) zircon age for dacitic lava from the southern IOG, the BIF conformably overlying the dacitic lava flow has not been dated. Thus, definite field evidence and precise geochronological data from the eastern, western and southern BIFs are of paramount geological importance to establish their mutual age relations. Clearly, a comprehensive systematic study of the volcanic stratigraphy and an understanding of the chemical evolution of lavas and their correlation in the three IOG basins (Fig. 1b) is of highest priority from regional, as well as Archaean, geological points of view (see Bose 2009).

Another controversial issue of the Precambrian stratigraphy of the Singhbhum crustal province is the temporal relationship between the SG and the IOG rocks. Interested readers should consult Mukhopadhyay (2001) for a detailed discussion on this issue. To summarize, one group of researchers considers the SG to be the basement of the IOG succession (Mukhopadhyay 1976; Banerji 1977; Banerjee 1982; Iyenger & Murthy 1982) while another group (Dunn 1929; Dunn & Dey 1942; Sarkar & Saha 1977, 1983) considers the SG to be younger. Subsequently, Saha *et al.* (1988) and Saha (1994) proposed that the SG phases I and II acted as the basement of IOG sedimentation (and, hence, are older than the IOG succession) while the SG phase III is intrusive into these rocks. If so, we believe that a precise U–Pb SHRIMP crystallization age of the SG phase III can constrain the age of the BIF if direct age determination is otherwise difficult. An attempt has been made to date the SG intrusive into the BIF (i.e. SG phase III) in the Deo River section (Paul *et al.* 1991), but the SG has yielded an errorchron age of 3145 $\pm$ 282 Ma because of its disturbed isotopic system. This reflects, however, the antiquity of the BIF (Paul *et al.* 1991; see also Mukhopadhyay 2001).

Effort has been made to compare the OMG and the eastern IOG (Gorumahisani–Badampahar) supracrustals. Acharyya (1993) compared the mafic enclaves of the OMG and the eastern IOG rocks and suggested that both the OMG and the eastern IOG supracrustals belong to the same succession. According to Acharyya *et al.* (2010*a*, *b*), the OMG supracrustals south of Holudpukur (cf. Saha 1994) are of lower metamorphic grade, and comprise pillow basalts, gabbros, ultramafics, ferruginous arenites and tuffaceous rocks that closely resemble the IOG assemblage. However, Mukhopadhyay (2001) questioned such a comparison because the OMTG shows a distinct intrusive relation with the OMG but no BIF raft has yet been reported from the OMTG. In addition, the sillimanite–garnet-bearing pelitic schists, which are an important constituent of the OMG, are absent within the IOG (Mukhopadhyay 2001, p. 315). A metamorphic discordance between the OMG and the IOG has been pointed out by Sarkar & Saha (1962).

As mentioned earlier, the maximum age of OMG sedimentation is around 3500 Ma (Goswami *et al.* 1995; Misra *et al.* 1999) (Table 1). No direct age data are yet available from the eastern IOG belt. Basu *et al.* (2008) reported an approximate 3400 Ma age for the western IOG belt and believed that all three IOG successions are of the same age. If we accept that the eastern IOG belt is approximately 3400 Ma old, then the OMG is definitely older than the eastern as well as the western IOG successions, and both the eastern and western IOG successions are broadly contemporaneous or slightly younger than the OMTG (Table 1). This possibly explains the absence of BIF enclaves within the OMTG as pointed out by Mukhopadhyay (2001). Earlier, Basu *et al.* (1996) reported an approximate 3350 Ma age from the zircons of the pelitic schists of the OMG and interpreted this age as the first Pb-loss event of the older zircons. But this approximate 3350 Ma Pb-loss event has not yet been documented from the IOG successions. The OMG consists of amphibolite-facies pelitic schists, quartz–magnetite–cummingtonite schists, quartzites, banded calc-gneiss, and para- and ortho-amphibolites (Mukhopadhyay 2001). Mukhopadhyay (2001, p. 309) speculated that the quartz–cummingtonite–magnetite schists probably represent metamorphosed iron formation. It thus appears that there are broad lithological similarities between OMG and IOG supracrustals, although their metamorphic grades are different. Mukhopadhyay *et al.* (2008) reported an approximate 3506 Ma age (see Table 1) from the dacitic lava of the southern IOG belt. Geochronological data (Goswami *et al.* 1995; Mishra *et al.* 1999; Mukhopadhyay *et al.* 2008) (Table 1) in combination with comparable lithological assemblages thus suggest that the southern IOG succession and the OMG are contemporaneous. A detailed sedimentological analysis, coupled with a U–Pb geochronological investigation, of the OMG and IOG successions is therefore essential to resolve these regional stratigraphic problems.

The Birtola Formation

An approximately 200 m-thick siliciclastic succession (Birtola Formation: Mahalik 1987) overlies the Iron Ore Group (*c.* 3400 Ma: Basu *et al.* 2008) and the Bonai Granite (*c.* 3163 Ma: Sengupta *et al.* 1991) (Fig. 2a). The Birtola Formation represents the lowermost lithostratigraphical unit of the three-tiered Darjing Group (Mahalik 1987, table 1; Kundu & Matin 2007). Rocks overlying the Birtola Formation have been intruded by the 2809 $\pm$ 12 Ma ($^{207}Pb/^{206}Pb$ age of zircons: see Bandyopadhyay *et al.* 2001) Tamparkola Granite.

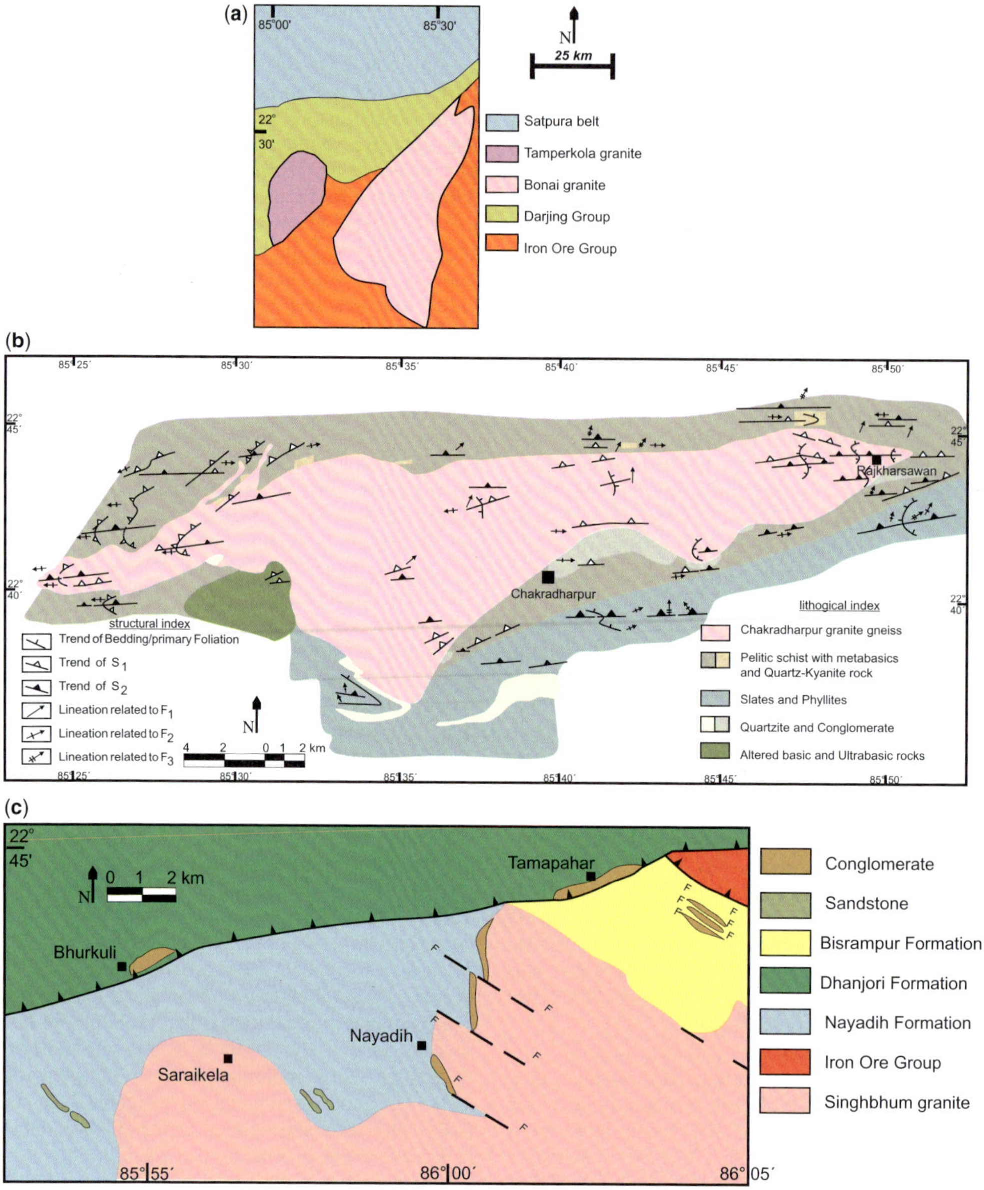

Fig. 2. (**a**) Geological map (after Saha *et al.* 2004) showing the distribution of the Darjing Group. (**b**) Geological map (after Bandyopadhyay 1981) in and around Rajkharsawan (Fig. 1a), showing unclassified conglomerates. (**c**) Geological map (modified after Bhattacharya & Mahapatra 2008) showing the location of the Nayadih Formation and its bounding units.

According to Bandyopadhyay *et al.* (2001), the Birtola Formation is thus older than 2800 Ma. Chaki *et al.* (2005) reported a Rb–Sr age of 2746 ± 144 Ma from the Tamparkola Granite. These authors suggested, however, based on their field observations, that the Tamparkola Granite possibly acted as the basement onto which the Darjing Group was deposited. Therefore, in their view, the Darjing Group cannot be older than about 2700 Ma (Chaki *et al.* 2005, p. 436). Thus, the minimum age of deposition of the Birtola Formation is ambiguous.

The Birtola succession starts with a thick conglomerate that grades upwards into coarse-grained feldspathic sandstone, which bears an original sedimentary texture (Mahalik 1987). Interbedded with this arenite is a greenish conglomerate with schistose sandy matrix. The conglomerate contains angular to well-rounded pebbles of BIF, quartzite and granite (Mahalik 1987, p. 77). From the lithological assemblages and the preserved sedimentary structures, four broad lithofacies have been recognized by Chakrabarti *et al.* (1998) in and around the Patasahi area where the Birtola Formation is well exposed. While the basal interlayered conglomerate and cross-bedded sandstone facies bears braided fluvial and intertidal signatures, a medium- to fine-grained, crudely laminated massive sandstone and a thick-bedded cross-stratified well-sorted sandstone facies contain evidence for dominant-wave processes, suggesting a highstand systems tract (Chakrabarti *et al.* 1998). The sandstones are characterized by well-rounded unaltered feldspars of different varieties. On the basis of their preliminary sedimentological studies, Chakrabarti *et al.* (1998) claimed that the Birtola siliciclastics possibly preserve the first record of a Late Archaean glaciation in the Indian subcontinent. Understandably, a detailed sedimentological verification is of paramount importance in order to justify their interpretation.

Unclassified siliciclastics of Late Archaean (?) age

There are some unclassified siliciclastic deposits in the Singhbhum crustal province of which the stratigraphic status and ages are very poorly known (Fig. 2b, c). Interesting conglomerate bodies are exposed in the Rajkharsawan–Chakradharpur area (Fig. 2b; Sarkar 1984). They are embedded within pelitic and mafic schists, and are locally directly overlying the Chakradharpur Granite Gneiss (Bandyopadhyay 1981). De (1957) presented some preliminary data on conglomerate deformation in the Rajkharsawan area. Initial investigation of this conglomeratic body in and around Udalkam village, south of the Rajkharsawan railway station, reveals that this is essentially a poorly sorted clast-supported conglomerate (Fig. 3a). The clasts are flattened owing to tectonic deformation, giving rise to pebble lineation. At places, this conglomerate contains folded clasts (Fig. 3b). A careful scrutiny of this conglomerate deposit further reveals that it grades into medium- to coarse-grained sandstone (quartzite). The conglomerate–sandstone association is thus made up of fining-upwards cycles (authors' unpublished data). The stratigraphic status of this conglomerate–sandstone–schist association is ambiguous. Mazumder *et al.* (2000) earlier considered this conglomeratic unit to be part of the Dhanjori succession. These authors further suspected that the conglomerate is possibly of glacial origin, but recent study revealed that this is not the case and that the conglomerates, unlike the lower Dhanjori basal conglomerate exposed in the type area (Singpura–Narwapahar sector: see Mazumder & Sarkar 2004, figs 2 & 4), contains BIF pebbles in addition to quartzite and slate pebbles. Thus, the Rajkharsawan conglomerates seem not to be part of the Dhanjori succession. The presence of BIF pebbles clearly shows that this conglomerate, the associated sandstones and the surrounding schists are definitely younger than the BIF of the IOG. It thus remains unclear whether the Rajkharsawan sediments are part of the Mesoarchaean or the Palaeoproterozoic. Detailed sedimentological and structural geological studies are essential in order to understand the provenance, depositional setting and the deformational history of this siliciclastic succession.

Bhattacharya & Mahapatra (2008, fig. 2a) have identified a poorly metamorphosed sedimentary succession (Nayadih Formation: Fig. 2c) that is less deformed than the IOG rocks; it skirts the western flank of the SG. Considering the less deformed and poorly metamorphosed nature of the succession and the occurrence of IOG chert pebbles in the basal conglomerate of this succession, an age younger than that of the Archaean IOG has been assigned to it (Bhattacharya & Mahapatra 2008, p. 304). Detailed sedimentary facies analysis of the Nayadih Formation is, however, yet to be done in order to identify the depositional environment and palaeogeography, and the precise sedimentation age of the Nayadih Formation is yet to be determined.

Newer Dolerite dyke swarm

An interesting dyke swarm, known as 'Newer Dolerite dykes' (Dunn & Dey 1942) in the Singhbhum stratigraphy (Saha *et al.* 1973; Saha 1994; Bose 2008) (Fig. 4), is spectacularly developed within the SG pluton (Dunn & Dey 1942; Saha 1949, 1994; Bose 2008; Mazumder & Saha 2009). This dyke swarm is also present in other granitoid bodies of the Singhbhum crustal province (Saha 1994; Mukhopadhyay 2001 and references therein). The dominant trend of these dykes is NNE–SSW to NE–SW, and a subsidiary trend is NW–SE to east–west (Dunn & Dey 1942; Mukhopadhyay 2001). The most common rock type is quartz dolerite (metamorphosed to amphibolite), with minor occurrences of norite and acidic rocks such as microgranite and syenodiorite (Saha 1994; Mukhopadhyay 2001). Associated with these dolerite dykes are some ultramafic dykes (Bose & Goles 1971; Saha *et al.* 1973; Saha 1994; Bose 2000, 2008; Roy *et al.* 2004). The readers may consult Bose

Fig. 3. (**a**) Clast-supported conglomerate near Udalkam village, Rajkharsawan (see Fig. 1a for the location of Rajkharsawan). (**b**) Folded pebbles within the Rajkharsawan conglomerate. The conglomerate suffered folding after lithification, as is obvious from the fold geometry of the pebbles and the matrix, and from the development of weak cleavage in the matrix parallel or subparallel to the axes of the folds.

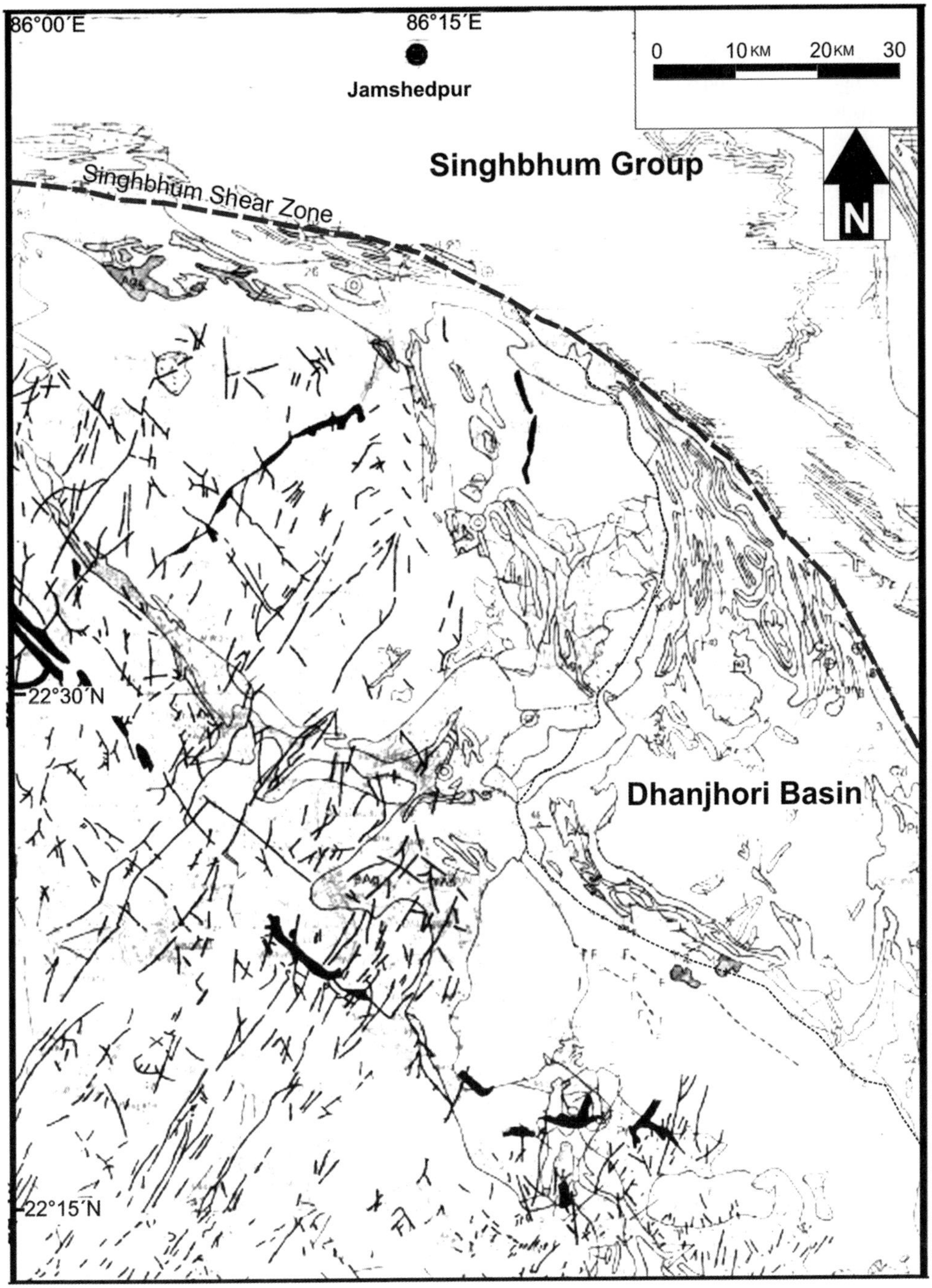

Fig. 4. Geological map showing the dolerite dykes of the Singhbhum crustal province (after Dunn & Dey 1942; Mandal *et al.* 2006).

(2008) for an overview of the petrology and geochemistry of the Newer Dolerite and associated ultramafic dykes. The dyke swarm is indicative of the termination of cratonization in the Singhbhum crustal province when the continental crust was sufficiently cooled down and became rigid enough to be fractured (Mukhopadhyay 2001; Bose 2009). Reliable geochronological data from the dolerite dyke swarm are lacking (K–Ar ages range from 923 to 2144 Ma: see Saha 1994 and references therein; Mukhopadhyay 2001). According to Mallick & Sarkar (1994), the available data (mostly K–Ar ages) apparently indicate three pulses of mafic intrusive activity, viz. 2100 ± 100, 1500 ± 100 and 1100 ± 200 Ma. In contrast, Roy *et al.* (2004) suggested that the ultramafic members of these swarms are 2613 ± 177 Ma (Rb–Sr isochron age) and claimed that these ultramafic members of the Newer Dolerite dyke swarm attest to the oldest stabilization event of all of the cratonic blocks of India.

The dyke swarm denotes a fracture system generated through north–south tectonic compression acting on a brittle cratonic nucleus. The fractures, filled by mafic magma, gradually evolved through nucleation, propagation and subsequent coalescence (see Mandal *et al.* 2006). However, it is hitherto unclear whether the mafic and ultramafic magmas were genetically related or whether these two compositionally different (mafic and ultramafic) sets of dykes represent two independent, temporally distinct magma sources and had different emplacement histories. It is therefore essential to test the genetic linkage between the mafic and ultramafic dykes through critical field and petrographical studies supplemented by trace element and REE (rare earth element) geochemistry. Palaeomagnetic studies by Verma & Prasad (1974) suggested three different episodes of dyke intrusion. The temporal variation of magma chemistry and their consequences in physical behaviour of the fracture systems should be critically examined through numerical experiments. It would be interesting to examine the behaviour of magma over a possible flattened large plume head (Bose 2009).

Gabbro–anorthosite ultramafic intrusives

A few gabbro–anorthosite ultramafic intrusives occur along the eastern margin of the SG pluton (Saha 1994). Petrogenetic studies of these intrusives reveal two magma types: one related to the ultramafic products and the other to gabbro–anorthosite emplacements (Deb & Chakraborty 1960; Mondal *et al.* 2006). The zircon ages of the Bangur Gabbro intruding the Baula Complex and its breccia matrix cluster around 3121 ± 3 Ma according to Augé *et al.* (2003), who estimated that stabilization of the cratonic nucleus and inception of normal magmatic processes occurred at about 3100 Ma.

Palaeoproterozoic stratigraphical record

The oval-shaped Archaean nucleus of Singhbhum (incorporating OMG, OMTG, SG and other related granitic bodies, and the IOG rocks; Fig. 1b), is enveloped by Proterozoic rocks both to the north and to the south (Fig. 1a) (Saha *et al.* 1988; Saha 1994; Mukhopadhyay 2001). The interested reader may consult Saha (1994) and Mukhopadhyay (2001) for a detailed account of the granitoids of Singhbhum – such as, for example, the Bonai Granite, the Chakradharpur Granite Gneiss, the Mayurbhanj Granite and the Soda Granite. The Proterozoic volcano-sedimentary succession occurring to the north with a deformation style that differs distinctly from that of the Archaean nucleus has been inferred to constitute the Singhbhum Mobile Belt or the North Singhbhum Mobile Belt (Saha 1994; Gupta & Basu 2000; Dasgupta 2004; Sengupta & Chattopadhyay 2004). The presence of orogens in Singhbhum had been anticipated by earlier geologists (Dunn & Dey 1942; Saha 1994 and references therein), although the elements of the orogen(s) and their boundaries remained unidentified (cf. Sengupta & Chattopadhyay 2004).

The Palaeoproterozoic deformed and metamorphosed (greenschist–amphibolite) Singhbhum supracrustal rocks (e.g. the Dhanjori, Chaibasa, Dhalbhum and Dalma formations; Fig. 1a), constituting the so-called North Singhbhum Mobile Belt (cf. Gupta & Basu 2000; Dasgupta 2004; Sengupta & Chattopadhyay 2004), are sandwiched between the SG batholith in the south and the Chottanagpur Gneissic Complex (CGC) in the north. They occur as a 200 km-long and 50–60 km-wide roughly east–west-trending curvilinear belt (Bose 1994; Saha 1994; Mukhopadhyay 2001).

The Singhbhum Shear Zone

A zone of intense shearing, the Singhbhum Shear Zone (SSZ), occurs close to the northern and eastern margins of the SG pluton (Ghosh *et al.* 1985; Ghosh & Sengupta 1987; Saha 1994; Joy & Saha 1998, 2000; Mukhopadhyay 2001) (Fig. 1a). A difference in opinion exists on various aspects of the SSZ, particularly on its lateral continuity, width and the extent of the rocks that it has affected (Joy & Saha 1998; Sengupta & Mukhopadhyay 2000; Sengupta & Chattopadhyay 2004; see Saha 1994 and references therein). The mesoscopic structures from the SSZ indicate that it is a major northwards-dipping (or northeastwards-dipping) shear zone with a southwards thrust movement (Ghosh & Sengupta 1987; Joy 1996). More recently, the idea of a more extensive effect of ductile shearing affecting rocks within the SSZ is promoted (Joy & Saha 1998, 2000; Sengupta & Mukhopadhyay 2000; Sengupta & Chattopadhyay 2004). Joy & Saha (2000) identified two zones with a strong L–S fabric within the Dhanjori Formation between the SSZ and the SG pluton, and named them the 'Jublatola Shear Zone' and the 'Rohinibera Shear Zone', respectively (Joy & Saha 2000, fig. 2). Contrary to the earlier interpretation (Sarkar & Saha 1962; Saha 1994), it has been established that the SSZ does not mark the interface between the Singhbhum Archaean nucleus and the Proterozoic North Singhbhum Mobile Belt/North Singhbhum Fold Belt (Blackburn & Srivastava 1994; Gupta & Basu 2000). The most dominant structures in the entire North Singhbhum Fold Belt are the east–west regional folds with a strong axial-planar fabric, which are the second-generation structures of the region (Sengupta & Chattopadhyay 2004, p. 28). It has been inferred further that the second deformation across both the fold belt and the SSZ has been contemporaneous and non-coaxial in both. All shear-sense indicators in the SSZ demonstrate an upwards transport of the northern rocks relative to the rocks in the south (Sengupta & Chattopadhyay 2004). The Rb–Sr whole-rock age of about 1600 Ma obtained from the Soda Granite (Sarkar *et al.* 1986; see also Krishna Rao *et al.* 1979: Pb/Pb age; Table 1) has been inferred to reflect the age of thrusting along the SSZ (Sengupta & Mukhopadhyay 2000). Sengupta & Mukhopadhyay (2000, p. 55) further suggested that the end of the thrusting event post-dates the eruption/consolidation age of the Dalma–Chandil volcano-sedimentary formations at approximately 1500 Ma. According to Sengupta & Chattopadhyay (2004), the SSZ represents a fold–thrust belt that affects the shallow-marine Chaibasa Formation; they also state that the SSZ is intracontinental in nature, possibly involving no major displacement. Mazumder *et al.* (2000) and Mazumder (2005) have established that the alluvial-fan/fluvial Dhanjori Formation (cf. Mazumder & Sarkar 2004) is conformably overlain by the Chaibasa Formation, which formed in a shallow-marine to shelf setting (Bose *et al.* 1997; Mazumder 2004; Mazumder & Arima 2005; Mazumder *et al.* 2006, 2009). The Dhanjori Formation is of Late Archaean–Palaeoproterozoic age (poorly constrained between 2600 and 2100 Ma: Acharyya *et al.* 2010*b*) and the Chaibasa Formation is of Palaeoproterozoic age (*c.* 2200 Ma: Sarkar *et al.* 1986). Understandably, shearing along the SSZ is a much younger geological event (*c.* 1600 Ma) than the formation of the units (>2100 Ma) affected by it. The U–Pb zircon SHRIMP ages from the Soda Granite and the precise K–Ar and/or Ar/Ar mineral ages of biotite and/or amphibole from the SSZ schists are essential to constrain the associated tectonothermal event(s) more precisely.

The Palaeoproterozoic supracrustal rocks are confined mostly to the north of the Archaean nucleus (the so-called North Singhbhum Fold Belt, see the earlier discussion), although roughly comparable and tentatively Palaeoproterozoic supracrustals (Mazumder *et al.* 2012) are located to the SE and SW of the Archaean nucleus as well (Fig. 1a). The Late Professor A. K. Saha and his colleagues (Sarkar & Saha 1962, 1977, 1983; Saha *et al.* 1988; Saha 1994) considered the SSZ as an interface (suture zone, see Sarkar & Saha 1977) between the Archaean Singhbhum cratonic mass and the Proterozoic Singhbhum Mobile Belt, and proposed a stratigraphic classification of the Singhbhum Precambrian rocks in which the lithostratigraphic units occurring along the periphery of the SG pluton were compartmentalized into those occurring: (1) to the south of the Singhbhum Shear Zone; and (2) to the north of the Singhbhum Shear Zone (Sarkar & Saha 1962; Saha *et al.* 1988; see also Saha 1994). This stratigraphic classification has become well accepted among the students and researchers of Singhbhum geology. More recently, the stratigraphic relationship between the Dhanjori and Chaibasa formations originally proposed by Saha and his colleagues (i.e. that the Chaibasa Formation is older than the Dhanjori Group: Saha *et al.* 1988) has been criticized (Sarkar & Deb 1971; Mukhopadhyay 1976; Gupta *et al.* 1980, 1985; Bose 1994). On the basis of field relationships and available geochronological data, it is now established that the Dhanjori Formation is older than the Chaibasa Formation (Mukhopadhyay 1976; Sarkar *et al.* 1986; Roy *et al.* 2002*a*; Acharyya *et al.* 2010*a*, *b*). Subsequent studies also clarified that the SSZ is not a plate boundary as proposed by Sarkar & Saha (1977). The Dhanjori Formation has been affected by shearing, and shows an excellent development of strong L–S tectonites in the quartzites, grits and metabasic rocks. Rocks within the SSZ show the same paragenesis as the Dhanjori Formation except for the presence of kyanite (Joy & Saha 2000). The SSZ has been interpreted as a fold–thrust belt that also affected the Chaibasa Formation (Sengupta & Mukhopadhyay 2000; Sengupta & Chattopadhyay 2004). Therefore the ambiguous stratigraphic classification of Saha and his colleagues (see Saha 1994; Mishra *et al.* 1999, table 1) should be treated with caution. The assumption that the lithostratigraphic units constituting the North Singhbhum Mobile Belt have roughly the same age (Bhattacharya & Mahapatra 2008, table1) is not supported by geochronological data (see Sarkar *et al.* 1986; Nelson *et al.* 2007; Reddy *et al.* 2009; Acharyya *et al.* 2010*b*) (see Table 1). We hope that, with the availability of more precise geochronological data, particularly those from the tentative Palaeoproterozoic successions occurring along the SW margins of the SG pluton, a new chronostratigraphic succession can be established.

The northern Palaeoproterozoic successions

The Palaeoproterozoic meta-sedimentary succession north of the Archaean Singhbhum nucleus is made up of the successively younger Dhanjori, Chaibasa, Dhalbhum, and Dalma formations (Fig. 1a). Earlier researchers assigned a 'Group' status for the Dhanjori rocks (Sarkar & Saha 1962; Saha *et al.* 1988; Saha 1994), and the Dhanjori Group was further subdivided into a Lower and an Upper Dhanjori Formation (Gupta *et al.* 1985). Subsequently, Mazumder (2002) assigned a 'Formation' status to the Dhanjoris, and the Lower Dhanjori Formation and the Upper Dhanjori Formation of Gupta *et al.* (1985) were classified as the Lower Dhanjori Member and the Upper Dhanjori Member, respectively (Mazumder & Arima 2004, 2009; Mazumder & Sarkar 2004; Mazumder 2005). The entirely terrestrial (alluvial-fan/fluvial) Dhanjori Formation is conformably overlain by the marine Chaibasa Formation. The Dhanjori–Chaibasa contact is therefore a sequence boundary (Mazumder 2005; Mazumder *et al.* 2012).

The Chaibasa Formation is entirely marine, and is made up of an alternation of fine-grained sandstones, heterolithic units (alternating sandstones and shales) and shales (Bhattacharya 1991; Bose *et al.* 1997; Bhattacharya & Bandyopadhyay 1998). The sandstones were formed in a subtidal setting, whereas the heterolithical units and the shales were formed in a shelf setting above and below the storm wave base, respectively (Bose *et al.* 1997; Mazumder 2005; Mazumder *et al.* 2009). The characteristics of the Chaibasa sedimentary facies are indicative of ongoing shallowing (Mazumder 2005).

In contrast to the underlying marine Chaibasa Formation, the overlying Dhalbhum Formation is characterized by medium- to coarse-grained, poorly sorted, compositionally immature sandstones (Mazumder 2003, 2005, fig. 10a). Sarkar & Saha (1962) determined the boundary between the Chaibasa and Dhalbhum formations on the basis of combined LANDSAT imagery and field data, although they could not correlate the boundary with any structural discontinuity in the field (see also Naha 1965, pp. 45–46). A drastic change in terms of texture and primary structures of sandstones, as well as in palaeocurrent direction, has, however, been used to differentiate the Dhalbhum Formation from the underlying Chaibasa Formation (cf. Naha & Ghosh 1960; Mazumder 2005). Compositional immaturity, a coarser character, poor sorting and fining-upwards cycles suggest that the Dhalbhum sandstones are of fluvial origin (Mazumder 2005). In addition, the presence of fine-grained well-sorted sandstones with large-scale cross-bedding and translatent strata (Mazumder 2005, fig. 10e; Mazumder *et al.* 2012) indicates that these are aeolian deposits (Simpson *et al.* 2004; Mazumder 2005). Thus, the Dhalbhum Formation consists of terrestrial deposits. The contact between the Chaibasa and the Dhalbhum formations is an unconformity (sequence boundary; cf. Posamantier & Vail 1988). The southerly Dhalbhum palaeocurrent trend close to the Chaibasa–Dhalbhum contact (Naha & Ghosh 1960; Naha 1965; Mazumder 2005) indicates that the sediments were derived from the north. It is interesting to note that, in contrast to the underlying Chaibasa Formation, the Dhalbhum Formation contains volcanic and volcaniclastic rocks, indicating a volcanic influence during sedimentation. Analysis of the provenance of these Palaeoproterozoic clastic rocks has not yet been carried out but is envisaged to be undertaken using heavy-mineral analysis.

The Dalma Formation conformably overlies the Dhalbhum Formation and is represented by a thick succession of mafic–ultramafic volcanic rocks with lenses of basic agglomerates (Chakraborti 1980; Gupta *et al.* 1980; Chakraborti & Bose 1985; Bose 1994; Gupta & Basu 2000; Mazumder 2005). The agglomerates are coarse-grained, characterized by angular basaltic fragments embedded in lava (Mazumder *et al.* 2012, fig. 11a) and are interlayered with the basalts (Gupta *et al.* 1980). Researchers have proposed continental rift (Dunn & Dey 1942; De 1964; Gupta *et al.* 1980), island arc (Naha & Ghosh 1960), back-arc (Bose *et al.* 1989; Bose 1994) and plume (Roy *et al.* 2002*b*; Mazumder 2005) models for the Dalma volcanism. The tectonic regime of the Dalma volcanism has been interpreted on the basis of inadequate geochemical data. The stratigraphic positions of most of the samples of Dalma volcanic rocks are unknown. This is the major impediment to infer the temporal and/or spatial change in geochemistry of the volcanics. Van Loon & Mazumder (2009) studied the volcaniclastic deposits occurring at the base of the Dalma Formation between Tatanagar and Chandil, and suggested that these are lahar (volcanic mudflow) deposits, but detailed facies analysis of the Dalma Formation is still to be carried out.

The overlying Chandil Formation includes quartzites, mica schists, carbonaceous phyllites, weakly metamorphosed acidic volcanic and volcaniclastic rocks (including vitric and lithic tuffs), and amphibolites (Bose 1994; Ray *et al.* 1996; Singh 1997, 1998; Sengupta *et al.* 2000; Mazumder 2005). Volcaniclastic rocks are abundant in the Chandil succession, which suggests volcanic influences on sedimentation. Preliminary studies reveal that both terrestrial and marine deposits occur within the formation (Singh 1997, 1998; Gupta & Basu 2000; Mazumder 2005). The sedimentary development and architecture are not clearly understood at present, and detailed sedimentary-facies and sequence-stratigraphic analyses remain to be performed (cf. Mazumder 2005).

The southern Palaeoproterozoic successions

The southern Palaeoproterozoic successions include the Ongarbira metasedimentary and metavolcanic rocks (Mukhopadhyay *et al.* 1990; Blackburn & Srivastava 1994), the Kolhan Group (Ghosh & Chatterjee 1994; Saha 1994; Mukhopadhyay *et al.* 2006), the Simlipal metavolcanic rocks (Sarkar & Saha 1962; Saha 1994) and the Jagannathpur mafic lava (Sarkar & Saha 1962; Banerjee 1982; Saha 1994). The Ongarbira volcano-sedimentary succession rests unconformably on a topographic low in the basement between the Chakradharpur Granite and the Singhbhum Granite (Blackburn & Srivastava 1994). Mafic (tholeiitic basalts–ultramafic pyroxenites) volcanic rocks, felsic volcaniclastics (tuffaceous metasediments) and low-grade pelites constitute the Ongarbira volcano-sedimentary succession (Blackburn & Srivastava 1994). The entire Ongarbira volcano-sedimentary succession closely resembles the Dalma–Chandil succession, and it has been claimed that both the Ongarbira and Dalma volcanic rocks occupy comparable structural horizons, and that the continuity between them is lost only because of attenuation and faulting on an intervening fold limb (Mukhopadhyay *et al.* 1990; see also Acharyya 2003, pp. 22–23).

The Kolhan Group (basal conglomerate–sandstone assemblage overlain by phyllitic shale and impersistent limestone) is almost undeformed and unmetamorphosed, and unconformably overlies the Singhbhum Granite and the IOG (Saha 1994; Mukhopadhyay 2001). The intercalated conglomerate–sandstone units, sheet sandstones and sandstone–siltstone alternations of the lower part of the Kolhan Group are inferred to have formed in a braided stream, on a beach and on a wave-dominated shelf (Ghosh & Chatterjee 1994; Bandopadhyay & Sengupta 2004; Bandopadhyay *et al.* 2010). Mukhopadhyay *et al.* (2006) studied the Kolhan Group in the Chaibasa–Jagannathpur area and inferred a Meso- to Neoproterozoic age (on the basis of lithostratigraphic and interpreted genetic correlations with other units), and interpreted a deep-water cratonic palaeogeography for the Kolhan Group. The age of this group is hitherto unknown.

The Simlipal and Jagannathpur volcano-sedimentary successions have been correlated with the Proterozoic Dhanjori volcano-sedimentary succession based solely on lithostratigraphy (cf. Sarkar & Saha 1962, 1977; Goodwin 1996, p. 338). Geochronological, sedimentological and stratigraphic analyses are essential to resolve the controversy regarding the depositional and tectonic setting of the southern Palaeoproterozoic litho-units and to correlate them with the northern Palaeoproterozoic successions (Mazumder *et al.* 2012).

Synthesis

The Singhbhum crustal province contains an extensive stratigraphic record across the Archaean–Proterozoic boundary (Mukhopadhyay 2001; Mazumder 2005) and therefore is of prime importance for understanding the Earth's evolution across this significant transition (Reddy & Evans 2009). It is at least possible if not likely that different cratons during this time span experienced significantly dissimilar geodynamic evolutions, as reflected in their variable supracrustal depositional records (e.g. Eriksson *et al.* 2011). While there were undoubtedly some global-scale events (e.g. the 'Great Oxidation Event' at *c.* 2.4–2.3 Ga; the first global glaciation at *c.* 2.4–2.2 Ga; and a possible 'magmatic shutdown' at *c.* 2.45–2.2 Ga; cf. Young 2004; Condie *et al.* 2009; Holland 2009), common to most preserved supracrustal cratonic records, many other facets of their evolution were apparently unique (Eriksson *et al.* 2011). The challenge thus lies in discriminating global and craton-specific events, and in this regard each ancient craton is of great importance. Unravelling the development of the Singhbhum crustal province thus has the potential to make a seminal contribution to these ongoing studies. Critical analyses of the stratigraphic record of Singhbhum in the light of recent geochronological findings indicate that there are several regional stratigraphic problems that deserve careful scrutiny. Here we point out those problematic issues of Singhbhum Precambrian stratigraphy that may motivate researchers to undertake further research.

- Provenance and sedimentary environment of OMG protoliths and their correlation with the IOG rocks.
- Age and correlation of the three IOG belts, and the IOG–OMTG relationship.
- Stratigraphic relationships between the Birtola, Nayadih and other tentative Archaean successions (e.g. Rajkharsawan siliciclastics, Fig. 3a, b), and their sedimentation age.
- Provenance characterization of the Singhbhum Group rocks and the Chandil Formation to understand the volcanic influence on sedimentation.
- Detailed geochemical (trace and HFSE element) characterization of the Dalma volcanic rocks, and their comparison with the Dhanjori and Ongarbira volcanic rocks.
- Age and possible correlation of the southern Palaeoproterozoic successions (e.g. the Kolhan, Ongarbira, Jagannathpur, Simlipal rock units) with northern Singhbhum equivalents.

We believe that the debate concerning the issues highlighted above need to be resolved in order to understand the relationship between the Singhbhum and other crustal fragments of India during the Mesoarchaean–Palaeoproterozoic transition.

The authors are grateful to Professors T. Bhattacharya, A.R. Basu and A. Basu for their critical comments and suggestions on an earlier version of the paper. Field trips and geoscience research in Singhbhum over the years were supported by the Japan Society for the Promotion of Science (2002–2004 to R. Mazumder and M. Arima), the Alexander von Humboldt Foundation (2005–2006 to R. Mazumder), the Ludwig-Maximilians University Munich (2006 to W. Altermann and R. Mazumder), University of Pretoria and National Research Foundation of South Africa (to P.G. Eriksson since 1999), the Indian Statistical Institute (2006–2008 to R. Mazumder), the Foundation Dr Schürmannfonds (The Netherlands, 2008, 2009, 2011 to A. J. van Loon), the Curtin University of Technology (2008 to S.M. Reddy), the University Grants Commission, Government of India (2008–2009 to L. Mallik), and the Department of Science and Technology, Government of India (2009–2010 to R. Mazumder and S. De). R. Mazumder is grateful to the participants of the UNESCO–IGCP 509 sponsored field workshop in Singhbhum in November 2009 for thought-provoking discussions, and to the Institute of Geoscience Research (TIGER), Department of Applied Geology, Curtin University of Technology, for a short-term Visiting Scientist position in September 2010 and for infrastructural facilities. This is an IGCP 509 contribution.

References

ACHARYA, S. 1984. Stratigraphic and structural evolution of the rocks of the iron ore basins in Singhbhum–Orissa iron ore province, India. *In*: *Crustal Evolution of the Indian Shield and its Bearing on Metallogeny, Seminar Volume*. Indian Society of Earth Science, Calcutta, 19–28.

ACHARYYA, S. K. 1993. Greenstones from Singhbhum craton, their Archaean character, oceanic crustal affinity and tectonics. *Proceedings National Academy of Science India, Section A*, **63**, 211–222.

ACHARYYA, S. K. 2003. A plate tectonic model for Proterozoic crustal evolution of Central Indian Tectonic Zone. *Gondwana Geological Magazine*, **7**, 9–31.

ACHARYYA, S. K., GUPTA, A. & ORIHASHI, Y. 2010*a*. New U–Pb zircon ages from Palaeo-Mesoarchaean TTG gneisses of the Singhbhum Craton, eastern India. *Geochemical Journal*, **44**, 81–88.

ACHARYYA, S. K., GUPTA, A. & ORIHASHI, Y. 2010*b*. Neoarchaean–Palaeoproterozoic stratigraphy of the Dhanjori basin, Singhbhum Craton, Eastern India: and recording of a few U–Pb zircon dates from its basal part. *Journal of Asian Earth Science*, **39**, 527–536, doi: 10.1016/j.jseaes.2010.04.023.

ASPLER, L. B. & CHIARENZELLI, J. R. 1998. Protracted breakup of Kenorland, a Neoarchean supercontinent? Geochronologic, tectonostratigraphic and sedimentologic evidence from the Paleoproterozoic. *Sedimentary Geology*, **120**, 75–104.

AUGÉ, T., COCHERE, A., GENNA, A., ARMSTRONG, R., GUERROT, C., MUKHARJEE, M. M. & PATRA, R. N. 2003. Age of the Baula PGE mineralization (Orissa, India) and its implications concerning the Singhbhum Archaean nucleus. *Precambrian Research*, **121**, 85–101.

AVASTHY, R. K. 1980. Stromatolites from Iron Ore Formation of Bonai–Keonjhar District, Orissa, India. *In*: *Stromatolites: Characteristics and Utility*. Geological Survey of India, Miscellaneous Publications, **44**, 54–56.

BANDOPADHYAY, P. C. & SENGUPTA, S. 2004. The Palaeoproterozoic supracrustal Kolhan Group in Singhbhum craton, India and the Indo-African supercontinent. *Gondwana Research*, **7**, 1228–1235.

BANDOPADHYAY, P. C., ERIKSSON, P. G. & ROBERTS, R. J. 2010. A vertic paleosol at the Archean–Proterozoic contact from the Singhbhum–Orissa craton, eastern India. *Precambrian Research*, **177**, 277–290.

BANDYOPADHYAY, P. K. 1981. Chakradharpur Granite Gneiss–A composite Batholith in the western part of the Singhbhum Shear Zone, Bihar. *Indian Journal of Earth Sciences*, **8**, 109–118.

BANDYOPADHYAY, P. K., CHAKRABARTI, A. K., DEOMURARI, M. P. & MISRA, S. 2001. 2.8 Ga old anorogenic granite-acid volcanic association from western margin of Singhbhum–Orissa craton, eastern India. *Gondwana Research*, **4**, 465–475.

BANERJI, A. K. 1977. On the Precambrian banded iron-formations and the manganese ores of the Singhbhum region, eastern India. *Economic Geology*, **72**, 90–98.

BANERJEE, P. K. 1982. *Stratigraphy, Petrology and Geochemistry of Some Precambrian Basic Volcanic and Associated Rocks of Singhbhum District, Bihar and Mayurbhanj and Keonjhar Districts, Orissa. Memoirs of the Geological Survey of India*, **111**.

BASU, A. R., BANDYOPADHYAY, P. K., CHAKRABORTI, R. & ZOU, H. 2008. Large 3.4 Ga Algoma type BIF in the Eastern Indian Craton. *In*: *Goldschmidt Conference 2008, Vancouver (Canada), Abstract Volume. Geochimica et Cosmochimica Acta*, **72**, (12S), A59.

BASU, A. R., SHARMA, M. & PREMO, W. R. 1996. U–Pb age of an older Metamorphic Group mica schist: earliest terrain of the eastern Indian Craton. *In*: SAHA, A. K. (eds) *Recent Researches in Geology and Geophysics of the Precambrians: Essays in Honour of Prof. Samarendra Nath Sarkar on his Seventyfith Birth Anniversary*. Recent Researches in Geology, **16**, 93–102.

BHATTACHARYA, H. N. 1991. A reappraisal of the depositional environment of the Precambrian metasediments around Ghatshila-Galudih, eastern Singhbhum. *Journal of the Geological Society of India*, **37**, 47–54.

BHATTACHARYA, H. N. & BANDYOPADHYAY, S. 1998. Seismites in a Proterozoic tidal succession, Singhbhum, Bihar, India. *Sedimentary Geology*, **119**, 239–252.

BHATTACHARYA, H. N. & MAHAPATRA, S. 2008. Evolution of the Proterozoic rift margin sediments–North Singhbhum Mobile Belt, Jharkhand–Orissa, India. *Precambrian Research*, **162**, 302–316.

BHATTACHARYA, H. N., CHAKRABORTY, I. & GHOSH, K. 2007. Geochemistry of some banded iron-formations of the Archaean supracrustals, Jharkhand–Orissa region, India. *Journal of Earth System Sciences*, **116**, 245–260.

BLACKBURN, W. H. & SRIVASTAVA, D. C. 1994. Geochemistry and tectonic significance of the Ongarbira metavolcanic rocks, Singhbhum district, India. *Precambrian Research*, **67**, 181–206.

BOSE, M. K. 1994. Sedimentation pattern and tectonic evolution of the Proterozoic Singhbhum basin in the eastern Indian shield. *Tectonophysics*, **231**, 325–346.

BOSE, M. K. 2000. Mafic–ultramafic magmatism in the eastern Indian craton – A review. *In*: *National Seminar Commemorating Dr M. S. Krishnan's Birth centenary (Proceedings Volume)*. Geological Survey of India, Special Publications, **55**, 227–258.

BOSE, M. K. 2008. Petrology and geochemistry of Proterozoic 'Newer Dolerite' and associated ultramafic dykes within Singhbhum granite pluton, eastern India. *In*: SRIVASTAVA, R. K., SIVAJI, C. & CHALAPATHI RAO, N. V. (eds) *Indian Dykes: Geochemistry, Geophysics, and Geochronology*. Narosa, New Delhi, 413–445.

BOSE, M. K. 2009. Precambrian Mafic Magmatism in the Singhbhum Craton, Eastern India. *Journal of the Geological Society of India*, **73**, 13–35.

BOSE, M. K. & GOLES, G. G. 1971. Chemical petrology of the ultramafic minor intrusions of Singhbhum, Bihar. *In*: *Second Symposium Upper Mantle Project*. National Geophysical Research Institute, Hyderabad, 305–326.

BOSE, M. K., CHAKRABARTI, M. K. & SAUNDERS, A. D. 1989. Petrochemistry of the lavas from Proterozoic Dalma volcanic belt, Singhbhum, eastern India. *Geologische Rundschau*, **78**, 633–648.

BOSE, P. K., MAZUMDER, R. & SARKAR, S. 1997. Tidal sandwaves and related storm deposits in the transgressive Protoproterozoic Chaibasa Formation, India. *Precambrian Research*, **84**, 63–81.

CHAKI, A., BHATTACHARYA, D., RAO, J. S., CHATURBEDI, A. K. & BAGCHI, A. K. 2005. Geochronology of the granitoids of Kunjar area, Sundargarh District, Orissa: implications to the regional stratigraphy. *Journal of the Geological Society of India*, **65**, 428–440.

CHAKRABARTI, A., BANDYOPADHYAY, P. K. & CHAKRABARTY, A. K. 1998. Rounded fresh feldspar in the Late Archaean nearshore deposits around Patasahi, Orissa: glacial activity? *Indian Journal of Earth Sciences*, **25**, 94–107.

CHAKRABORTI, M. K. 1980. On the pyroclastic rocks of Dalma volcanic sequence, Singhbhum, Bihar. *Indian Journal of Earth Sciences*, **7**, 216–222.

CHAKRABORTI, M. K. & BOSE, M. K. 1985. Evaluation of the tectonic setting of Precambrian Dalma volcanic belt, eastern India using trace element data. *Precambrian Research*, **28**, 253–268.

CHAKRABORTY, K. L. 1996. A comprehensive geological review of the Precambrian banded iron formation (BIF) of Bihar and Orissa, eastern India. *Indian Journal of Geology*, **68**, 211–236.

CHAKRABORTY, K. L. & MAJUMDER, T. 1986. Geological aspects of the banded Iron Formation of Bihar and Orissa. *Journal of the Geological Society of India*, **28**, 109–133.

CONDIE, K. C. 1989. *Plate Tectonics and Crustal Evolution*, 3rd edn. Pergamon, Oxford.

CONDIE, K. C. 1997. *Plate Tectonics and Crustal Evolution*, 4th edn. Butterworth–Heinmann, Oxford.

CONDIE, C. K., O'NEILL, C. & ASTER, R. 2009. Evidence and implications for a widespread magmatic shutdown for 250 My on Earth. *Earth and Planetary Science Letters*, **282**, 294–298.

DASGUPTA, S. 2004. Modelling ancient orogens – an example from North Singhbhum Mobile Belt. *In*: *Proceedings of the Workshop on (IGCP-453): Uniformitarianism Revised Edition on Orogens of India*. Geological Survey of India, Special Publications, **84**, 33–42.

DASGUPTA, H. C., SAMBASIVA RAO, V. V. & KRISNA, C. 1999. Chemical environment of deposition of ancient iron- and manganese-rich sediments and cherts. *Sedimentary Geology*, **125**, 83–98.

DE, A. 1957. Conglomerate deformation in the western part of the Singhbhum Shear Zone. Proceedings of the Indian Science Congress Abstract, 168–169.

DE, A. 1964. Precambrian Dalma Dhanjori volcanicity of E. India and its stratigraphic significance. *Proceedings of the International Geological Congress*, **10**, 59–70.

DEB, S. & CHAKRABORTY, K. L. 1960. Trend of differentiation in the gabbro-anorthosite suite of rocks, Nausahi. *Proceedings of the National Institute of Sciences*, **26A**, 427.

DUNN, J. D. 1929. The Geology of north Singhbhum including parts of Ranchi and Manbhum districts. *Memoirs of the Geological Survey of India*, **54**, 1–66.

DUNN, J. D. 1940. The stratigraphy of South Singhum. *Memoirs of the Geological Survey of India*, **63**, 303–369.

DUNN, J. D. & DEY, A. K. 1942. The Geology and Petrology of Eastern Singhbhum and surrounding areas. *Memoirs of the Geological Survey of India*, **69**, 281–456.

ERIKSSON, P. G., CATUNEANU, O., NELSON, D. R., MUELLER, W. U. & ALTERMANN, W. 2004. Towards a synthesis. *In*: ERIKSSON, P. G., ALTERMANN, W., NELSON, D. R., MUELLER, W. U. & CATUNEANU, O. (eds) *The Precambrian Earth: Tempos and Events*. Elsevier, Amsterdam, 739–769.

ERIKSSON, P. G., LENHARDT, N., WRIGHT, D. T., MAZUMDER, R. & BUMBY, A. J. 2011. Late Neoarchaean–Palaeoproterozoic supracrustal basin-fills of the Kaapvaal craton: relevance of the supercontinent cycle, the 'Great Oxidation Event' and'Snowball Earth'? *Marine and Petroleum Geology*, **28**, 1385–1401.

ERIKSSON, P. G., MAZUMDER, R., CATUNEANU, O., BUMBY, A. J. & OUNTSCHE ILONDO, B. 2006. Precambrian continental freeboard and geological evolution: a time perspective. *Earth-Science Reviews*, **79**, 165–204.

ERIKSSON, P. G., MAZUMDER, R., SARKAR, S., BOSE, P. K., ALTERMANN, W. & VAN DER MERWE, R. 1999. The 2.7–20 Ga volcano-sedimentary record of Africa, India and Australia: evidence for global and local changes in sea level and continental freeboard. *Precambrian Research*, **97**, 269–302.

GHOSH, G. & MUKHOPADHYAY, J. 2007. Reappraisal of the structure of the Western Iron Ore Group, Singhbhum craton, eastern India: implications for the exploration of BIF-hosted iron ore deposits. *Gondwana Research*, **12**, 525–532.

GHOSH, G., KUMARI, S., PATIL, S. K., MUKHOPADHYAY, J. & ARIJIT, R. 2010. Superposed deformation fabrics in the Precambrian metabasic rocks of the Iron Ore Group, Singhbhum craton, Eastern India: evidences from anisotropy of magnetic susceptibility studies. *Journal of Structural Geology*, **32**, 249–261.

GHOSH, S. K. & CHATTERJEE, B. K. 1994. Depositional mechanism as revealed from grain size measures of the Paleoproterozoic Kolhan siliciclastics, Keonjhar district, Orissa, India. *Sedimentary Geology*, **89**, 181–196.

GHOSH, S. K. & SENGUPTA, S. 1987. Progressive development of structures in a ductile shear zone. *Journal of Structural Geology*, **9**, 277–287.

GHOSH, S. K., SARKAR, S. C. & SENGUPTA, S. 1985. *A Field Guide for Ghatshila and Neighbouring Mineral Belt*. COSIP-ULP in Geology. Jadavpur University, Kolkata.

GOODWIN, A. M. 1996. *Principles of Precambrian Geology*. Academic Press, London.

GOSWAMI, J. N., MISHRA, S., WIEDENBECK, M., RAY, S. L. & SAHA, A. K. 1995. $^{207}Pb/^{206}Pb$ ages from the OMG, the oldest recognized rock unit from Singhbhum–Orissa Iron Ore craton, E. India. *Current Science (Bangalore)*, **69**, 1008–1012.

GRANT, P. R., MURTY, V. N. & SENGUPTA, S. 1980. The first record of stromatolites from the Koira Group (Iron Ore Series) Precambrian of Bihar–Orissa, India. *In*: *Stromatolites: Characteristics and Utility*. Geological Survey of India, Miscellaneous Publications, **44**, 49–53.

GUPTA, A. & BASU, A. 2000. North Singhbhum Proterozoic mobile belt Eastern India—a review. *In*: *National Seminar Commemorating Dr M. S. Krishnan's Birth centenary (Proceedings Volume)*. Geological Survey of India, Special Publications, **55**, 195–226.

GUPTA, A., BASU, A. & GHOSH, P. K. 1980. The Proterozoic ultramafic and mafic lavas and tuffs of the Dalma greenstone belt, Singhbhum, eastern India. *Canadian Journal of Earth Sciences*, **17**, 210–231.

GUPTA, A., BASU, A. & GHOSH, P. K. 1982. Ultramafic volcaniclastics of the Precambrian Dalma volcanic belt, Singhbhum, Eastern India. *Geological Magazine*, **119**, 505–510.

GUPTA, A., BASU, A. & SINGH, S. K. 1985. Stratigraphy and petrochemistry of Dhanjori greenstone belt, eastern India. *Quarterly Journal of the Geological, Mineralogical and Metallurgical Society of India*, **57**, 248–263.

HOLLAND, H. D. 2009. Why the atmosphere became oxygenated: a proposal. *Geochimica et Cosmochimica Acta*, **73**, 5241–5255.

IYENGER, S. V. P. & MURTHY, Y. G. K. 1982. The evolution of the Archaean Proterozoic crust in parts of Bihar and Orissa, eastern India. *Geological Survey of India Record*, **112**, 1–5.

IYENGER, S. V. P., CHANDY, K. C. & NARAYANASWAMY, R. 1981. Geochronology and Rb–Sr systematics of the igneous rocks of the Simlipal complex, Orissa. *Indian Journal of Earth Sciences*, **8**, 61–65.

JOY, S. 1996. *Kinematics of Deformation and Quartz c-Axis Fabric in a Part of Singhbhum Shear Zone and its Footwall, Singhbhum District, Bihar*. PhD thesis, University of Calcutta.

JOY, S. & SAHA, D. 1998. Influence of micaceous impurity on dynamically recrystallized quartz c-axis fabric in L–S tectonites from the Singhbhum Shear Zone and its footwall, Eastern India. *Journal of Structural Geology*, **20**, 1509–1520.

JOY, S. & SAHA, D. 2000. Dynamically recrystallised quartz c-axis fabrics in greenschist facies quartzites, Singhbhum shear zone and its footwall, eastern India – influence of high fluid activity. *Journal of Structural Geology*, **22**, 777–793.

KRISHNA RAO, N., AGARWAL, S. K. & RAO, G. V. U. 1979. Lead isotopic ratio of uraninites and the age of uranium mineralisation in Singhbum Shear Zone, Bihar. *Journal of the Geological Society of India*, **20**, 24–27.

KUNDU, A. & MATIN, A. 2007. Sedimentary and tectonic environment of deposition of the Archaean Birtola Formation of Singhbhum Orissa craton, Sundergarh District, Orissa. *Indian Journal of Geology*, **79**, 22–34.

MAHALIK, N. K. 1987. Geology of rocks lying between Gangpur Group and Iron Ore Group of the Horse Shoe Syncline in North Orissa. *Indian Journal of Earth Sciences*, **14**, 73–83.

MAJUMDER, T. 1985. Environment and petrogenesis of the Precambrian Banded Iron Formation of the Iron Ore Group, eastern India. *Geological Survey of Finland Bulletin*, **331**, 201–213.

MAJUMDER, T. & CHAKRABORTY, C. 1977. Primary sedimentary structures in the Banded Iron Formation of Orissa. *Sedimentary Geology*, **19**, 287–300.

MALLICK, A. K. & SARKAR, A. 1994. Geochronology and geochemistry of mafic dykes from the Precambrian of Keonjhar, Orissa. *Indian Minerals*, **48**, 13–24.

MANDAL, N., MITRA, A. K., MISRA, S. & CHAKRABORTY, C. 2006. Is the outcrop topology of dolerite dykes of the Precambrian Singhbhum craton fractal? *Journal of Earth System Sciences*, **115**, 643–660.

MAZUMDER, R. 2002. *Sedimentation History of the Dhanjori and Chaibasa Formations, Eastern India and its Implications*. PhD thesis, Jadavpur University, Kolkata.

MAZUMDER, R. 2003. Correlations between the Eastern Block of the North China Craton and the South Indian Block of the Indian Shield: an Archaean to Palaeoproterozoic link–comment. *Precambrian Research*, **127**, 379–380.

MAZUMDER, R. 2004. Implications of lunar orbital periodicities from Chaibasa tidal rhythmite of late Paleoproterozoic age. *Geology*, **32**, 841–844.

MAZUMDER, R. 2005. Proterozoic sedimentation and volcanism in the Singhbhum crustal province, India and their implications. *Sedimentary Geology*, **176**, 167–193.

MAZUMDER, R. & ARIMA, M. 2004. Proterozoic events in the Singhbhum crustal province, India. *Gondwana Research*, **7/4S**, 1343–1345.

MAZUMDER, R. & ARIMA, M. 2005. Tidal rhythmites and their implications. *Earth Science Reviews*, **69**, 79–95.

MAZUMDER, R. & ARIMA, M. 2009. Implication of mafic magmatism in an intracontinental rift setting: a case study from the Palaeoproterozoic Dhanjori Formation, Singhbhum crustal province, India. *Journal of Geology*, **117**, 455–466.

MAZUMDER, R. & SAHA, D. 2009. *Palaeoproterozoic of Singhbhum Crustal Province*. Indian Statistical Institute, Kolkata.

MAZUMDER, R. & SARKAR, S. 2004. Sedimentation history of the Palaeoproterozoic Dhanjori Formation, Singhbhum, India and its implications. *Precambrian Research*, **130**, 267–287.

MAZUMDER, R., BOSE, P. K. & SARKAR, S. 2000. A commentary on the tectono-sedimentary record of ther pre-2.0 Ga continental growth of India vis-a-vis Pre-Gondwana Afro-Indian supercontinent. *Journal of African Earth Sciences*, **30**, 201–217.

MAZUMDER, R., ERIKSSON, P. G., DE, S., BUMBY, A. J. & LENHARDT, N. 2012. Palaeoproterozoic sedimentation on the Singhbhum Craton: global context and comparison with Kaapvaal. *In*: MAZUMDER, R. & SAHA, D. (eds) *Palaeoproterozoic of India*. Geological Society, London, Special Publications, **365**, 49–74, http://dx.doi.org/10.1144/SP365.4.

MAZUMDER, R., RODRIGUEZ-LOPEZ, J. A., ARIMA, M. & VAN LOON, A. J. 2009. Palaeoproterozoic seismites (fine-grained facies of the Chaibasa Fm., E. India) and their soft-sediment deformation structures. *In*: REDDY, S. M., MAZUMDER, R., EVANS, D. A. & COLLINS, A. S. (eds) *Palaeoproterozoic Supercontinent and its Global Evolution*. Geological Society, London, Special Publications, **323**, 301–318.

MAZUMDER, R., VAN LOON, A. J. & ARIMA, M. 2006. Soft-sediment deformation structures in the Earth's oldest seismites. *Sedimentary Geology*, **186**, 19–26.

MISHRA, S., DEOMURARI, M. P., WIEDENBECK, M., GOSWAMI, J. N., RAY, S. & SAHA, A. K. 1999. $^{207}Pb/^{206}Pb$ zircon ages and the evolution of the Singhbhum Craton, eastern India: an ion microprobe study. *Precambrian Research*, **93**, 139–151.

MONDAL, S. K., FREI, R. & RIPLEY, E. M. 2007. Os isotope systematics of Mesoarchaean chromite-PGE deposits in the Singhbhum Craton (India): implications for the evolution of lithospheric mantle. *Chemical Geology*, **244**, 391–408.

MONDAL, S. K., RIPLEY, E. M., LI, C. & FREI, R. 2006. The genesis of Archaean chromitites from Nuasahi and Sukinda massifs in the Singhbhum craton, India. *Precambrian Research*, **148**, 45–66.

MUKHOPADHYAY, D. 1976. Present status of Precambrian Stratigraphy of Singhbhum – the problems and prospects. *Indian Journal of Earth Sciences*, **3**, 208–219.

MUKHOPADHYAY, D. 1988. Precambrian of Eastern Indian Shield – Perspective of the Problems. *Memoirs of the Geological Society of India*, **8**, 1–12.

MUKHOPADHYAY, D. 2001. The Archaean nucleus of Singhbhum: the present state of knowledge. *Gondwana Research*, **4**, 307–318.

MUKHOPADHYAY, D., BHATTACHARYA, T., CHAKRABORTY, T. & DEY, A. K. 1990. Structural pattern in the Precambrian rocks of Sonua-Lotapahar region, North Singhbhum, eastern India. *Proceedings of the Indian Academy of Science (Earth and Planetary Sciences)*, **99**, 249–268.

MUKHOPADHYAY, J., BEUKES, N. J., ARMSTRONG, R. A., ZIMMERMANN, U., GHOSH, G. & MEDDA, R. A. 2008. Dating the oldest Greenstone in India: a 3.51 Ga precise U–Pb SHRIMP Zircon Age for Dacitic Lava of the Southern Iron Ore Group, Singhbhum Craton. *Journal of Geology*, **116**, 449–461.

MUKHOPADHYAY, J., GHOSH, G., NANDI, A. K. & CHAUDHURI, A. K. 2006. Depositional setting of the Kolhan Group: its implications for the development of a Meso to Neoproterozoic deep-water basin on the South Indian craton. *South African Journal of Geology*, **109**, 183–192.

MURTHY, V. N. & ACHARYA, S. 1975. Lithostratigraphy of the Precambrian rocks around Koira, Sundargarh District, Orissa. *Journal of the Geological Society of India*, **16**, 55–68.

NAHA, K. 1965. Metamorphism in relation to stratigraphy, structure and movements in parts of east Singhbhum, Eastern India. *Quarterly Journal of the Geological, Mineralogical, Metallargical Society of India*, **37**, 41–88.

NAHA, K. & GHOSH, S. K. 1960. Archaean palaeogeography of E. and N. Singhbhum. *Geological Magazine*, **97**, 436–439.

NELSON, D. R., BHATTACHARYA, H. N., MISRA, S., DASGUPTA, N. & ALTERMANN, W. 2007. New SHRIMP U–Pb zircon dates from the Singhbhum craton, Jharkhand–Orissa region, India. Abstract. *In*: BANERJEE, S. (ed.) *International Conference on Precambrian Sedimentation & Tectonics & the Second GPSS Meeting*. Indian Institute of Technology, Bombay.

NOFFKE, N. 2010. *Geobiology: Microbial Mats in Sandy Deposits from the Archaean Era to Today*. Springer, Berlin.

PAUL, D. K., MUKHOPADHYAY, D., PYNE, T. K. & BISHOI, P. K. 1991. Rb–Sr age of granitoid in the Deo River section, Singhbhum and its relevance to the age of iron formation. *Indian Minerals*, **45**, 51–56.

POSAMANTIER, H. W. & VAIL, P. R. 1988. Eustatic controls on clastic deposition 1. Sequence and systems tract models. *In*: WILGUS, C. K., ROSS, C. A. & POSAMENTIER, H. (eds) *Sea Level Changes: An Integrated Approach*. SEPM, Special Publications, **42**, 109–124.

RAI, K. L., SARKAR, S. N. & PAUL, P. R. 1980. Primary depositional and diagenetic features in the banded iron formation and associated iron ore deposits of Noamundi, Singhbhum district, Bihar, India. *Mineral Deposita*, **15**, 189–200.

RAO, V. V. S. & DASGUPTA, H. C. 1995. Clay mineralogy of the argillites associated with banded iron formations from the Noamundi basin, eastern India: a clue to its chemical environment of deposition. *Indian Minerals*, **49**, 31–44.

RAY, K. K., GHOSH-ROY, A. K. & SENGUPTA, S. 1996. Acid volcanic rocks between the Dalma volcanic belt and Chotanagpur gneissic complex, east Singhbhum and Purulia districts of Bihar and West Bengal. *Indian Mineralogist*, **50**, 1–8.

REDDY, S. M., CLARKE, C. & MAZUMDER, R. 2009. Temporal constraints on the evolution of the Singbhum Crustal Province from U–Pb SHRIMP data. *In*: SAHA, D. & MAZUMDER, R. (eds) *Paleoproterozoic Supercontinents and Global Evolution. International Conference on Paleoproterozoic Supercontinents and Global Evolution Abstract Volume*. International Association for Gondwana Research, Conference Series, **9**, 17–18.

REDDY, S. M. & EVANS, D. A. D. 2009. Palaeoproterozoic supercontinents and global evolution:correlations from core to atmosphere. *In*: REDDY, S. M., MAZUMDER, R., EVANS, D. A. D. & COLLINS, A. (eds)

Palaeoproterozoic Supercontinents and Global Evolution. Geological Society, London, Special Publications, **323**, 1–26.

Roy, A., Sarkar, A., Jeyakumar, S., Aggrawal, S. K. & Ebihara, M. 2002*a*. Sm–Nd age and mantle characteristics of the Dhanjori volcanic rocks, Eastern India. *Geochemical Journal*, **36**, 503–518.

Roy, A., Sarkar, A., Jeyakumar, S., Aggrawal, S. K. & Ebihara, M. 2002*b*. Mid-Proterozoic plume-related thermal event in Eastern Indian craton: evidence from trace elements, REE geochemistry and Sr–Nd isotope systematics of basic-ultrabasic intrusives from Dalma Volcanic Belt. *Gondwana Research*, **5**, 133–146.

Roy, A., Sarkar, A., Jeyakumar, S., Aggrawal, M., Ebihara, M. & Satoh, H. 2004. Late Archaean mantle metasomatism below eastern Indian cratons: evidence from trace elements, REE geochemistry and Sr–Nd–O isotope systematics of ultramafic dykes. *Proceedings of the Indian Academy of Science (Earth and Planetary Science)*, **113**, 649–666.

Saha, A., Basu, A. R., Garzione, C., Bandyopadhyay, P. K. & Chakrabarti, A. 2004. Geochemical and petrological evidence for subduction accretion processes in the Archean Eastern Indian Craton. *Earth and Planetary Science Letters*, **220**, 91–106.

Saha, A. K. 1949. Dolerite dykes and sills around Chaibasa. *Quarterly Journal of the Geological Mining Metallurgical Society of India*, **21**, 77–83.

Saha, A. K. 1994. *Crustal Evolution of Singhbhum–North Orissa, eastern India*. Memoirs of the Geological Society of India, **27**.

Saha, A. K., Ray, S. L. & Sarkar, S. N. 1988. Early history of the Earth: evidence from the Eastern Indian Shield. *In*: Mukhopadhyay, D. (ed.) *Precambrian of the Eastern Indian Shield*. Memoirs of the Geological Society of India, **8**, 13–37.

Saha, A. K., Sankaran, A. V. & Bhattacharyya, T. K. 1973. Geochemistry of Newer Dolerite suite of intrusions within Singhbhum granite. *Journal of the Geological Society of India*, **14**, 324–346.

Sahu, N. K. & Mukherjee, M. M. 2001. Spinifex textured komatiite from Badampahar–Gorumahishini Schist belt, Mayurbhanj Dist, Orissa. *Journal of the Geological Society of India*, **57**, 529–534.

Sarkar, B. 1989. Algal Stromatolites and associated microbiota of the Precambrian Banded Iron Formation in Noamundi basin, eastern India. *Himalayan Geology*, **13**, 29–37.

Sarkar, S. C. 1984. *Geology and Ore Mineralisation of the Singhbhum Copper–Uranium Belt, Eastern India*. Jadavpur University, Calcutta.

Sarkar, S. C. & Deb, M. 1971. Dhanjori basalts and some related rocks. *Quarterly Journal of the Geological, Mineralogical, and Metallurgical Society of India*, **43**, 29–37.

Sarkar, S. N. & Saha, A. K. 1962. A revision of the Precambrian stratigraphy and tectonics of Singhbhum and adjacent region. *Quarterly Journal of the Geological, Mineralogical and Metallurgical Society of India*, **34**, 97–136.

Sarkar, S. N. & Saha, A. K. 1977. The present status of the Precambrian stratigraphy, tectonics and geochronology of Singhbhum–Keonjhar–Mayurbhanj region, eastern India. *Indian Journal of Earth Sciences*, **4**, 37–65.

Sarkar, S. N. & Saha, A. K. 1983. Structure and tectonics of the Singhbhum–Orissa Iron Ore Craton, eastern India. *In*: *Recent Researches in Geology: Structure and Tectonics of Precambrian Rocks*. Hindustan Publishing Corporation, New Delhi, 1–25.

Sarkar, S. N., Ghosh, D. K. & Lambert, R. J. St. 1986. Rubidium–Strontium and lead isotopic studies of the Soda granites from Musaboni area, Singhbhum Copper Belt. *In*: Sarkar, S. N. (ed.) *Geology and Geochemistry of Sulphide Ore Bodies and Associated Rocks in Musaboni and Rakha Mines Section in the Singhbhum Copper Belt*. Diamond Jubilee Monograph. Indian School of Mines, Dhanbad, 101–110.

Sengupta, S. & Chattopadhyay, B. 2004. Singhbhum Mobile Belt – How far it fits an Ancient Orogen. *In*: *Proceedings of the Workshop on (IGCP-453): Uniformitarianism Revised Edition on Orogens of India*. Geological Survey of India, Special Publications, **84**, 24–31.

Sengupta, S. & Mukhopadhyay, P. K. 2000. Sequence of Precambrian events in the eastern Indian craton. *Geological Survey of India Special Publication*, **57**, 49–56.

Sengupta, S., Acharyya, S. K. & De Smith, J. B. 1997. Geochemistry of Archaean volcanic rocks from Iron Ore Supergroup, Singhbhum, eastern India. *Proceedings of the Indian Academy of Science (Earth Planetary Science)*, **106**, 327–342.

Sengupta, S., Paul, D. K., De Laeter, J. R., McNaughton, N. J., Bandyopadhyay, P. K. & De Smeth, J. B. 1991. Mid-Archaean evolution of the eastern Indian craton: geochemical and isotopic evidence from Bonai pluton. *Precambrian Research*, **49**, 23–37.

Sengupta, S., Paul, D. K., Bishui, P. K., Gupta, S. N., Chakraborty, R. & Sen, P. 1994. Geochemical and Rb–Sr isotopic study of Kuilapal granite and Arkasani granophyre from the eastern Indian craton. *Indian Minerals*, **48**, 77–88.

Sengupta, S., Sarkar, G., Ghosh-Roy, A. K., Bhaduri, S. K., Gupta, S. N. & Mandal, A. 2000. Geochemistry and Rb–Sr Geochronology of acid tuffs from the northern fringe of the Singhbhum craton and their significance in the Precambrian evolution. *Indian Minerals*, **54**, 43–56.

Sharma, M., Basu, A. R. & Ray, S. L. 1994. Sm–Nd isotopic and geochemical study of the Archaean tonalite amphibolite association from the eastern Indian Craton. *Contributions to Mineralogy and Petrology*, **117**, 45–55.

Simpson, E. L., Alkmin, F. F., Bose, P. K., Bumby, A. J. & Eriksson, K. A. 2004. Sedimentary dynamics of Precambrian aeolianites. *In*: Eriksson, P. G., Altermann, W., Nelson, D. R., Mueller, W. U. & Catuneanu, O. (eds) *The Precambrian Earth: Tempos and Events*. Developments in Precambrian Geology, **12**, 642–657.

Singh, S. P. 1997. Geochemistry of acid volcanics of the Dalma Group, Singhbhum, Eastern India. *Journal of the Geological Society of India*, **49**, 437–441.

Singh, S. P. 1998. Precambrian stratigraphy of Bihar — an overview. *In*: Paliwal, B. S. (ed.) *The Indian Precambrian*. Scientific Publishers, Jodhpur, 376–408.

TAIT, J., ZIMMERMANN, U., MIYAZAKI, T., PRESNYAKOV, S., CHANG, Q., MUKHOPADHYAY, J. & SERGEEV, S. 2011. Possible juvenile Palaeoarchaean TTG magmatism in eastern India and its constraints for the evolution of the Singhbhum craton. *Geological Magazine*, **148**, 340–347.

VAN LOON, A. J. & MAZUMDER, R. 2009. A Palaeoproterozoic (~1.6 Ga) Lahar Deposit (Dalma Formation, Eastern India). *In*: SAHA, D. & MAZUMDER, R. (eds) *Paleoproterozoic Supercontinents and Global Evolution. International Conference on Paleoproterozoic Supercontinents and Global Evolution Abstract Volume*. International Association for Gondwana Research, Conference Series, **9**, 4.

VERMA, R. K. & PRASAD, S. N. 1974. Paleomagnetic study and chemistry of Newer Dolerites from Singhbhum, Bihar, India. *Canadian Journal of Earth Sciences*, **11**, 1043–1054.

VOHRA, C. P., DASGUPTA, S., PAUL, D. K., BISHOI, P. K., GUPTA, S. N. & GUHA, S. 1991. Rb–Sr chronology and petrochemistry of granitoids from the southeastern part of the Singhbhum craton, Orissa. *Journal of the Geological Society of India*, **38**, 5–22.

YOUNG, G. M. 2004. Earth's two great Precambrian glaciations: aftermath of the 'Snowball Earth' hypothesis. *In*: ERIKSSON, P. G., ALTERMANN, W., NELSON, D. R., MUELLER, W. U. & CATUNEANU, O. (eds) *The Precambrian Earth: Tempos and Events*. Elsevier, Amsterdam, 440–448.

Palaeoproterozoic sedimentation on the Singhbhum Craton: global context and comparison with Kaapvaal

R. MAZUMDER[1], P. G. ERIKSSON[2]*, S. DE[1], A. J. BUMBY[2] & N. LENHARDT[2]

[1]*Geological Studies Unit and Fluvial Mechanics Laboratory, Indian Statistical Institute, 203 B.T. Road, Kolkata 700 108, India*

[2]*Department of Geology, University of Pretoria, Pretoria 0002, South Africa*

**Corresponding author (e-mail: pat.eriksson@up.ac.za)*

Abstract: The Singhbhum Craton has a limited Palaeoproterozoic supracrustal record, which suggests a three-part history, comprising: a long period of high freeboard and palaeosol formation on granitoids; subsequent rift-related mafic–ultramafic volcanism and subordinate sedimentation (*c.* 2.25–2.1 Ga: Dhanjori and Jagannathpur basin-fills; possibly also Simlipal, Malangtoli and Ongarbira basin-fills), which overlapped locally with mafic soil formation; and a major regression at around 2.0 Ga. Following a long hiatus, the approximately 1.6 Ga Dhalbhum–Dalma succession was laid down, probably under continental conditions. This rather truncated record stands in contrast to the chronologically and geographically much more widespread supracrustal basin-fills of the Kaapvaal Craton, and there appears to be an overall poor comparison between these two early Precambrian crustal blocks. However, on Kaapvaal, three analogous events are identified: widespread approximately 2.2 Ga mafic volcanism, followed by a well-developed palaeosol and a major transgression prior to 2.05 Ga. The three shared events between the two cratons are compatible with the postulate of a global, approximate 2.45–2.2 Ga shutdown of magmatic and tectonic geodynamics, with the origin of the triumvirate directly reflecting its resumption again after about 2.2 Ga. We recognize here that a large diversity of views on Singhbhum's geodynamic history exists, predicated on a lack of precise geochronology and commonly poor outcrops, and the current hypotheses are presented with these factors in mind.

The Palaeoproterozoic period (2500–1600 Ma) is one fraught with interest as it most probably post-dates the onset of supercontinentality globally, although this was apparently not a pervasive process and might even have been diachronous on the planetary scale (e.g. Eriksson *et al.* 2011). In addition, this time interval encompasses evidence for significant palaeo-atmospheric turmoil (Fig. 1), although possibly not the permanent change envisaged in the 'Great Oxidation Event' (e.g. Karhu & Holland 1996) hypothesis (Eriksson *et al.* 2009*a*; Frauenstein *et al.* 2009; Frei *et al.* 2009; Voegelin *et al.* 2010). Many and, perhaps, a majority of Precambrian geologists might agree that a gradual growth of continental crust, reaching an impersistent and probably diachronous high approximately across the Archaean–Proterozoic boundary, is pertinent. Similarly, many accept that the dominance of mantle–thermal processes in the Hadean did not persist thereafter, and that it changed somehow (with the Trendall 2002 'plughole' model providing one hypothesis: see also De Wit 1998) to the plate-tectonically-dominated Earth familiar to all. The Palaeoproterozoic essentially follows the inferred peak in crustal growth and post-dates the onset of Phanerozoic-style plate tectonics (e.g. Eriksson & Catuneanu 2004) but may have been characterized by a 'global magmatic lull' from about 2.45 to 2.2 Ga, including also global tectonic quiescence (Condie *et al.* 2009). This hypothesis is one proposed on a global scale, with obviously local exceptions occurring during the period.

Mantle plumes (also superplumes and superplume events: see Condie 2001, 2004*a*, *b* for terminology; compare with alternative terms/definitions of Ernst *et al.* 2004) were also a feature of the Palaeoproterozoic period, being exemplified by direct proxies such as komatiites, picrites, flood basalts and large dyke swarms (Isley & Abbott 1999; Ernst & Buchan 2002), and indirect evidence in the form of peaks in banded iron formation (BIF) occurrence, shallow-marine sediments, black shales and the Chemical Index of Alteration (CIA; cf. Nesbitt & Young 1982) (Condie 2004*a*). Analysis of the plume record through these inferred proxies indicates several major superplume events during the Palaeoproterozoic that preferentially affected particular cratonic terrains: 2.45 Ga (India and the North China Craton), 2.1–2.15 Ga (Guiana Shield and west Africa), 2.0–1.9 Ga (global) and

From: Mazumder, R. & Saha, D. (eds) 2012. *Palaeoproterozoic of India*. Geological Society, London, Special Publications, **365**, 51–76. http://dx.doi.org/10.1144/SP365.4

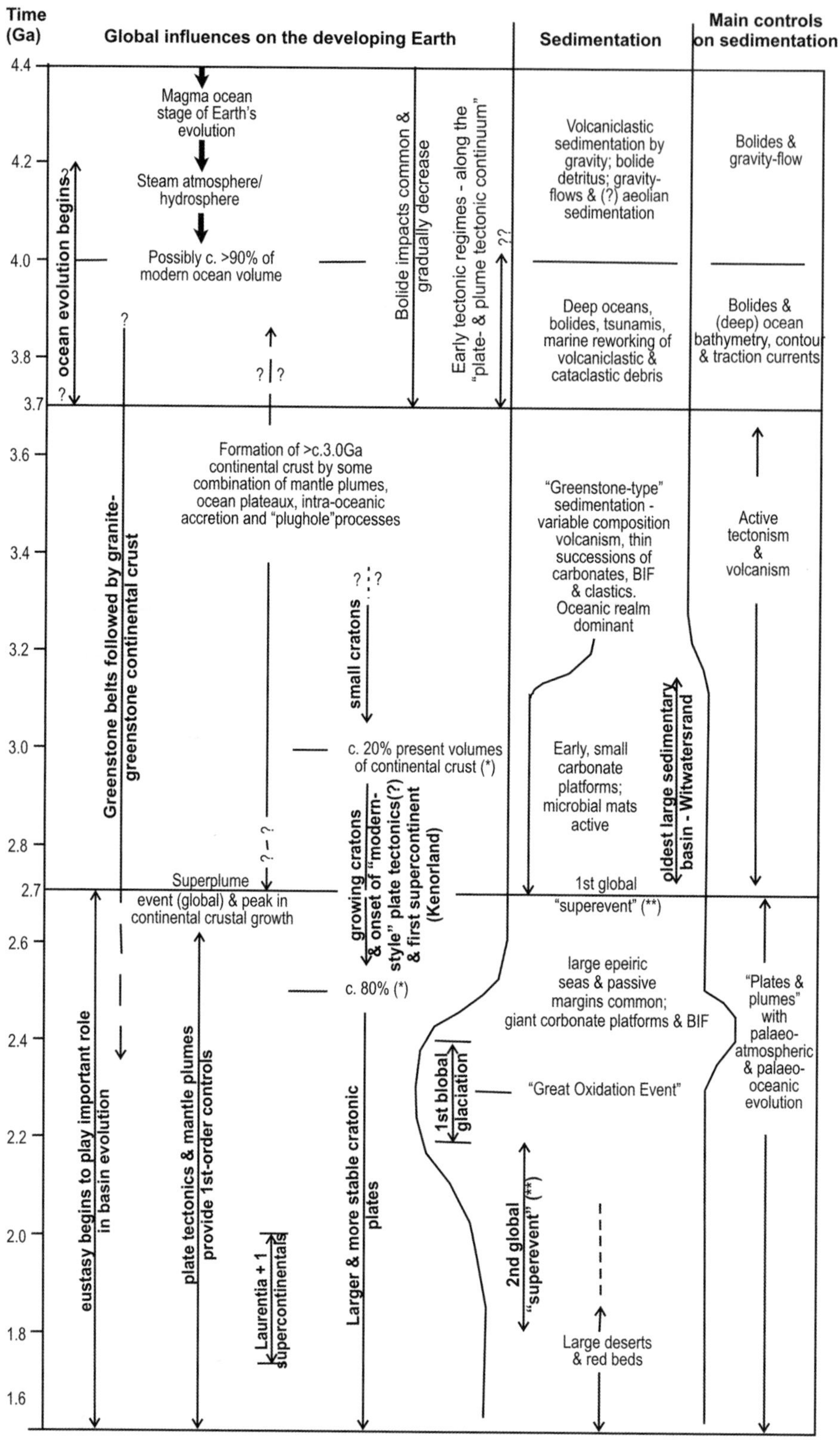
Time (Ga)
Global influences on the developing Earth
Sedimentation
Main controls on sedimentation
4.4
4.2
4.0
3.8
3.7
3.6
3.4
3.2
3.0
2.8
2.7
2.6
2.4
2.2
2.0
1.8
1.6
Magma ocean stage of Earth's evolution
Steam atmosphere/ hydrosphere
Possibly c. >90% of modern ocean volume
ocean evolution begins
Bolide impacts common & gradually decrease
Early tectonic regimes - along the "plate- & plume tectonic continuum"
Volcaniclastic sedimentation by gravity; bolide detritus; gravity-flows & (?) aeolian sedimentation
Bolides & gravity-flow
Deep oceans, bolides, tsunamis, marine reworking of volcaniclastic & cataclastic debris
Bolides & (deep) ocean bathymetry, contour & traction currents
Formation of >c.3.0Ga continental crust by some combination of mantle plumes, ocean plateaux, intra-oceanic accretion and "plughole"processes
Greenstone belts followed by granite-greenstone continental crust
small cratons
c. 20% present volumes of continental crust (*)
"Greenstone-type" sedimentation - variable composition volcanism, thin successions of carbonates, BIF & clastics. Oceanic realm dominant
Active tectonism & volcanism
Early, small carbonate platforms; microbial mats active
oldest large sedimentary basin - Witwatersrand
Superplume event (global) & peak in continental crustal growth
growing cratons & onset of "modern-style" plate tectonics(?) & first supercontinent (Kenorland)
1st global "superevent" (**)
c. 80% (*)
large epeiric seas & passive margins common; giant corbonate platforms & BIF
"Plates & plumes" with palaeo-atmospheric & palaeo-oceanic evolution
eustasy begins to play important role in basin evolution
plate tectonics & mantle plumes provide 1st-order controls
Larger & more stable cratonic plates
1st global glaciation
"Great Oxidation Event"
2nd global "superevent" (**)
Laurentia + 1 supercontinentals
Large deserts & red beds

1.75–1.70 Ga (southern Baltica and in Laurentia) (Condie 1992, 1998, 2004*a*; Eriksson *et al.* 2004). Two globally relevant superplume events were postulated by Condie (e.g. 2004*a*): one at approximately 2.7 Ga and the other at around 1.9 Ga.

A general greenhouse palaeo-atmosphere is envisaged by most workers, prior to about 2.4 Ga, when it possibly collapsed and cooling occurred, including an approximately 2.4–2.2 Ga planetary-scale glaciation event (Young 2004; Lowe & Tice 2007). The latter has been interpreted within the 'snowball Earth hypothesis' (SEH) (e.g. Kirschvink 1992; Hoffman *et al.* 1998), and related to palaeomagnetic evidence for low palaeolatitudes (e.g. Harland & Bidgood 1959; Harland 1964; Embleton & Williams 1986; Schmidt *et al.* 1991; Schmidt & Williams 1995; Sohl *et al.* 1999) and glaciation near sea level (Young 2004). However, opponents of the SEH point out a significant number of arguments against this model (Williams 1975, 1986, 1998, 2004, 2008; Young 2004; Williams *et al.* 2008), and present alternative explanations for this glacial event including genesis within rift settings related to supercontinental break-up (e.g. summarized by Eriksson & Altermann 2004) or possible tectonic and magmatic quiescence from about 2.45 to 2.2 Ga (Condie *et al.* 2009). The veracity of palaeomagnetic data from rocks older than about 1.8 Ga has also been questioned (e.g. Meert 2002; Pesonen *et al.* 2003).

It is the intention of this paper to examine briefly the geological evolution (with emphasis on depositional basins) of the Singhbhum Craton during the Palaeoproterozoic, and to compare this with the general framework outlined above and with a globally relevant example of a well-studied craton, namely Kaapvaal in southern Africa, which is among the oldest preserved such terranes known (e.g. Eriksson *et al.* 2009*b*). The latter craton is briefly described prior to Singhbhum, as it well illustrates many of the global processes discussed in this introductory section, and in order to provide a comparative framework for the Singhbhum Craton's inferred evolution. This paper is meant to complement and build upon another paper within this volume (Mazumder *et al.* 2012) devoted to a detailed treatment of the supracrustal record of the Singhbhum Craton. We will use this second manuscript as a source of reference to support our comparison between the northern and southern Singhbhum successions, and their possible relationship with global-scale events and those identified on Kaapvaal. The focus of the current paper is thus distinct from Mazumder *et al.* (2012).

Kaapvaal Craton: a brief summary of Palaeoproterozoic evolution

Figure 2 provides a very brief summary of the evolution of the Kaapvaal Craton from about 3.1 to 1.6 Ga, covering its development from post-cratonization through the earliest known large sedimentary basin, the Witwatersrand–Pongola depository (*c.* 3.1–2.8 Ga), the subsequent Ventersdorp (superplume) volcano-sedimentary deposits (*c.* 2714–2709 Ma), the Transvaal basins (*c.* 2.66–2.05 Ga), the Bushveld Complex intrusion (*c.* 2060 Ma) and the succeeding Waterberg–Soutpansberg rift basins (*c.* <2.05–1.7 Ga) (e.g. Eriksson *et al.* 2006*a*). The nucleus of the craton had formed by around 3.1 Ga through initial thin-skinned thrusting within intra-oceanic and arc settings, and the succeeding amalgamation of such terranes, with accompanying granitoid magmatism (de Wit *et al.* 1992). Accretion of composite terranes from the north and west followed between approximately 3.23 and 2.65–2.7 Ga (e.g. Zeh *et al.* 2009 and references therein).

Evolution of the Witwatersrand–Pongola Basin was essentially coeval with these Neoarchaean cratonization processes, with sediment derived from both older nuclear and younger accreted crust as well as from pulses of granitic intrusions (Robb & Meyer 1995; Frimmel 2005). A complex flexural foreland basin model is commonly applied to this depository, with the Witwatersrand Supergroup having been deposited within the foredeep and the Pongola Supergroup in the back-bulge part of the system, with a flexural forebulge separating the two (Beukes & Cairncross 1991; Catuneanu 2001 and references therein). The two foreland sub-basins were characterized by initial volcanic rocks, with subsequent thermal subsidence in the foredeep accommodating a starved basin succession there, with the coeval Pongola back-bulge sub-basin deposits being of shallow-water affinity (Eriksson *et al.* 1981, 2005; Stanistreet & McCarthy 1991; Robb & Meyer 1995; Catuneanu 2001). Fluvial sandstones and lesser conglomerates thereafter formed the famous upper Witwatersrand auriferous deposits within a now stuffed basin-type foredeep depository (Robb & Meyer 1995; Els 1998*a*, *b*; Catuneanu 2001).

Following deformation and partial erosion of the Witwatersrand strata, the Kaapvaal Craton was affected directly by the approximately 2.7 Ga global

Fig. 1. Schematic summary chart of Earth evolution from about 4.4–1.6 Ga, showing inferred major changes and events, as well as their possible impact on the sedimentary record for the Palaeoproterozoic period within a broader Precambrian framework (modified after Eriksson *et al.* 2007).

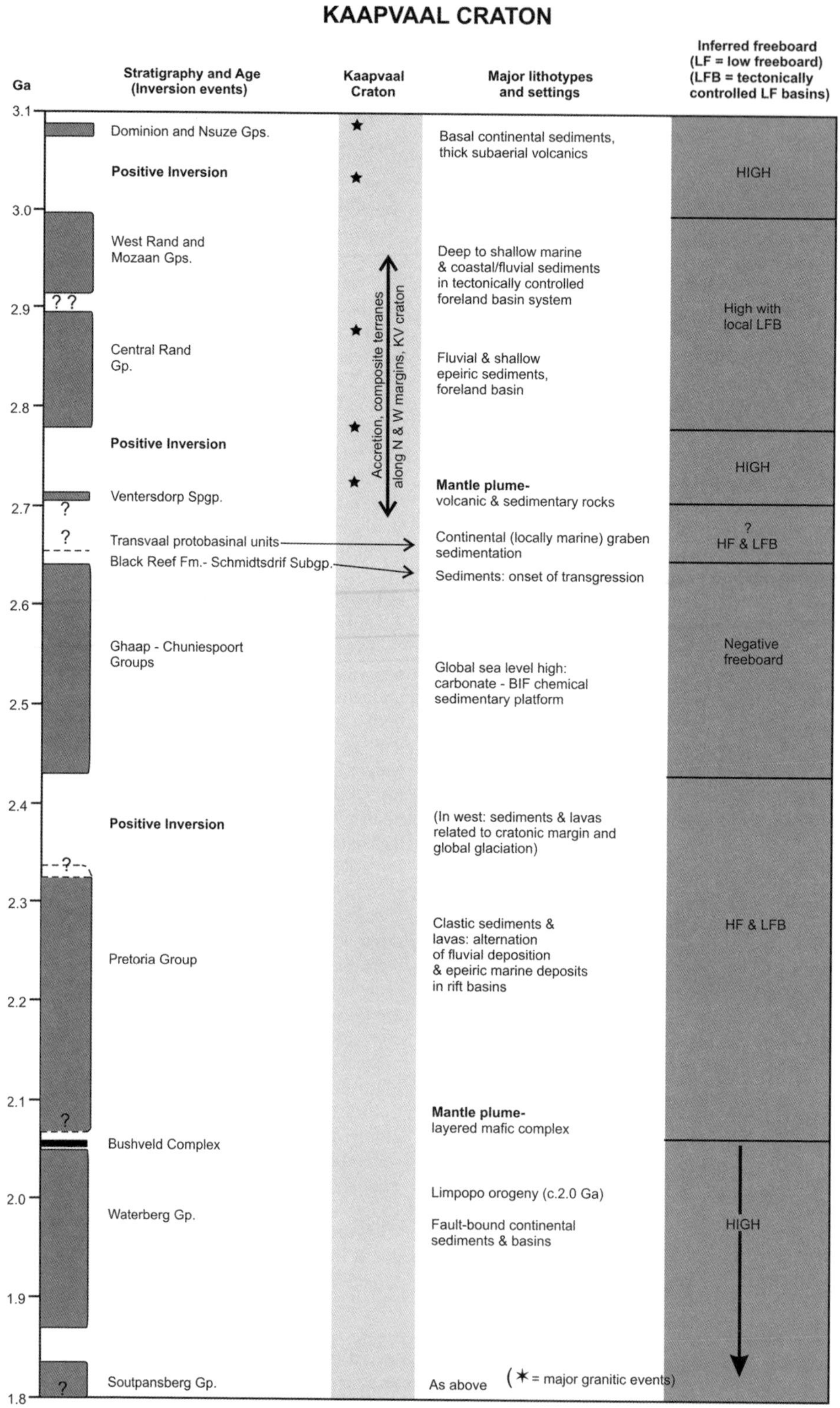
KAAPVAAL CRATON
Ga
Stratigraphy and Age (Inversion events)
Kaapvaal Craton
Major lithotypes and settings
Inferred freeboard (LF = low freeboard) (LFB = tectonically controlled LF basins)
3.1
3.0
2.9
2.8
2.7
2.6
2.5
2.4
2.3
2.2
2.1
2.0
1.9
1.8
Dominion and Nsuze Gps.
Basal continental sediments, thick subaerial volcanics
Positive Inversion
HIGH
West Rand and Mozaan Gps.
Deep to shallow marine & coastal/fluvial sediments in tectonically controlled foreland basin system
High with local LFB
Accretion, composite terranes along N & W margins, KV craton
Central Rand Gp.
Fluvial & shallow epeiric sediments, foreland basin
Positive Inversion
HIGH
Ventersdorp Spgp.
Mantle plume- volcanic & sedimentary rocks
Transvaal protobasinal units
Continental (locally marine) graben sedimentation
? HF & LFB
Black Reef Fm.- Schmidtsdrif Subgp.
Sediments: onset of transgression
Ghaap - Chuniespoort Groups
Global sea level high: carbonate - BIF chemical sedimentary platform
Negative freeboard
Positive Inversion
(In west: sediments & lavas related to cratonic margin and global glaciation)
Pretoria Group
Clastic sediments & lavas: alternation of fluvial deposition & epeiric marine deposits in rift basins
HF & LFB
Mantle plume- layered mafic complex
Bushveld Complex
Limpopo orogeny (c.2.0 Ga)
Waterberg Gp.
Fault-bound continental sediments & basins
HIGH
Soutpansberg Gp.
As above
(★ = major granitic events)

superplume event (Condie 2004*a*), with the formation of the 2714–2709 Ma (Armstrong *et al.* 1991) Ventersdorp Supergroup (Fig. 2). This volcano-sedimentary succession encompassed: lower flood basalts (erupted from ponded plume magma beneath thinned crust underlying the Witwatersrand depository); medial fault-bounded continental sedimentary and bimodal volcanic lithologies; uppermost units related to thermal subsidence and the widespread deposition of volcano-sedimentary beds as the inferred plume responsible for this supergroup's formation subsided (van der Westhuizen *et al.* 1991; Eriksson *et al.* 2002 references in both).

The approximately 2.66–2.05 Ga Transvaal Supergroup, preserved within three basins on the Kaapvaal Craton (Transvaal (TB) and Griqualand West (GB) basins in South Africa, and the Kanye Basin (KB) in Botswana), succeeded the relatively short-lived Ventersdorp plume event (Catuneanu & Eriksson 1999, 2002; Eriksson *et al.* 2001, 2006*b* references therein). Four major subdivisions of Transvaal stratigraphy are recognized (in ascending order): 'protobasinal' (a non-genetic use of the term; only in TB) rocks; Black Reef Formation (TB and KB; Vryburg Formation equivalent in GB); a carbonate-BIF platformal succession (Chuniespoort, Ghaap and Taupone groups, respectively: TB, GB, KB); clastic sedimentary–lesser volcanic succession of the Pretoria Group (TB and Segwagwa Group equivalent in KB) and an attenuated correlate in GB, the Postmasburg Group (Eriksson *et al.* 2006*b*) (Fig. 3).

The protobasinal successions occur as discrete, fault-bounded depositories, filled by varying proportions of bimodal volcanic to sedimentary rock types, with the latter tending to increase in maturity upwards from alluvial to basinal deposits, within individual basin-fill successions (Hartzer 1995; Eriksson *et al.* 2001). These relatively small basin-fills have been interpreted as reflecting either extensional or strike-slip basins preserved across a wide (*c.* 200 km) zone of rifting, which might have been related to late Ventersdorp events (e.g. Olsson *et al.* 2010). These basins are succeeded unconformably by the thin (generally *c.* 30–60 m) sheet sandstones of the Black Reef Formation, which also transgressed onto adjacent basement lithologies (Eriksson *et al.* 2006*b*). Equivalent deposits mark the base of the Transvaal succession in the KB; the formation is interpreted as basal fluvial deposits that grade up into transgressive marine beds (e.g. Button 1973; Key 1983; Henry *et al.* 1990; Els *et al.* 1995). The basal unit of the Transvaal Supergroup in the GB is the Vryburg Formation, generally considered to be approximately coeval with the (undated) Black Reef lithologies (e.g. Eriksson *et al.* 2006*b*) (Fig. 2). The Vryburg unconformably overlies Ventersdorp rocks, is mostly about 100–300 m thick, and consists of clastic sedimentary, lesser carbonate sedimentary beds and uppermost 2642 ± 3 Ma (Walraven & Martini 1995) basaltic–andesitic lavas (Eriksson *et al.* 2006*b* and references therein). This succession is ascribed to a range of depositional conditions, from fluvial–marginal marine to deeper oceanic settings (Beukes 1979; Altermann & Siegfried 1997).

A regional unconformity ushered in through tilting and base-level fall was followed by the advance of an epeiric sea that eventually covered much of the craton, and which accommodated a thick package of stromatolitic carbonate rocks (*c.* 1200 m in TB and >2.5 km in GB), followed by BIF (*c.* 640 m in TB) and uppermost clastic and chemical sedimentary rocks (Koegas Subgroup of GB, *c.* 1100 m-thick Duitschland Formation of TB) as subsidence continued (Altermann & Siegfried 1997; Eriksson *et al.* 2001, 2006*b*) (Fig. 3). This chemical sedimentary platform environment persisted for over 200 Ma (from *c.* 2.65–2.64 to 2.40 Ga: Trendall *et al.* 1990; Knoll & Beukes 2009), with deposition initially in the SW part of the GB. At approximately 2550 Ma, a major transgression occurred and the carbonate platform expanded over the rest of the GB and the basins now preserved as the TB and KB; palaeoenvironmental conditions encompassed a spectrum from deep carbonate platform to exposed peritidal flats (e.g. Eriksson & Altermann 1998). Deeper epeiric conditions concomitant with a second major transgression at about 2500 Ma drowned the carbonate platform and BIF were deposited across a wide region, including all three preservational basins (Altermann & Nelson 1998). The uppermost Chuniespoort–Ghaap–Taupone groups' clastic–chemical sedimentary successions are ascribed to a final regressive withdrawal of the epicontinental sea, from approximately NE to SW, off the craton (e.g. Eriksson *et al.* 2005).

In contrast to the conformably-based Koegas Subgroup of GB, the Duitschland Formation is unconformably-based, very localized in occurrence in the NE of the TB, has a highly irregular geometry and tends to be thicker where removal of underlying BIF of the Chuniespoort Group has been greater (Potgieter 1992; Hälbich *et al.* 1993; Bekker *et al.* 2001; Frauenstein *et al.* 2009). It is thus intrinsically related to an erosive/weathering hiatus separating the Chuniespoort and Pretoria groups in the TB (Eriksson *et al.* 2001). Where best developed,

Fig. 2. Schematic summary of geohistory for the Kaapvaal Craton supracrustal rocks, from about 3.1 to 1.8 Ga. Inferred continental freeboard conditions are shown in far right column (modified after Eriksson *et al.* 2006*a*).

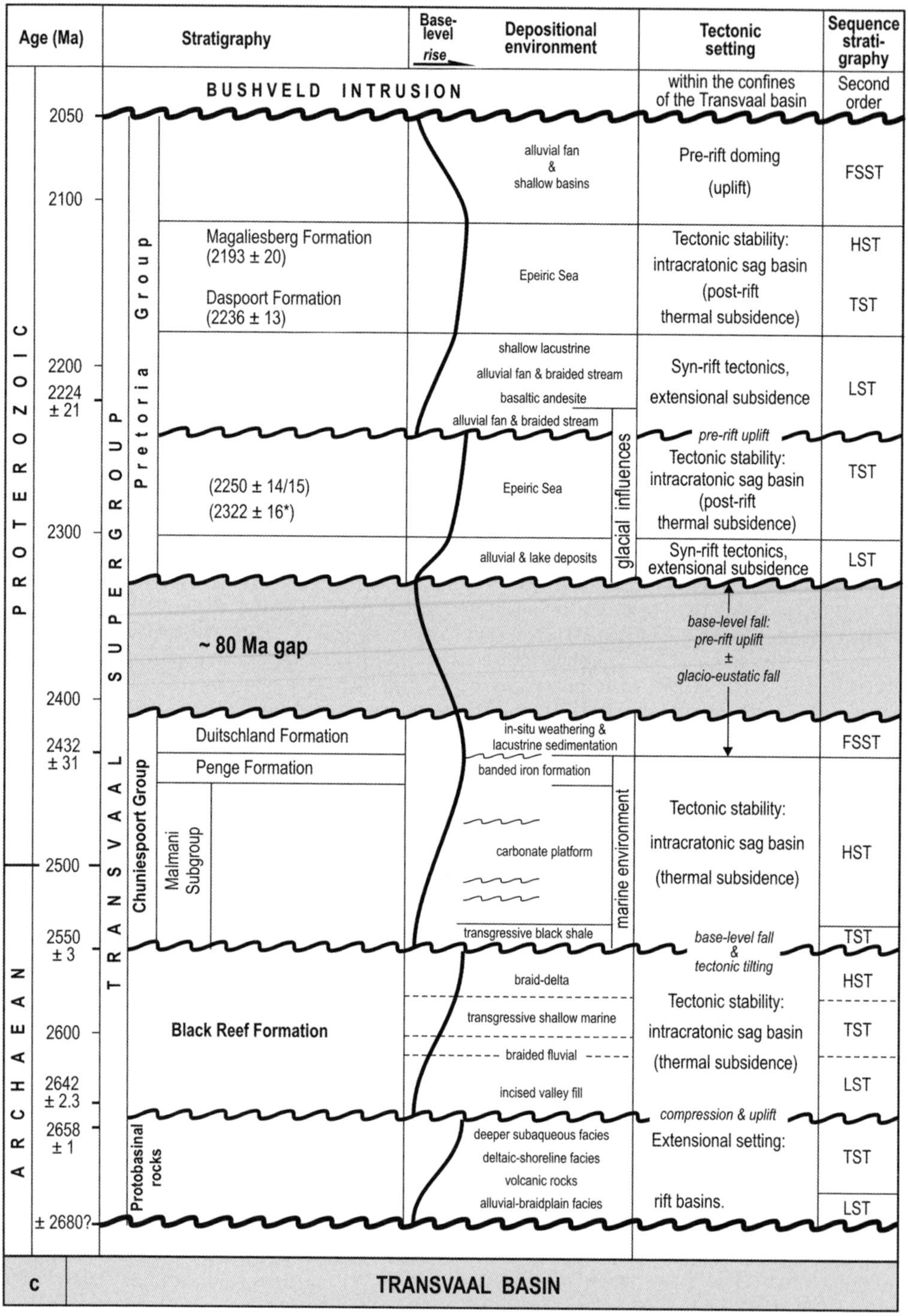

Fig. 3. Schematic geodynamic history chart for the Transvaal Supergroup, shown only for the Transvaal Basin (one of three preserved), illustrating lithostratigraphy, chronology, inferred tectonic settings and depositional palaeoenvironments, as well as interpreted sequence stratigraphy (modified after Catuneanu & Eriksson 1999). The maximum age for protobasinal rocks of approximately 2.68 Ga (Olsson *et al.* 2010) is speculative. Age data for Pretoria Group sandstones (shown in brackets: Mapeo *et al.* 2006) are from detrital zircons and represent maximum depositional

the up to 1.1 km-thick Duitschland encompasses: (1) a basal glacigenic diamictite and succeeding conglomerate–quartzite unit; (2) an inferred deep-water black shale, marl, carbonate interbed unit; (3) a shallow-water quartzite–carbonate–shale facies, with a second diamictite and conglomerate–quartzite couplet (Bekker *et al.* 2001; Frauenstein *et al.* 2009). Eriksson *et al.* (2001) related the Duitschland deposition to the weathering and erosion of carbonate rocks in the south of the TB, and to the basal Rooihoogte Formation (karst-fill deposits) of the succeeding Pretoria Group. Both of these formations and the inferred glacial diamictite at the base of the Duitschland lie chronologically within the approximately 80 Ma (Eriksson & Reczko 1995; or even 200 Ma in the KB: Mapeo *et al.* 2006) hiatus preceding the main Pretoria Group deposition. The tillite would thus equate to the approximately 2.4–2.2 Ga global glaciation event discussed in the first section of this paper (cf. Young 2004; Lowe & Tice 2007).

The Pretoria Group (TB; closely correlated with the Segwagwa Group in KB) consists of a 6–7 km-thick succession with a sheet-like geometry of predominant argillites (inferred epeiric marine deposits), lesser sandstones (ascribed to fluvial deposition) and two major volcanic intervals (e.g. Eriksson *et al.* 2006*b*). The age range is bracketed by 2316 $\pm$ 7 Ma (Re–Os: Hannah *et al.* 2004) near the base of the group, and a maximum age of less than 2058 $\pm$ 0.8 Ma (Buick *et al.* 2001) determined for the Bushveld Complex that intrudes this succession (Fig. 3). A less complete succession in the GB includes the Makganyene glacial formation, thought by a majority of workers to be equivalent in age to the basal epeiric shales of the Pretoria Group in the TB and KB (see, however, Moore *et al.* 2001, who placed it chronologically in the sub-Pretoria hiatus).

Deposition of the Pretoria Group ceased at some time prior to the Bushveld intrusion, and regional-scale compressive deformation of the Transvaal basin-fill followed (Bumby *et al.* 1998; Eriksson *et al.* 1998), which may have been related, at least partially, to the approximately 2.2–1.8 Ga Eburnean supercontinental assembly event that affected many of the current cratons of Africa. The Bushveld Complex, covering an area of about 61 000 km^2, comprises the layered mafic rocks of the Rustenburg Suite, felsic rocks of the lopolithic Lebowa Granite Suite (centrally emplaced within the marginal, inward-dipping sheet-like mafic lithologies) and localized occurrences of the Rashoop Granophyre Suite; the maximum thickness of the complex is approximately 11 km (Von Gruenewaldt & Harmer 1993; Eriksson *et al.* 1995 and references therein). Associated with the Bushveld intrusives into the upper Pretoria Group, are the essentially coeval Rooiberg Felsites (with bimodal volcanism and inferred crustal melts: Buchanan *et al.* 2002, 2004), up to about 5 km thick with an estimated eruptive volume of possibly 300 000 km^3 (Twist 1985; Cawthorn & Walraven 1997). Basal Rooiberg lavas overlie the Pretoria Group unconformably and were detached from the main Rooiberg succession (which now forms the roof of the Bushveld) by this major intrusion (e.g. Eriksson *et al.* 1995 and references therein).

Kaapvaal appears to have become part of a small supercontinent at about 2.0 Ga through amalgamation with the Zimbabwe Craton along the Limpopo Mobile Belt (LMB) over a protracted and complex geodynamic history spanning some 700 Ma (e.g. Barton *et al.* 2006; Boshoff *et al.* 2006; Zeh *et al.* 2007, 2008; Chudy *et al.* 2008; Millonig *et al.* 2008; Perchuk *et al.* 2008; van Reenen *et al.* 2008; Gerdes & Zeh 2009). A much earlier amalgamation event, at approximately 2.67–2.61 Ga, occurred when the small and exotic terrane of the Central Zone of the LMB accreted onto the Kaapvaal Craton (e.g. Zeh *et al.* 2009 and references therein). The concomitant far-field tectonic effects may have influenced the evolution of the Transvaal protobasinal depositories (Eriksson *et al.* 2005). Later collision of this assembled Kaapvaal–Central Zone terrane with the Zimbabwe Craton occurred at around 2.0 Ga (e.g. Kamber *et al.* 1995; Holzer *et al.* 1998; Kröner *et al.* 1999) (Fig. 2).

The approximately 2.0 Ga events in the Limpopo belt, the intrusion of the Bushveld Complex at about 2.06 Ga (Fig. 2) and the Eburnean assembly of African cratons provided the geodynamic framework for deposition within the Waterberg basins on Kaapvaal; these continental depositories are bound by fundamental Archaean structures and were related to active synsedimentary tectonism (Callaghan 1987; Eriksson *et al.* 1996). The Waterberg succession counts amongst the earliest appearances globally of red beds (*sensu stricto*; related to free oxygen in Earth's palaeo-atmosphere) and large erg deposits (Eriksson & Cheney 1992; Eriksson & Simpson 1998). Deposition occurred within a larger Main Basin (with early northern and southern sub-basins, and filled by 12 formations) and a smaller Middelburg Basin (one formation), between about 2.06 and 1.88 Ga (South African Committee for Stratigraphy 1980; Jansen 1982; Walraven & Hattingh 1993; Eglington & Armstrong

Fig. 3. (*Continued*) ages for sampled sandstones; age at the base of the Pretoria Group (shown with *) from Hannah *et al.* (2004); remaining age data from references in Eriksson *et al.* (2001).

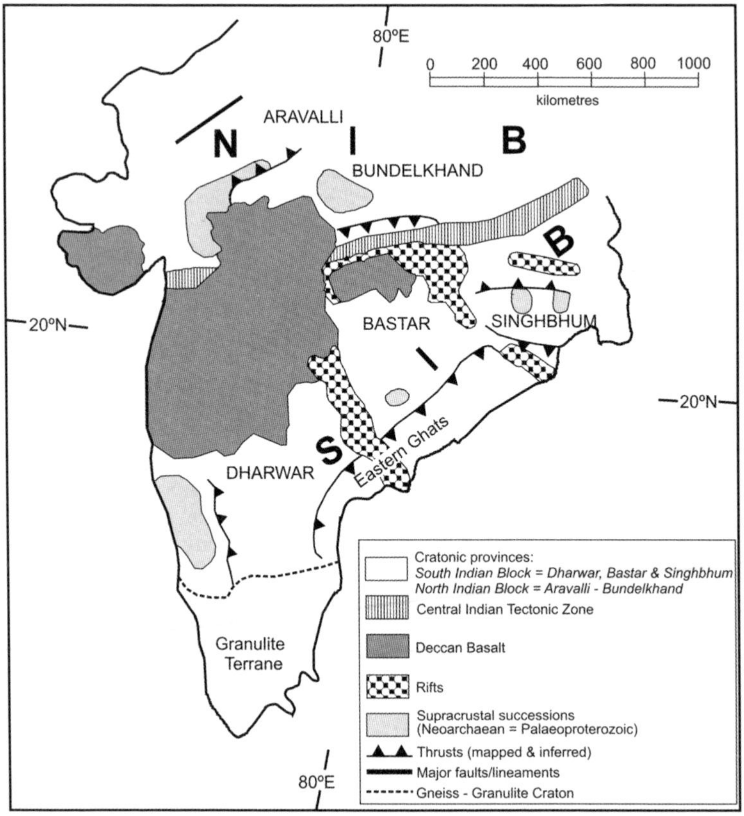

Fig. 4. Crustal architecture of Peninsular India, showing the major cratons making up the South Indian Block (SIB: Dharwar, Bastar and Singhbhum cratons) and the North Indian Block (NIB: Bundelkhand and Aravalli cratons). Major exposures of Neoarchaean–Palaeoproterozoic supracrustal rocks, Deccan Basalts and major Phanerozoic rifts shown as well. Modified after Eriksson *et al.* (2006*a*).

2004; Hanson *et al.* 2004). Sedimentation systems were largely fluvial, with subordinate fan and lacustrine settings, and a prominent palaeo-desert deposit (Callaghan *et al.* 1991; Eriksson *et al.* 2000, 2009*a*; Simpson *et al.* 2002, 2004) (Fig. 2). The Soutpansberg Group, also a fault-bounded basin encompassing essentially continental sedimentary rocks similar to the Waterberg depositories, is thought to be a bit younger than these basins, and occurs to the ENE of the Main Waterberg Basin (Barker *et al.* 2006).

Singhbhum Craton: Palaeoproterozoic evolution

The Singhbhum Craton forms part of the South Indian Block within the crustal architecture of Peninsular India (Fig. 4). Palaeoproterozoic volcano-sedimentary successions of the Singhbhum Craton comprise a northern and a southern group, the former lying to the north, and the latter lying both to the east and west of the Archaean granitoid core of the craton (Fig. 5) (see also Mazumder *et al.* 2012). Those successions occurring to the north of the Archaean Singhbhum Granitoid (Fig. 6) provide important insights into the understanding of the extant mantle and surface processes, crust–mantle interactions, and the tectonic framework of sedimentation (e.g. Bose 1994; Mazumder 2005 and references therein; see also Mazumder *et al.* 2009). Supracrustal successions to the south (i.e. those lying west and east of the Singhbhum Granitoid; Fig. 5) have been less intensely studied. Earlier studies focused on various aspects of the Palaeoproterozoic geology of the Singhbhum crustal province but a comprehensive account of the Palaeoproterozoic geological evolution of the Singhbhum is still awaited (see, however, Mazumder *et al.* 2012), and only a brief overview will be attempted here. Geochronological data available for this terrane are limited, particularly for the Palaeoproterozoic supracrustal rocks (Table 1).

Supracrustal successions north of the Singhbhum Granitoid

The Archaean Singhbhum Granite (Saha 1994; Mishra *et al.* 1999) is overlain by the siliciclastic metasedimentary and mafic metavolcanic rocks of the 2.1 Ga (Sm–Nd age from upper Dhanjori mafic rocks: see Roy *et al.* 2002*a*) Dhanjori Formation (Fig. 7). The Dhanjori rocks are deformed and metamorphosed to greenschist facies (Gupta *et al.* 1985; Saha 1994; Joy & Saha 2000). This formation is made up of two members: phyllites, quartzites and thin conglomerate comprise the lower member, whereas volcanic and volcaniclastic rocks along with subordinate quartzites and phyllites characterize the upper member (Gupta *et al.* 1985; Mazumder & Sarkar 2004; Mazumder 2005; Mazumder & Arima 2009). The lower member starts with a lenticular basal conglomerate, and is overall fining-upwards, as is the upper volcano-sedimentary member of the Dhanjori Formation (Mazumder & Sarkar 2004, fig. 2). The Dhanjori Formation has been interpreted as a terrestrial (dominantly fluvial) volcano-sedimentary succession (Mazumder & Sarkar 2004; Mazumder 2005; Mazumder *et al.* 2012). Basal conglomerate (Fig. 8a), coarse-grained sandstone (Fig. 8b) and shale represent the distal fringe of a basal alluvial-fan complex, interpreted as reflecting sieve, sheet flood, mass flow, and alluvial channel and overbank deposits. Clastic sedimentary rocks within the rest of the formation are entirely constituted by fining-upwards fluvial cycles (Mazumder & Sarkar 2004). An event of basin tilting and volcanic eruption, encompassing profuse volcanic rocks and accompanying volcaniclastic deposits, intervened during fluvial sedimentation of the Upper Member, although the general fluvial depositional framework remained essentially unaltered. The geochemical characteristics of the upper Dhanjori volcanic rocks indicate that these are basaltic komatiites (Mazumder & Arima 2009). These geochemical data, in combination with the sedimentological and stratigraphic characteristics of the interbedded metasedimentary rocks, strongly support an intracontinental rift setting for Dhanjori volcanism and sedimentation (Mazumder & Arima 2009).

The Chaibasa Formation sharply overlies the Dhanjori Formation, and a transgressive lag demarcates their contact (Bose *et al.* 1997; Mazumder 2005; Mazumder *et al.* 2012). Lithologically, the Chaibasa is characterized by the interbedding of compositionally and texturally mature fine-grained sandstones (Fig. 9a), a heterolithic (fine sandstone/siltstone–mudstone; Fig. 9b) facies and shales (Fig. 9c), with variable bed thicknesses (Bhattacharya 1991; Bose *et al.* 1997). The Chaibasa sandstones are interpreted to have formed in a subtidal setting, and the heterolithic facies in a tide–storm interactive offshore palaeogeography between the fair-weather and storm wave base. The shales are thought to reflect a distal offshore setting below the storm wave base and probably represent distal turbidite deposits (Bose *et al.* 1997; Mazumder 2005). The Chaibasa succession is thus seen as being generally transgressive, with intermittent punctuations caused by short-term lowstands during which sandstones were emplaced (Mazumder *et al.* 2000; Mazumder 2005). The shallow-marine subtidal sandstones bear an excellent record of sandwave migration and provide a rare opportunity to unlock the only unambiguous Late Palaeoproterozoic lunar tidal rhythm (Mazumder 2004, 2005; Mazumder & Arima 2005). The Chaibasa Formation bears numerous and varied soft-sediment deformation structures, several of which show characteristics that indicate earthquakes as a trigger (Bhattacharya & Bandyopadhaya 1998; Mazumder *et al.* 2006, 2009). The inferred Chaibasa seismites represent the oldest unambiguous seismites inferred to date (Mazumder *et al.* 2006).

The Dhalbhum Formation, which succeeds the Chaibasa, comprises phyllites, quartzites (Fig. 10a, b) and shales at the base with common ultramafic–mafic intrusive bodies, followed by a largely tuffaceous (Fig. 10c) upper part; thin mafic–ultramafic lava flows occur throughout the succession (Gupta *et al.* 1980, 1982). An overall terrestrial (fluvial–aeolian) palaeogeography is envisaged for the Dhalbhum sedimentation (Simpson *et al.* 2004; Mazumder 2005). The Chaibasa–Dhalbhum boundary therefore represents an unconformity (within the sequence stratigraphic paradigm; cf. Posamentier *et al.* 1988; Mazumder 2005) (Fig. 7). The change from shallow-marine upper Chaibasa deposits to fully continental Dhalbhum lithologies is postulated to reflect crustal uplift under the influence of the early stages of a mantle plume (Eriksson *et al.* 1999; Mazumder *et al.* 2000; Mazumder 2005). A plume influence (see Kent 1991) is also supported by the change in freeboard conditions from neutral-negative (Chaibasa) to terrestrial and positive during Dhalbhum deposition, and also by the presence of komatiitic rocks and bimodal volcanics (Mazumder 2005; Eriksson *et al.* 2006*a*).

The Dalma Formation is a thick sequence of mafic–ultramafic volcanic rocks containing lenses of basic agglomerates (Fig. 11a) and minor quartzites that conformably overlies the Dhalbhum Formation (Bhattacharyya & Bhattacharyya 1970); pillow basalts are common in the lower parts of the Dalma succession (Fig. 11b) (Chakraborti 1980; Gupta *et al.* 1980; Chakraborti & Bose 1985; Bose

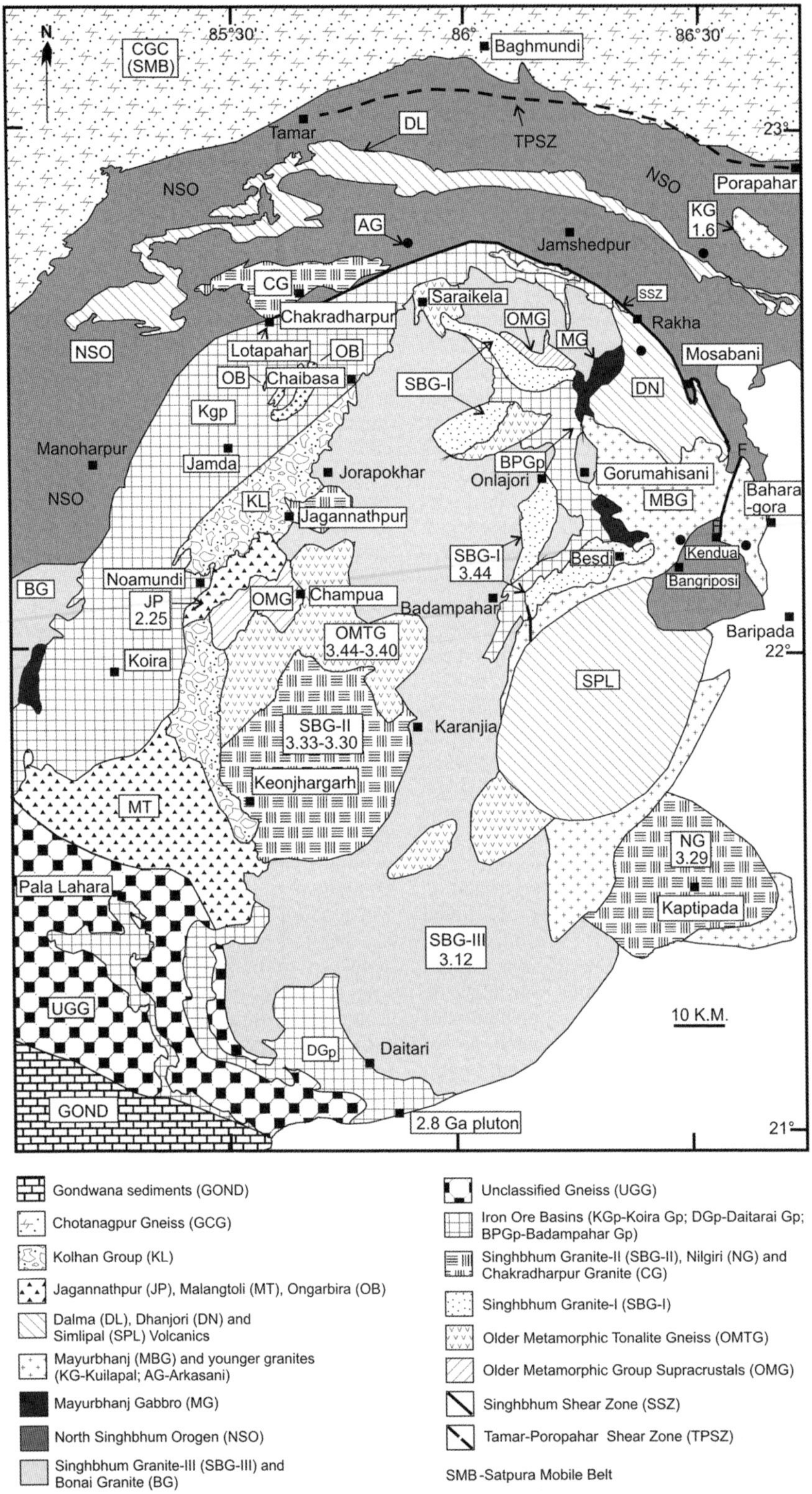
N
85°30′
86°
86°30′
23°
22°
21°
CGC (SMB)
Baghmundi
Tamar
DL
TPSZ
NSO
Porapahar
KG 1.6
AG
Jamshedpur
CG
Saraikela
SSZ
Rakha
Chakradharpur
OMG
MG
Mosabani
Lotapahar
OB
Chaibasa
SBG-I
DN
Kgp
Manoharpur
Jamda
Jorapokhar
BPGp
Onlajori
Gorumahisani
MBG
Bahara -gora
KL
Jagannathpur
F
SBG-I 3.44
Besdi
Kendua
Bangriposi
BG
Noamundi
JP 2.25
Champua
Badampahar
Baripada
OMTG 3.44-3.40
Koira
SPL
SBG-II 3.33-3.30
Karanjia
Keonjhargarh
MT
NG 3.29
Pala Lahara
Kaptipada
SBG-III 3.12
UGG
10 K.M.
DGp
Daitari
GOND
2.8 Ga pluton
Gondwana sediments (GOND)
Chotanagpur Gneiss (GCG)
Kolhan Group (KL)
Jagannathpur (JP), Malangtoli (MT), Ongarbira (OB)
Dalma (DL), Dhanjori (DN) and Simlipal (SPL) Volcanics
Mayurbhanj (MBG) and younger granites (KG-Kuilapal; AG-Arkasani)
Mayurbhanj Gabbro (MG)
North Singhbhum Orogen (NSO)
Singhbhum Granite-III (SBG-III) and Bonai Granite (BG)
Unclassified Gneiss (UGG)
Iron Ore Basins (KGp-Koira Gp; DGp-Daitarai Gp; BPGp-Badampahar Gp)
Singhbhum Granite-II (SBG-II), Nilgiri (NG) and Chakradharpur Granite (CG)
Singhbhum Granite-I (SBG-I)
Older Metamorphic Tonalite Gneiss (OMTG)
Older Metamorphic Group Supracrustals (OMG)
Singhbhum Shear Zone (SSZ)
Tamar-Poropahar Shear Zone (TPSZ)
SMB -Satpura Mobile Belt

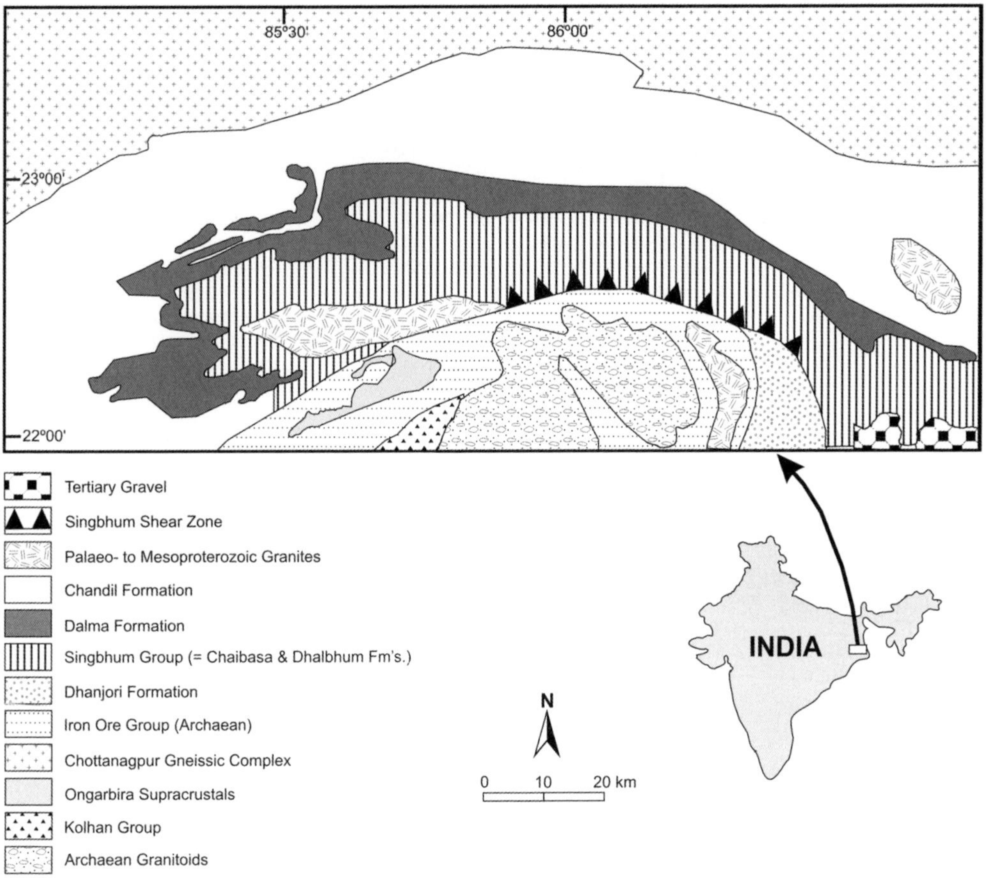

Fig. 6. Geological sketch map of the northern supracrustal successions of the Singhbhum Craton. Modified after Mazumder (2005).

1994; Singh 1997, 1998; Sengupta *et al.* 2000). The Dalma volcanic rocks form an arcuate belt along the NW margin of the Singhbhum Craton (Fig. 6) and their origin is still debated. An island arc setting has been suggested for the Dalma volcanism (Naha & Ghosh 1960), whereas Bose *et al.* (1989) and Bose (1994) proposed a back-arc setting. In contrast, De (1964) suggested that the Dalma Formation represents continental volcanism. A plume-related genesis is vigorously argued by Roy (1998; see also Roy *et al.* 2002*b*), an idea also supported by Mazumder & Arima (2004) and Mazumder (2005) who proposed an intracontinental rift setting for the volcanism. The age of the Dalma rocks is uncertain (discussion in Mazumder *et al.* 2012); gabbro-pyroxenitic intrusives into these volcanic lithologies are dated at 1619 ± 38 Ma (Rb–Sr whole rock: Roy *et al.* 2002*b*). These intrusives exhibit trace element and rare earth element (REE) signatures that support a comagmatic relationship with the Dalma volcanic rocks (Acharyya 2003*a*, *b*).

The Dalma Volcanic Belt is separated from the Chotanagpur Gneissic Complex, which forms the northern part of the Singhbhum Craton (Fig. 6), by a belt of metasedimentary and metavolcanics rocks known as the Chandil Formation (Ray *et al.*

Fig. 5. Simplified geological sketch map of the Singhbhum Craton and its surrounds, modified after Misra (2006) and Ramakrishnan & Vaidyanadhan (2008). Note the 'northern' Singhbhum supracrustals lying north of the centrally placed Singhbhum Granitoid core (and illustrated in Fig. 6), and those termed 'southern' Singhbhum supracrustals here, which lie to both east (Simlipal Volcanics) and west (Kolhan Group; Jagannathpur, Malangtoli and Ongarbira volcanics) of the granitoid core.

Table 1. *Relevant ages from the Singhbhum Craton and a summary of Mesoarchaean to Palaeoproterozoic–Mesoproterozoic geodynamic events (modified after Mazumder* et al. *2012)*

Event	Age and data source	Geochronological method
Thermal events (1500–1400 Ma)	1487 ± 34 Ma: Sengupta *et al.* (2000)	Rb–Sr age
	1547 ± 20 Ma: Saha (1994)	K–Ar age
Crystallization of Chandil Rhyolite and *c.* 1600 Ma events	1628.5 ± 4.3 Ma: Reddy *et al.* (2009)	U–Pb SHRIMP concordant age
	1619 ± 38 Ma: Roy *et al.* (2002*b*)	Rb–Sr WR age
	c. 1580 Ma: Krishna Rao *et al.* (1979)	Pb/Pb age
	1677 ± 11 Ma: Sarkar *et al.* (1986)	Rb–Sr, Pb/Pb ages
	1638 ± 38 Ma: Sengupta *et al.* (1994)	Rb–Sr age
Thermal events (2200–2000 Ma)	*c.* 2200 Ma: Sarkar *et al.* (1986)	Rb–Sr and Pb/Pb
	2072 ± 106 Ma: Roy *et al.* (2002*a*)	Sm–Nd age
	c. 2008 Ma: Iyenger *et al.* (1981)	Rb–Sr age
	c. 1960 Ma: Vohra *et al.* (1991)	Rb–Sr age
Thermal events *c.* 2800 Ma (emplacement of Temperkola Granite; emplacement of main phase of Mayurbhanj Granite)	*c.* 2800 Ma: Bandyopadhyay *et al.* (2001)	$^{207}Pb/^{206}Pb$ age
	c. 2800 Ma: Acharyya *et al.* (2010)	U–Pb zircon age
Metamorphic event	3241 ± 7 Ma: Mishra *et al.* (1999)	$^{207}Pb/^{206}Pb$ age
Emplacement of Singhbhum Granitoid phase III	3288 ± 8 Ma: Reddy *et al.* (2009)	U–Pb SHRIMP concordant age
Emplacement of Singhbhum Granitoid phase II	3328 ± 7 Ma: Mishra *et al.* (1999)	$^{207}Pb/^{206}Pb$ age
Thermal Pb-loss event	*c.* 3350 Ma: Basu *et al.* (1996)	U–Pb age
Emplacement of Older Metamorphic Tonalite Gneiss (OMTG)	3437 ± 9 Ma: Mishra *et al.* (1999)	$^{207}Pb/^{206}Pb$ age
	3448 ± 19 Ma: Acharyya *et al.* (2010)	U–Pb zircon age
Older Metamorphic Group (OMG) sedimentation	*c.* 3500 Ma: Mishra *et al.* (1999)	$^{207}Pb/^{206}Pb$ age
Dacitic volcanism	3506.8 ± 2.3 Ma: Mukhopadhyay *et al.* (2008) (oldest Indian greenstone succession)	
Crust formation including a significant granitoid component	*c.* 3600 Ma: Mishra *et al.* (1999) (detrital OMG zircons)	$^{207}Pb/^{206}Pb$ age

1996; Mazumder 2005). Bose (1994) first pointed out that the lithological association of these supracrustals occurring to the north of the Dalma volcanic belt (Bose 1994, his sub-basin II) is markedly different from those occurring to the south of the Dalma volcanic belt. The Chandil succession is made of metapelites with intercalated lenticular quartzites, metabasic, volcaniclastic and minor carbonate rocks (Bhattacharya 1992; Bose 1994), and is post-Dalma in age (Bhattacharyya & Bhattacharyya 1970; Mazumder 2005; Nelson *et al.* 2007). A single sample of Chandil Formation rhyolite from the base of the unit contains a single, largely concordant, igneous zircon population that records an age of 1629 ± 5 Ma (SHRIMP (Sensitive High Resolution Ion Microprobe): Reddy *et al.* 2009). Preliminary sedimentological investigation suggests that the Chandil quartzites are terrestrial deposits (Mazumder 2005). Gupta & Basu (2000), however, proposed that the psammopelitic rhythmites bearing flaser–wavy–lenticular bedding in the Lawa area were formed in an intertidal–subtidal setting. The mode of the Chandil sequence-building is not clearly understood at present and detailed sedimentological studies remain to be done.

Southern supracrustal successions east and west of the Singhbhum Granitoid

As noted above, Palaeoproterozoic supracrustals also occur to the south of the Singhbhum Shear Zone (Mazumder *et al.* 2012) (Fig. 5). The precise ages of these supracrustals are poorly constrained (Table 1). They include the Kolhan Group (Ghosh & Chatterjee 1994; Saha 1994; Mukhopadhyay

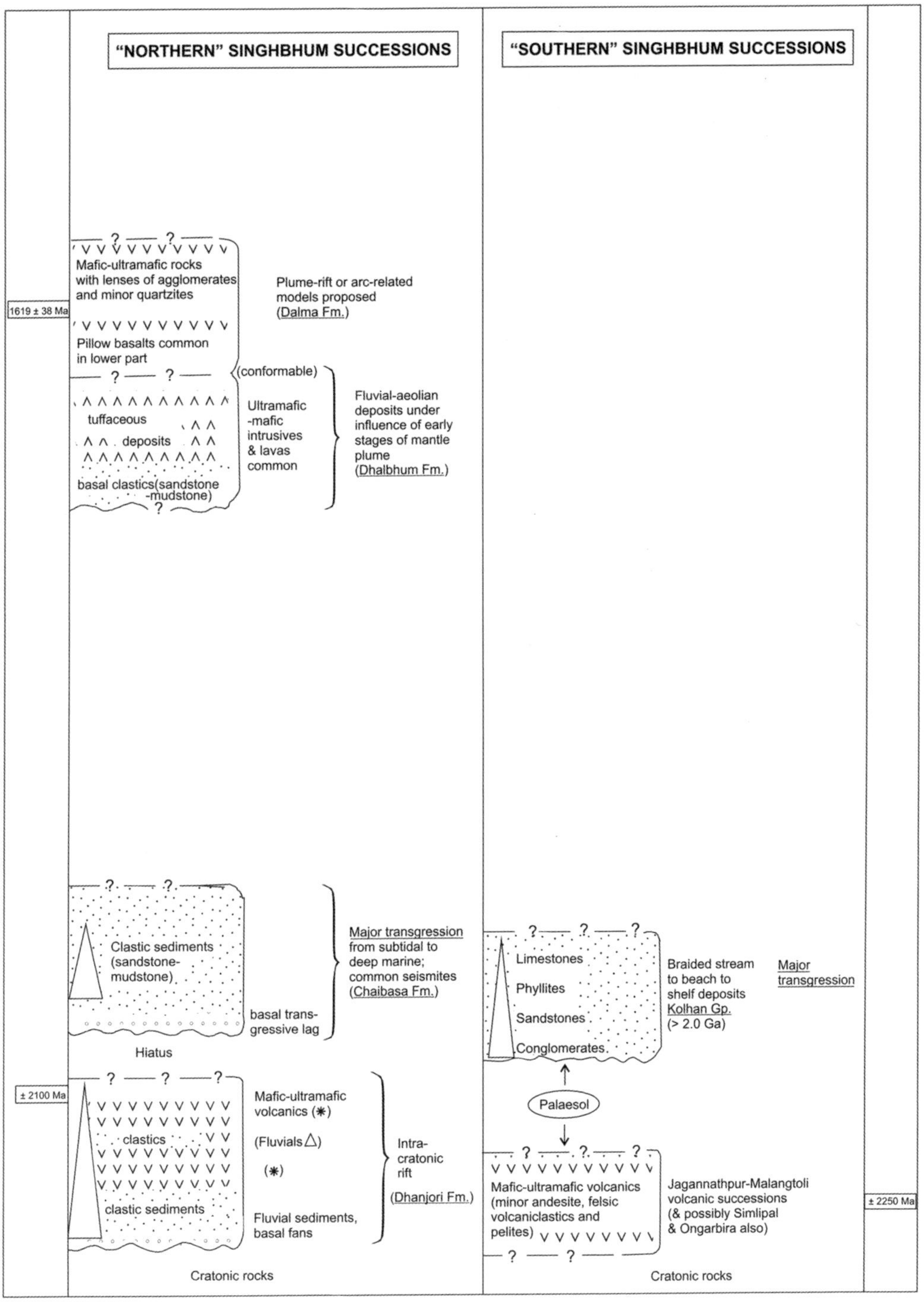

Fig. 7. Schematic geohistory sketch showing major lithostratigraphic units for both the 'southern' (right-hand column) and 'northern' (left-hand column) Singhbhum Palaeoproterozoic volcano-sedimentary supracrustal successions. Note the minimal age date for the Dhanjori and Dalma formations, the Kolhan Group and Jagannathpur Volcanics, and their inferred depositional settings and possible tectonic frameworks. The term cratonic rocks refers to cratonic basement lithologies (gneissic–granitic rocks).

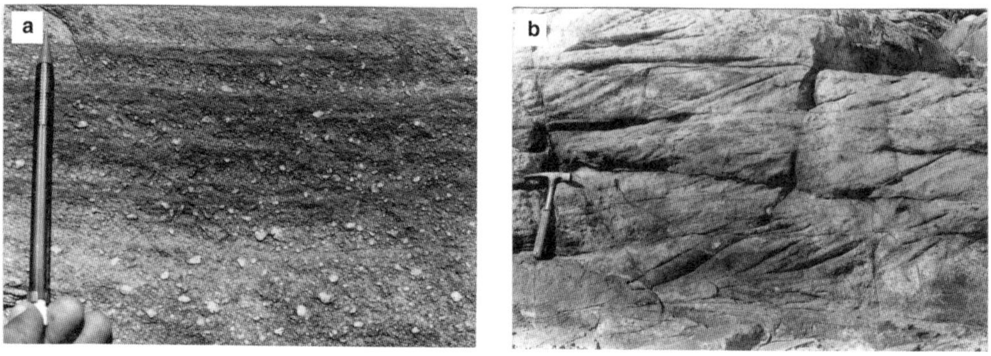

Fig. 8. Dhanjori Formation: (**a**) basal conglomerate, containing small pebbles of quartzite, vein quartz and granite; (**b**) coarse-grained, poorly sorted Dhanjori sandstone, showing cross-stratification.

et al. 2006), Ongarbira metasedimentary and metavolcanic rocks (Gupta *et al.* 1981; Mukhopadhyay *et al.* 1990; Blackburn & Srivastava 1994), Simlipal metavolcanic rocks (Sarkar & Saha 1962; Saha 1994), and the Jagannathpur and Malangtoli mafic lavas (Sarkar & Saha 1962; Banerjee 1982; Saha 1994) (Fig. 5). The Kolhan Group (basal conglomerate–sandstone assemblage overlain by phyllitic

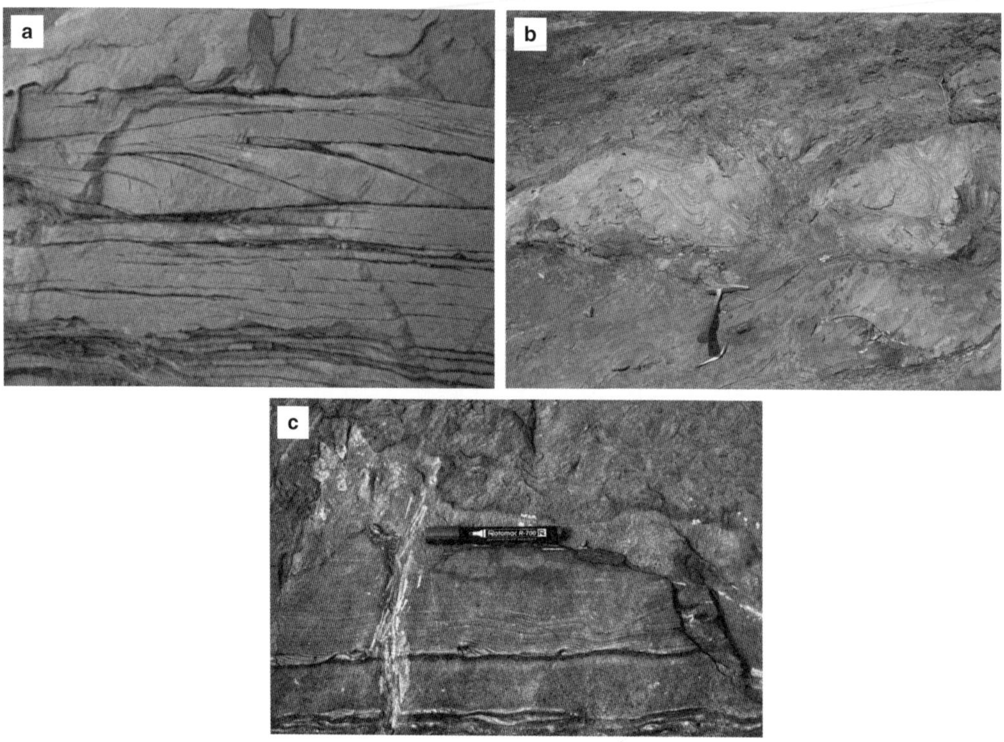

Fig. 9. Chaibasa Formation: (**a**) medium- to fine-grained cross-stratified texturally and compositionally mature sandstone facies; (**b**) heterolithic facies; note the penecontemporaneously folded quartzite interbands within garnetiferous quartz mica schist, reflecting the alteration of pre-metamorphic regular alternation of sandstone and mudstone interbeds; the hammer is for scale; (**c**) Chaibasa Shale facies with intercalated thin beds of very fine sandstone (light-coloured); the sandstone laminae show either horizontal or ripple cross-lamination followed by draping and horizontal laminations. The sandstone beds have sharp lower contacts but gradational upper contacts with the overlying mudstones (the pen length is 10 cm). Note that the original shale has become replaced by schist during metamorphism (and sandstones are generally recrystallized to quartzite).

Fig. 10. Dhalbhum Formation: (**a**) trough cross-bedded sandstones; (**b**) aeolian quartzites (locally recrystallized sandstones) with very large-scale cross-bed foresets of inversely graded sandstones; the inversely graded laminae are visible at the base of the large foreset in the centre of the photograph (the person is for scale); locality is Tiraldih, west of Tatanagar; (**c**) Dhalbhum felsic tuff occurring close to the Dhalbhum–Dalma contact (location is Dimna lake, near Tatanagar; the hammer length is 42 cm).

shale and impersistent limestone) is almost undeformed and unmetamorphosed, and unconformably overlies the Singhbhum Granite and the Iron Ore Group (cf. Saha 1994; Mukhopadhyay 2001). The intercalated conglomerate–sandstones, sheet sandstones and sandstone–siltstone alterations of the lower part of the Kolhan Group are inferred to have formed in braided stream, beach and wave-dominated shelf settings (Ghosh & Chatterjee 1994; Bandopadhyay & Sengupta 2004; Bandopadhyay *et al.* 2010). Mukhopadhyay *et al.* (2006), who studied the Kolhan Group in the Chaibasa–Jagannathpur area (Fig. 5), inferred a Mesoproterozoic–Neoproterozoic age (on the basis of lithostratigraphic and interpreted genetic correlations with other units) and interpreted the palaeoenvironment as having comprised a deep-water cratonic setting. The marginal-marine Kolhan Group in an area further to the south of that studied by Mukhopadhyay *et al.* (2006), in the general Keonjhargarh region (Fig. 5), overlies a Palaeoproterozoic palaeosol (Keonjhargarh palaeosol, of white shale with quartz, muscovite and pyrophyllite, whose tentative age is between 2.0 and 2.25 Ga: see Bandopadhyay *et al.* 2010 for discussion) (Fig. 12). The Ongarbira volcano-sedimentary succession rests unconformably on a basement topographic low between the Chakradharpur Granite and the Singhbhum Granite (Blackburn & Srivastava 1994) (Fig. 5). Mafic (tholeiitic basalts) to ultramafic (pyroxenite) volcanic rocks, felsic volcaniclastics (tuffaceous metasediments) and low-grade pelites constitute the Ongarbira succession (Gupta *et al.* 1981; Mukhopadhyay *et al.* 1990; Blackburn & Srivastava 1994; Mazumder 2005). Mukhopadhyay *et al.* (1990) claimed that both the Ongarbira and Dalma volcanic rocks occupy comparable structural horizons, and that continuity between them is lost because of attenuation and faulting on a fold limb. The Simlipal and Jagannathpur volcano-sedimentary successions have been correlated with the Proterozoic Dhanjori volcano-sedimentary succession solely based on lithostratigraphy (cf. Sarkar & Saha 1962, 1977; Goodwin 1996, p. 338). The Jagannathpur Volcanics are thought to be cogenetic with the Malangtoli Volcanics; the former are dated at approximately 2250 Ma (Pb/Pb, whole rock) and comprise relatively undeformed mafic–ultramafic flows with

Fig. 11. Dalma Formation: (**a**) agglomerates; note the poorly sorted, angular clasts of basalt; (**b**) highly weathered pillow basalts.

Fig. 12. General view of the approximately 10 m-thick Keonjhargarh palaeosol overlying a saprolitic substrate (the parked tractor on the quarry floor is for scale) and succeeded by horizontal beds of Kolhan Group sandstone. Photograph kindly provided by P. Bandopadhyay.

minor andesitic and basanitic compositions (Ramakrishnan & Vaidyanadhan 2008). Bandopadhyay *et al.* (2010) suggested that these mafic lavas might have provided a younger weathered component to the longer-lived granitoid-derived palaeosol beneath the Kolhan Group. The Simlipal Basin is large (150 by 60 km; Fig. 5), and is filled by a succession comprising basal quartz-pebble conglomerate–orthoquartzite–feldspathic quartzite, overlain by three horizons of spilitic basalts alternating with cross-bedded orthoquartzite layers. The lava flows exhibit basal breccias, with tuffs and tuffaceous sediments at their tops; their composition is tholeiitic to basaltic–andesitic (Ramakrishnan & Vaidyanadhan 2008). Age determinations on the intrusive Mayurbhanj Granitoid range widely, 3100–2200 Ma, and thus do not constrain the upper limit of Simlipal deposition in any meaningful way (Ramakrishnan & Vaidyanadhan 2008). Sedimentological data from all of these tentative 'southern' Palaeoproterozoic supracrustal volcano-sedimentary successions (excluding the Kolhan Group) are limited, and much detailed work remains to be carried out to clarify the variable viewpoints on their genesis.

Discussion

Comparison of Palaeoproterozoic successions on the Singhbhum Craton

In the virtual absence of any reliable age data from the Palaeoproterozoic successions of this craton, recourse must be had to examining the lithostratigraphic records (see also Mazumder *et al.* 2012) from the northern and southern parts of the craton, using also inferred major geodynamic and eustatic events as possible indicators of a shared history. Obviously, this is a partially speculative process given the current stage of knowledge, and this caveat will apply equally when comparing Singhbhum overall with the inferred Palaeoproterozoic evolution of the Kaapvaal Craton. Lacking accurate geochronological data for these purposes leaves the choice of either doing nothing at all or of an attempted examination of possible geological history, as we present here in the full assumption that future data might invalidate much of what can currently be proposed. The dearth of chronological data (Table 1) is exacerbated by commonly poor field exposures.

In addition to problems inherent in the discussed lack of geochronological data and good exposure, we recognize the potential difficulty raised by our geodynamic comparisons arising from divergent views on the Singhbhum shear zone. A separate paper in this volume (Mazumder *et al.* 2012), which describes the geology of the Singhbhum Craton in great detail, also addresses more controversial aspects such as the possible age and the geodynamic interpretation of this shear zone, and how these then relate to viewpoints on stratigraphy and stratigraphic comparisons, such as we make in this section of the paper. In particular, the variable models of Singhbhum Shear Zone geohistory impact directly on inferred correlations between northern and southern Singhbhum supracrustal successions. We thus rely here on readers having consulted Mazumder *et al.* (2012) prior to perusing this paper.

In the northern successions, the association of fluvial cycles and mafic–ultramafic volcanics within the approximately 2.1 Ga Dhanjori Formation, together ascribed to an intracratonic rift, can be considered as possible equivalents to the 'southern' approximately 2250 Ma Jagannathpur–Malangtoli (and possibly also the Simlipal and even Ongarbira) successions, also mainly comprised of mafic–ultramafic volcanic rocks and lesser clastic sedimentary lithologies (Fig. 7). Together, they may reflect an overall essentially intracratonic volcanically dominated rift geodynamic setting at about 2.2–2.1 Ga, which occurred at several places on the craton (cf. Mazumder & Arima 2009). Interestingly, in both northern and southern parts of Singhbhum, these possibly approximately coeval volcanic-rift deposits are overlain, following hiatuses, by sedimentary rocks consistent with major transgressions (Fig. 7). Eriksson *et al.* (2006*a*) postulated that continental freeboard for the Singhbhum Craton was at an approximate maximum prior to these transgressions, which might thus reflect a major eustatic or local drowning event. In the case of northern Singhbhum, the clastic sedimentary rocks of the Chaibasa Formation (no age estimate) overlie an interpreted transgressive lag and are ascribed to a vertical transition from subtidal to deep-marine conditions, with common seismites suggesting a continuing active tectonic subsidence influence (cf. Bose *et al.* 1997; Mazumder 2005; Mazumder *et al.* 2006, 2012). In the case of the southern Singhbhum transgression, the vertical trend in the more than approximately 2.0 Ga Kolhan Group deposits is inferred to reflect a change from braided streams through beach to shelf conditions (cf. Ghosh & Chatterjee 1994; Bandopadhyay & Sengupta 2004; Bandopadhyay *et al.* 2010).

However, any suggestion of a possible major transgressive episode that developed across several parts of the Singhbhum Craton and an overall shared geodynamic history in the approximate 2.2–2.0 Ga period, must be questioned on the basis of the widespread and well-preserved, thick Keonjhargarh palaeosol (age estimated to be *c.* 2.0–2.25 Ga but also overlies 3.32 Ga cratonic rocks in many places) beneath the Kolhan Group (Bandopadhyay *et al.* 2010). There is no equivalent palaeosol known or preserved within the hiatus of unknown length between the Dhanjori and Chaibasa formations (Fig. 7). Analogous palaeosol horizons overlie approximately 3.33 Ga Banded Gneissic Complex rocks in both the Aravalli and Bundelkhand cratons (North Indian Block, NIB, see Fig. 4), which many workers suggest were not even then joined to the South Indian Block (SIB) that includes the Singhbhum Craton (see the discussion in Bandopadhyay *et al.* 2010). It can be speculated that continued active tectonism in northern Singhbhum related to seismite development in the Chaibasa succession might have led to the erosive removal of any possible equivalent palaeosol developed there. Interestingly, the Kolhan Group–Keonjhargarh palaeosol contact shows soft-sediment deformation, implying that the transgression there occurred while the soil was still soft and moist following rainfall, with no discernable hiatus separating these two disparate deposits (Bandopadhyay *et al.* 2010). The transgressive lag at the base of the Chaibasa Formation transgression (Fig. 7) would support possible erosion of any palaeosol developed there and in more general terms, during transgression, development of the wave ravinement surface can be responsible for significant erosion (up to *c.* 15–20 m: Catuneanu 2006).

The two younger units in northern Singhbhum, the Dhalbhum and Dalma formations (with a conformable contact between them), have no known equivalents in southern Singhbhum (Fig. 7). The inferred poorly defined age of the Dalma ($>1619 \pm 38$ Ma; Rb–Sr, whole rock: Roy *et al.* 2002*b*) suggests a major hiatus separating these two conformable formations from the preceding Chaibasa. The Dhalbhum–Dalma succession may have been related to a plume origin (e.g. Mazumder 2005).

Comparison of Palaeoproterozoic successions on the Singhbhum Craton with those from Kaapvaal

It would appear that earlier sedimentary basins on the Kaapvaal Craton (*c.* 3.1–2.8 Ga Witwatersrand, *c.* 2.7 Ga Ventersdorp and *c.* 2.66–2.4 Ga lower Transvaal (carbonate–BIF platform) basins) have no equivalents on the Singhbhum Craton. The so-called Younger Iron Ore Sequence (cf. the 'Western Iron Ore Group') of Singhbhum does comprise a chemical sedimentary covering of predominant BIF and some carbonates of inferred continental platform character, analogous thus to the lower Transvaal (Ghaap–Chuniespoort–Taupone groups) carbonate–BIF platform, but the Indian deposits are estimated to be considerably older, at around 3.3–3.1 Ga (Misra 2006; Ramakrishnan & Vaidyanadhan 2008). While Kaapvaal was dominated by a shallow epeiric platform setting from about 2.66 to 2.04 Ga, and probably low to negative freeboard conditions (Eriksson *et al.* 2006*a*), it is possible that Singhbhum was characterized by elevated freeboard in order to allow the thick and widespread Keonjhargarh palaeosol to develop above the Singhbhum Granitoid basement.

The upper part of the Transvaal Supergroup basin-fill, the Pretoria Group, that was deposited from about 2316 ± 7 Ma (Re–Os; Hannah *et al.* 2004) to less than 2058 ± 0.8 Ma (Buick *et al.* 2001), and which is related to relatively high freeboard and a two-stage rift basin geodynamic history (Catuneanu & Eriksson 1999; Eriksson *et al.* 2006*a*, *b*) (Fig. 3), has a rather truncated possible equivalent in Singhbhum, in the approximately 2.2–2.1 Ga Dhanjori Formation intracratonic rift volcano-fluvial deposits (Fig. 7). Within the Pretoria Group succession, two major clastic epeiric transgressions are identified (Fig. 3) (Eriksson *et al.* 2008), with only the second one, the Silverton–Magaliesberg transgression (>2.05 Ga), possibly being analogous to the Kolhan–Chaibasa transgression (>*c.* 2.0 Ga) in Singhbhum. The Pretoria Group is up to 6–7 km thick and covers a wide area with sheet-like sedimentary beds, in contrast to the much more spatially restricted possible Singhbhum equivalents. However, the 2222 ± 13 Ma (Pb/Pb: Cornell *et al.* 1996) Hekpoort flood basalts (Fig. 3) in the Pretoria Group are chronologically close to the approximately 2250 Ma Jagannathpur lavas of southern Singhbhum (Figs 5 & 7). Also, both the Hekpoort and the Jagannathpur lavas appear to have developed, or at least contributed to, thick weathering mantles; respectively, the less than 2.22 Ga Hekpoort (e.g. Driese 2004) and the less than 2.25 Ga Keonjhargarh (Bandopadhyay *et al.* 2010) palaeosols. Both palaeosols are also succeeded by major transgressive facies associations in the two cratons (e.g. Eriksson *et al.* 2001, 2006*a*). It should be noted that the Keonjhargarh palaeosol appears to be derived from an older granitoid contribution and a younger mafic (Jagannathpur) component, and, thus, in totality, post-dated the latter lavas (Bandopadhyay *et al.* 2010).

In contrast, major geodynamic events on Kaapvaal, such as the 2058 ± 0.8 Ma (Buick *et al.* 2001) intrusion of the enormous Bushveld Complex (probably plume-related: cf. Hatton 1995), the approximately 2.0 Ga Limpopo mobile belt orogenesis and demonstrable glacial deposits in the approximately 2.4–2.2 Ga interval within the Pretoria Group have no known equivalents in Singhbhum.

The Kaapvaal–Singhbhum comparison: validity, supercontinentality and global v. local events

The significant differences in geodynamic evolution and in the chronology, geometry and inferred depositional conditions of the supracrustal successions on Kaapvaal and Singhbhum do not readily support a common Palaeoproterozoic history nor a shared supercontinental heritage (e.g. compare Figs 2 & 3 with Fig. 7). Validating the supercontinental paradigm in the earlier part of the Precambrian is, in any case, fraught with difficulty, reflecting long and complex geological histories for possible daughter fragments from the earlier postulated supercontinents, the debated unreliability of palaeomagnetic data older than about 1.8 Ga and a number of suggested reconstructions (e.g. Rogers 1996; Meert 2002; Rogers & Santosh 2002, 2009). A fairly strong case can be made for a Neoarchaean amalgamation of the North American cratons in the Kenorland supercontinent, and even for an ‘expanded Kenorland’ encompassing also the Baltic and Siberian shields (e.g. Aspler & Chiarenzelli 1998). However, for Kaapvaal, despite a divergence of views in the literature, a recent analysis of possible cratons that might have been joined to Kaapvaal suggests that this craton probably enjoyed an independent plate history during the Palaeoproterozoic (Eriksson *et al.* 2009*b*).

The only possibly common large-scale geodynamic events that can be inferred from the relatively poorly dated supracrustal Palaeoproterozoic record of Kaapvaal and the even more geochronologically impoverished record of Singhbhum are:

(1) an approximate 2.2 (−2.1?) Ga major mafic (−ultramafic in Singhbhum) volcanic event;
(2) well-developed palaeosols above these inferred correlatable deposits;
(3) major transgressions above the palaeosols, at about 2.1–2.0 Ga.

Bearing in mind that these inferred commonalities between Kaapvaal and Singhbhum are essentially exceptions within two disparate supracrustal records, and that the two cratons being joined in a Palaeoproterozoic supercontinent cannot be argued on any reliable basis, what possible connection might there be? A possible solution might lie in the revolutionary idea of Condie *et al.* (2009) that there may have been a global-scale magmatic shutdown between about 2.45 and 2.2 Ga, characterized by a concomitant overall tectonic stagnation, and that this major event may also have been directly and (partly?) responsible for the first widespread glaciation and atmospheric oxygenation events in that same time period.

If this recent postulate has veracity, then it could be expected that resumption of the thermal–tectonic engine driving Earth evolution at approximately 2.2 Ga would have been marked by a widespread mantle thermal event, possibly even of global compass, resulting in widespread mafic (–ultramafic) magmatism in various forms, and seen in examples such as the approximately 2.2–2.25 Ga Hekpoort (Kaapvaal) and Jagannathpur and Dhanjori (both Singhbhum) lavas. Resumption of widespread mafic magmatism would also have driven enhanced supplies of greenhouse gases to the palaeo-atmosphere resulting in global warming (cf. Condie *et al.* 2009; Eriksson *et al.* 2009*a*) and, thus, also enhanced weathering. This could explain the approximately less than 2.2 Ga Hekpoort (cf. Driese 2004) and less than 2.25 Ga Keonjhargarh palaeosols, as well as inferred possible equivalents in the NIB (Bandopadhyay *et al.* 2010). Resumption of widespread magmatic and tectonic processes on a global scale from about 2.2 Ga would also logically have driven enhanced freeboard (cf. Eriksson *et al.* 2006*a*) for cratonic terranes (easing palaeosol formation), but probably also caused eustatic rise (due to rapid mid-ocean ridge growth once more) thus drowning many cratonic margins where freeboard may not have been high enough. This might explain the major transgressions inferred for both the Singhbhum and Kaapvaal cratons at around 2.1–2.0 Ga.

In contrast, the Kaapvaal Craton, which does not appear to have formed part of a Palaeoproterozoic supercontinent, enjoyed a much more chronologically and geographically widespread set of intracratonic basins (*c.* 3.1–2.8 Ga Witwatersrand, *c.* 2.7 Ga Ventersdorp and *c.* 2.66–2.05 Ga Transvaal), as well as unique events such as the approximately 2.05 Ga Bushveld Complex, the approximately 2.0 Ga Limpopo Orogeny when the Zimbabwe Craton became joined to it, as well as at least two glacial events. The only common elements between Kaapvaal and Singhbhum thus appear to be the approximately 2.2 Ga mafic volcanic units, their subsequent palaeosols and, finally, approximately 2.1–2.0 Ga major transgressive clastic sedimentary units. This triumvirate of possibly shared events can be accounted for by the global-scale postulate of Condie *et al.* (2009) of an approximately 2.45–2.2 Ga shutdown of magmatic and tectonic geodynamic systems, and the expected set of events to follow such a unique period in Earth history. It is noticeable that, although the postulated shared geodynamic events between Kaapvaal and Singhbhum are exceptions within a scenario of geological disparity in terms of their Palaeoproterozoic supracrustal records, these relatively minor exceptions all support the Condie *et al.* (2009) global magmatic shutdown model. Presumably, the effects of the latter inferred global event were such that, although their preserved remnants might have become muted in preserved geological records dominated by the products of craton-specific events, their widespread imprint nevertheless remains visible through a spectrum of localized geological histories.

Conclusions

The Palaeoproterozoic supracrustal record of the Singhbhum Craton in India is fairly restricted, with evidence for possible high freeboard conditions that stimulated palaeosol development up till around 2.25–2.2 Ga. Following this, mafic–ultramafic volcanism (lesser clastic sedimentary rocks) associated with intracratonic rifting at approximately 2.25–2.1 Ga produced a set of localized basin-fills in Singhbhum (e.g. Dhanjori, Jagannathpur, and possibly also Simlipal, Malangtoli and Ongarbira). Weathered material from these mafic volcanics probably contributed to palaeosol development also. Subsequent major transgressions at about 2.0 Ga deposited littoral–deep-marine deposits above these successions, preserved in the Kolhan Group–Chaibasa Formation association. A long hiatus ensued before deposition of the Dhalbhum and conformably succeeding Dalma formations in Singhbhum.

P. G. Eriksson, A. J. Bumby and N. Lenhardt acknowledge research funding from the National Research Foundation, South Africa, and the University of Pretoria. R. Mazumber acknowledges financial support from JSPS (Japan), the AvH Foundation (Germany) and the Department of Science and Technology, Government of India. T. Rajab is thanked for the drafting of figures. We appreciate the comments and improvements offered by reviewers U. Zimmermann and J. Mukhopadhyay.

References

Acharyya, S. K. 2003*a*. The nature of Mesoproterozoic Central Indian Tectonic Zone with exhumed and reworked older granulites. *Gondwana Research*, **6**, 197–214.

Acharyya, S. K. 2003*b*. A plate tectonic model for Proterozoic Crustal evolution of Central Indian Tectonic Zone. *Gondwana Geological Magazine*, **7**, 9–31.

Acharyya, S. K., Gupta, A. & Orihashi, Y. 2010. Neoarchaean–Palaeoproterozoic stratigraphy of the Dhanjori basin, Singhbhum Craton, Eastern India: and recording of a few U–Pb zircon dates from its basal part. *Journal of Asian Earth Science*, **7**, 9–31, doi: 10.1016/j.jseaes.2010.04.023.

ALTERMANN, W. & NELSON, D. R. 1998. Sedimentation rates, basin analysis and regional correlations of three Neoarchaean and Palaeoproterozoic sub-basins of the Kaapvaal craton as inferred from precise U–Pb zircon ages from volcaniclastic sediments. *Sedimentary Geology*, **120**, 225–256.

ALTERMANN, W. & SIEGFRIED, H. P. 1997. Sedimentology, facies development and type-section of an Archaean shelf-carbonate platform transition, Kaapvaal Craton, as deduced from a deep borehole core at Kathu, South Africa. *Journal of African Earth Sciences*, **24**, 391–410.

ARMSTRONG, R. A., COMPSTON, W., RETIEF, E. A., WILLIAMS, I. S. & WELKE, H. J. 1991. Zircon ion microprobe studies bearing on the age and evolution of the Witwatersrand Triad. *Precambrian Research*, **53**, 243–266.

ASPLER, L. B. & CHIARENZELLI, J. R. 1998. Protracted breakup of Kenorland, a Neoarchean supercontinent? Geochronologic, tectonostratigraphic and sedimentologic evidence from the Paleoproterozoic. *Sedimentary Geology*, **120**, 75–104.

BANDOPADHYAY, P. C. & SENGUPTA, S. 2004. The Paleoproterozoic supracrustal Kolhan Group in Singhbhum Craton, India and the Indo-African supercontinent. *Gondwana Research*, **7**, 1228–1235.

BANDOPADHYAY, P. C., ERIKSSON, P. G. & ROBERTS, R. J. 2010. A vertic paleosol at the Archean–Proterozoic contact from the Singhbhum–Orissa craton, eastern India. *Precambrian Research*, **177**, 277–290.

BANDYOPADHYAY, P. K., CHAKRABARTI, A. K., DEOMURARI, M. P. & MISRA, S. 2001. 2.8 Ga old anorogenic granite–acid volcanic association from western margin of Singhbhum–Orissa craton, eastern India. *Gondwana Research*, **4**, 465–475.

BANERJEE, P. K. 1982. Stratigraphy, petrology and geochemistry of some Precambrian basic volcanic and associated rocks of Singhbhum District, Bihar and Mayurbhanj and Keonjhar districts, Orissa. *Memoirs of the Geological Survey of India*, **111**, 54.

BARKER, O. B., BRANDL, G., CALLAGHAN, C. C., ERIKSSON, P. G. & VAN DER NEUT, M. 2006. The Soutpansberg and Waterberg Groups and the Blouberg Formation. *In*: JOHNSON, M. R., ANHAEUSSER, C. R. & THOMAS, R. J. (eds) *The Geology of South Africa*. Geological Society of South Africa, Johannesburg and Council for Geoscience, Pretoria, 301–318.

BARTON, J. M. JR., KLEMD, R. & ZEH, A. 2006. The Limpopo belt: a result of Archean to Proterozoic, Turkic-type orogenesis? *In*: REIMOLD, W. U. & GIBSON, R. (eds) *Processes on the Early Earth*. Geological Society of America, Boulder, CO, Special Publications, **405**, 315–332.

BASU, A. R., SHARMA, M. & PREMO, W. R. 1996. U–Pb age of an older Metamorphic Group mica schist: earliest terrain of the eastern Indian Craton. *In*: SAHA, A. K. (eds) *Recent Researches in Geology and Geophysics of the Precambrians: Essays in Honour of Prof. Samarendra Nath Sarkar on his Seventyfifth Birth Anniversary*. Recent Researches in Geology, **16**, 93–102.

BEKKER, A., KAUFMAN, A. J., KARHU, J. A., BEUKES, N. J., SWART, Q. D., COETZEE, L. L. & ERIKSSON, K. A. 2001. Chemostratigraphy of the Paleoproterozoic Duitschland Formation, South Africa: implications for coupled climate change and carbon cycling. *American Journal of Science*, **301**, 261–285.

BEUKES, N. J. 1979. Litostratigrafiese onderverdeling van die Schmidtsdrif-Subgroep van die Ghaap-Groep in Noord-Kaapland. *Transactions of the Geological Society of South Africa*, **82**, 313–327.

BEUKES, N. J. & CAIRNCROSS, B. 1991. A lithostratigraphic-sedimentological reference profile for the Late Archaean Mozaan Group, Pongola Sequence: application to sequence stratigraphy and correlation with the Witwatersrand Supergroup. *South African Journal of Geology*, **94**, 44–69.

BHATTACHARYA, H. N. 1991. A reappraisal of the depositional environment of the Precambrian metasediments around Ghatshila-Galudih, eastern Singhbhum. *Journal of the Geological Society of India*, **37**, 47–54.

BHATTACHARYA, H. N. & BANDYOPADHAYA, S. 1998. Seismites in a Proterozoic tidal succession, Singhbhum, Bihar, India. *Sedimentary Geology*, **119**, 239–252.

BHATTACHARYA, S. 1992. Evolution of Singhbhum mobile belt by continental rifting and implications of the geochemistry of Purulia amphibolites. *Indian Journal of Earth Sciences*, **19**, 9–17.

BHATTACHARYYA, D. S. & BHATTACHARYYA, T. K. 1970. Geological and geophysical investigations of a basaltic layer in Archaean terrain of eastern India. *Geological Society of America Bulletin*, **81**, 3073–3078.

BLACKBURN, W. H. & SRIVASTAVA, D. C. 1994. Geochemistry and tectonic significance of the Ongarbira metavolcanic rocks, Singhbhum district, India. *Precambrian Research*, **67**, 181–206.

BOSE, M. K. 1994. Sedimentation pattern and tectonic evolution of the Proterozoic Singhbhum basin in the eastern Indian shield. *Tectonophysics*, **231**, 325–346.

BOSE, M. K., CHAKRABARTI, M. K. & SAUNDERS, A. D. 1989. Petrochemistry of the lavas from Proterozoic Dalma volcanic belt, Singhbhum, eastern India. *Geologische Rundschau*, **78**, 633–648.

BOSE, P. K., MAZUMDER, R. & SARKAR, S. 1997. Tidal sandwaves and related storm deposits in the transgressive Protoproterozoic Chaibasa Formation, India. *Precambrian Research*, **84**, 63–81.

BOSHOFF, R., VAN REENEN, D. D., SMIT, C. A., PERCHUK, L. L., KRAMERS, J. D. & ARMSTRONG, R. 2006. Geologic history of the Central Zone of the Limpopo Complex: the West Alldays area. *Journal of Geology*, **114**, 699–716.

BUCHANAN, P. C., REIMOLD, W. U., KOEBERL, C. & KRUGER, F. J. 2002. Geochemistry of intermediate to siliceous volcanic rocks of the Rooiberg Group, Bushveld Magmatic Province, South Africa. *Contributions to Mineralogy and Petrology*, **144**, 131–143.

BUCHANAN, P. C., REIMOLD, W. U., KOEBERL, C. & KRUGER, F. J. 2004. Rb–Sr and Sm–Nd isotopic compositions of the Rooiberg Group, South Africa: early Bushveld-related volcanism. *Lithos*, **29**, 373–388.

BUICK, I. S., MAAS, R. & GIBSON, R. 2001. Precise U–Pb titanite age constraints on the emplacement of the Bushveld Complex, South Africa. *Journal of the Geological Society, London*, **158**, 3–6.

BUMBY, A. J., ERIKSSON, P. G. & VAN DER MERWE, R. 1998. Compressive deformation in the floor rocks to the Bushveld Complex (South Africa): evidence from the Rustenburg Fault Zone. *Journal of African Earth Sciences*, **27**, 307–330.

BUTTON, A. 1973. *A Regional Study of the Stratigraphy and Development of the Transvaal Basin in the Eastern and Northeastern Transvaal*. PhD thesis, University of the Witwatersrand, Johannesburg.

CALLAGHAN, C. C. 1987. *The Geology of the Waterberg Group in the Southern Portion of the Waterberg Basin*. M.Sc. thesis, University of Pretoria, Pretoria.

CALLAGHAN, C. C., ERIKSSON, P. G. & SNYMAN, C. P. 1991. The sedimentology of the Waterberg group in the Transvaal, South Africa: an overview. *Journal of African Earth Sciences*, **13**, 121–139.

CATUNEANU, O. 2001. Flexural partitioning of the Late Archaean Witwatersrand foreland system, South Africa. *Sedimentary Geology*, **141–142**, 95–112.

CATUNEANU, O. 2006. *Principles of Sequence Stratigraphy*. Elsevier, Amsterdam.

CATUNEANU, O. & ERIKSSON, P. G. 1999. The sequence stratigraphic concept and the Precambrian rock record: an example from the 2.3–2.1 Ga Pretoria Group, Kaapvaal craton. *Precambrian Research*, **97**, 215–251.

CATUNEANU, O. & ERIKSSON, P. G. 2002. Sequence stratigraphy of the Precambrian Rooihoogte-Timeball Hill rift succession, Transvaal Basin, South Africa. *Sedimentary Geology*, **147**, 71–88.

CAWTHORN, R. G. & WALRAVEN, F. 1997. *The Bushveld Complex: A Time to Fill and a Time to Cool*. Economic Geology Research Unit Information Circular, University of the Witwatersrand, Johannesburg.

CHAKRABORTI, M. K. 1980. On the pyroclastic rocks of Dalma volcanic sequence, Singhbhum, Bihar. *Indian Journal of Earth Sciences*, **7**, 216–222.

CHAKRABORTI, M. K. & BOSE, M. K. 1985. Evaluation of the tectonic setting of Precambrian Dalma volcanic belt, eastern India using trace element data. *Precambrian Research*, **28**, 253–268.

CHUDY, T., ZEH, A., GERDES, A., KLEMD, R. & BARTON, J. M. JR. 2008. Paleoarchean (3.3 Ga) mafic magmatism and Paleoproterozoic (2.02 Ga) amphibolite-facies metamorphism in the Central Zone of the Limpopo Belt: new geochronological, petrological and geochemical constraints from metabasic and metapelitic rocks from the Venetia area. *South African Journal of Geology*, **111**, 387–408.

CONDIE, K. C. (ed.) 1992. *Proterozoic Crustal Evolution*. Elsevier, Amsterdam.

CONDIE, K. C. 1998. Episodic continental growth and supercontinents: a mantle avalanche connection? *Earth and Planetary Science Letters*, **163**, 97–108.

CONDIE, K. C. 2001. *Mantle Plumes and their Record in Earth History*. Cambridge University Press, Cambridge.

CONDIE, K. C. 2004*a*. Precambrian superplume events. *In*: ERIKSSON, P. G., ALTERMANN, W., NELSON, D. R., MUELLER, W. U. & CATUNEANU, O. (eds) *The Precambrian Earth: Tempos and Events*. Developments in Precambrian Geology, **12**, 163–173.

CONDIE, K. C. 2004*b*. Supercontinents and superplume events: distinguishing signals in the geological record. *Physics of the Earth and Planetary Interiors*, **146**, 319–332.

CONDIE, K. C., O'NEILL, C. & ASTER, R. 2009. Evidence and implications for a widespread magmatic shutdown for 250 My on Earth. *Earth and Planetary Science Letters*, **282**, 294–298.

CORNELL, D. H., SCHÜTTE, S. S. & EGLINGTON, B. L. 1996. The Ongeluk Basaltic Andesite Formation in Griqualand West, South Africa: submarine alteration in a 2222 Ma Proterozoic sea. *Precambrian Research*, **79**, 101–124.

DE, A. 1964. Precambrian Dalma Dhanjori volcanicity of E. India and its stratigraphic significance. *Proceedings of the International Geological Congress*, **10**, 59–70.

DE WIT, M. J. 1998. On Archean granites, greenstones, cratons and tectonics: does the evidence demand a verdict? *Precambrian Research*, **91**, 181–226.

DE WIT, M. J., ROERING, C. *ET AL*. 1992. Formation of an Archaean continent. *Nature*, **357**, 553–562.

DRIESE, S. G. 2004. Pedogenic translocation of Fe in modern and ancient vertisols and implications for interpretations of the Hekpoort paleosol (2.25 Ga). *Journal of Geology*, **112**, 543–560.

EGLINGTON, B. M. & ARMSTRONG, R. A. 2004. The Kaapvaal Craton and adjacent orogens, southern Africa: a geochronological database and overview of the geological development of the craton. *South African Journal of Geology*, **107**, 13–32.

ELS, B. G. 1998*a*. The auriferous late Archaean sedimentation systems of South Africa: unique palaeoenvironmental conditions? *Sedimentary Geology*, **120**, 205–224.

ELS, B. G. 1998*b*. The question of alluvial fans in the auriferous Archaean and Proterozoic successions of South Africa. *South African Journal of Geology*, **101**, (1), 17–25.

ELS, B. G., VAN DEN BERG, W. A. & MAYER, J. J. 1995. The Black Reef Quartzite Formation in the western Transvaal: sedimentological and economic aspects, and significance for basin evolution. *Mineralium Deposita*, **30**, 112–123.

EMBLETON, B. J. J. & WILLIAMS, G. E. 1986. Low palaeolatitudes of deposition for late Precambrian periglacial varvites in South Australia: implications for palaeoclimatology. *Earth and Planetary Science Letters*, **79**, 419–430.

ERIKSSON, K. A. & SIMPSON, E. L. 1998. Controls on spatial and temporal distribution of Precambrian aeolianites. *Sedimentary Geology*, **120**, 275–294.

ERIKSSON, K. A., TURNER, B. R. & VOS, R. G. 1981. Evidence of tidal processes from the lower part of the Witwatersrand Supergroup, South Africa. *Sedimentary Geology*, **29**, 309–325.

ERIKSSON, P. G. & ALTERMANN, W. 1998. An overview of the geology of the Transvaal Supergroup dolomites (South Africa). *Environmental Geology*, **36**, 179–188.

ERIKSSON, P. G. & ALTERMANN, W. 2004. Commentary, chapter 5: evolution of the hydrosphere and atmosphere. *In*: ERIKSSON, P. G., ALTERMANN, W., NELSON, D. R., MUELLER, W. U. & CATUNEANU, O. (eds) *The Precambrian Earth: Tempos and Events*. Elsevier, Amsterdam, 505–511.

ERIKSSON, P. G. & CATUNEANU, O. 2004. A commentary on Precambrian plate tectonics. *In*: ERIKSSON, P. G.,

Altermann, W., Nelson, D. R., Mueller, W. U. & Catuneanu, O. (eds) *The Precambrian Earth: Tempos and Events*. Elsevier, Amsterdam, 201–213.

Eriksson, P. G. & Cheney, E. S. 1992. Evidence for the transition to an oxygen-rich atmosphere during the evolution of red beds in the Lower Proterozoic sequences of southern Africa. *Precambrian Research*, **54**, 257–269.

Eriksson, P. G. & Reczko, B. F. F. 1995. The sedimentary and tectonic setting of the Transvaal Supergroup floor rocks to the Bushveld complex. *Journal of African Earth Sciences*, **21**, 487–504.

Eriksson, P. G., Altermann, W., Catuneanu, O., van der Merwe, R. & Bumby, A. J. 2001. Major influences on the evolution of the 2.67–2.1 Ga Transvaal basin, Kaapvaal craton. *Sedimentary Geology*, **141–142**, 205–231.

Eriksson, P. G., Altermann, W. & Hartzer, F. J. 2006*b*. The Transvaal Supergroup and its precursors. *In*: Johnson, M. R., Anhaeusser, C. R. & Thomas, R. J. (eds) *The Geology of South Africa*. Geological Society of South Africa, Johannesburg and Council for Geoscience, Pretoria, 237–260.

Eriksson, P. G., Banerjee, S., Catuneanu, O., Sarkar, S., Bumby, A. J. & Mtimkulu, M. N. 2007. Prime controls on Archaean–Palaeoproterozoic sedimentation: change over time. *Gondwana Research*, **12**, 550–559.

Eriksson, P. G., Banerjee, S. *et al.* 2009*b*. A Kaapvaal craton debate: nucleus of an early small supercontinent or affected by an enhanced accretion event? *Gondwana Research*, **15**, 354–372.

Eriksson, P. G., Bose, P. K., Catuneanu, O., Sarkar, S. & Banerjee, S. 2008. Precambrian clastic epeiric embayments: examples from South Africa and India. *In*: Pratt, B. R. & Holmden, C. (eds) *Dynamics of Epeiric Seas*. Geological Association of Canada, Special Paper, **48**, 119–136.

Eriksson, P. G., Catuneanu, O., Nelson, D. R., Mueller, W. U. & Altermann, W. 2004. Towards a synthesis. *In*: Eriksson, P. G., Altermann, W., Nelson, D. R., Mueller, W. U. & Catuneanu, O. (eds) *The Precambrian Earth: Tempos and Events*. Elsevier, Amsterdam, 739–769.

Eriksson, P. G., Condie, K. C. *et al.* 2002. Late Archaean superplume events: a Kaapvaal–Pilbara perspective. *Journal of Geodynamics*, **34**, 207–247.

Eriksson, P. G., Els, B. G. & Bumby, A. J. 2005. Kaapvaal craton: changing first- and second-order controls on sea level from *c.* 3 to 2.0 Ga. *Sedimentary Geology*, **176**, 121–148.

Eriksson, P. G., Hattingh, P. J. & Altermann, W. 1995. An overview of the geology of the Transvaal Sequence and Bushveld Complex, South Africa. *Mineralium Deposita*, **30**, 98–111.

Eriksson, P. G., Mazumder, R., Catuneanu, O., Bumby, A. J. & Ountsche Ilondo, B. 2006*a*. Precambrian continental freeboard and geological evolution: a time perspective. *Earth-Science Reviews*, **79**, 165–204.

Eriksson, P. G., Mazumder, R., Sarkar, S., Bose, P. K., Altermann, W. & van der Merwe, R. 1999. The 2.7–.20 Ga volcano-sedimentary record of Africa, India and Australia: evidence for global and local changes in sea level and continental freeboard. *Precambrian Research*, **97**, 269–302.

Eriksson, P. G., Rautenbach, C. J. de W., Wright, D. T., Bumby, A. J., Catuneanu, O., Mostert, P. & van der Neut, M. 2009*a*. Possible evidence for episodic epeiric marine and fluvial sedimentation (and implications for palaeoclimatic conditions), *c.* 2.3–1.8 Ga, Kaapvaal craton, South Africa. *Palaeogeography, Palaeoclimatology, Palaeoecology*, **273**, 153–173.

Eriksson, P. G., Reczko, B. F. F., Corner, B. & Jenkins, S. L. 1996. The Kanye axis, Kaapvaal craton, southern Africa: a postulated Archaean crustal architectural element inferred from three-dimensional basin modelling of the lower Transvaal Supergroup. *Journal of African Earth Sciences*, **22**, 223–233.

Eriksson, P. G., Rigby, M. J., Bandopadhyay, P. C. & Steenkamp, N. C. 2011. The Kaapvaal craton (South Africa): no evidence for a supercontinental affinity prior to 2.0 Ga? *International Geology Review*, **53**, 1312–1330.

Eriksson, P. G., Simpson, E. L., Eriksson, K. A., Bumby, A. J., Steyn, G. L. & Sarkar, S. 2000. Muddy roll-up structures in siliciclastic interdune beds of the ca. 1.8 Ga Waterberg Group, South Africa. *Palaios*, **15**, 177–183.

Eriksson, P. G., van der Merwe, R. & Bumby, A. J. 1998. The Palaeoproterozoic Woodlands Formation of eastern Botswana – northwestern South Africa: lithostratigraphy and relationship with Transvaal Basin inversion structures. *Journal of African Earth Sciences*, **27**, 349–358.

Ernst, R. E. & Buchan, K. L. 2002. Maximum size and distribution in time and space of mantle plumes: evidence from large igneous provinces. *Journal of Geodynamics*, **34**, 309–342.

Ernst, R. E., Buchan, K. L. & Prokoph, A. 2004. Large igneous province record through time. *In*: Eriksson, P. G., Altermann, W., Nelson, D. R., Mueller, W. U. & Catuneanu, O. (eds) *The Precambrian Earth: Tempos and Events*. Developments in Precambrian Geology, **12**, 173–180.

Frauenstein, F., Veizer, J., Beukes, N., van Niekerk, H. S. & Coetzee, L. L. 2009. Transvaal Supergroup carbonates: implications for Paleoproterozoic $\delta^{18}O$ and $\delta^{13}C$ records. *Precambrian Research*, **175**, 149–160.

Frei, R., Gaucher, C., Poulton, S. W. & Canfield, D. E. 2009. Fluctuations in Precambrian atmospheric oxygenation recorded by chromium isotopes. *Nature*, **461**, 250–253.

Frimmel, H. E. 2005. Archaean atmospheric evolution: evidence from the Witwatersrand gold fields, South Africa. *Earth-Science Reviews*, **70**, 1–46.

Gerdes, A. & Zeh, A. 2009. Zircon formation versus zircon alteration: new insights from combined U–Pb and Lu–Hf *in situ* LA-ICP-MS analyses, and consequences for the interpretation of Archean zircon from the Limpopo Belt. *Chemical Geology*, **261**, 230–243.

Ghosh, S. K. & Chatterjee, B. K. 1994. Depositional mechanism as revealed from grain-size measures of the Paleoproterozoic Kolhan siliciclastics, Keonjhar district, Orissa, India. *Sedimentary Geology*, **89**, 181–196.

Goodwin, A. M. 1996. *Principles of Precambrian Geology*. Academic Press, London.

Gupta, A. & Basu, A. 2000. North Singh bhum Proterozoic mobile belt, eastern India – a review. *In*: *National Seminar Commemorating Dr M. S. Krishnan's Birth centenary (Proceedings Volume)*. Geological Survey of India, Special Publications, **55**, 195–226.

Gupta, A., Basu, A. & Ghosh, P. K. 1980. The Proterozoic ultramafic and mafic lavas and tuffs of the Dalma greenstone belt, Singhbhum, eastern India. *Canadian Journal of Earth Sciences*, **17**, 210–231.

Gupta, A., Basu, A. & Ghosh, P. K. 1982. Ultramafic volcaniclastics of the Precambrian Dalma volcanic belt, Singhbhum, Eastern India. *Geological Magazine*, **119**, 505–510.

Gupta, A., Basu, A. & Singh, S. K. 1985. Stratigraphy and petrochemistry of Dhanjori greenstone belt, eastern India. *Quarterly Journal of the Geological, Mineralogical and Metallurgical Society of India*, **57**, 248–263.

Gupta, A., Basu, A. & Srivastava, D. C. 1981. Mafic and ultramafic volcanism of Ongarbira greenstone belt, Singhbhum, Bihar. *Journal of the Geological Society of India*, **22**, 593–596.

Hälbich, I. W., Scheepers, R., Lamprecht, D. F., van Deventer, J. L. & De Kock, N. J. 1993. The Transvaal-Griqualand West banded iron formation: geology, genesis, iron exploitation. *Journal of African Earth Sciences*, **16**, 63–120.

Hannah, J. L., Bekker, A., Stein, H. J., Markey, R. J. & Holland, H. D. 2004. Primitive Os and 2316 Ma age for marine shale: implications for Paleoproterozoic glacial events and the rise of atmospheric oxygen. *Earth and Planetary Science Letters*, **225**, 43–52.

Hanson, R. E., Gose, W. A. *et al*. 2004. Paleoproterozoic intraplate magmatism and basin development on the Kaapvaal Craton: age, paleomagnetism and geochemistry of ~1.93 to ~1.87 Ga post-Waterberg dolerites. *South African Journal of Geology*, **107**, 233–254.

Harland, W. B. 1964. Critical evidence for a great Infra-Cambrian glaciation. *Geologische Rundschau*, **54**, 45–61.

Harland, W. B. & Bidgood, D. E. T. 1959. Paleomagnetism in some Norwegian sparagmites and the late-Precambrian ice age. *Nature*, **184**, 1860–1862.

Hartzer, F. J. 1995. Transvaal Supergroup inliers: geology, tectonic development and relationship with the Bushveld Complex, South Africa. *Journal of African Earth Sciences*, **21**, 521–547.

Hatton, C. J. 1995. Mantle plume origin for the Bushveld and Ventersdorp magmatic provinces. *Journal of African Earth Sciences*, **21**, 571–577.

Henry, G., Clendenin, C. W. & Charlesworth, E. G. 1990. Depositional facies of the Black Reef Quartzite Formation in the eastern Transvaal. *In*: *Proceedings of the 23rd GeoCongress Conference (Abstract Volume)*. Geological Society of South Africa, Cape Town, 234–237.

Hoffman, P. F., Kaufman, A. J., Halverson, G. P. & Schrag, D. P. 1998. A Neoproterozoic Snowball Earth. *Science*, **281**, 1342–1346.

Holzer, L., Frei, R., Barton, J. M., Jr. & Kramers, J. D. 1998. Unraveling the record of successive high grade events in the Central Zone of the Limpopo Belt using Pb single phase dating of metamorphic minerals. *Precambrian Research*, **87**, 87–115.

Isley, A. E. & Abbott, D. H. 1999. Plume-related mafic volcanism and the deposition of banded iron formation. *Journal of Geophysical Research*, **B104**, 15461–15477.

Iyenger, S. V. P., Chandy, K. C. & Narayanaswamy, R. 1981. Geochronology and Rb–Sr systematics of the igneous rocks of the Simlipal complex, Orissa. *Indian Journal of Earth Sciences*, **8**, 61–65.

Jansen, H. 1982. *The geology of the Waterberg Basin in the Transvaal, Republic of South Africa*. Geological Survey of South Africa, Memoirs, **71**.

Joy, S. & Saha, D. 2000. Dynamically recrystallised quartz c-axis fabrics in greenschist facies quartzites, Singhbhum shear zone and its footwall, eastern India – influence of high fluid activity. *Journal of Structural Geology*, **22**, 777–793.

Kamber, B. S., Blenkinsop, T. G., Villa, I. M. & Dahl, P. S. 1995. Proterozoic transpressive deformation in the Northern Marginal Zone, Limpopo Belt, Zimbabwe. *Journal of Geology*, **100**, 490–508.

Karhu, J. A. & Holland, H. D. 1996. Carbon isotopes and the rise of atmospheric oxygen. *Geology*, **24**, 867–870.

Kent, R. 1991. Lithospheric uplift in eastern Gondwana: evidence for a long-lived mantle plume system? *Geology*, **19**, 19–23.

Key, R. M. 1983. *The Geology of the Area Around Gaborone and Lobatse, Kweneng, Kgatleng, Southern and South East Districts*. Geological Survey of Botswana District, Memoirs, **5**.

Kirschvink, J. L. 1992. Late Proterozoic low-latitude global glaciation: the snowball Earth. *In*: Schopf, J. W. & Klein, C. (eds) *The Proterozoic Biosphere: A Multidisciplinary Study*. Cambridge University Press, Cambridge, 51–52.

Knoll, A. H. & Beukes, N. J. 2009. Introduction: initial investigations of a Neoarchean shelf margin-basin transition (Transvaal Supergroup, South Africa). *Precambrian Research*, **169/1–4**, 1–14.

Krishna Rao, N., Agarwal, S. K. & Rao, G. V. U. 1979. Lead isotopic ratio of uraninites and the age of uranium mineralisation in Singhbum Shear Zone, Bihar. *Journal of the Geological Society of India*, **20**, 124–127.

Kröner, A., Jaeckel, P., Brandl, G., Nemchin, A. A. & Pidgeon, R. T. 1999. Single zircon ages for granitoid gneisses in the Central Zone of the Limpopo Belt, Southern Africa and geodynamic significance. *Precambrian Research*, **93**, 299–337.

Lowe, D. R. & Tice, M. M. 2007. Tectonic controls on atmospheric, climatic, and biological evolution 3.5–2.4 Ga. *Precambrian Research*, **158**, 177–197.

Mapeo, R. B. M., Armstrong, R. A., Kampunzu, A. B., Modisi, M. P., Ramokate, L. V. & Modie, B. N. J. 2006. A ca. 200 Ma hiatus between the Lower and Upper Transvaal Groups of southern Africa: SHRIMP U–Pb detrital zircon evidence from the Segwagwa Group, Botswana: implications for Palaeoproterozoic glaciations. *Earth and Planetary Science Letters*, **244**, 113–132.

Mazumder, R. 2004. Implications of lunar orbital periodicities from Chaibasa tidal rhythmite of late Paleoproterozoic age. *Geology*, **32**, 841–844.

MAZUMDER, R. 2005. Proterozoic sedimentation and volcanism in the Singhbhum crustal province, India and their implications. *Sedimentary Geology*, **176**, 167–193.

MAZUMDER, R. & ARIMA, M. 2004. Proterozoic events in the Singhbhum crustal province, India. *Gondwana Research*, **7**, (4S), 1343–1345.

MAZUMDER, R. & ARIMA, M. 2005. Tidal rhythmites and their implications. *Earth-Science Reviews*, **69**, 79–95.

MAZUMDER, R. & ARIMA, M. 2009. Implication of mafic magmatism in an intracontinental rift setting: a case study from the Palaeoproterozoic Dhanjori Formation, Singhbhum crustal province, India. *Journal of Geology*, **117**, 455–466.

MAZUMDER, R. & SARKAR, S. 2004. Sedimentation history of the Palaeoproterozoic Dhanjori Formation, Singhbhum, India and its implications. *Precambrian Research*, **130**, 267–287.

MAZUMDER, R., BOSE, P. K. & SARKAR, S. 2000. A commentary on the tectono-sedimentary record of ther pre-2.0 Ga continental growth of India vis-a-vis Pre-Gondwana Afro-Indian supercontinent. *Journal of African Earth Sciences*, **30**, 201–217.

MAZUMDER, R., RODRIGUEZ-LOPEZ, J. A., ARIMA, M. & VAN LOON, A. J. 2009. Palaeoproterozoic seismites (fine-grained facies of the Chaibasa Fm., E. India) and their soft-sediment deformation structures. *In*: REDDY, S. M., MAZUMDER, R., EVANS, D. A. & COLLINS, A. S. (eds) *Palaeoproterozoic Supercontinent and its Global Evolution*. Geological Society, London, Special Publications, **323**, 301–318.

MAZUMDER, R., VAN LOON, A. J. & ARIMA, M. 2006. Soft-sediment deformation structures in the Earth's oldest seismites. *Sedimentary Geology*, **186**, 19–26.

MAZUMDER, R., VAN LOON, A. J., MALLIK, L., REDDY, S. M., ARIMA, M., ALTERMANN, W., ERIKSSON, P. G. & DE, S. 2012. Mesoarchaean–Palaeoproterozoic stratigraphic record of the Singhbhum crustal province, eastern India: a synthesis. *In*: MAZUMDER, R. & SAHA, D. (eds) *Palaeoproterozoic of India*. Geological Society, London, Special Publications, **365**, 29–47, http://dx.doi.org/10.1144/SP365.3.

MEERT, J. G. 2002. Paleomagnetic evidence for a Paleo-Mesoproterozoic supercontinent. *Gondwana Research*, **5**, 207–215.

MILLONIG, L., ZEH, A., GERDES, A. & KLEMD, R. 2008. Late Archaean high-grade metamorphism in the Central Zone of the Limpopo Belt (South Africa): petrological and geochronological evidence from the Bulai Pluton. *Lithos*, **103**, 333–351.

MISHRA, S., DEOMURARI, M. P., WIEDENBECK, M., GOSWAMI, J. N., RAY, S. & SAHA, A. K. 1999. 207Pb/206Pb zircon ages and the evolution of the Singhbhum Craton, eastern India: an ion microprobe study. *Precambrian Research*, **93**, 139–151.

MISRA, S. 2006. Precambrian chronostratigraphic growth of Singhbhum–Orissa Craton, Eastern Indian Shield: an alternative model. *Journal of the Geological Society of India*, **67**, 356–378.

MOORE, J. M., TSIKOS, H. & POLTEAU, S. 2001. Deconstructing the Transvaal Supergroup, South Africa: implications for Palaeoproterozoic palaeoclimate models. *Journal of African Earth Sciences*, **33**, 437–444.

MUKHOPADHYAY, D. 2001. The Archean nucleus of Singhbhum: the present state of knowledge. *Gondwana Research*, **4**, 307–318.

MUKHOPADHYAY, D., BHATTACHARYA, T., CHAKRABORTY, T. & DEY, A. K. 1990. Structural pattern in the Precambrian rocks of Sonua-Lotapahar region, North Singhbhum, eastern India. *Proceedings of the Indian Academy of Science (Earth and Planetary Sciences)*, **99**, 249–268.

MUKHOPADHYAY, J., GHOSH, G. & NANDI, A. K. 2006. Depositional setting of the Kolhan Group: its implications for the development of a Meso to Neoproterozoic deep-water basin on the South Indian craton. *South African Journal of Geology*, **109**, 183–192.

MUKHOPADHYAY, J., BEUKES, N. J., ARMSTRONG, R. A., ZIMMERMANN, U., GHOSH, G. & MEDDA, R. A. 2008. Dating the oldest greenstone in India: a 3.51-Ga precise U–Pb SHRIMP zircon age for dacitic lava of the southern Iron Ore Group, Singhbhum craton. *Journal of Geology*, **116**, 449–461.

NAHA, K. & GHOSH, S. K. 1960. Ghosh, Archaean palaeogeography of E. and N. Singhbhum. *Geological Magazine*, **97**, 436–439.

NELSON, D. R., BHATTACHARYA, H. N., MISRA, S., DASGUPTA, N. & ALTERMANN, W. 2007. New SHRIMP U–Pb zircon dates from the Singh bhum craton, Jharkhand–Orissa region, India. *In*: *International Conference on Precambrian Sedimentation and Tectonics and 2nd GPSS Meeting, IIT, Mumbai (Bombay) (Abstracts)*. Indian Institute of Technology, Bombay, **47**.

NESBITT, H. W. & YOUNG, G. M. 1982. Early Proterozoic climates and plate motions inferred from major element chemistry of lutites. *Nature*, **299**, 715–717.

OLSSON, J. R., SÖDERLUND, U., KLAUSEN, M. B. & ERNST, R. E. 2010. U–Pb baddeleyite ages linking major Archean dyke swarms to volcanic-rift forming events in the Kaapvaal craton (South Africa), and a precise age for the Bushveld Complex. *Precambrian Research*, **183**, 490–500.

PERCHUK, L. L., VAN REENEN, D. D., VARLAMOV, D. A., VAN KAL, S. M., TABATABAEIMANESH & BOSHOFF, R. 2008. P–T record of two high-grade metamorphic events in the Central Zone of the Limpopo Complex, South Africa. *Lithos*, **103**, 70–105.

PESONEN, L. J., ELMING, S.-A. ET AL. 2003. Palaeomagnetic configuration of continents during the Proterozoic. *Tectonophysics*, **375**, 289–324.

POSAMENTIER, H. W., JERVEY, M. T. & VAIL, P. R. 1988. Eustatic controls on clastic deposition I: conceptual framework. *In*: WILGUS, C. K., HASTINGS, B. S., KENDALL, C. G. ST. C., POSAMENTIER, H. W., ROSS, C. A. & VAN WAGONER, J. C. (eds) *Sea Level Changes: An Integrated Approach*. SEPM, Special Publications, **42**, 110–124.

POTGIETER, G. J. 1992. *Tektonisme langs die noordoostelike rand van die Bosveldkompleks, Suid-Afrika*. PhD thesis, University of Pretoria.

RAMAKRISHNAN, M. & VAIDYANADHAN, R. 2008. *Geology of India, Volume 1*. Geological Society of India, Bangalore.

RAY, K. K., GHOSH-ROY, A. K. & SENGUPTA, S. 1996. Acid volcanic rocks between the Dalma volcanic belt and Chotanagpur gneissic complex, east Singhbhum and Purulia districts of Bihar and West Bengal. *Indian Mineralogist*, **50**, 1–8.

Reddy, S. M., Clarke, C. & Mazumder, R. 2009. Temporal constraints on the evolution of the Singbhum Crustal Province from U–Pb SHRIMP data. *In*: Saha, D. & Mazumder, R. (eds) *Paleoproterozoic Supercontinents and Global Evolution. International Conference on Paleoproterozoic Supercontinents and Global Evolution Abstract Volume*. International Association for Gondwana Research, Conference Series, **9**, 17–18.

Robb, L. J. & Meyer, F. M. 1995. *The Witwatersrand Basin, South Africa: Geological Framework and Mineralization Processes*. Economic Geology Research Unit Information Circular, **293**. University of the Witwatersrand, Johannesburg.

Rogers, J. J. W. 1996. A history of continents in the past three billion years. *Journal of Geology*, **104**, 91–107.

Rogers, J. J. W. & Santosh, M. 2002. Configuration of Columbia, a Mesoproterozoic supercontinent. *Gondwana Research*, **5**, 5–22.

Rogers, J. J. W. & Santosh, M. 2009. Tectonics and surface effects of the supercontinent Columbia. *Gondwana Research*, **15**, 373–380.

Roy, A. 1998. *Isotopic Evolution of Precambrian Mantle – A Case Study of Singhbhum–Orissa Craton*. PhD thesis, Indian School of Mines, Dhanbad.

Roy, A., Sarkar, A., Jeyakumar, S., Aggrawal, S. K. & Ebihara, M. 2002*a*. Sm–Nd age and mantle characteristics of the Dhanjori volcanic rocks, Eastern India. *Geochemical Journal*, **36**, 503–518.

Roy, A., Sarkar, A., Jeyakumar, S., Aggrawal, S. K. & Ebihara, M. 2002*b*. Mid-Proterozoic plume-related thermal event in Eastern Indian craton: evidence from trace elements, REE geochemistry and Sr–Nd isotope systematics of basic–ultrabasic intrusives from Dalma Volcanic Belt. *Gondwana Research*, **5**, 133–146.

Saha, A. K. 1994. *Crustal Evolution of Singhbhum–North, Orissa, Eastern India*. Geological Society of India, Memoirs, **27**.

Sarkar, S. N. & Saha, A. K. 1962. A revision of the Precambrian stratigraphy and tectonics of Singhbhum and adjacent region. *Quarterly Journal of the Geological, Mineralogical and Metallurgical Society of India*, **34**, 97–136.

Sarkar, S. N. & Saha, A. K. 1977. The present status of the Precambrian stratigraphy, tectonics and geochronology of Singhbhum–Keonjhar–Mayurbhanj region, eastern India. *Indian Journal of Earth Sciences*, **4**, 37–65.

Sarkar, S. N., Ghosh, D. K. & Lambert, R. J. St. 1986. Rubidium–Strontium and lead isotopic studies of the soda granites from Musaboni area, Singhbhum Copper Belt. *In*: Sarkar, S. N. (ed.) *Geology and Geochemistry of Sulphide Ore Bodies and Associated Rocks in Musaboni and Rakha Mines Section in the Singhbhum Copper Belt*. Diamond Jubilee Monograph. Indian School of Mines, Dhanbad, 101–110.

Schmidt, P. W. & Williams, G. E. 1995. The Neoproterozoic climatic paradox: equatorial palaeolatitude for Marinoan glaciation near sea level in South Australia. *Earth and Planetary Science Letters*, **134**, 107–124.

Schmidt, P. W., Williams, G. E. & Embleton, B. J. J. 1991. Low palaeolatitude of Late Proterozoic glaciation: early timing of remanence in haematite of the Elatina Formation, South Australia. *Earth and Planetary Science Letters*, **105**, 355–367.

Sengupta, S., Paul, D. K., Bishui, P. K., Gupta, S. N., Chakraborty, R. & Sen, P. A. 1994. Geochemical and Rb–Sr isotopic study of Kuilapal granite and Arkasoni granophyre from the eastern Indian craton. *Indian Mineralogist*, **48**, 77–88.

Sengupta, S., Sarkar, G., Ghosh-Roy, A. K., Bhaduri, S. K., Gupta, S. N. & Mandal, A. 2000. Geochemistry and Rb–Sr Geochronology of acid tuffs from the northern fringe of the Singhbhum craton and their significance in the Precambrian evolution. *Indian Mineralogist*, **54**, 43–56.

Simpson, E. L., Eriksson, K. A., Eriksson, P. G. & Bumby, A. J. 2002. Eolian dune degradation and generation of massive sandstone bodies in the Paleoproterozoic Makgabeng Formation, Waterberg Group, South Africa. *Journal of Sedimentary Research*, **72**, 40–45.

Simpson, E. L., Eriksson, K. A., Kuklis, C. A., Eriksson, P. G., Bumby, A. J. & van Jaarsveld, C. F. 2004. Saline pan deposits from the ~1.8 Ga Makgabeng Formation, Waterberg Group, South Africa. *Sedimentary Geology*, **163**, 279–292.

Singh, S. P. 1997. Geochemistry of acid volcanics of the Dalma Group, Singhbhum, Eastern India. *Journal of the Geological Society of India*, **49**, 437–441.

Singh, S. P. 1998. Precambrian stratigraphy of Bihar – an overview. *In*: Paliwal, B. S. (ed.) *The Indian Precambrian*. Scientific Publishers, Jodhpur, 376–408.

Sohl, L. E., Christie-Blick, N. & Kent, D. V. 1999. Paleomagnetic polarity reversals in Marinoan (ca. 600 Ma) glacial deposits of Australia: implications for the duration of low-latitude glaciation in Neoproterozoic time. *Geological Society of America Bulletin*, **111**, 1120–1139.

SOUTH AFRICAN COMMITTEE FOR STRATIGRAPHY. 1980. *In*: Kent, L. E. (ed.) *Stratigraphy of South Africa, Part 1*. Handbook Series, **8**. Geological Survey of South Africa, Pretoria.

Stanistreet, I. G. & McCarthy, T. S. 1991. Changing tectono-sedimentary scenarios relevant to the development of the Late Archaean Witwatersrand Basin. *Journal of African Earth Sciences*, **13**, 65–82.

Trendall, A. F. 2002. The significance of iron-formation in the Precambrian stratigraphic record. *In*: Altermann, W. & Corcoran, P. L. (eds) *Precambrian Sedimentary Environments: A Modern Approach to Depositional Systems*. International Association of Sedimentologists, Special Publications, **44**, 33–66.

Trendall, A. F., Compston, W. *et al.* 1990. *Chronological comparison of the volcano-sedimentary sequences of the Kaapvaal and Pilbara cratons between 3.2 and 2.4 Ga*. *In*: Glover, J. E. & Ho, S. E. (eds) *Third International Archaean Symposium, Perth, Australia. Extended Abstracts Volume*. Geoconferences, Perth, WA, 81–83.

Twist, D. 1985. Geochemical evolution of the Rooiberg silicic lavas in the Loskop Dam area, Southeastern Bushveld. *Economic Geology*, **80**, 1153–1165.

van der Westhuizen, W. A., de Bruiyn, H. & Meintjies, P. G. 1991. The Ventersdorp Supergroup:

an overview. *Journal of African Earth Sciences*, **13**, 83–105.

van Reenen, D. D., Boshoff, R., Smit, C. A., Perchuk, L. L., Kramers, J. D., McCourt, S. & Armstrong, R. A. 2008. Geochronological problems related to polymetamorphism in the Limpopo Complex, South Africa. *Gondwana Research*, **14**, 644–662.

Voegelin, A. R., Nägler, T. F., Beukes, N. J. & Lacassie, J. P. 2010. Molybdenum isotopes in late Archean carbonate rocks: implications for early earth oxygenation. *Precambrian Research*, **182**, 70–82.

Vohra, C. P., Dasgupta, S., Paul, D. K., Bishoi, P. K., Gupta, S. N. & Guha, S. 1991. Rb–Sr chronology and petrochemistry of granitoids from the southeastern part of the Singhbhum craton, Orissa. *Journal of the Geological Society of India*, **38**, 5–22.

Von Gruenewaldt, G. & Harmer, R. E. 1993. Tectonic setting of Proterozoic layered intrusions with special reference to the Bushveld Complex. *In*: Condie, K. C. (ed.) *Proterozoic Crustal Evolution*. Elsevier, Amsterdam, 181–213.

Walraven, F. & Hattingh, E. 1993. Geochronology of the Nebo granite, Bushveld Complex. *South African Journal of Geology*, **96**, 31–41.

Walraven, F. & Martini, J. 1995. Zircon Pb-evaporation age determinations of the Oak tree formation, Chuniespoort Group, Transvaal Sequence: implications for Transvaal-Griqualand West basin correlations. *South African Journal of Geology*, **98**, 58–67.

Williams, G. E. 1975. Late Precambrian glacial climate and the Earth's obliquity. *Geological Magazine*, **112**, 441–465.

Williams, G. E. 1986. Precambrian permafrost horizons as indicators of palaeoclimate. *Precambrian Research*, **32**, 233–242.

Williams, G. E. 1998. Precambrian tidal and glacial clastic deposits: implications for Precambrian Earth–Moon dynamics and palaeoclimate. *Sedimentary Geology*, **120**, 55–74.

Williams, G. E. 2004. The paradox of Proterozoic glaciomarine deposition, open seas and strong seasonality near the Palaeo-Equator: global implications. *In*: Eriksson, P. G., Altermann, W., Nelson, D. R., Mueller, W. U. & Catuneanu, O. (eds) *The Precambrian Earth: Tempos and Events*. Elsevier, Amsterdam, 448–459.

Williams, G. E. 2008. Proterozoic (pre-Ediacaran) glaciation and the high obliquity, low-latitude ice, strong seasonality (HOLIST) hypothesis: principles and tests. *Earth-Science Reviews*, **87**, 61–93.

Williams, G. E., Gostin, V. A., McKirdy, D. M. & Preiss, W. V. 2008. The Elatina glaciation, late Cryogenian (Marinoan Epoch), South Australia: sedimentary facies and palaeoenvironments. *Precambrian Research*, **163**, 307–331.

Young, G. M. 2004. Earth's two great Precambrian glaciations: aftermath of the 'Snowball Earth' hypothesis. *In*: Eriksson, P. G., Altermann, W., Nelson, D. R., Mueller, W. U. & Catuneanu, O. (eds) *The Precambrian Earth: Tempos and Events*. Elsevier, Amsterdam, 440–448.

Zeh, A., Gerdes, A. & Barton, J. M. Jr. 2009. Archean accretion and crustal evolution of the Kalahari Craton – the zircon age and Hf isotope record of granitic rocks from Barberton/Swaziland to the Francistown Arc. *Journal of Petrology*, **50**, 933–966.

Zeh, A., Gerdes, A., Klemd, R. & Barton, J. M. Jr. 2007. Archaean to Proterozoic crustal evolution in the central zone of the Limpopo Belt (South Africa–Botswana): constraints from combined U–Pb and Lu–Hf isotope analyses of zircon. *Journal of Petrology*, **48**, 1605–1639.

Zeh, A., Gerdes, A., Klemd, R. & Barton, J. M. Jr. 2008. U–Pb and Lu–Hf isotope record of detrital zircon grains from the Limpopo Belt – evidence for crustal recycling at the Hadean to early-Archean transition. *Geochimica et Cosmochimica Acta*, **72**, 5304–5329.

Native iron in the Palaeoproterozoic Chaibasa Formation: primary or secondary?

DALLAS H. ABBOTT[1]*, RAJAT MAZUMDER[2] & DEE BREGER[3]

[1]*Lamont-Doherty Earth Observatory of Columbia University, 61 Route 9W, Palisades, NY 10964, USA*

[2]*Geological Studies Unit and Fluvial Mechanics Laboratory, Indian Statistical Institute, 203, B.T. Road, Kolkata 700108, India*

[3]*Micrographic Arts, 49 Dunham Pond Road, Greenfield Center, NY 12833, USA*

**Corresponding author (e-mail: dallas@ldeo.columbia.edu)*

Abstract: Native Fe in the Chaibasa Formation may be the oldest native Fe ever found. The Chaibasa Formation contains pre-1.6 Ga offshore shelf sediments from the Singhbhum Craton in India. The sediments are light green in colour, consistent with a low content of organic matter and a metamorphic grade of greenschist–lower amphibolite facies. The native Fe grains are enclosed in fibrous mica. The native Fe occurs with magnetite crystals, evidence of relatively oxidizing conditions in the Chaibasa Formation during diagenesis. (Mn,Fe)S, a compound characteristic of reducing conditions, is found within the native Fe. The native Fe contains small amounts of native Si. Most grains of native Fe have rough edges. One bulbous grain has smooth edges and an internal wavy structure that is suggestive of a drop of melted iron. The melting temperature of pure Fe is 1535 °C. We also found native Fe with small amounts of carbon in it that appears to have a plate-like, pearlitic structure. Pearlite typically forms at temperatures of 723–500 °C. All of these temperatures are too high to be achieved during greenschist-facies metamorphism. As a result, we infer that the native Fe is primary rather than secondary. Our work and that of others suggests a primary, impact origin for the native Fe.

Our initial purpose in examining the Chaibasa Formation was to look for impact ejecta (Mazumder 2004). Because marine shales often have lower deposition rates than sandstones, we chose a low-grade, metamorphosed marine shale in order to examine the maximum amount of geological time. We used the magnetic susceptibility of the shale and sandstone facies as a tool in our study. Our idea was that impact ejecta might have a higher magnetic susceptibility (Abbott & Hagstrum 2004). We found native Fe, native Si and (Fe,Mn)S in one of the samples that had a higher magnetic susceptibility. Some of the native Fe appears melted and some has a pearlitic structure. In addition, the shale facies contains well-preserved, fine veins of FeTi oxide that would be recrystallized during high-grade metamorphism. In this paper, we argue that these features and associations attest to a primary origin of the native Fe.

Geological setting

The Chaibasa Formation is part of the Palaeoproterozoic supracrustal sequence of the Singhbhum Craton in eastern India (inset, Fig. 1) (Mazumder 2005). The Chaibasa Formation is 6–8 km thick and is composed of clastic sediments. It is part of the Palaeoproterozoic Singhbhum Group of rocks (Saha 1994) in eastern India. The Chaibasa Formation is underlain by the Dhanjori Formation and overlain by the Dhalbhum Formation (Fig. 1). There are no direct radiometric age dates from the Chaibasa Formation. The immediately underlying Dhanjori volcanics have a Sm–Nd isochron age of 2.1 Ga (Roy *et al.* 2002*a*). Sm–Nd isochron ages from basic–ultrabasic intrusives into the lavas of the Dalma Formation are 1.6 Ga (Roy *et al.* 2002*b*). Thus, the Chaibasa Formation is of Palaeoproterozoic age (Fig. 1) (see also Mazumder 2005 for an overview of the Proterozoic stratigraphy of the Singhbhum Group). Lithologically, the Chaibasa Formation is characterized by the interbanding of precursor sandstones, heterolithic (sandstone–shale) and shales. The sandstones formed in a tide–storm interactive subtidal setting (Bose *et al.* 1997; Bhattacharya & Bandyopadhyay 1998) (see Fig. 2). The heterolithic material formed in an offshore setting above the storm wave base and the shale facies formed in an offshore setting below the wave base (Bose *et al.* 1997; Mazumder 2005). The Chaibasa Formation sharply

From: Mazumder, R. & Saha, D. (eds) 2012. *Palaeoproterozoic of India*. Geological Society, London, Special Publications, **365**, 77–90. http://dx.doi.org/10.1144/SP365.5

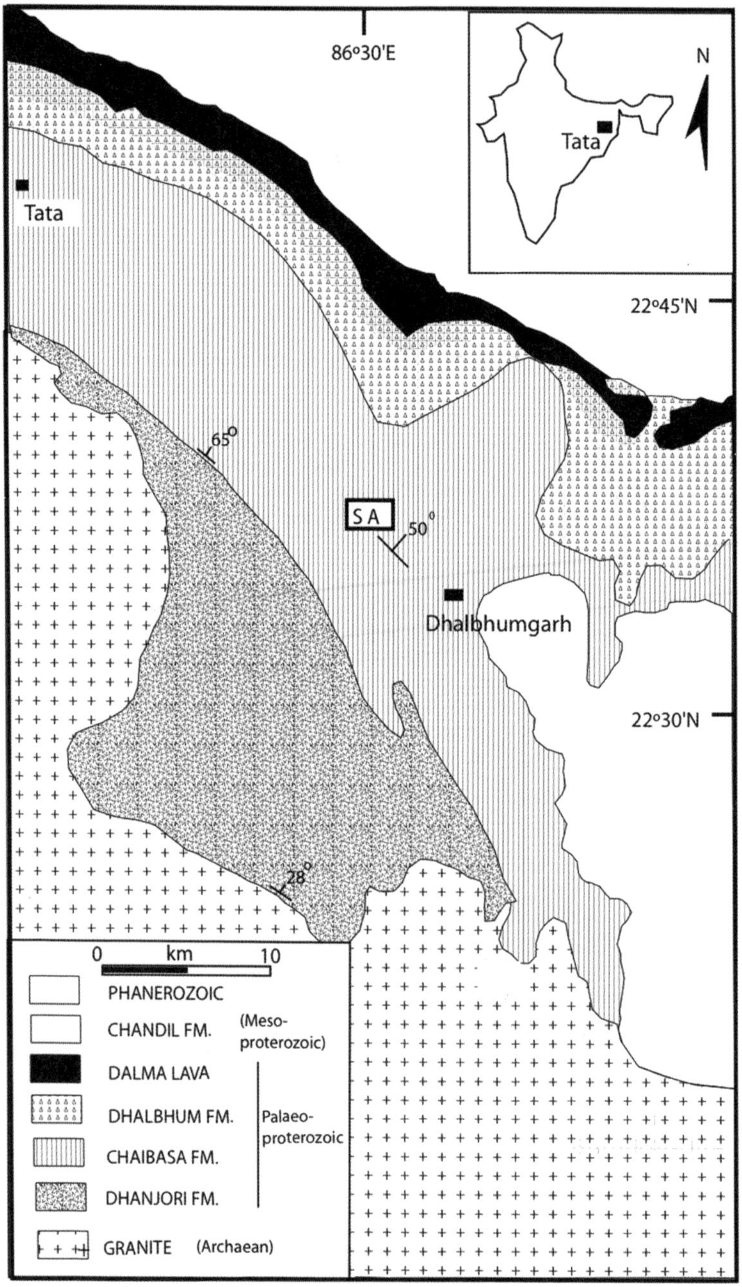

Fig. 1. Simplified geological map showing the disposition of the Chaibasa Formation and its bounding litho-units (after Saha 1994). SA is the study area that is shown in detail in Figure 2.

overlies the Dhanjori Formation and a transgressive lag demarcates their contact. The basal part of the Chaibasa Formation immediately overlying the transgressive lag deposit represents a transgressive systems tract, while offshore shales that sharply overlie the shallow-marine sandstones upsection represent marine flooding surfaces (Mazumder 2005). These rocks are deformed and metamorphosed to greenschist (locally amphibolite) facies (Naha 1965).

Methods

Hand samples were collected with a rock hammer from outcrops with good preservation and exposure. The whole rocks are soft and friable owing to tropical weathering and their high mica content. They are usually light green in colour. The hand samples were subsequently broken up into centimetre- to millimetre-sized pieces. Their magnetic susceptibility was measured using a Bartington magnetic susceptibility meter. Sample preparation involved further breaking up of the rocks so that the pieces were small enough to fit into the small plastic boxes used in measurements of magnetic susceptibility. The samples were purposely not ground up as later analyses on the rock fragments needed to be performed. The amount of rock that was put into each box was weighed and its susceptibility then measured three times. A hand magnet was used to separate out the magnetic grains. The magnetic grains were then made into thin sections with a microprobe-grade polish.

A petrographical microscope was used to look at the polished thin sections of rocks from sites 27A–27G. Some of the samples had magnetic grains that were relatively large – hundreds of micrometres in size. Those polished sections had a high proportion of highly altered groundmass and not much material with a good polish. Because material that lacks a good polish will not analyse well on the microprobe, polished thin sections from sites 27A, 27D and 27G were selected for a more detailed examination on a scanning electron microscope (SEM) and X-ray microanalyser (EDX) system.

The thin sections intended for SEM–EDX investigation were coated with a conductive layer of Au–Pd and examined in an FEI XL30 ESEM at 15 kV in high vacuum mode. An EDAX energy-dispersive X-ray analysis (EDX) microanalyser attached to the SEM was used to qualitatively analyse and chemically map the brighter grains. Thin sections for microprobe investigation were coated with C and quantitatively analysed on a Cameca SX-100 electron microprobe.

Results

Magnetic susceptibility of the Chaibasa Formation

The magnetic susceptibilities of our samples from the Chaibasa Formation are bi-modal. Four of the samples have a moderate susceptibility of 3 cgs units or less. All of these samples are from the Chaibasa sandstone facies. The remaining seven samples have extremely high susceptibilities of 90 cgs units or higher. All of these samples are from the Chaibasa shale facies. In our studies of Phanerozoic impact ejecta layers and volcanic ash layers, we have seldom found magnetic susceptibilities that are as high as those of the samples from the shale facies. In general, we have had most success at finding volcanic ash or impact ejecta if the magnetic susceptibility of the layer was about 8–10 cgs units or higher. Thus, the high susceptibility samples from the Chaibasa shale facies are good candidates for a more detailed examination.

Analytical results

The samples with the highest magnetic susceptibility in the Chaibasa Formation are 27A–27G (Fig. 3 & Table 1). These samples are all greenschist-facies mica schists. When we looked at samples of coarsely broken mica schist, it was found that they were full of magnetic material. A hand-held magnet was used to separate out the magnetic grains and the magnetic grains had been made into polished sections.

The polished thin section from 27D (22°34′23.106′N, 86°29′24.3744′E) had the highest proportion of relatively unaltered grains with a good polish. These grains appeared shiny in reflected light. Most of the shiny opaque grains were magnetite. The magnetite grains appeared crystalline, with straight edges that formed triangular, square and octahedral outlines. These straight-edged grains had an uneven level of polish owing to the lack of cleavage of the magnetite. There was a subset of shiny opaque grains with uneven outlines, as if they were fractured fragments. The edges of the shiny, uneven grains were often composed of a dark fibrous groundmass, suggesting that the fractured appearance was a primary feature and not a function of the later breakage of the rock (Figs 4 & 5).

The grains with uneven outlines also appeared brighter than the magnetite grains when they were imaged in the SEM's backscatter mode. These brighter grains had a yellowish tinge (characteristic of Fe) under the optical microscope and a much more even polish. The brighter grains are pure Fe with no oxygen (Figs 4 & 5). Because the backscatter signal from mineral grains is proportional to the average atomic number of the material, the native Fe and magnetite present differing grey levels when imaged in backscatter mode.

The most interesting of the smaller native Fe grains is sample 27D-2. The left-hand side of the grain has a bulbous shape and the right-hand side is broken. This bulbous grain could not be transported a long distance as a detrital grain. In addition, the overall shape of the grain and the wavy structure near the bottom left of the grain are suggestive of a deforming drop of pure Fe melt.

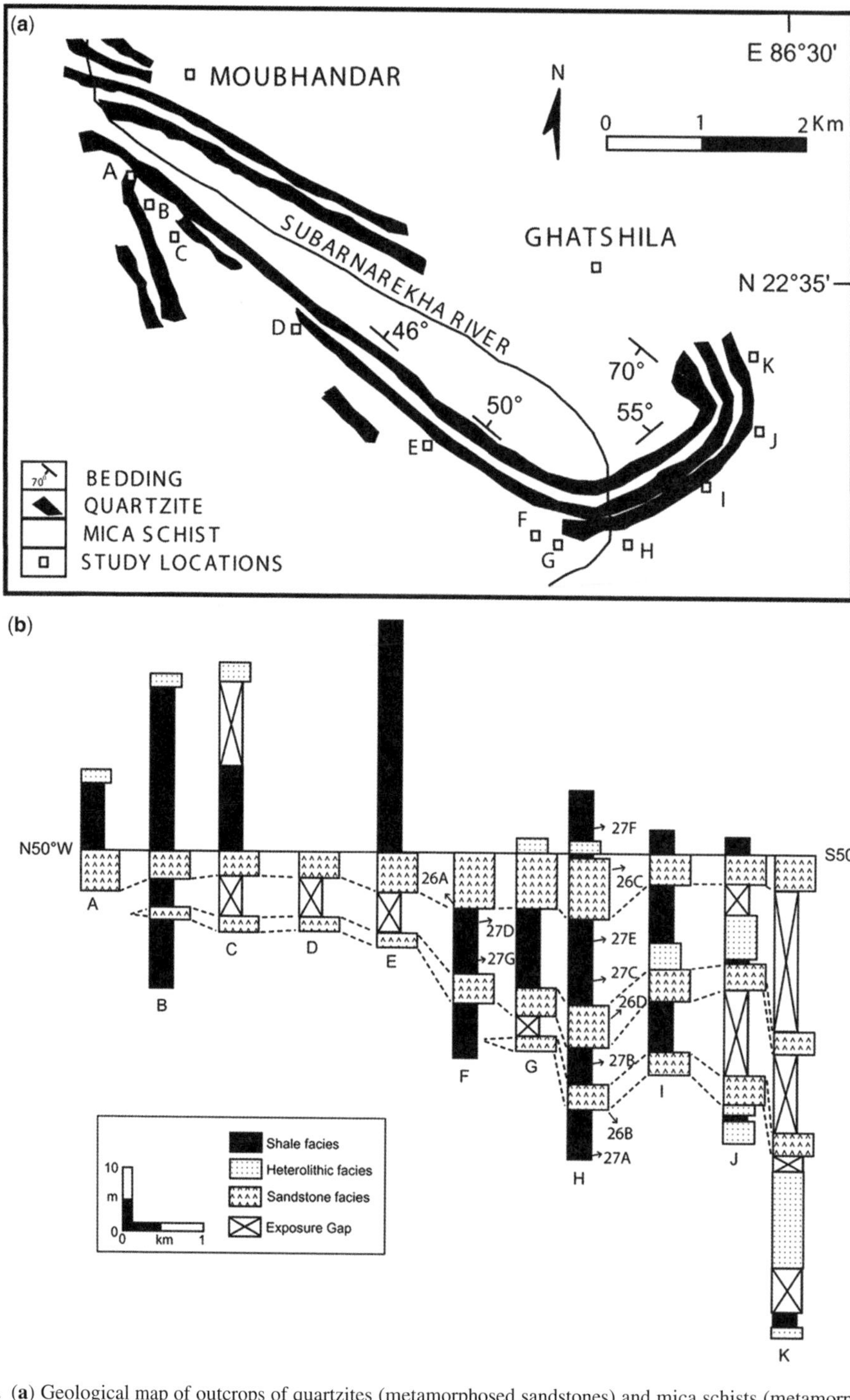

Fig. 2. (**a**) Geological map of outcrops of quartzites (metamorphosed sandstones) and mica schists (metamorphosed shales) of the Chaibasa Formation near the town of Ghatshila (modified after Naha 1965). The sediments are metamorphosed into the greenschist facies and sometimes the amphibolite facies. The letters designate the locations of the cross-sections. (**b**) Lateral and vertical facies changes in the Chaibasa Formation at locations A–K (modified after Bose *et al.* 1997). Our sampling locations are labelled with numbers 26A–27G.

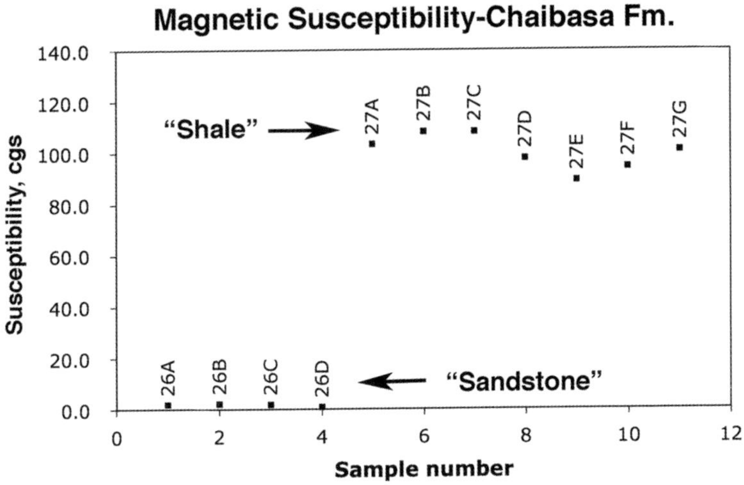

Fig. 3. Plot of magnetic susceptibility measurements on metamorphosed equivalents of shale and sandstone from the Chaibasa Formation. The final numbers on the plot and in Table 1 are the mean and standard deviation of the susceptibility measurements.

Visual comparisons of native Fe and magnetite grains – images and element maps

Figure 5 shows an SEM backscatter image of a native Fe grain (right) next to a magnetite grain (on the left). In backscatter mode, the compound with a higher average atomic number appears brighter. The native Fe is brighter than the magnetite. The polishing patterns provide another visual clue for discriminating between magnetite and native Fe. Magnetite has a hardness of 7.5–8 and no cleavage (Deer *et al.* 1966), and polishes poorly, producing an irregular surface in our thin sections. Native Fe has a hardness of 4–5 and polishes more easily, producing a smoother surface.

The element maps show the distinct difference between the oxygen content of the native Fe and the magnetite grains. On the element map for oxygen, the native Fe shows up as a black area, consistent with no oxygen. In contrast, the magnetite is bright yellow.

Microprobe analyses

We performed microprobe analyses of the components of grain 21 (Figs 6, 7 and Tables 2 & 3). All of the microprobe analyses included the same group of common rock-forming elements: Ca, Na, K, Ti, Fe, Mg, Al, Si, Mn, Cr, Ni and S. (Analyses that are not reported in any given table were all less than 0.1% for that particular element.) In addition, analyses for C were performed. C analyses are much less accurate than the other analyses, as all samples are coated with C, but the results for C can be compared for relative amounts. The microprobe totals with C reported as CO_2 and Fe reported as FeO are all significantly higher than 100%. This is expected, as the standard assumption in microprobe analyses is that O is present but is not directly analysed for. As a result, the presence of native elements and/or abundant C will produce microprobe totals that exceed 100%.

Discussion of microprobe analyses

When the microprobe analysis totals were recalculated to consider native Fe instead of FeO and with no CO_2, the totals in Table 2 are reduced to totals of between 101 and 97%. These totals are within the typical range for our standards (Table 4). However, this leaves no room for the C that we know is present from the X-ray analyses of similar samples coated in Au–Pd (Fig. 8). We therefore tried to also include the effect of native Si. When this was done, all of our totals were less than 100%, the expected result if C is present at levels of around 1–3%.

The presence of SiO_2 instead of native Si was checked by plotting Si v. the 'missing mass' ('O by difference' in Table 2). It was found that the two are negatively correlated; that is, there is more missing mass when there is less Si rather than more missing mass when there is more Si. The best-fit equation is oxygen = 5.36–2.41Si, with an R^2 of 0.5. Therefore, the Si must be present as native Si dissolved in the native Fe rather than as SiO_2.

The microprobe totals for the dark areas (Table 3) are too high if S, Fe and Mn are assumed to be present

Table 1. *Magnetic susceptibility of samples from the Chaibasa Formation*

Sample no.	Magnetic susceptibility (cgs)	Standard deviation
26A	2.01	0.04
26B	2.26	0.18
26C	1.92	0.05
26D	0.98	0.01
27A	103.39	0.69
27B	108.27	0.61
27C	108.21	0.33
27D	97.77	0.87
27E	89.09	0.32
27F	94.40	0.83
27G	100.84	0.46

as oxides. If the analyses are recalculated with no oxygen, the totals are in a normal range of 97–98%. Therefore, the dark areas in our images are most likely to be a solid solution of (Fe,Mn)S with minor Al, Si, Ni, Ca, Ti and Cr. This composition is generally associated with reducing conditions.

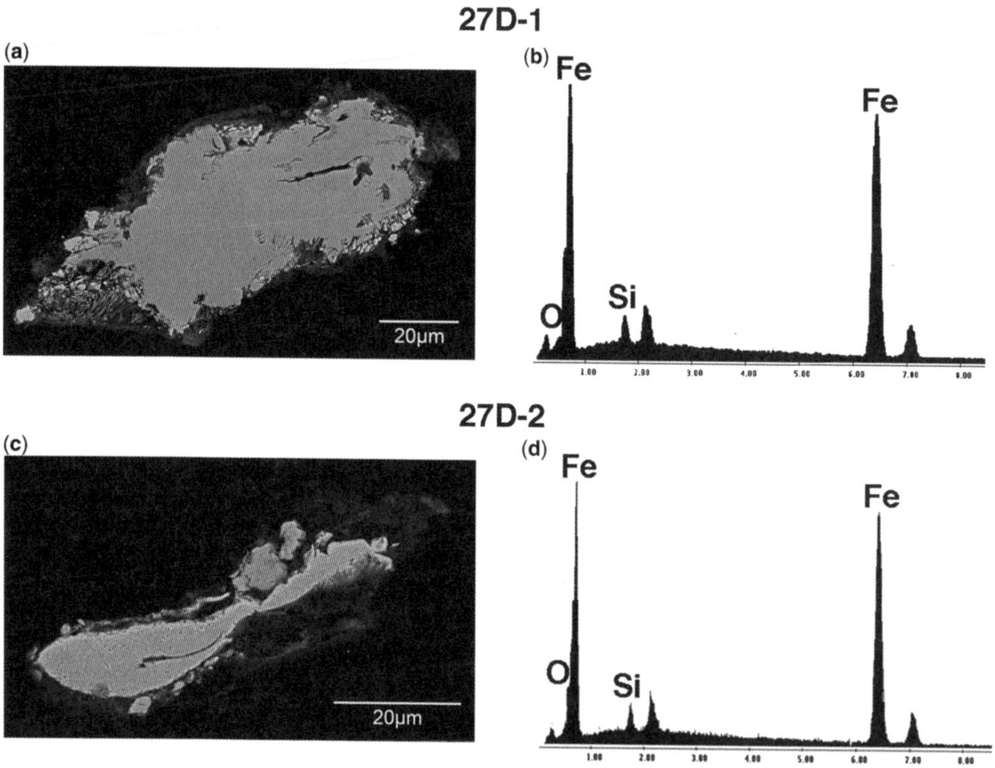

Fig. 4. Native iron grains from layer 27D in the Chaibasa Formation. (**a**) SEM backscatter image of grain 27D-1. Note that the edges of the grain show fibrous material that is another mineral. The edges of the grain are rough. (**b**) EDX spectrum from grain 27D-1. The predominant spectral peak is iron, with minor silica and oxygen. (**c**) SEM backscatter image of grain 27D-2. The left half of the native Fe grain has a bulbous shape, and is completely surrounded by silicate material. There is also a region of silicate within the grain that forms a wavy line. The right half of the grain is broken. (**d**) EDX spectrum from grain 27D-2. Again, the predominant spectral peak is iron. Ni was not detectable under our analytical conditions. The unlabelled peak represents the applied conductive coating.

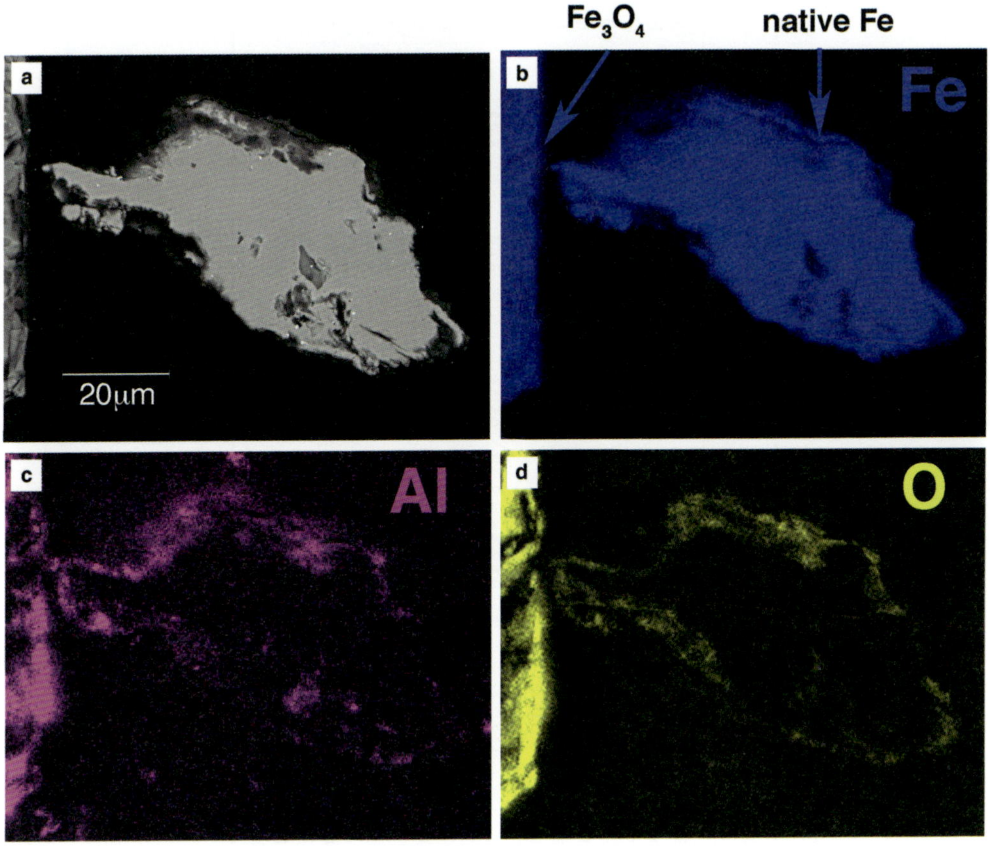

Fig. 5. Images and element maps of a native Fe grain (right) and a magnetite grain (left). (**a**) SEM backscatter image. (**b**) Fe element map. (**c**) Al element map. (**d**) O element map.

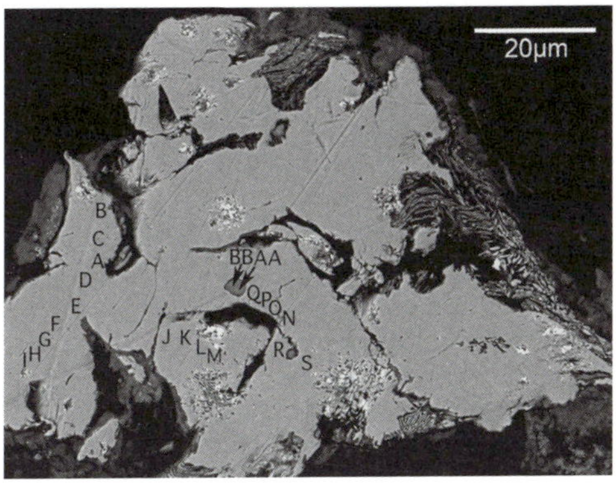

Fig. 6. Grain 21. SEM Backscatter image. Native Fe appears smooth and light grey. Dark grey rounded areas are (Fe,Mn)S. White areas are Au–Pd relicts from an earlier coating of the sample before new polishing. Dark textured areas on the right-hand side contain Fe and O. Locations of microprobe analyses are labelled with letters. Native Fe analyses are labelled from A to Q. Analyses of (Fe,Mn)S are labelled AA and BB.

Supporting evidence from SEM image and element maps

The native Fe is relatively bright in a backscatter image compared to rocky or other grains that contain oxygen (Fig. 7a). Both the native Fe and the (Mn,Fe)S show up as black on the oxygen map, consistent with the absence of oxygen in both minerals. The border of grain 21 contains oxygen but does not contain significant Al or Si (not shown). Based on the Fe map, it contains oxidized Fe.

Inclusions and textures of the native Fe

Some of the larger native Fe grains contain inclusions of native Fe with interesting textures (Fig. 8). In some cases, the native Fe locally has a striped appearance. When these were analysed, the light stripes had an apparently identical composition to the darker groundmass. Both contain minor C in qualitative X-ray analyses of Au–Pd-coated grains

Table 2. *Microprobe analyses of native Fe*

Al_2O_3	SiO_2	FeO	MnO	NiO	CO_2	Cr_2O_3	Total: all	Total: CO_2	Total: All elements*	Total: All elements†	O (by difference)	C from CO_2	Label
0.15	3.08	124.7	0.26	0.11	11.9	0.04	140.4	128.4	100.7	99.0	0.96	3.25	A
0.35	3.30	124.3	0.14	0.14	12.2	0.06	140.6	128.5	100.8	99.0	0.97	3.32	B
0.10	3.36	123.9	0.08	0.19	16.8	0.09	144.6	127.9	100.3	98.5	1.52	4.58	C
0.33	3.14	122.1	0.14	0.00	24.3	0.11	150.3	126.0	98.8	97.1	2.86	6.62	D
0.11	3.28	124.0	0.21	0.07	15.2	0.06	143.0	127.8	100.2	98.4	1.58	4.14	E
0.19	3.30	124.7	0.25	0.14	17.4	0.11	146.4	128.9	101.1	99.4	0.61	4.76	F
0.29	3.55	124.2	0.28	0.15	9.9	0.03	138.5	128.5	100.9	99.0	1.01	2.71	G
0.75	3.30	125.2	0.12	0.08	10.7	0.00	140.2	129.5	101.6	99.8	0.15	2.93	H
2.59	3.56	116.1	0.48	0.12	19.9	0.07	143.0	123.1	97.3	95.4	4.64	5.43	I
0.15	2.99	123.3	0.11	0.03	17.3	0.13	144.2	126.9	99.4	97.8	2.16	4.71	J
0.13	3.06	123.4	0.06	0.15	22.1	0.07	149.1	127.0	99.5	97.8	2.16	6.04	K
0.08	2.89	123.4	0.00	0.09	25.3	0.13	151.9	126.6	99.1	97.6	2.39	6.91	L
0.39	2.68	123.3	0.18	0.12	20.3	0.01	147.1	126.8	99.3	97.9	2.11	5.54	M
0.34	2.03	123.3	0.00	0.00	17.9	0.20	143.9	126.0	98.5	97.4	2.58	4.88	N
0.19	2.08	123.5	0.10	0.00	22.5	0.11	148.6	126.0	98.6	97.4	2.55	6.15	O
0.18	2.04	122.5	0.04	0.02	24.9	0.17	149.9	125.0	97.7	96.6	3.36	6.79	P
0.42	2.02	122.8	0.14	0.13	28.1	0.14	154.0	125.8	98.5	97.4	2.61	7.68	Q
0.14	2.91	123.8	0.17	0.13	12.5	0.10	140.0	127.4	99.9	98.3	1.69	3.42	R
0.20	3.24	122.4	0.17	0.16	21.2	0.01	147.4	126.2	99.0	97.2	2.77	5.79	S

*FeO as Fe native.
†SiO_2 as Si native and Fe native.

Table 3. *Microprobe analyses of (Fe,Mn)S*

Al_2O_3	SiO_2	FeO	MnO	NiO	SO_2	TiO_2	Cr_2O_3	Total with oxygen	Total number of oxygen	Label on figure
0.12	0.33	24.37	61.17	0.03	61.24	0.72	0.76	148.79	98.23	AA
0.63	0.95	51.97	42.39	0.00	44.48	0.40	0.51	141.40	97.06	BB

(Fig. 8). The striped region in Figure 8 is truncated on the upper side by an irregular area that appears to be eating away at its edge.

The striped area most closely resembles pearlite, a plate-like structure that forms in Fe alloys with more than 1% C. Pearlite typically forms when C-bearing Fe is heated to temperatures above 723–500 °C (Shiflet 2008). The pearlite is then cooled rapidly compared to typical geological processes. For example, Zhang & Kelly (2009) described an optimal cooling rate of 2 °C min^{-1} used for the production of pearlite from steel heated and annealed at over 1000 °C. So far as known, pearlite is not a structure that could be produced by the slow heating and cooling of geological metamorphism. In addition, this part of the Chaibasa Formation appears to be of greenschist facies, and thus could never have experienced metamorphic temperatures above 500 °C. Therefore the appearance of pearlite is suggestive of a primary origin for the native Fe. The carbon may have acted as a reducing agent, helping to preserve the native Fe.

This grain also contains inclusions of (Mn,Fe)S with some C. (Mn,Fe)S typically forms under reducing conditions and is rarely found in sediments (Lepland & Stevens 1998).

Other components of the Chaibasa Formation

The Chaibasa Formation at site 27D also contains a well-preserved iron–titanium oxide rock (Fig. 9). The rock is highly fragmented on the micrometre scale, with extremely sharp vein edges. The veins are straight but trend in multiple directions. The veins contain less oxygen and more Ti than the rest of the rock. The straightness of the veins and the sharpness of their contacts argue for their formation from a rock that was already lithified and was relatively cold. The earlier veins are larger and are cross-cut by smaller, later veins. The multiple directions, clear contacts and tiny sizes of the veins are inconsistent with deep burial or high-grade metamorphism of the rock.

Discussion

Origin of native Fe

The occurrence of native Fe in shale is fairly unusual. In this case, native Fe is found in metamorphosed shale that does not have a high content of organic matter. If the original shale had been reducing because of the inclusion of organic matter, it would be expected that the organic matter would have had sufficient sulphur to convert the iron into pyrite. No pyrite was found in the shale facies, only magnetite. Thus, we infer that conditions during deposition and later diagenesis of the shale were generally oxidizing.

Fibrous-looking muscovite–biotite schist surrounds the native iron. Because the rock is of greenschist facies, we infer that the muscovite mica with a modest biotite component was originally an illite–smectite mixed layer clay. As clay minerals are

Table 4. Microprobe analyses of standards

Na_2O	Al_2O_3	SiO_2	MgO	FeO	MnO	NiO	CaO	K_2O	TiO_2	Total	Comment
0.40	35.63	42.84	0.11	0.38	0.02	0.00	19.35	0.00	0.00	98.73	anorthite std 1
0.45	34.97	42.44	0.08	0.48	0.01	0.01	19.54	0.01	0.00	98.00	anorthite std 2
0.53	35.56	42.20	0.10	0.43	0.09	0.08	19.55	0.00	0.01	98.57	anorthite std 3
0.00	0.01	40.29	56.09	3.31	0.09	0.04	0.04	0.00	0.02	99.96	ol que 1*
0.00	0.00	40.41	55.75	3.18	0.09	0.00	0.02	0.01	0.00	99.45	ol que 2†
2.53	13.91	39.36	12.93	10.87	0.17	0.00	10.31	2.08	5.03	97.17	kak hb 1‡
0.03	0.02	28.96	0.23	69.05	2.29	0.10	0.03	0.02	0.00	100.72	rkfa 1§

*Olivine 1.
†Olivine 2.
‡Hornblende 1.
§Fayalite 1.

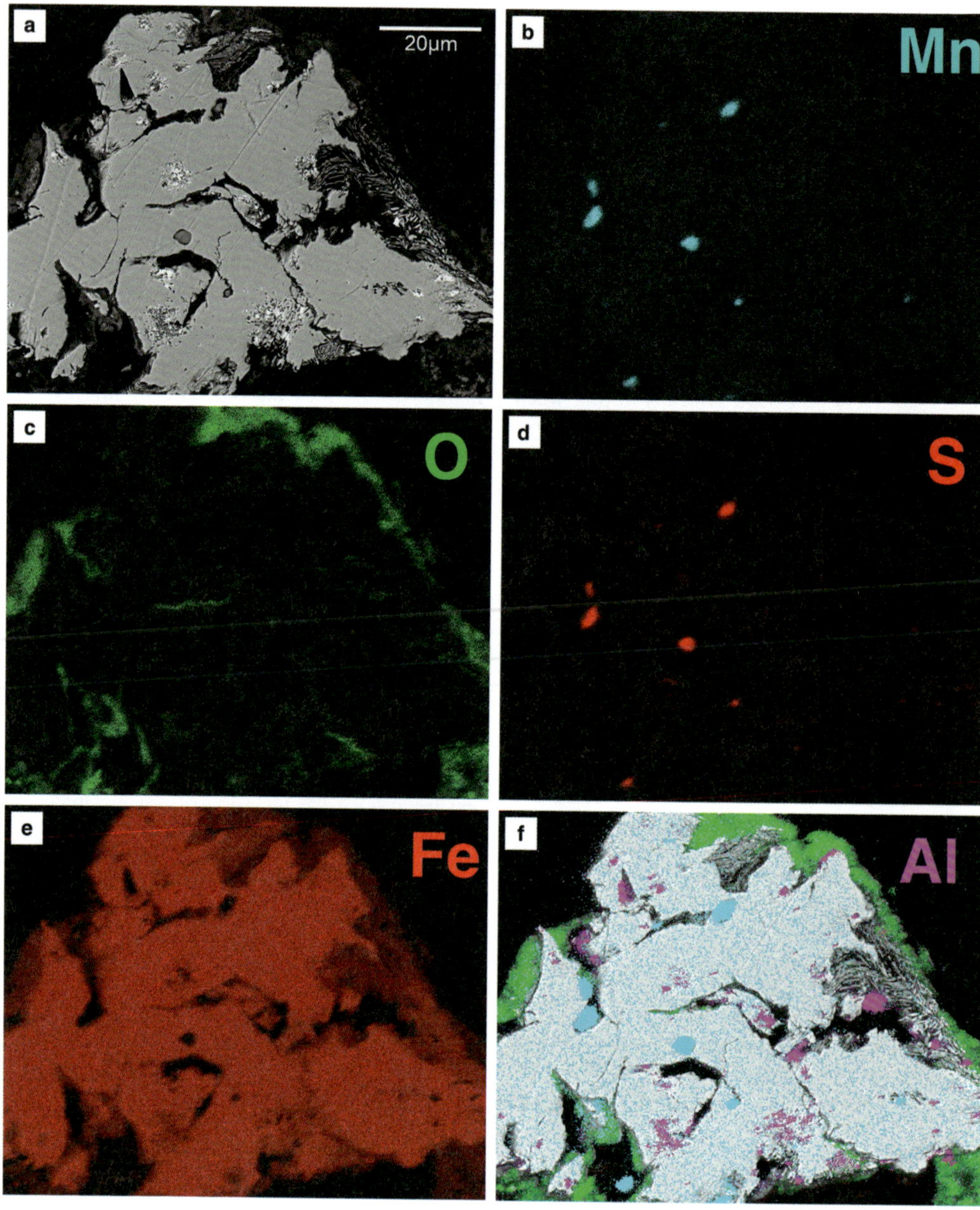

Fig. 7. Grain 21. (**a**) SEM backscatter image of grain 21. (**b**) Element map of Mn. (**c**) Element map of O. (**d**) Element map of S. (**e**) Element map of Fe. (**f**) Element map of Al (red), Mn (cyan) and O (green) overlain on a backscatter image of grain 21.

extremely impermeable, the clay may have acted as a seal on the native Fe, thus preventing later oxidation of the native element. There are abundant, well-formed magnetite crystals in the same layer as the native Fe. If the magnetite were detrital in origin, the edges of the grains would be rounded. So because they have straight edges, we conclude that the magnetite crystals formed during metamorphism, and that metamorphic conditions were, in fact, oxidizing rather than reducing.

In contrast to the magnetite, the native Fe grains have rough edges or elongated bulbous shapes and cannot have been transported any significant distance as detrital grains. The bulbous grains

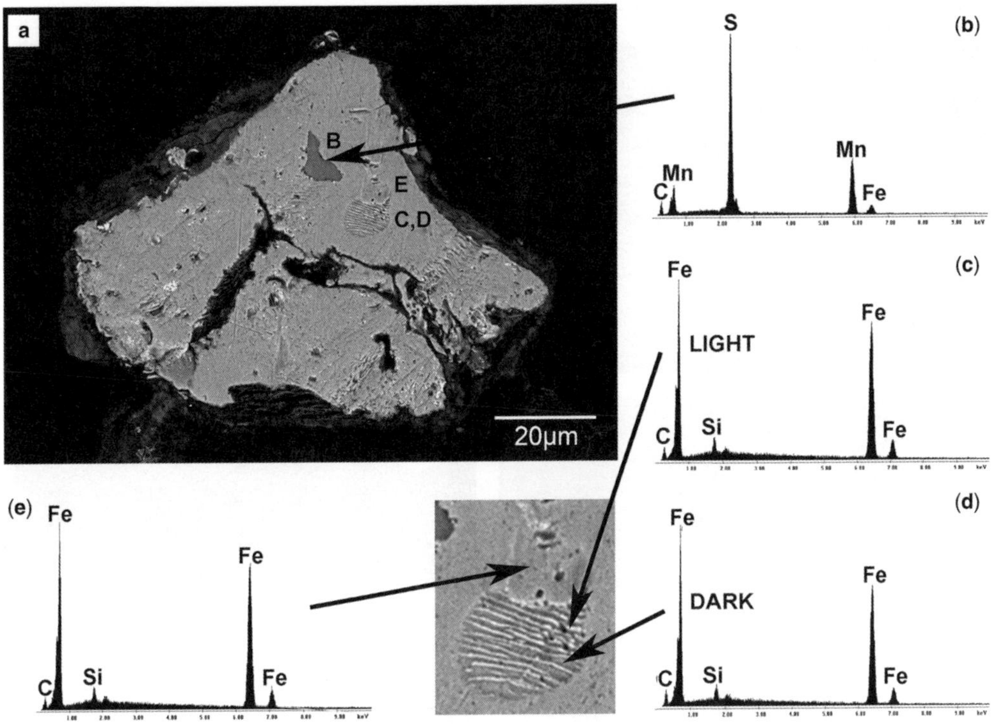

Fig. 8. (**a**) SEM backscatter image of grain 16. The grey groundmass is native Fe with native Si and C. (**b**) EDX analysis of the darker area of (Mn,Fe)S with a trace amount of C. (**c**) EDX analysis of light stripes on the area of the enlarged inset of grain 16. (**d**) EDX analysis of the darker groundmass on an area of the enlarged inset of grain 16. There is little apparent difference in composition between C and D owing to the large surface area (wider than the width of the stripes) being analysed in a qualitative way. (**e**) EDX analysis of melted-appearing area truncating the edge of the striped region. The carbon content is less than that of points C and D.

appear melted. The melting temperature of pure Fe given on the periodic table of elements is 1535 °C. Although these temperatures are locally present within komatiites, they are not found in sedimentary rocks. We thus infer that the native Fe grains were deposited immediately after they had formed. Therefore, the native Fe must have a local, primary origin. If so, it is the oldest native Fe ever found.

Independent supporting data

Subsequent to our initial work on, and finding of, native Fe in the Chaibasa Formation (Abbott *et al.* 2006), other scientists used Mossbauer spectroscopy to confirm the presence of native Fe (Chandra *et al.* 2010). Chandra *et al.* (2010) also found cohenite $(FeNi)_3C$ with Co as a trace impurity at the parts per million (ppm) level. Cohenite is typically found in iron meteorites and is diagnostic of strongly reducing conditions (Rubin 1977). They also found a reduction of 4 GPa in the pressure of the bcc (body centred cubic) to hcp (hexagonal close packed) martensitic phase transition. They attributed this reduction to the nucleation of stacking faults during shock metamorphism. Their finding of cohenite and a 4 GPa change in the pressure of the hcp phase transition provides support for the impact hypothesis for the origin of the Chaibasa Formation (Chandra *et al.* 2010).

Preservation of native Fe in sediments for millions of years

The native Fe found in the Chaibasa Formation has been preserved for over 1.6 Ga. It may seem unlikely that native Fe could remain unoxidized for such a long period but there is some precedent. Ordovician sandstones contain alloys of metallic Fe and Ni that represent fragmented iron meteorites (Schmitz *et al.* 2001). Native metal globules are also associated with the Permo-Triassic boundary at Graphite Peak, Antarctica and at Meishan, China (Basu *et al.* 2003). The native metals from the Permo-Triassic boundary are interpreted as bits of

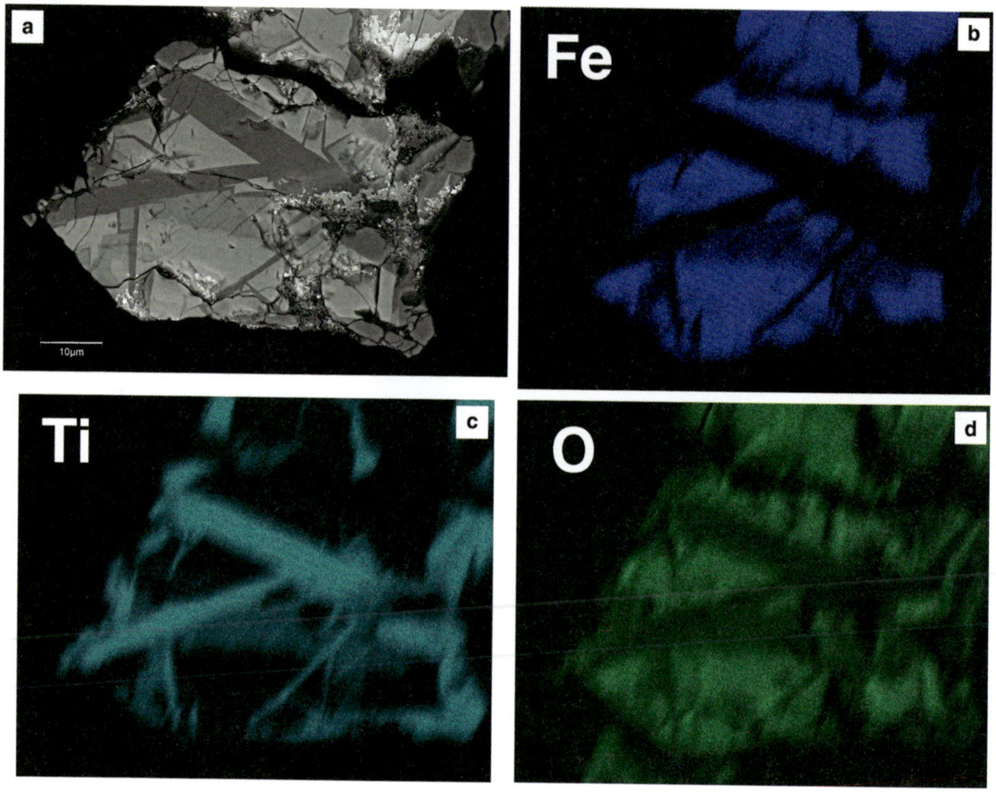

Fig. 9. (**a**) SEM backscatter image of Ti–Fe oxide. (**b**) Element map of Fe content. (**c**) Element map of Ti content. (**d**) Element map of O content.

impactor that have been preserved for over 250 Ma. Permo-Triassic sandstones from Japan contain pure iron nuggets (Erwin *et al.* 1998). Less than 1% of the metal nuggets are enriched in Cr and Ni; most metal nuggets consist of pure Fe with no appreciable Ni.

It is known that levels of atmospheric oxygen were lower during the Precambrian than at present (Holland 1984). However, there are a few reduced compounds (e.g. uraninite) that are preserved as detrital grains. These reduced grains date to the middle Archaean rather than to the Proterozoic. Atmospheric oxygen levels were sufficient to oxidize iron in sea water by 2.3 Ga (Rye & Holland 1998); thus, we cannot call upon a reduced atmosphere (or reduced sea water) as the source of the reduction of the iron. Instead, the native Fe must have formed before the grains were deposited in the sediment.

The fact that native Fe is not commonly found in sediments of any geological age (Ulff-Moller 1990) suggests that conditions for the Chaibasa samples were somewhat unusual. We suggest that two conditions (beyond some dissolved C) needed to be present. The first is that the sediments were deposited rapidly in a thick layer. The second is that the sediments were not disturbed by bioturbation. The second condition is easy to meet in Precambrian sediments as bioturbation is absent. Rapid burial is more difficult to prove but could be inferred from the presence of soft-sediment deformation in the Chaibasa Formation (Bhattacharya & Bandyopadhaya 1998; Mazumder *et al.* 2006, 2009). The Chaibasa Formation consists of at least 6 km of alternating shale and sandstone facies. As sandstone is much more permeable and much more dense than shale, a sequence of rapidly deposited shales will tend to deform over time owing to the combination of water retention in the low-permeability shale and sediment loading from the overlying dense sandstone layers. As Precambrian shales tend to be finely laminated, the fine laminations will be deformed by the rapid deposition of heavy sandstones on top of weak, rapidly deposited shales.

Although we do not know the conditions of deposition of the Ordovician meteorite fragments, the native metals from the Permo-Triassic boundary meet one or both of these conditions. Although nearly all Phanerozoic sediments are heavily bioturbated, the mass extinction at the Permo-Triassic boundary killed all of the bioturbating organisms. As a result, most early Triassic sediments are laminated. At Meishan, the Permo-Triassic boundary is bracketed by numerous ash layers (Erwin *et al.* 1998); indeed, the boundary clay is covered by an ash layer. As ash layers are geologically instantaneous deposits, the ash is prima facie evidence for rapid burial of the pressure–temperature (P–T) boundary layer at this location. Thus, where we have sufficient knowledge, the preservation of native metals in sedimentary sequences is associated both with rapid burial and unbioturbated sediments.

Conclusions

The native Fe in the Chaibasa Formation is most probably primary and not secondary. It occurs with (Fe,Mn)S, a mineral that is characteristic of reducing conditions. It also contains pearlite-like structures that resemble those found in quenched steel. It may contain a few melted drops of native Fe. Because the temperatures required to form pearlite are at least 500 °C, and the Chaibasa Formation is greenschist–lower-amphibolite grade, the pearlite must be primary. The low permeability of the shale facies, the lack of bioturbation in Precambrian sediments and the rapid deposition of the shale facies all worked together to preserve the native Fe for 1.6 Ga. The exact conditions that produced the native Fe are still unknown. Other workers have found cohenite at the ppm level within the native Fe; cohenite is typically found in meteorites.

Because we have not yet found shocked minerals, impact remains implied but not yet proven. Ongoing work should solve the puzzle.

We thank C. Mandeville at the Museum of Natural History in New York for assistance with the electron microprobe. The SEM and EDX investigations were carried out in the Materials Characterization Facility of the Drexel Nanotechnology Institute, Drexel University. R. Mazumder acknowledges financial support from the Alexander von Humboldt Foundation, Germany and the Indian Statistical Institute at Kolkata. We thank L. Cannone for the preparation of thin sections.

References

Abbott, D. H. & Hagstrum, J. 2004. Strategies for finding the record of Archean impact events. *In*: Eriksson, P. G., Altermann, W., Nelson, D. R., Mueller, W. U. & Catuneanu, O. (eds) *The Precambrian Earth: Tempos and Events*. Developments in Precambrian Geology, **12**, 45–62.

Abbott, D. H., Mazumder, R. & Breger, D. 2006. Native iron in the Chaibasa Formation, India: result of a pre-1.6 Ga impact? *In*: *Proceedings of the 37th Annual Lunar and Planetary Science Conference, March 13–17, 2006, League City, Texas.* Lunar and Planetary Institute, Houston, TX, Abstract 1889.

Basu, A., Pataev, M. I., Poreda, R. J., Jacobsen, S. B. & Becker, L. 2003. Chondritic meteorite fragments associated with the Permian–Triassic boundary in Antarctica. *Science*, **302**, 1388–1392.

Bhattacharya, H. N. & Bandyopadhaya, S. 1998. Seismites in a Proterozoic tidal succession, Singhbhum, Bihar, India. *Sedimentary Geology*, **119**, 239–252.

Bose, P. K., Mazumder, R. & Sarkar, S. 1997. Tidal sandwaves and related stormwave deposits in the transgressive Protoproterozoic Chaibasa Formation, India. *Precambrian Research*, **84**, 63–81.

Chandra, U., Sharma, P., Parthasarathy, G. & Sreedhar, B. 2010. ^{57}Fe Mossbauer spectroscopy and electrical resistivity studies on naturally occurring native iron under high pressures up to 9.1 GPa. *American Mineralogist*, **95**, 870–875.

Deer, W. A., Howie, R. A. & Zussman, J. 1966. *An Introduction to the Rock-forming Minerals*. Longmans, London.

Erwin, D. H., Bowring, S. A. & Martin, M. 1998. Constraining causes of the end-Permian mass extinction with geochronology. Geological Society of America, Abstracts with Programs, **21**, 310, Abstract 50105.

Holland, H. D. 1984. *The Chemical Evolution of the Atmosphere and Oceans*. Princeton University Press, Princeton, NJ.

Lepland, A. & Stevens, R. L. 1998. Manganese authigenesis in the Landsort Deep, Baltic Sea. *Marine Geology*, **151**, 1–25.

Mazumder, R. 2004. Late Paleoproterozoic meteorite impact? Insights from the Chaibasa Shale, eastern India. Geological Society of America, Abstracts with Programs, **36**, (5), 207.

Mazumder, R. 2005. Proterozoic sedimentation and volcanism in the Singhbhum crustal province, India and their implications. *Sedimentary Geology*, **176**, 167–193.

Mazumder, R., Rodriguez-Lopez, J. A., Arima, M. & van Loon, A. J. 2009. Palaeoproterozoic seismites (fine-grained facies of the Chaibasa Fm., E. India) and their soft-sediment deformation structures. *In*: Reddy, S. M., Mazumder, R., Evans, D. A. & Collins, A. S. (eds) *Palaeoproterozoic Supercontinent and its Global Evolution*. Geological Society, London, Special Publications, **323**, 301–318.

Mazumder, R., van Loon, A. J. & Arima, M. 2006. Soft-sediment deformation structures in the Earth's oldest seismites. *Sedimentary Geology*, **186**, 19–26.

Naha, K. 1965. Metamorphism in relation to stratigraphy, structure, and movements in parts of east Singhbhum, Eastern India. *Quarterly Journal of the Geological, Mineralogical and Metallurgical Society of Eastern India*, **37**, 41–88.

Roy, A. B., Sarkar, A., Jeyakumar, S., Agrawal, S. K. & Ebihara, M. 2002*a*. Mid-Proterozoic plume-related thermal event on the Eastern Indian Craton: evidence from trace elements, REE geochemistry, and Sm–Nd isotopic systematics of basic–ultrabasic intrusives from the Dalma volcanic belt. *Gondwana Research*, **5**, 133–146.

Roy, A. B., Sarkar, A., Jeyakumar, S., Agrawal, S. K. & Ebihara, M. 2002*b*. Sm–Nd age and mantle characteristics of the Dhanjori volcanic rocks. *Geochemical Journal*, **36**, 503–518.

Rubin, A. E. 1977. Mineralogy of meteorite groups. *Meteoritics and Planetary Science*, **32**, 231–247.

Rye, R. & Holland, H. D. 1998. Paleosols and the evolution of atmospheric oxygen: a critical review. *American Journal of Science*, **298**, 621–672.

Saha, A. K. 1994. *Crustal Evolution of Singhbhum-North Orissa, eastern India.* Geological Society of India Memoir, **27**.

Schmitz, B., Tassinari, M. & Peucker-Ehrenbrink, B. 2001. A rain of ordinary chondritic meteorites in the Early Ordovician. *Earth and Planetary Science Letters*, **194**, 1–15.

Shiflet, G. J. 2008. *Pearlite, Encyclopedia of Materials: Science and Technology*. Elsevier, Amsterdam, 6783–6787.

Ulff-Moller, F. 1990. Formation of native iron in sediment-contaminated magma: I. A case study of the Hanekammen Complex on Disko Island, West Greenland. *Geochimica et Cosmochimica Acta*, **54**, 57–70.

Zhang, M. X. & Kelly, P. M. 2009. The morphology and formation mechanism of pearlite in steels. *Materials Characterization*, **60**, 545–554.

Stability of chloritoid + biotite-bearing assemblages in some metapelites from the Palaeoproterozoic Singhbhum Shear Zone, eastern India and their implications

NANDINI SENGUPTA

Department of Geology, University of Calcutta, 35 Ballygunge Circular Road, Kolkata 700019, India (e-mail: nandinisg@yahoo.com)

Abstract: The arcuate Singhbhum Shear Zone (SSZ) of the East Indian Shield area, fringing the Archaean Singhbhum Craton, exposes a mélange-like ensemble of polymetamorphosed and highly tectonized siliciclastic sediments, mafic–ultramafic extrusive, tourmaline-rich rocks, magnetite–apatite rocks, granites and Cu–Fe–U sulphide ores of Meso- to Palaeoproterozoic age (1.5–1.77 Ga). Metapelites from two localities in the SSZ developed the enigmatic assemblage chloritoid–biotite–garnet–chlorite, which are the first reported from India and the eighth in world occurrence. Textural studies and algebraic analyses of the phase compositions in the KFMASH system indicate the operation of the reaction, chloritoid + biotite → garnet + chlorite + H_2O and this reaction has a negative slope in pressure–temperature ($P-T$) space. Quantitative geothermobarometry and pseudosections in the NCMnKFMASHO system indicate that this reaction occurred during prograde metamorphism that culminated at 6.3 ± 1 kbar and 490 ± 40 °C. The stability of the chloritoid + biotite is also sensitive to the MnO content of the bulk rock, and to the chemical potential of H_2O and oxygen (μ_{H_2O} and μ_{O_2}, respectively), of the ambient fluid phase. Thrusting of continental crust in a collisional setting is invoked to explain the peak metamorphic temperatures and the clockwise $P-T$ trajectory is construed from the petrological study of the chloritoid–biotite schist.

Petrological characterization of metapelitic rocks occurring in the fold belts peripheral to one or more cratonic blocks is used as a common marker to decipher the complex geological history and tectonic setting in many terrains, particularly those experiencing medium-grade metamorphism (Harte & Hudson 1979; Labotka 1980, 1981; Wang & Spear 1991; Ganguly *et al.* 2000; Goscombe *et al.* 2003; Mahato *et al.* 2008; Dasgupta *et al.* 2009; Sharma 2010). Although a large number of studies on the petrology of medium-grade pelites across the world have been published, there exist only seven localities where pelites record the four AFM phases garnet–chlorite–chloritoid–biotite (Droop & Harte 1995). Evolution of pelites with these four phases is a subject of considerable debate owing to contrasting topologies proposed for the assemblage garnet–muscovite–chlorite–biotite–chloritoid–quartz–vapour in the KFMASH (K_2O–FeO–MgO–Al_2O_3–SiO_2–H_2O) system (Droop & Harte 1995). The KFMASH topology of Harte & Hudson (1979), which is supported by Droop & Harte (1995), assumes a steep positive pressure–temperature ($P-T$) slope of the KFMASH univariant reaction (Als, St): garnet + chlorite + muscovite = chloritoid + biotite + quartz + H_2O with garnet + chlorite placed on the lower-temperature side of the reaction. The thermodynamically computed KFMASH topology of Spear & Cheney (1989), although supporting the positive slope of the reaction (Als, St), places the garnet + chlorite + H_2O on the higher-temperature side of the reaction.

The Singhbhum Shear Zone that separates the Archaean Singhbhum Craton from the Palaeo- to Meso-Proterozoic North Singhbhum Fold Belt (Fig. 1) exposes chloritoid–biotite schist containing the enigmatic mineral assemblage garnet–chlorite–chloritoid–biotite. The stability and rarity of this particular assemblage in nature cannot be explained with the help of the topological arrangement described in existing KFMASH petrogenetic grids. Other than studying the characteristics of stability of chloritoid + biotite v. garnet + chlorite, this paper attempts to:

(a) to decipher the paragenetic relationship between chloritoid and the associated minerals present in the metapelitic rocks of the studied area;
(b) to place constraints on the pressure–temperature conditions under which such enigmatic mineral assemblage formed;
(c) to construct an alternative schematic KFMASH petrogenetic grid to explain the stability and mutual relationship between chloritoid and the associated minerals in $P-T$ space;
(d) to evaluate the control to the bulk chemical composition of metapelites, and the chemical

From: Mazumder, R. & Saha, D. (eds) 2012. *Palaeoproterozoic of India*. Geological Society, London, Special Publications, **365**, 91–116. http://dx.doi.org/10.1144/SP365.6

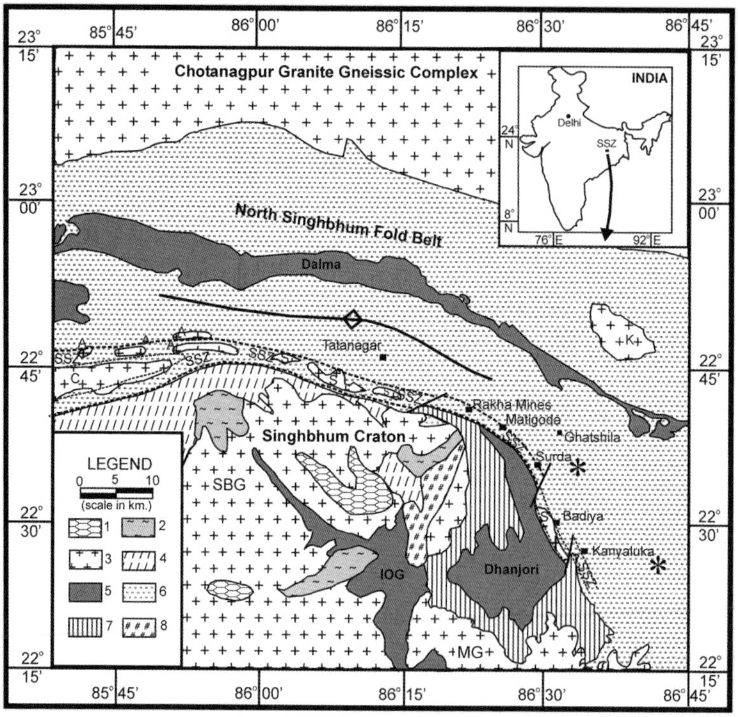

Fig. 1. Geological map showing the distribution of stratigraphic units in parts of eastern India after Saha (1994): 1, Older Metamorphic Group Granite; 2, Older Metamorphic Tonalitic Gneiss; 3, Granite (A, Arkasani Granite Gneiss and Granophyre; C, Chakradharpur Granite Gneiss; K, Kuilapal Granite; MG, Mayurbhanj Granite; S, Soda Granite; SBG, Singhbhum Granite); 4, Iron Ore Group shale, tuff and phyllite; 5, Basic lava and intrusives (Dalma, Dalma Group of volcanics; Dhanjori, Dhanjori Group of metabasic rocks; IOG, Iron Ore Group volcanics); 6, pelites and psammopelites of the Singhbhum Group; 7, Dhanjori Group metasediments; 8, Proterozoic gabbro-anorthosite; SSZ, Singhbhum Shear Zone; the locations marked with a star exposes the four-phase assemblage chloritoid–biotite–garnet–chlorite.

potentials of H_2O and O_2 (μ_{H_2O} and μ_{O_2}, respectively) on the stability of the mineral assemblage;

(e) to understand the effect of the construed P–T path in the light of an existing tectonic model.

Geological setting

The North Singhbhum Fold Belt (NSFB), the arc-shaped belt of Palaeo- to Meso-Proterozoic age, is sandwiched between the 3–3.4 Ga old Singhbhum Craton containing granitoids and minor supracrustal rocks in the south and the Meso- to Neo-Proterozoic Chotanagpur Granite Gneiss Complex in the north (Fig. 1) (Sharma *et al.* 1994; Sarkar 2000; Sengupta *et al.* 2005; Mahato *et al.* 2008; Chatterjee *et al.* 2010). The NSFB, which is dominated by metasediments, contains two mafic–ultramafic suites interstratified with metasedimentary rocks: the Dhanjori Formation to the south and the Dalma volcanics at the axial part of the belt (Fig. 1). The more than 150 km-long arc-shaped Singhbhum Shear Zone (SSZ) of variable width (*c.* 0.5–5.0 km) separates the NSFB from the Singhbhum Craton. The SSZ, which is developed along the northern boundary of the Singhbhum Craton and northeastern fringe of the Dhanjori Formation exposes an ensemble of laterally discontinuous lithologies including kyanite–quartzite and kyanite–mica schist, metapelitic and psammopelitic rocks, amphibole + chlorite ± talc schist, apatite–magnetite rock, conglomerate, tourmalinite and mylonitized granite resembling a tectonic mélange (Sarkar 2000; Sengupta *et al.* 2005). These rocks are highly tectonized with extensive development of mylonites and phyllonites with typical S–C fabric and various shear markers (Sarkar 1984; Ghosh & Sengupta 1987, 1990; Mukhopadhyay & Deb 1995; Sengupta *et al.* 2005). Kinematic shear indicators in the SSZ indicate the thrusting of the rocks of the NSFB over the Singhbhum Craton (Fig. 1). Important economic deposits of Fe, Cu, U and other non-metallic minerals, such as kyanite and apatite, are present along this zone (Sarkar 2000).

Limited geochronological data suggest that sediments of the NSFB were deposited after 2.2 Ga (Mahato *et al.* 2008), which was followed by mafic magmatism of the Dhanjori Formation at around 2.1 Ga (Roy *et al.* 2002; Mahato *et al.* 2008). Subsequently, in the SSZ, the supracrustal rocks underwent multiple episodes of folding and shearing over a protracted period from about 1.7 to 1.3 Ga in response to regional compression that acted perpendicular to the trend of the SSZ (Ghosh & Sengupta 1990; Mukhopadhyay & Deb 1995; Sarkar 2000; Sengupta *et al.* 2005; Mahato *et al.* 2008). Vestiges of approximately 0.8 Ga age have recently been documented from a few places in the eastern part of the NSFB (Chatterjee *et al.* 2010). Detailed structural and limited petrological studies of the rocks of the SSZ have identified three major phases of deformation (D_1, D_2 and D_3) and two phases of metamorphism (M_1 and M_2), with a major phase of shearing broadly coeval with the D_2 event (Bandyopadhyay 2003; Sengupta *et al.* 2005). The approximately 1.5–1.8 Ga U–Pb date of uraninite is believed to be the time of occurrence of the major shearing event and, hence, the time that the D_2 deformation occurred in the SSZ (Rao & Rao 1983). The M_1 metamorphism in the SSZ is marked by the formation of porphyroblasts (e.g. garnet, chloritoid, staurolite and kyanite) in metapelites and psammopelites that developed syn- to post-tectonically with the regional schistosity and/or shear foliation (S_2) related to the D_2 deformation. M_2 is marked by retrogressive metamorphism and is correlated with the D_3 deformational event. Upper-greenschist- to amphibolite-facies metamorphic conditions have been inferred from the assemblages of the porphyroblastic minerals in pelitic and psammopelitic rocks of the SSZ (Sarkar 2000; Bandyopadhyay 2003; Sengupta *et al.* 2005).

Chloritoid–biotite schist has been found to occur in two localities of the SSZ (Surda and Kanyaluka: Fig. 1) and is the subject of the present communication. The colour of chloritoid–biotite schist varies from pale to dark grey depending on the proportions of garnet, chloritoid, magnetite, muscovite, chlorite and biotite. The rock exposed near Kanyaluka is dark grey in colour owing to the large proportion of magnetite (Fig. 2a). The rocks occurring in the close vicinity to the chloritoid–biotite schist include kyanite–quartzite, kyanite–mica schist, sheared granite and thinly banded psammopelitic schist. A set of schistosity defined by the alternate banding of muscovite and biotite is the dominant planar fabric of the studied area (Fig. 2a). The schistosity, which is phyllonitic in places, marks the S_2 foliation plane. Because of intense shearing, sets of earlier schistosity (S_1) and primary bedding planes (S_0) are rarely observed on the meso-scale. Commonly, the S_2 fabric bent around nodular aggregates of porphyroblastic chloritoid, quartz and magnetite (Fig. 2b). In places, randomly oriented needles of chloritoid occur in the schistose micaceous matrix (Fig. 2c). Porphyroblastic garnet has an euhedral outline and appears to grow under static condition (Fig. 2d).

Mineral assemblages and petrography

Several samples of chloritoid–biotite schist have been collected from two areas, namely Surda and Kanyaluka, in the SSZ (Fig. 1). For quantitative geothermobarometry and to place constraints on the stability of the rare chloritoid + biotite-bearing assemblages, samples from the psammopelites exposed in adjoining areas that share a common metamorphic evolution with the chloritoid–biotite schist are also chosen for the present study. After routine microscopic study of all the collected samples, representative thin sections have been studied:

- to recognize any variation in the chloritoid-bearing assemblages present in the studied area;
- to understand mutual textural relationships between chloritoid and other KFMASH minerals;
- to select samples for detailed chemical analyses from the chloritoid–biotite schist and psammopelitic schist.

Minerals present in the chloritoid–biotite schist and in the psammopelitic schist are listed in Table 1. On the basis of mineralogy, three associations, namely A (Cld + Bt + Grt + Chl), B (Cld + Bt + Grt), C (Cld + Bt + Chl), are recognized in the chloritoid–biotite schist. In all the three associations, flakes of biotite and muscovite define a prominent schistosity (S_2) (Fig. 3a). Laths of chlorite athwart the schistosity are common. The same schistosity is also traceable and is the prominent planar fabric in the adjoining psammopelitic schist. Garnet (Association A and B) and chloritoid (Association A, B and C) form porphyroblasts in chloritoid–biotite schist. Commonly, chloritoid blades are flattened parallel to the schistosity and bear the impression of deformation manifested by undulose extinction, and the development of kinks (Fig. 3b–e) and fractures perpendicular to the length of grain (Fig. 3c). The schistosity swerves around the chloritoid porphyroblasts in places (Fig. 3d). Rarely, inclusions of quartz in chloritoid define an internal fabric (S_i-trail) that is parallel with the external schistosity (S_e) defined by mica (Fig. 3c). In places, twinned

Fig. 2. (**a**) Outcrop photograph of schistosity (S_2) in chloritoid–biotite schist defined by alternate muscovite-rich and biotite-rich banding. (**b**) Outcrop photograph of schistosity swerves around nodular chloritoid porphyroblasts. (**c**) Outcrop photograph of randomly oriented needles of chloritoid in schistose micaceous matrix. (**d**) Outcrop photograph of euhedral garnet porphyroblasts in chloritoid–biotite schist.

chloritoid grains that impinge into the foliation include polygonal aggregates of quartz (Fig. 3e). Garnet porphyroblasts commonly show an euhedral outline and appear to grow over the schistosity defined by mica and the elongated chloritoid (Fig. 3c, d, f). The pervasive schistosity abruptly stops at the euhedral faces of garnet (Fig. 3c, d, f). Locally, garnet porphyroblasts contain inclusions of chloritoid laths (Fig. 4a). Volumetrically minor chlorite flakes athwart the schistosity, cross-cut

Table 1. *Mineral associations in the chloritoid–biotite schist and the psammopelitic schist*

Rock type	Association	Sample no.	Area	AFM phases	Accessory minerals
Chloritoid–biotite schist	A	L5A (1–6)	Surda	Grt + Chl + Cld + Bt	Tur + Ilm + Zrn
		K3CG (1–8)	Kanyaluka	Grt + Chl + Cld + Bt	Mag + Ilm + Tur + Ap + Zrn
		24C (1–1a)	Kanyaluka	Grt + Chl + Cld + Bt	Mag + Ilm + Tur + Ap + Zrn
	B	L5X (1–3)	Surda	Grt + Cld + Bt	Mag + Ilm + Ap + Zrn
		K3 (1–6)	Kanyaluka	Grt + Cld + Bt	Mag + Ilm + Ap + Zrn
	C	K5 (1–2)	Kanyaluka	Cld + Bt + Chl	Ap + Mag + Zrn
Psammo-pelitic schist	D	H2q (1–8)	Matigoda	Grt + Bt + Chl + Plag	Ep + Tur + Mag + Ilm
		H2m (1–4)	Matigoda	Grt + Bt + Chl + Plag	Ep + Tur + Mag + Ilm
		G2 (1–5)	Rakha Mines	Grt + Bt + Chl + Plag	Ep + Zo + Tur + Mag + Ilm
	E	16A (1–5)	Badiya	Grt + Bt + Plag	Ap + Ep + Tur + Mag + Ilm + Sulphides

Note: All of the associations include quartz and muscovite; the mineral abbreviations used in this communication are after Kretz (1983).

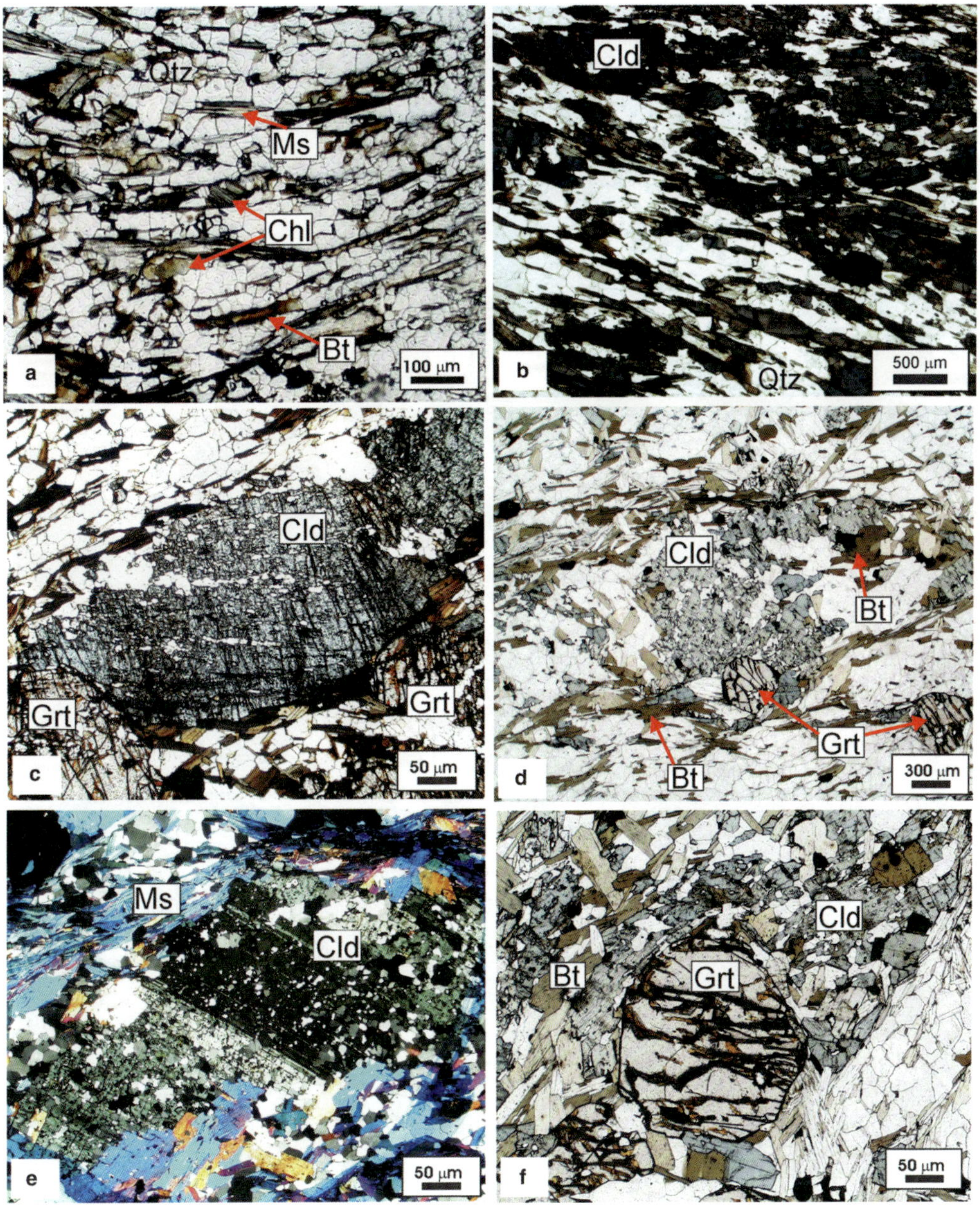

Fig. 3. (**a**) Photomicrograph of needles of muscovite and biotite, together with the elongated polygonal grains of quartz, define the schistosity (S_2) and laths of chlorite in places athwart the schistosity. (**b**) Photomicrograph of deformed and fractured grains of chloritoid in parallelism with the S_2. (**c**) Photomicrograph of quartz inclusions defining the S_i-trail within chloritoid is parallel to S_2. (**d**) Photomicrograph of S_2 swerves around rounded porphyroblasts of chloritoid. (**e**) Photomicrograph (in cross-polarized light) of chloritoid grains showing lamellar twining and also bearing a S_i-trail parallel to S_2. (**f**) Photomicrograph of an euhedral grain of garnet overprinted upon schistosity defined by chloritoid. Photomicrographs (a), (b), (c), (d) and (f) are taken in plane-polarized light. All of the mineral abbreviations used in this paper are after Kretz (1983).

both by biotite laths (Fig. 4b–d) and chloritoid blades (Fig. 4c). In places, small prisms of chloritoid are engulfed within chlorite flakes (Fig. 4d). Magnetite occurs in two modes. Commonly, euhedral and subhedral magnetite grains are scattered in the matrix, and show intergrowth with chloritoid

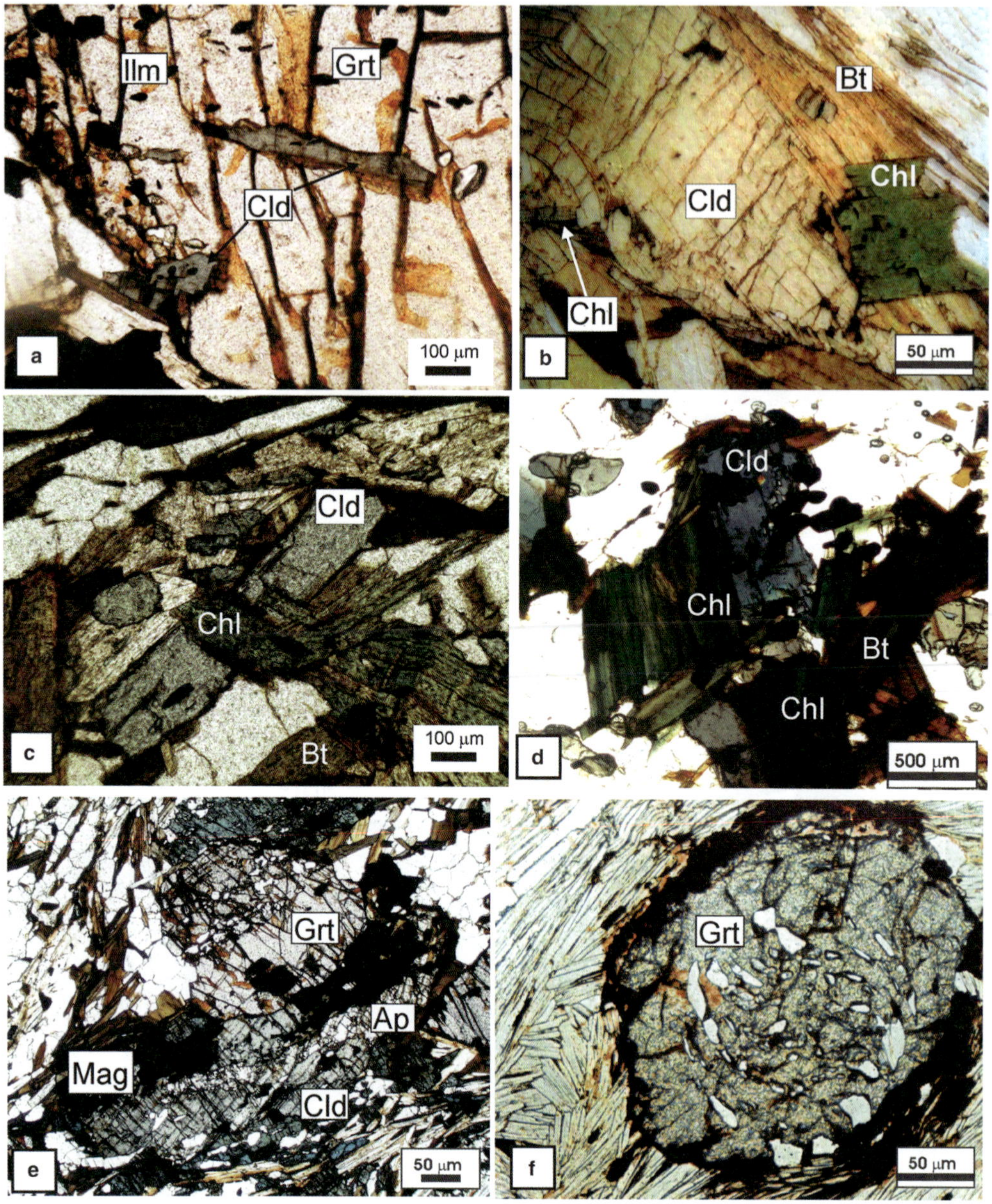

Fig. 4. (**a**) Photomicrograph showing inclusions of chloritoid and ilmenite within garnet. (**b**) Laths of chlorite cross-cuts biotite needles. (**c**) Photomicrograph of laths of chlorite cross-cutting a chloritoid blade. (**d**) Photomicrograph of a grain of chloritoid engulfed within flakes of chlorite. (**e**) Photomicrograph showing two sets of shear foliation (*S–C* fabric) around the garnet porphyroblast: one set is defined by needles of biotite running in the NW corner of the photomicrograph; and another set is defined by white mica, biotite and chloritoid blades aligned in the lower half of the photomicrograph. Veins of apatite parallel to the shear foliation passes through the large blade of chloritoid, and magnetite is aligned along the boundary between garnet and chloritoid. (**f**) Photomicrograph of a relict fold defined by earlier schistosity (S_1) in a matrix with a similar pattern to that of the sigmoid S_i-trail within the core of garnet; the rim of garnet is devoid of inclusions. All of the above photomicrographs are taken in plane-polarized light.

(Fig. 4e). Magnetite, in the other mode, occurs as inclusions in garnet and chloritoid porphyroblasts. Apatite occurs in the matrix either as isolated grains or as cluster of smaller grains. Locally, veins of apatite cross-cut chloritoid porphyroblasts (Fig. 4e). The textural features indicate that chloritoid grew broadly syntectonic to the schistosity, whereas garnet and chlorite formed under static conditions after the schistosity-forming stress withdrew or significantly waned out. From this, it follows that chloritoid and biotite are paragenetically older than both garnet and chlorite.

In the psammopelitic schist, muscovite, biotite, chlorite (when present) define the schistosity (S_2). In places, folded traces of an earlier schistosity (S_1) defined by muscovite is observed (Fig. 4f). Quartz and plagioclase form polygonal aggregates in the matrix. Garnet occurs as porphyroblasts. The core of garnet is studded with quartz inclusions that define a sigmoidal pattern (Fig. 4f). The rims of the garnet porphyroblasts show a euhedral outline and are practically devoid of inclusion (Fig. 4f). Like the chloritoid–biotite schist, the external schistosity sharply abuts against the euhedral face of garnet. This feature suggests that garnet grew in two stages. The growth of garnet started at the time of formation of the schistosity (the core), whereas the final growth of garnet (the rim) outlasted deformation and, hence, post-dated the schistosity.

Mineral chemistry

The compositions of minerals in the studied rocks were determined with a CAMECA SX51 electron probe microanalyser at the Mineralogisch Petrologisches Institut, Bonn University. Wavelength Dispersive Spectrometry (WDS) analyses were performed using an accelerating voltage of 15 kV, a current of 5.10 A and a 1–2 μm-diameter electron beam. For muscovite and plagioclase, the electron beam was broadened to 5–10 μm (diameter) to avoid alkali loss. The following well-calibrated standards were used: F = BaF_2; Na, Ca and Si = jadeite–diopside glass; Mg = MgO; Al = Al_2O_3; Fe = pure Fe; K = sanidine; Ti = TiO_2; Cr = pure Cr; and Mn = Mn with a Fe-mixture. The raw data were processed using the PAP procedure (Pouchou & Pichoir 1984). At least three points from each individual grain and at least five grains were analysed from each thin section for minerals that are not chemically zoned. At least five pairs of core and rim from each grain were analysed to detect the presence of chemical zoning. In some sections, garnet porphyroblasts were traversed at the interval of 50–100 μm to check the compositional variation or zoning from core to rim. Representative compositions of the minerals are presented in Tables 2–7.

None of the minerals in the chloritoid–biotite schist shows significant compositional zoning. Chloritoid in all the chloritoid–biotite schist is extremely Fe-rich (Mg concentration $X_{Mg} = Mg/(Mg + Fe^{2+}) = 0.077–0.141$) with chloritoid in Association C being most magnesian (X_{Mg} goes up to 0.141: Table 2). The levels of Fe^{3+} in the chloritoid analyses that were recalculated following the procedure of Chopin *et al.* (1992) are low (0.00–0.07 apfu (atoms per formula unit)).

Biotite in the chloritoid–biotite schist is more magnesian (X_{Mg}: 0.363–0.410, Table 3) compared to the coexisting chloritoid. The TiO_2 content of biotite was found to vary between 0.50 and 1.72 wt% (Table 3), while biotite in the ilmenite-bearing samples has a higher TiO_2 content (>1 wt%) than the ilmenite-free samples (<0.6 wt%). Biotite in psammopelitic schist is more magnesian (X_{Mg}: 0.42–0.51) and titanian (TiO_2: 1.71–2.95 wt%, Table 3) than that of chloritoid–biotite schist.

Garnet in chloritoid–biotite schist is rich in almandine (77.0–94.9 mol%, Table 4). Garnet in all the associations from chloritoid–biotite schist (A & B) shows a very low concentration of grossular and spessartine molecules (sum of the two never exceeds the range 1.21–5.21 mol%); barring one sample of Association A (#L5A/2), which contains garnet with a higher content of spessartine (3.89–4.94 mol%) and grossular (11.81–12.70 mol%). Garnet is slightly less magnesian ($X_{Mg} = Mg/(Mg + Fe^{2+})$: 0.036–0.065) than the coexisting chloritoid. In contrast to chloritoid–biotite schist, the garnet core in the psammopelitic schist has a distinctly higher spessartine (11.01–27.63 mol%) and grossular (9.56–15.81 mol%) content. Strongest compositional zoning in garnet is noted in one set of samples (#H2q/1), where the spessartine content drops significantly from the core to the rim of garnet (core 12.67 mol%; rim 1.79 mol%). No significant variation in terms of grossular content was, however, noted from the core to the rim of garnet (Table 4).

In chloritoid–biotite schist, X_{Mg} of chlorite shows insignificant variation (0.367–0.400). Following the trends of biotite and chloritoid, chlorite in association C is marginally more magnesian (X_{Mg}: 0.373–400) than chlorite in Association A (X_{Mg}: 0.367–0.375). Chlorite in psammopelitic schist is distinctly more magnesian (X_{Mg}: 0.424–0.488) than the chlorite in chloritoid–biotite schist (Table 5). In psammopelitic schist, both primary chlorite and secondary retrograde chlorite are present, but there is no compositional variation among them.

Muscovite shows a considerable amount of phlogopite–annite, celadonite, pyrophyllite, margarite and paragonite. Together, they vary within the ranges of 21.16–30.51 and 29.95–34.44 mol% in chloritoid–biotite schist and psammopelitic schist, respectively. Muscovite in psammopelite has distinctly higher paragonite (11.86 ± 4.47 mol%)

Table 2. *Representative analyses of chloritoid from Associations A–C*

Association	A				B				C	
Sample no.	L5A/1		K3CG/1		L5X/1		K3/1		K5/1	
	core	rim	core	rim	core	rim	core	rim	core	rim
SiO_2	24.91	24.72	23.94	23.77	23.97	23.67	24.04	23.83	25.21	24.57
TiO_2	0.01	0.01	0.10	0.25	0.02	0.00	0.25	0.36	0.01	0.03
Al_2O_3	39.94	40.06	41.69	40.87	40.37	40.45	41.02	41.25	41.44	40.78
Cr_2O_3	0.00	0.00	0.02	0.05	0.06	0.00	0.03	0.01	0.05	0.03
FeO	24.97	25.17	25.44	25.95	26.04	25.98	25.91	25.64	22.96	24.06
MnO	0.23	0.16	0.04	0.03	0.02	0.03	0.05	0.08	0.08	0.14
MgO	2.08	2.06	1.44	1.49	1.77	1.75	1.22	1.35	2.11	2.02
CaO	0.00	0.00	0.01	0.01	0.01	0.03	0.01	0.02	0.03	0.07
Na_2O	0.03	0.03	0.14	0.11	0.00	0.00	0.12	0.11	0.12	0.13
K_2O	0.01	0.00	0.02	0.02	0.00	0.03	0.01	0.02	0.01	0.01
Total	92.19	92.22	92.84	92.55	92.26	91.94	92.66	92.67	92.02	91.84
Si	2.063	2.049	1.976	1.976	1.996	1.980	1.994	1.974	2.062	2.033
Ti	0.001	0.001	0.006	0.016	0.001	0.000	0.016	0.022	0.000	0.002
Al	3.900	3.914	4.057	4.006	3.964	3.990	4.011	4.030	3.996	3.979
Cr	0.000	0.000	0.001	0.003	0.004	0.000	0.002	0.001	0.004	0.002
Fe^{2+}	1.694	1.709	1.756	1.804	1.780	1.789	1.797	1.777	1.570	1.665
Fe^{3+}*	0.035	0.035	0.000	0.000	0.034	0.028	0.000	0.000	0.000	0.000
Mn	0.016	0.011	0.003	0.002	0.001	0.002	0.003	0.005	0.005	0.010
Mg	0.257	0.255	0.177	0.185	0.220	0.218	0.150	0.166	0.257	0.249
Ca	0.000	0.000	0.001	0.001	0.001	0.003	0.001	0.002	0.003	0.006
Na	0.005	0.005	0.023	0.018	0.000	0.000	0.020	0.018	0.019	0.021
K	0.001	0.000	0.002	0.003	0.000	0.003	0.001	0.003	0.001	0.001
Σcation	7.972	7.979	8.002	8.014	8.001	8.013	7.994	7.998	7.918	7.968
X_{Mg}	0.132	0.130	0.091	0.093	0.110	0.109	0.077	0.086	0.141	0.130

Note: *Fe^{3+} recalculation after Chopin *et al.* (1992), $X_{Mg} = Mg/(Mg + Fe^{2+})$; cation recalculation on 12(O).

than muscovite in chloritoid–biotite schist (7.53 ± 1.66 mol%) (Table 6).

Plagioclase composition that is restricted to psammopelite varies with in An_{19}–An_{29}, although no compositional zoning was noted in a single grain (Table 7).

Thermobarometric calculation

For quantitative temperature estimates of the studied rocks, a number of geo-thermometers (Table 4) were applied to the mineral assemblages in chloritoid–biotite schist and psammopelitic schist in the adjoining localities. Pressure was computed using the assemblage garnet–biotite–plagioclase–quartz (GBPQ: Wu *et al.* 2004) and garnet–muscovite–plagioclase–quartz (GMPQ: Wu & Zhao 2006) in psammopelitic schist. To minimize the problem of kinetic inhibition, estimates were carried out using ferromagnesian minerals that were either touching each other or occurred within 10–15 μm distance of each other. Barring one sample (#H2q/1: Association D) consisting of zoned garnet (Table 4), the compositions of minerals remain virtually constant within individual samples irrespective of the micro-domains (i.e. biotite touching or separated by inert minerals such as quartz). This feature suggests that the kinetic barrier to Fe–Mg diffusion among the ferromagnesian minerals was minimal within the domains of chemical analyses.

The estimated temperature clusters within the range of 490 ± 40 °C (Table 8). The estimates yielded by the garnet–chlorite thermometry in the chloritoid–biotite schist in samples with the prefix #K3CG have not been considered within this cluster because the extreme Fe-rich composition of garnet in these samples is outside the composition range for which the garnet–chlorite thermometers are calibrated (Grambling 1990).

The estimated pressure values fall in narrow spreads of 6.78 ± 1.0 kbar (GBPQ) and 5.77 ± 0.90 kbar (GMPQ) (Table 8). A pressure of about 6.3 ± 1.1 kbar is obtained if all of the pressure estimates from the GBPQ and GMPQ barometers are combined. In view of the foregoing analyses, it is presumed that the metamorphism in the studied areas culminated at about 6.3 ± 1.1 kbar and

Table 3. *Representative analyses of biotite*

Association	A		B		C		D		E	
Sample no.	L5A/1	K3CG/1	K3/1	L5X/1	K5/1	G2/1	H2q/1	H2m/1	16A/1	16A/2
SiO_2	34.95	34.99	35.69	34.86	35.36	37.26	35.76	36.04	37.99	38.66
TiO_2	1.72	1.10	1.15	1.31	0.50	1.97	2.95	1.91	1.71	1.75
Al_2O_3	18.20	18.60	16.86	17.50	19.69	18.31	16.91	17.73	16.74	16.20
Cr_2O_3	0.02	0.10	0.01	0.00	0.02	0.07	0.08	0.05	0.03	0.00
FeO	23.10	23.49	24.36	22.93	22.22	18.44	20.88	19.94	19.33	20.15
MnO	0.02	0.10	0.00	0.04	0.02	0.01	0.01	0.06	0.20	0.12
MgO	7.40	7.82	7.28	8.96	8.30	10.86	8.53	9.61	9.71	9.45
CaO	0.02	0.00	0.01	0.04	0.10	0.16	0.06	0.10	0.40	0.45
Na_2O	0.17	0.01	0.14	0.20	0.42	0.11	0.09	0.13	0.09	0.09
K_2O	9.79	9.47	9.65	9.99	8.92	8.80	9.82	9.52	9.22	8.24
Total	95.40	95.68	95.15	95.83	95.55	95.99	95.09	95.08	95.42	95.11
Si	5.434	5.418	5.590	5.408	5.421	5.559	5.519	5.517	5.748	5.847
Ti	0.201	0.128	0.135	0.153	0.058	0.221	0.342	0.220	0.195	0.199
Al	3.336	3.395	3.113	3.201	3.559	3.221	3.077	3.199	2.986	2.889
Cr	0.002	0.012	0.001	0.000	0.002	0.008	0.010	0.006	0.004	0.000
Fe	3.004	3.042	3.191	2.975	2.849	2.301	2.695	2.553	2.446	2.549
Mn	0.003	0.013	0.000	0.005	0.003	0.001	0.001	0.008	0.026	0.015
Mg	1.715	1.805	1.699	2.072	1.897	2.415	1.962	2.192	2.190	2.130
Ca	0.003	0.000	0.002	0.007	0.016	0.026	0.010	0.017	0.065	0.073
Na	0.052	0.003	0.043	0.060	0.125	0.032	0.027	0.039	0.026	0.026
K	1.942	1.871	1.928	1.977	1.745	1.675	1.933	1.858	1.780	1.590
Cation total	15.692	15.687	15.703	15.857	15.675	15.459	15.576	15.609	15.465	15.318
X_{Mg}	0.363	0.372	0.347	0.410	0.400	0.512	0.421	0.462	0.472	0.455
End members (mol%)*										
Ti–biotite	10.06	6.40	6.77	7.64	2.88	11.06	17.12	10.98	9.73	9.95
Talc–minnesotite	0.32	6.31	1.46	0.00	6.52	14.66	1.98	5.14	9.69	19.18
Muscovite	7.91	7.10	8.97	6.05	8.88	5.54	6.26	6.75	9.65	8.77
Eastonite–siderophyllite	22.69	26.48	17.22	18.34	31.26	27.89	17.25	22.34	17.42	19.24
Wonesite	2.59	0.15	2.13	3.01	6.24	1.59	1.35	1.95	1.32	1.32
Phlogopite–annite	56.44	53.56	63.45	64.96	44.22	39.25	56.05	52.85	52.19	41.54

Note: *End-member recalculation after Holdaway *et al.* (1988); cation recalculation on 22(O).

Table 4. *Representative analyses of garnet*

Association	A				A	B				D				E	
Sample no.	L5A/2		K3CG/2		24C/1a	L5X/3		K3/1		H2 q/1		H2 q/2		16A/1	
	core	rim	core	rim		core	rim	core	rim	core	rim	core	rim	core	rim
SiO_2	37.01	36.91	35.81	36.24	36.26	35.86	35.66	36.84	37.02	37.96	36.50	36.48	37.21	35.82	35.79
TiO_2	0.05	0.03	0.00	0.04	0.01	0.00	0.00	0.01	0.00	0.14	0.06	0.00	0.04	0.07	0.01
Al_2O_3	21.08	21.09	20.90	20.94	20.43	21.30	21.14	20.93	21.09	21.78	21.66	20.74	20.73	20.62	20.84
Cr_2O_3	0.00	0.02	0.04	0.00	0.01	0.04	0.05	0.03	0.00	0.05	0.04	0.06	0.05	0.04	0.00
FeO	33.69	34.41	40.25	41.18	40.75	39.52	40.46	40.16	39.77	30.33	35.30	31.28	31.99	23.37	24.16
MnO	2.13	1.67	0.22	0.13	0.27	0.44	0.50	0.15	0.12	5.30	0.77	4.76	3.60	12.36	11.59
MgO	1.32	1.22	0.84	0.93	1.32	1.27	1.17	1.24	1.20	1.47	1.89	1.45	1.61	1.29	1.36
CaO	4.34	4.00	1.20	0.32	0.58	1.29	1.43	0.29	0.94	3.16	3.33	3.97	3.92	5.59	5.68
Na_2O	0.00	0.01	0.02	0.01	0.00	0.02	0.03	0.00	0.00	0.30	0.14	0.04	0.00	0.05	0.04
K_2O	0.00	0.02	0.02	0.00	0.01	0.00	0.00	0.00	0.00	0.00	0.00	0.00	0.00	0.00	0.00
Total	99.62	99.38	99.30	99.80	99.64	99.75	100.44	99.65	100.15	100.49	99.70	98.78	99.15	99.21	99.47
Si	6.003	6.006	5.925	5.963	5.977	5.888	5.849	6.030	6.023	6.056	5.914	5.982	6.050	5.879	5.858
Ti	0.006	0.004	0.000	0.005	0.001	0.000	0.000	0.001	0.000	0.017	0.008	0.000	0.005	0.009	0.001
Al	4.031	4.046	4.077	4.062	3.970	4.124	4.088	4.039	4.046	4.096	4.137	4.010	3.974	3.990	4.022
Cr	0.000	0.003	0.005	0.000	0.001	0.005	0.006	0.004	0.000	0.007	0.005	0.008	0.006	0.005	0.000
Fe^{2+}	4.569	4.682	5.570	5.667	5.589	5.426	5.550	5.497	5.412	4.047	4.782	4.290	4.330	3.203	3.307
Fe^{3+}*	0.000	0.000	0.000	0.000	0.028	0.000	0.000	0.000	0.000	0.000	0.000	0.000	0.020	0.005	0.000
Fe^{total}	4.569	4.682	5.570	5.667	5.618	5.426	5.550	5.497	5.412	4.047	4.782	4.290	4.350	3.208	3.307
Mn	0.293	0.230	0.031	0.018	0.038	0.062	0.069	0.021	0.017	0.716	0.106	0.661	0.496	1.718	1.607
Mg	0.318	0.296	0.207	0.228	0.324	0.311	0.286	0.304	0.292	0.349	0.457	0.354	0.390	0.316	0.332
Ca	0.754	0.697	0.213	0.057	0.102	0.227	0.251	0.050	0.164	0.540	0.577	0.698	0.683	0.983	0.996
Na	0.000	0.003	0.006	0.003	0.000	0.006	0.010	0.000	0.000	0.093	0.045	0.013	0.000	0.016	0.013
K	0.000	0.004	0.004	0.000	0.002	0.000	0.000	0.000	0.000	0.000	0.000	0.000	0.000	0.000	0.000
ΣCation	15.975	15.970	16.039	16.003	16.034	16.051	16.109	15.947	15.954	15.922	16.030	16.015	15.955	16.123	16.136
X_{Mg}	0.065	0.060	0.036	0.039	0.055	0.054	0.049	0.052	0.051	0.079	0.087	0.076	0.082	0.090	0.091
End members (mol%)															
Almandine	77.00	79.28	92.51	94.92	92.33	90.03	90.14	93.61	91.97	71.60	80.75	71.46	73.40	51.49	52.98
Pyrope	5.36	5.02	3.44	3.82	5.36	5.17	4.65	5.17	4.96	6.17	7.72	5.90	6.61	5.07	5.31
Grossular	12.70	11.81	3.53	0.95	1.69	3.77	4.08	0.85	2.78	9.56	9.75	11.62	11.58	15.81	15.96
Spessartine	4.94	3.89	0.51	0.31	0.62	1.03	1.13	0.36	0.29	12.67	1.79	11.01	8.40	27.63	25.74
Andradite	0.00	0.00	0.00	0.00	0.71	0.00	0.00	0.00	0.00	0.00	0.00	0.00	0.49	0.13	0.00

Note: *Fe^{3+} recalculation after Droop (1987) and cation recalculation on 24(O).

Table 5. *Representative chemical analyses of chlorite*

Association	A			C		D			
Sample no.	L5A/1	L5A/2	K3CG/2	K5/1	K5/2	H2q/2	H2q/3	G2/1	G2/2
SiO_2	22.99	22.83	22.75	22.76	22.98	23.84	25.19	23.83	23.66
TiO_2	0.08	0.13	0.33	0.11	0.02	0.24	0.10	0.07	0.09
Al_2O_3	21.51	22.03	22.49	22.50	23.38	21.94	21.61	22.21	21.84
Cr_2O_3	0.02	0.01	0.00	0.05	0.05	0.02	0.04	0.09	0.04
FeO	30.16	29.73	30.33	29.34	29.50	28.25	25.07	27.88	27.47
MnO	0.06	0.05	0.29	0.08	0.11	0.12	0.20	0.04	0.01
MgO	9.49	9.59	9.68	10.57	9.56	11.11	12.74	12.35	12.25
CaO	0.08	0.02	0.00	0.15	0.08	0.14	0.13	0.01	0.01
Na_2O	0.08	0.04	0.44	0.14	0.12	0.15	0.17	0.01	0.01
K_2O	0.11	0.03	0.02	0.01	0.01	0.12	0.22	0.01	0.00
Total	84.58	84.47	86.33	85.72	85.81	85.93	85.46	86.50	85.38
Si	5.183	5.138	5.021	5.020	5.062	5.212	5.431	5.166	5.191
Ti	0.014	0.022	0.055	0.018	0.003	0.040	0.016	0.011	0.015
Al	5.717	5.845	5.851	5.852	6.073	5.655	5.493	5.676	5.649
Cr	0.004	0.002	0.000	0.009	0.008	0.003	0.007	0.015	0.007
Fe^{2+}	5.508	5.453	5.306	5.212	5.283	4.923	4.300	4.951	4.934
Fe^{3+}*	0.177	0.142	0.288	0.198	0.151	0.241	0.219	0.103	0.105
Fe^{total}	5.685	5.594	5.594	5.410	5.435	5.164	4.519	5.054	5.039
Mn	0.011	0.010	0.055	0.015	0.021	0.023	0.036	0.007	0.002
Mg	3.189	3.217	3.182	3.474	3.140	3.621	4.094	3.990	4.006
Ca	0.019	0.005	0.000	0.036	0.020	0.034	0.030	0.002	0.002
Na	0.035	0.018	0.188	0.060	0.050	0.062	0.071	0.004	0.004
K	0.032	0.009	0.006	0.003	0.003	0.033	0.060	0.003	0.000
Σcation	19.888	19.859	19.951	19.897	19.815	19.846	19.758	19.929	19.916
X_{Mg}	0.367	0.371	0.375	0.400	0.373	0.424	0.488	0.446	0.448

Note: *Fe^{3+} recalculation after Zane *et al.* (1998), $X_{Mg} = Mg/(Mg + Fe^{2+})$; cation calculation on 28(O).

490 ± 40 °C. The presence of kyanite in the adjoining kyanite–quartzite and kyanite–mica schist is consistent with the inferred $P–T$ values for the peak metamorphism in the studied areas.

Mineral reactions in chloritoid–biotite schist

In the absence of definitive mineral inclusion, mineral reactions that stabilized chloritoid in the studied rocks remain unclear. Nevertheless, the following KFMASH ($K_2O–FeO–MgO–Al_2O_3–SiO_2–H_2O$) model reaction seems to be a distinct possibility:

$$\text{Ms} + \text{Chl} \longrightarrow \text{Cld} + \text{Bt} + \text{Qtz} + H_2O. \quad (1)$$

For chlorite-free rock, textural relations described before are consistent with the KFMASH reaction,

$$\text{Cld} + \text{Bt} + \text{Qtz} \longrightarrow \text{Grt} + \text{Ms} + H_2O. \quad (2)$$

Reactions (1) and (2) are consistent with the chemographical projections in the AFM plane (Fig. 5a, b).

Textural features demonstrate that garnet and chlorite were developed after the formation of biotite and chloritoid (Figs 3d, f & 4a–d), which by implication suggests following tie-line switching reaction involving the AFM phases in the system KFMASH (Fig. 5c):

$$\text{Cld} + \text{Bt} \longrightarrow \text{Grt} + \text{Chl}. \quad (3)$$

H_2O is always placed on the right-hand side of reaction (3) – that is, with garnet + chlorite – except in the balanced reaction given by Labotka 1980 (Table 9). Mineral compositions of #K3CG/1 have been used to balance the reaction as non-KFMASH components are minimal in the rock (Table 9). Ignoring the small concentrations of the non-KFMASH components, the balanced reaction places H_2O on the same side as garnet and chlorite:

$$10.58\text{Cld} + 4.81\text{Bt} + 9.21\text{Qtz} = 3.97\text{Grt} + 2.40\text{Chl} + 4.81\text{Ms} + H_2O. \quad (3)$$

The implication of this reaction is discussed in the following section.

Table 6. *Representative analyses of muscovite*

Association	A				B			C	D				E	
Sample no.	L5A/1	L5A/2	K3CG/1	K3CG/1	L5X/1	K3/1	K3/2	K5/2	G2/1	G2/4	H2q/1	H2m/1	16A/1	16A/1
SiO_2	47.82	48.02	46.86	47.26	46.54	47.93	47.51	46.71	47.43	47.37	47.33	46.85	47.66	46.89
TiO_2	0.27	0.28	0.08	0.24	0.28	0.41	0.24	0.03	0.20	0.24	0.30	0.55	0.54	0.70
Al_2O_3	34.16	33.93	35.35	34.26	35.34	33.00	33.62	35.57	34.25	34.22	34.62	33.94	32.10	31.90
Cr_2O_3	0.02	0.00	0.06	0.02	0.04	0.02	0.04	0.02	0.02	0.00	0.06	0.06	0.05	0.00
FeO	2.32	2.41	1.76	2.77	3.22	3.02	2.94	2.34	2.95	2.50	2.32	2.69	4.44	4.46
MnO	0.03	0.02	0.04	0.00	0.00	0.07	0.03	0.07	0.05	0.01	0.01	0.04	0.04	0.03
MgO	0.73	0.67	0.92	0.74	0.80	1.16	0.93	0.70	0.68	1.23	0.96	0.99	1.24	1.18
CaO	0.03	0.00	0.03	0.02	0.01	0.03	0.02	0.04	0.19	0.03	0.01	0.09	0.02	0.04
Na_2O	0.79	0.80	0.50	0.47	0.54	0.57	0.53	0.50	1.06	1.28	1.20	1.02	0.41	0.57
K_2O	10.49	10.51	10.49	10.40	9.69	10.08	10.16	10.18	9.04	9.26	9.46	9.86	9.25	10.19
Total	96.66	96.64	96.09	96.18	96.46	96.29	96.02	96.16	95.87	96.14	96.27	96.09	95.75	95.96
Si	6.299	6.327	6.191	6.263	6.144	6.345	6.305	6.173	6.276	6.251	6.235	6.218	6.360	6.296
Ti	0.027	0.028	0.008	0.024	0.028	0.041	0.024	0.003	0.020	0.024	0.030	0.055	0.054	0.071
Al	5.305	5.270	5.506	5.352	5.500	5.150	5.260	5.542	5.343	5.323	5.377	5.311	5.050	5.049
Cr	0.002	0.000	0.006	0.002	0.004	0.002	0.004	0.002	0.002	0.000	0.006	0.006	0.005	0.000
Fe	0.256	0.266	0.195	0.307	0.356	0.334	0.326	0.259	0.326	0.276	0.256	0.299	0.495	0.501
Mn	0.003	0.002	0.004	0.000	0.000	0.008	0.003	0.008	0.006	0.001	0.001	0.004	0.005	0.003
Mg	0.143	0.132	0.180	0.146	0.157	0.229	0.184	0.138	0.134	0.242	0.188	0.196	0.247	0.236
Ca	0.004	0.000	0.004	0.003	0.001	0.004	0.003	0.006	0.027	0.004	0.001	0.013	0.003	0.006
Na	0.202	0.204	0.129	0.121	0.138	0.146	0.136	0.128	0.272	0.327	0.307	0.263	0.106	0.148
K	1.763	1.767	1.769	1.758	1.632	1.702	1.720	1.716	1.526	1.559	1.590	1.670	1.575	1.746
Σcation	14.003	13.996	13.993	13.976	13.961	13.962	13.967	13.974	13.931	14.007	13.992	14.034	13.899	14.056
End members (mol%)*														
Ti–muscovite	1.34	1.39	0.40	1.20	1.39	2.04	1.20	0.15	1.00	1.19	1.49	2.75	2.71	3.53
Phlogopite–annite	1.46	1.11	4.03	4.59	9.26	4.97	4.99	5.71	4.93	5.77	4.31	3.93	10.28	7.64
Celadonite	14.24	15.13	6.26	7.68	0.00	11.21	9.35	2.54	7.24	7.38	7.79	10.20	3.55	10.40
Pyrophyllite	1.24	1.34	4.20	5.75	9.35	6.80	6.54	6.94	8.32	5.41	4.58	2.06	15.20	4.85
Paragonite	10.09	10.22	6.45	6.04	6.91	7.32	6.82	6.41	13.60	16.37	15.33	13.13	5.30	7.42
Margarite	0.21	0.00	0.21	0.14	0.07	0.21	0.14	0.28	1.35	0.21	0.07	0.64	0.14	0.29
Muscovite	71.43	70.82	78.44	74.60	73.03	67.45	70.96	77.97	63.57	63.66	66.44	67.31	62.82	65.88

Note: *End-member recalculation after Holdaway *et al.* (1988); cation recalculation on 22(O).

Table 7. *Representative analyses of plagioclase (Pl) and ilmenite (Ilm)*

Mineral	Pl						Ilm					
Association	D				C		A		B	D		E
Sample no.	G2/1	G2/2	H2q/1	H2q/2	16A/1	16A/2	L5A/1	24C/1 (incl-Cld)	K3/1 (incl-Grt)	G2/1	H2q/1	16A/1
SiO_2	62.16	61.59	62.94	62.26	62.98	62.50	0.02	0.04	0.04	0.13	0.03	0.05
TiO_2	0.00	0.00	0.00	0.02	0.00	0.00	54.00	55.67	52.23	54.74	52.81	53.94
Al_2O_3	23.51	23.90	23.41	23.63	23.31	24.11	0.00	0.05	0.05	0.01	0.00	0.04
Cr_2O_3	0.03	0.02	0.00	0.00	0.00	0.00	0.01	0.06	0.05	0.05	0.06	0.00
FeO	0.15	0.19	0.36	0.08	0.10	0.07	44.97	43.92	47.05	44.24	45.08	41.82
MnO	0.00	0.02	0.00	0.00	0.04	0.00	0.31	0.06	0.05	0.40	1.39	3.81
MgO	0.00	0.00	0.00	0.01	0.00	0.00	0.06	0.13	0.19	0.10	0.01	0.00
CaO	4.22	5.20	4.75	4.84	4.53	5.71	0.01	0.00	0.01	0.00	0.02	0.00
Na_2O	9.92	9.24	9.13	8.90	8.81	7.65	0.04	0.02	0.15	0.02	0.01	0.02
K_2O	0.06	0.05	0.05	0.11	0.07	0.06	0.06	0.00	0.00	0.00	0.01	0.11
Total	100.05	100.21	100.64	99.85	99.84	100.10	99.49	99.96	99.82	99.68	99.43	99.80
Si	2.759	2.734	2.773	2.762	2.788	2.758	0.001	0.001	0.001	0.003	0.001	0.001
Ti	0.000	0.000	0.000	0.001	0.000	0.000	1.021	1.038	0.993	1.028	1.005	1.018
Al	1.230	1.251	1.216	1.236	1.216	1.254	0.000	0.002	0.001	0.000	0.000	0.001
Cr	0.001	0.001	0.000	0.000	0.000	0.000	0.000	0.001	0.001	0.001	0.001	0.000
Fe	0.006	0.007	0.013	0.003	0.004	0.003	0.946	0.911	0.995	0.924	0.954	0.877
Mn	0.000	0.001	0.000	0.000	0.001	0.000	0.007	0.001	0.001	0.008	0.030	0.081
Mg	0.000	0.000	0.000	0.001	0.000	0.000	0.002	0.005	0.007	0.004	0.000	0.000
Ca	0.201	0.247	0.224	0.230	0.215	0.270	0.000	0.000	0.000	0.000	0.001	0.000
Na	0.854	0.795	0.780	0.766	0.756	0.655	0.002	0.001	0.007	0.001	0.001	0.001
K	0.003	0.003	0.003	0.006	0.004	0.003	0.002	0.000	0.000	0.000	0.000	0.004
Σcation	5.054	5.039	5.010	5.005	4.984	4.943	1.980	1.960	2.008	1.969	1.994	1.983
X_{Ab}	0.810	0.763	0.777	0.769	0.779	0.708						
X_{An}	0.190	0.237	0.223	0.231	0.221	0.292						

Note: Cation recalculation of Pl and Ilm, respectively, on 8(O) and 3(O); incl: included within porphyroblast.

Table 8. *Temperature (°C) and pressure (kbar) estimated from representative mineral compositions*

Sample no.*	Analysis no.	T^1 (°C)	T^2 (°C)	T^3 (°C)	T^4 (°C)	T^5 (°C)	T^6 (°C)	T^7 (°C)	Sample no.†	Analysis no.	T^1 (°C)	T^2 (°C)	T^3 (°C)	T^4 (°C)	T^5 (°C)	T^6 (°C)	T^7 (°C)	P^1 (kbar)	P^2 (kbar)
L5A/1	1 rim	510	511	530	506	484	489	480	H2 q/1	1 rim	514	522	530	506	466	474	464	7.51	4.98
	1 core	530	516	542	517	519	514	509		1 core	458	460	490	482	435	448	438	6.04	3.44
L5A/2	1 rim	498	514	533	503	487	491	482	H2 q/3	2 rim	502	513	522	558	533	524	521	7.90	5.90
	1 core	514	516	540	513	509	507	501		2 core	476	480	503	552	530	522	519	5.11	6.53
K3CG/	1 rim	472	488	475	495	408	423	415	H2 q/4	1 rim	494	490	534	530	518	514	508	7.51	6.26
1	1 core	482	495	488	487	389	405	399		1 core	452	463	490	509	490	493	485	5.93	5.84
K3CG/	1 rim	455	477	480	481	417	432	422	H2 q/5	3 rim	483	490	510	519	511	509	503	7.26	4.09
2	1 core	474	489	497	484	424	438	428		3 core	440	453	485	496	461	470	460	4.72	6.47
K3CG/	2 rim	459	483	486	470	391	407	400	H2m/1	1 rim	554	538	593	546	548	534	533	7.90	6.53
3	2 core	506	509	513	502	426	440	430		1 core	501	513	558	513	463	472	462	7.92	7.01
24C/1	1 rim	478	490	488	490	439	452	442	H2m/2	4 rim	531	536	588	542	525	519	514	7.67	6.43
	1 core	500	506	503	503	471	478	469		4 core	453	470	530	498	389	405	399	6.49	5.96
24C/1a	2 rim	482	493	488	474	453	464	454	G2/1	1 rim	487	495	516	479	480	475	508	6.46	6.49
	2 core	494	502	493	479	459	469	459		1 core	482	490	518	476	493	496	488	6.47	6.31
L5X/1	1 rim	443	468	468	500				G2/1a	2 rim	480	475	503	475	492	495	486	5.09	5.89
	1 core	449	475	476	505					2 core	485	480	510	480	480	465	490	5.26	6.04
L5X/2	1 rim	428	459	455	477				G2/3	2 rim	487	492	513	465	500	470	478	5.77	6.01
	1 core	417	450	444	470					2 core	483	490	512	463	496	470	475	5.96	6.00
L5X/3	2 rim	413	448	445	489				G2/5	4 rim	475	486	498	470	465	474	464	6.53	4.97
	2 core	427	462	460	499					4 core	498	512	522	485	472	480	465	6.27	3.77
K3/1	1 rim	462	482	477	446				16A/1	1 rim	463	475	495	495				7.37	5.33
	1 core	481	496	494	457					1 core	464	479	472	484				7.18	5.66
K3/2	2 rim	466	485	475	474				16A/2	2 rim	469	480	480	487				7.98	6.59
	2 core	479	495	490	480					2 core	461	470	476	489				8.46	6.09
K3/3	1 rim	456	472	474	469				16A/4	1 rim	469	480	477	527				7.56	5.66
	1 core	490	498	491	470					1core	469	480	492	527				7.37	5.67
Average		472	488	489	486	448	458	449			482	489	512	502	487	485	483	6.76	5.77
Standard deviation		30	20	26	18	42	35	36			25	22	31	28	37	30	32	1.05	0.88

Note: *Chloritoid–biotite schist; †Psammopelitic schist; T^{1-3}, Grt–Bt thermometers; T^1, Ganguly & Saxena (1984); T^2, Perchuk & Lavrentieva (1983); T^3, Holdaway (2000); T^4, Grt–Ms thermometer (Hynes & Forest 1988); T^{5-7}, Grt–Chl thermometers; T^5, Dickenson & Hewitt (1986); T^6, Grambling (1990); T^7, Ghent *et al.* (1987); P^1, Grt–Bt–Pl–Qtz barometer (Wu *et al.* 2004); P^2, Grt–Ms–Pl--Qtz barometer (Wu & Zhao 2006).

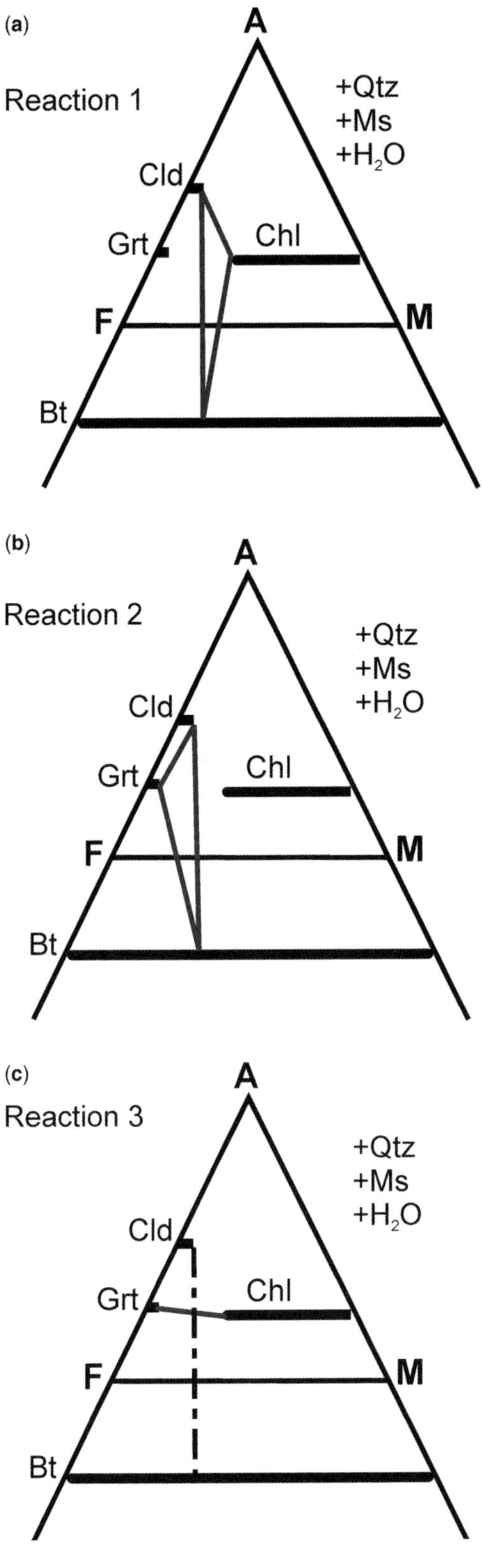

Fig. 5. Chemographic projections of the silicate phases representing different mineral reactions in chloritoid–biotite schist: (**a**) Ms + Chl → Cld + Bt + Qtz + H_2O – (1); (**b**) Cld + Bt + Qtz → Grt + Ms + H_2O – (2); and (**c**) Cld + Bt + Qtz → Grt + Chl + Ms + H_2O – (3).

Stability of chloritoid + biotite v. garnet + chlorite

Petrogenetic grid in the KFMASH system

In the six-component KFMASH system, the phases garnet–chloritoid–aluminosilicates–staurolite–biotite–chlorite–quartz–muscovite–H_2O ($P = 9$) has a variance of -1. If the three minerals – namely, quartz, muscovite and H_2O – are considered to be omnipresent, there will be a maximum of six invariant points [Cld], [Grt], [Als], [St], [Bt] and [Chl]. The mineral assemblages of the univariant reactions related to the [Als] invariant point are widespread in medium-grade metapelites. Therefore, understanding the topological relations around this invariant point is crucial in interpreting the observed mineralogy and their mutual relations in the low- to medium-grade metapelites. Different workers suggested contrasting topological phase relations around the [Als] invariant point (Albee 1965; Harte & Hudson 1979; Labotka 1981; Spear & Cheney 1989; Powell & Holland 1990; Wang & Spear 1991; Droop & Harte 1995). The nature and orientation of the (Als, St) univariant reaction that involves the four AFM phases garnet, chlorite, chloritoid and biotite is the main source of dispute between different workers.

The published topology can be divided into two basic types. Figure 6 describes the contrasting topological relations around the [Als] invariant point. The topology presented in Figure 6a was initially proposed by Albee (1965, 1972) from the reaction textures of a suite of metapelites, and was subsequently supported by Harte & Hudson (1979) and Droop & Harte (1995). In this topological arrangement, the univariant reaction (Als, St) has a positive slope, and the restricted P–T stability field with chloritoid + biotite is placed on the low-pressure side of the reaction (Als, St) below the [Als] invariant point (*c*. 6 kbar, Fig. 6a). The internally consistent thermodynamic database of Powell & Holland (1990) does not stabilize the [Als] invariant point in the KFMASH system. However, if Fe_2O_3 is incorporated into this system, the [Als] invariant point stabilizes and the observed topological relations are similar to that of Harte & Hudson (1979) and Droop & Harte (1995).

Using the internally consistent thermodynamic database, Spear & Cheney (1989) proposed a drastically different topology around the [Als] invariant

Table 9. *Balanced (Als, St) KFMASH reactions deduced from the chemical composition of the mineral phases in different areas*

Area	Balanced reaction	Source
South Brittany, France	10Bt + 23.8Cld + 17.85Qtz = 8.6Grt + 2.73Chl + 10Ms + 1.96H_2O	Triboulet (1983)
Tri-state, USA	1.82Bt + 4.04Cld + 10Qtz = 2.85Grt + 0.70Chl + 3.43Ms + 0.88H_2O	Wang & Spear (1991)
Tauern Window	(i) 1.71Bt + 4.40Cld + 3.8Qtz = 1.9Grt + Chl + 1.71Ms + 0.40H_2O	Droop & Harte (1995)
	(ii) 1.56Bt + 4.20Cld + 3.88Qtz = 1.86Grt + Chl + 1.56Ms + 0.20H_2O	
Vermont, USA	1.32Bt + 5.21Cld + 3.95Qtz = 1.39Grt + Chl + 1.32Ms + 1.21H_2O	Albee (1965)
Funeral Mt, California	1.63Bt + 1.79Cld + 2.88Qtz + 0.42H_2O = 1.11Grt + Chl + 1.63Ms	Labotka (1980)
Kanyaluka, SSZ	4.81Bt + 10.58Cld + 9.21Qtz = 3.97Grt + 2.40Chl + 4.81Ms + H_2O	Present study

point (Fig. 6b). They showed that because of the strong and opposite curvature of the univariant reactions (Als, St) and (Als, Cld), two [Als] invariant points stabilize, one at high pressure (14.8 kbar, 580 °C) and another at low pressure (6.5 kbar, 545 °C). An important feature of this grid is that the (Als, St) reaction is relatively insensitive to pressure and the garnet + chlorite is stabilized at a higher temperature compared to chloritoid + biotite (Fig. 6b).

In favour of the topological relationships proposed by the Spear & Cheney (1989) topology around the [Als] invariant point, Wang & Spear (1991) presented the following arguments.

- The appearance of chloritoid + biotite in metapelites at pressures as high as 7 kbar in SE Tauern Window, Austria can be explained using the grid of Spear & Chenny (1989), but is inconsistent with the grid proposed by Harte & Hudson (1979). This is because the latter allows only low-pressure (<6 kbar) stability of the chloritoid + biotite-bearing assemblage in the system KFMASH.
- The mass-balance calculations by least-squares method using the phases of the reaction (Als, St) in the Mn–Ca–Na-poor rocks show H_2O to be present with garnet + chlorite in most places (Table 9). This is consistent with higher entropy and, therefore, high-temperature stability of garnet + chlorite compared to chloritoid + biotite.
- The textural features and the phase compositional attributes of the four-phase assemblages (garnet–chlorite–chloritoid–biotite) from a Barrovian sequence of a tri-state area in the

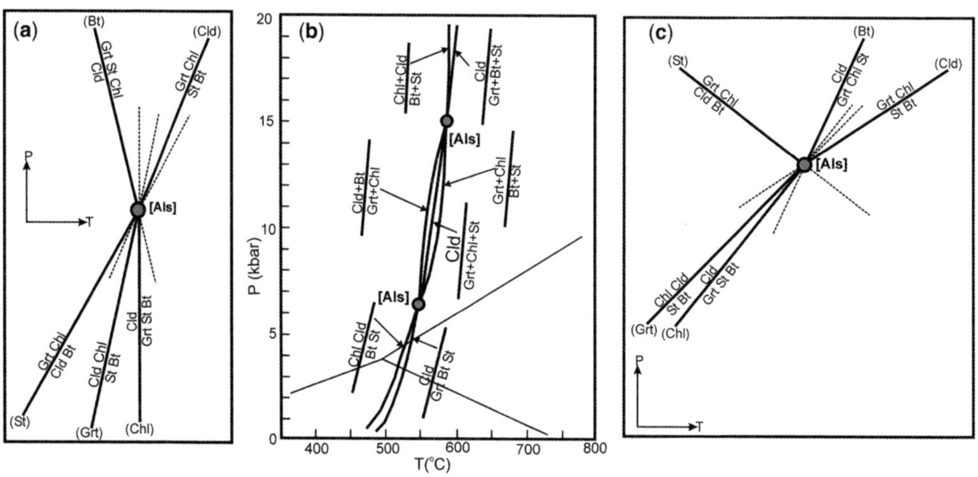

Fig. 6. Geometrical arrangements of the univariant reactions around the aluminosilicate-absent [Als] invariant point in the system KFMASH. (**a**) Wang & Spear (1991), (**b**) Harte & Hudson (1979) and (**c**) present study. Note that the arrangements depicted in both (b) and (c) are similar, and both satisfy the Schreinemakers' principals.

Taconic Range (Wang & Spear 1991) show the low-temperature stability of chloritoid + biotite compared to garnet + chlorite in Mn-poor rocks. This is consistent with the topology presented by Spear & Cheney (1989) but inconsistent with that of Harte & Hudson (1979) (Fig. 6a, b). However, the enlarged stability field of garnet + chlorite for the Mn-rich bulk composition explains the stability of garnet + chlorite at lower temperature rather than that of chloritoid + biotite (Wang & Spear 1991). The enlargement of the stability field of garnet-bearing assemblages was first demonstrated by Ganguly (1968), which was further substantiated by Loomis (1986)

In support of the topological relation of Harte & Hudson (1979), Droop & Harte (1995) presented the following arguments.

- In most of the natural rocks, chloritoid + biotite are found to stabilize at or below 7 kbar pressure. At higher pressures, garnet + chlorite are the dominant minerals at temperature below the staurolite-in isograd. This is consistent with the topology of Harte & Hudson (1979), which predicts stability of chloritoid + biotite below $P[_{\mathrm{Als}}]$ of around 7 kbar, but is inconsistent with the arrangements proposed by Spear & Cheney (1989), which predicts chloritoid + biotite to occur over a large range of pressures (Fig. 6b).
- Where present, the zone containing the garnet + chlorite zone is generally found to develop at lower grade relative to zones with chloritoid + biotite. This observation can be readily explained by the geometrical arrangements of Harte & Hudson (1979) but not by those of Spear & Cheney (1989). Furthermore, Droop & Harte (1995) cited the crucial observation of Labotka (1980) to contradict Wang & Spear (1991). Labotka (1980) described a garnet + chlorite assemblage close to the garnet-in isograd in Mn-poor metapelites from the Funeral Mountain, California, where garnet + chlorite assemblage gave away directly to staurolite-bearing assemblages without any intermediate chloritoid + biotite zone.
- The chloritoid + biotite zone is generally absent from the typical Barrovian sequences. When present, chloritoid + biotite-bearing assemblages develop only in aluminous rocks that plot in the high-Al side of the critical garnet–chlorite join in AFM plots (Droop & Harte 1995). This observation is inconsistent with the arrangements presented by Spear & Cheney (1989) because this topological arrangement cannot explain any bulk compositional restriction for the stability of chloritoid + biotite in low-grade metapelites.

A careful scrutiny of the published literature reveals the following points.

- Table 9 presents the stoichiometrically balanced reaction involving Cld, Bt, Grt and Chl in rocks with bulk compositions effectively represented by the KFMASH system. It is clear from Table 9 that H_2O, in most places, coexists with garnet + chlorite, which suggests that chloritoid + biotite is present in the lower temperature of the reaction. This feature is in agreement with the topology of Spear & Chenny (1989). But in the reaction provided by Labotka (1980), H_2O is placed with chloritoid + biotite. A careful scrutiny of his data reveals that these four phases never coexist in one sample. It is, therefore, not clear whether the compositions of minerals used to balance the reaction are of equilibrium assemblage or not.
- Table 10 presents the $P–T$ conditions of metamorphism and composition of garnet (if present) in low- to medium-grade metapelites containing chloritoid + biotite. It is evident from Table 10 that chloritoid + biotite-bearing assemblages do not occur at pressures much above 7 kbar. This feature is corroborated by the topology of Droop & Harte (1995).
- Natural examples presented in Table 10 demonstrate that garnet + chlorite appears early in a progressive metamorphic sequence, only when the bulk composition of the rock is rich in Ca and Mn (reflected in the garnet composition). This is consistent with the observation of Wang & Spear (1991). However, the absence of chloritoid + biotite in one Ca–Mn-poor metapelite occurrence described by Labotka (1981) merits discussion at this stage. From the muscovite + plagioclase-bearing assemblages, Labotka (1981) calculated a range of X_{H_2O} values (0.32–0.83) for pelitic rocks, which indicates a significant dilution of H_2O by non-hydrous fluid species. If the concentration of H_2O is diluted, then garnet + chlorite will be preferred over chloritoid + biotite, as H_2O is present with garnet + chlorite in the stoichiometrically balanced reaction (Table 9). This possibly explains the non-appearance of chloritoid + biotite in the Ca–Mn-poor and andalusite-bearing rocks of Paramint Mountain, California (Labotka 1981). This then follows that garnet + chlorite will develop prior to chloritoid + biotite if the rock is rich in Ca and Mn, and/or equilibrates with fluids other than pure H_2O. This contention is supported by stability of chloritoid + biotite in the Ca–Mn-poor metapelites and the absence of this assemblage in the Ca–Mn-rich psammopelites of the SSZ.

Table 10. *The occurrence of the assemblage Cld + Bt + Grt + Chl*

Locality	Stable Al-silicate	Garnet $X_{Ca} + X_{Mn}$	P (kbar)	T (°C)	Source	Remarks
Dalradian, Scotland	–	–	*c.* 7.5	550–600	Baltatzis & Wood (1977)	+Ab ±Pg
Naxos, Greece*	Ky, Sill	–	7–9	490–530	Andriessen *et al.* (1979)	+Gln +Ep +Ab +Pg, no Grt
South Brittany, France	–	0.22	6 ± 0.5	500–540	Triboulet (1983)	
Hida terrane, Central Japan	–	0.16–0.18	5	540	Hiroi (1983)	+St
Yilgarn Block, Australia	And	0.07–0.16	4.2	530–560	Bickle & Archibald (1984)	+St ±Crd ±And ±Pl
Mt Moosilauke, New Hampshire, USA	And	0.08	3.5–4	520	Dickenson (1988)	
Eastern Rhodope Zone, northern Greece	Sill	0.09	11	550	Mposkos (1989)	+St +Pg
Truchas Range, New Mexico, USA	Ky, And, Sill	0.06	4	500–550	Grambling (1983)	+St +Rt
Tri-state, USA	Ky	0.03	4.4–4.8	490–530	Wang & Spear (1991)	
Stonehaven, Scotland	–	–	*c.* 5	500 ± 25	Droop & Harte (1995)	
SE Tauern Window, Austria	Ky	0.15–0.22	7 ± 1	475 ± 25	Droop & Harte (1995)	
Dutches County, NY, USA	Ky	0.17	4	475–480	Whitney *et al.* (1996)	+Pl
Hlinsko region, Bohemian massif, Czech Republic	And	0.22–0.28	4	530–570	Pitra & Guiraud (1996)	+St +And
Island of Kos, Greece*	–	0.03	2	600	Kalt *et al.* (1998)	+Crd +St +Pl, no Chl
Moldanubicum, Bohemian massif, Czech Republic*	And	–	4.2 ± 1.9	<500	Babûrek (1998)	No Chl and Grt
Ayakhtinsk aureole, Siberia*	And		3.2 ± 0.3	450	Likhanov *et al.* (2001)	No Chl and Grt
Mungyong coalfield, central South Korea*	And	–	1–2	–	Ahn & Nakamura (2000)	+And, no Grt
Bushveld Complex, South Africa*	And	–	3.5	<450	Waters & Lovegrove (2002)	+St +Crd, no Grt
The SSZ, eastern India	Ky	0.01–0.17	6.5 ± 0.5	480 ± 40	Present study	

Note: *Indicates the absence of chlorite and/or garnet; the localities written in bold font record the assemblage stable without staurolite.

It is, therefore, evident from the foregoing analyses that none of the existing grids can explain all of the following features:

- the restriction of chloritoid + biotite at pressures below 7 kbar;
- the formation of chloritoid + biotite prior to garnet + chlorite in Ca–Mn-poor rocks of the studied area;
- the occurrence of H_2O on the same side as garnet + chlorite in #K3CG/1 where the concentration of non-KFMASH components are minimal.

With this backdrop, a new petrogenetic grid in parts of the KFMASH system has been developed integrating the extant petrological, phase chemical and experimental data (Figs 6c & 7). In this grid, the KFASH invariant point [Bt, Chl] has been considered to be stable together with [Bt, Grt], [Cld, Grt], [Cld, Chl], [Grt, Als] and [Chl, Als]. Slopes of most of the univariant KFASH reactions are taken from the experimental data of Ganguly (1972), and from Harte & Hudson (1979) and Spear & Cheney (1989) (Fig. 7). The slopes of the rest (marked with a star in Fig. 7) have been construed from mutual topological relationships with the other reactions. Owing to the large uncertainties of the thermodynamic data of the phases, the KFMASH topology was derived from the KFASH reaction nets following Schreinemaker's principals (Fig. 7). This approach stabilized three KFMASH invariant points namely, [Cld], [St] and [Bt]. The controversial univariant reaction (Als, St) assumes a moderate negative slope with chloritoid + biotite on the low-pressure side (Figs 6c & 7). Weak to strong negative P–T slope is not uncommon for dehydration reactions above about 5 kbar (fig. 3 of Ganguly 1972; fig. 9.12 of Spear 1993).

Barring the slopes of the KFMASH univariant reactions (Als, St) and (Als, Bt), topology of the proposed grid is in agreement with the topology predicted by Harte & Hudson (1979) (Fig. 6a). Both the topologies are, however, correct on the basis of Schreinemaker's principals. The new grid depicted in Figure 7 can explain: (a) the restriction of chloritoid + biotite at lower pressures and

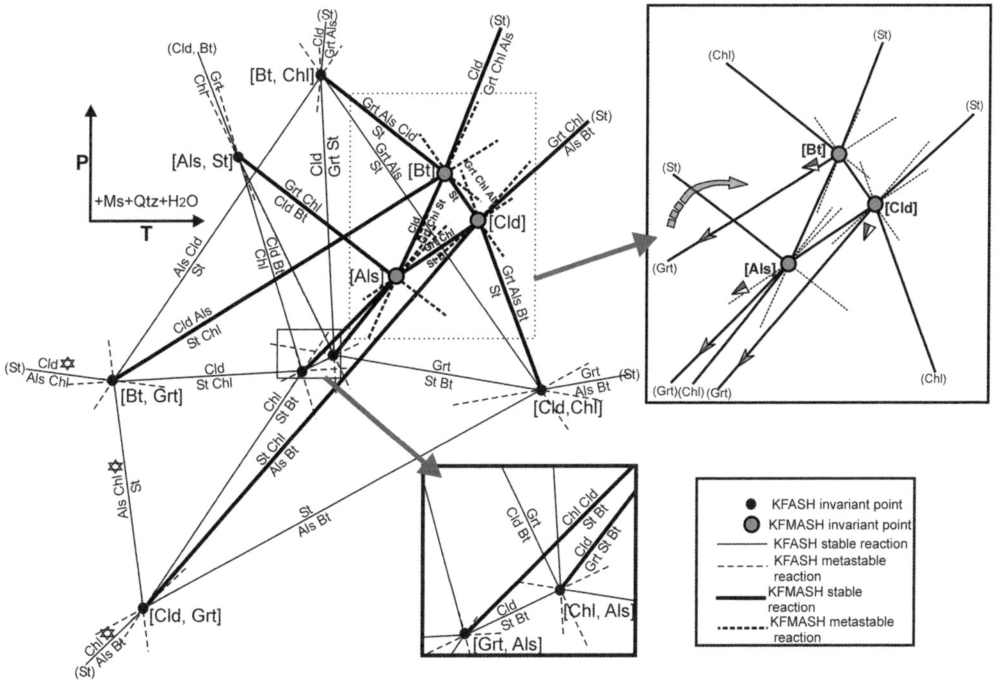

Fig. 7. Schematic partial petrogenetic grid in the KFASH (thin line) and KFMASH (thick lines) systems involving garnet, staurolite, chloritoid, biotite, chlorite, muscovite, aluminosilicate, quartz and H_2O. Note that the (Als, St) reaction has a negative slope with chloritoid + biotite on the low-P, low-T side of the reaction. The inset in the NE corner of the figure shows the likely P–T path of the studied rocks. The arrowheads depict the direction along which the KFMASH invariant points will shift due to the addition of Mn (filled arrowhead) and the reduced activity of H_2O (half-filled arrowhead).

temperatures compared to garnet + chlorite; and (b) the formation of garnet + chlorite from chloritoid + biotite during the progressive metamorphism of Mn-poor metapelites, as is documented in this study. Garnet is a strong sink for Mn (and also for Ca). Therefore, in Mn-rich metapelite, the KFMASH invariant points [Cld], [Als] and [Bt] will be displaced towards lower $P-T$ conditions along their corresponding garnet-absent univariant reactions (Fig. 7). The effect of the reduction of H_2O will be the same as the effect of Mn in bulk rock composition (Fig. 7). The net effect of an increase in Mn and a reduction of H_2O will be to reduce the $P-T$ field of chloritoid + biotite and to restrict the assemblage at very low pressure (Fig. 7). This probably explains the rarity of the chloritoid + biotite and preponderance of garnet + chlorite in low- to medium-grade metapelites. The influences of the non-KFMASH components also explain the early development of chloritoid + biotite in Mn–Ca-poor metapelite and garnet + chlorite in Mn–Ca-rich psammopelitic schist in the SSZ.

Pseudosections in parts of the Na_2O–CaO–MnO–K_2O–FeO–MgO–Al_2O_3–SiO_2–H_2O–O (NCMnKFMASHO) system

With the availability of an internally consistent thermodynamic dataset and the activity–composition relationships of solid-solution phases, it is now possible to compute phase diagrams numerically that portray mineralogical changes in a given bulk composition in response to the changing pressure, temperature and chemical potentials of H_2O and O_2. For this purpose, the computer program PERPLE_X 07 (Connolly 2005) is used along with the internally consistent thermodynamic dataset of Holland & Powell (1998, updated in 2004). Activity–composition relationships used for different solid-solution phases are: garnet (Ganguly *et al.* 1996), staurolite (Holland & Powell 1998), chlorite (Holland *et al.* 1998), biotite and muscovite (Powell & Holland 1999), and plagioclase (Newton *et al.* 1980). Use of the garnet solution model of Holland & Powell (1998) does not change the topology of the calculated phase diagrams.

Considering the bulk compositions and the mineralogy of chloritoid–biotite schist (bulk compositions were determined using X-ray fluorescence (XRF) at the University of Bonn, Germany), modelling was performed on the NCMnKFMASHO system. The value of H_2O was arbitrarily fixed at 3 wt% for the construction of $P-T$ pseudosections; an enhanced H_2O value does not change the overall topological relationships. TiO_2 was excluded from the system as it was effectively fixed in ilmenite and occurs only in very small concentrations in phyllosilicates (Tables 3, 6 & 7). It was expected that discarding TiO_2 from the system component would not have any major impact on the relative phase relations (Wei & Powell 2004; Jerabek *et al.* 2008). For the $P-T$ pseudosections, the bulk composition of two samples of chloritoid–biotite schist, viz. #L5A/1 and #K3CG/1, were chosen (Fig. 8). In phase diagrams, both of the samples were seen to develop the critical assemblage garnet–chlorite–chloritoid–biotite at a pressure above 6 kbar and at temperatures of between 450–525 and 450–550 °C in #L5A/1 and #K3CG/1, respectively (Fig. 8). These $P-T$ values are in good agreement with the $P-T$ values of approximately 6.3 ± 1 kbar and 490 ± 40 °C estimated using quantitative geothermobarometry. It is intriguing to note that the assemblage in which garnet–chlorite–chloritoid–biotite are stable forms a narrow band with a conspicuous negative $P-T$ slope. In both of the $P-T$ pseudosections, chloritoid + biotite- and garnet + chlorite-bearing pentavariant assemblages occur on the low- and high-pressure sides of this quadrivariant band, respectively (Fig. 8a, b). The dispositions of these quadri- and pentavariant assemblages are in good agreement with the negative slope of the KFMASH univariant reaction (Als, St), with garnet + chlorite on the high-pressure side of the univariant reaction (Fig. 7).

To demonstrate the control of bulk Mn, μ_{H_2O} and μ_{O_2} in metamorphic fluid on the stability of the critical assemblage garnet–chlorite–chloritoid–biotite, the following diagrams are computed using the bulk compositions of sample #K3CG/1 as the monitor:

- isothermal–isobaric μ_{H_2O} v. μ_{O_2} diagrams at different temperatures (Fig. 9b–d);
- T (°C) v. MnO diagram, where the temperature was varied along a geothermal gradient (24 °C km^{-1}) that passes through the $P-T$ box (*c.* 6.3 ± 1 kbar and 490 ± 40 °C) computed for the peak pressure and temperature of the area (Fig. 9a). To construct this diagram the concentration of MnO was varied from 0.00 to 1.00 wt%, keeping the concentration of other oxides in #K3CG/1 constant.

In the T (°C) v. MnO diagram, the garnet–chlorite–chloritoid–biotite-bearing assemblage has a wide stability field between 450 and 500 °C, covering the entire range of MnO (Fig. 9a). The stability field of the assemblage shrinks and is replaced with garnet–chloritoid–biotite at higher levels of MnO (>0.50 wt%) (Fig. 9a). This four-phase assemblage, therefore, stabilizes at a higher temperature range (475–500 °C) in rocks with a MnO-rich bulk composition (Fig. 9a). This feature also explains the relatively higher thermometric value yielded

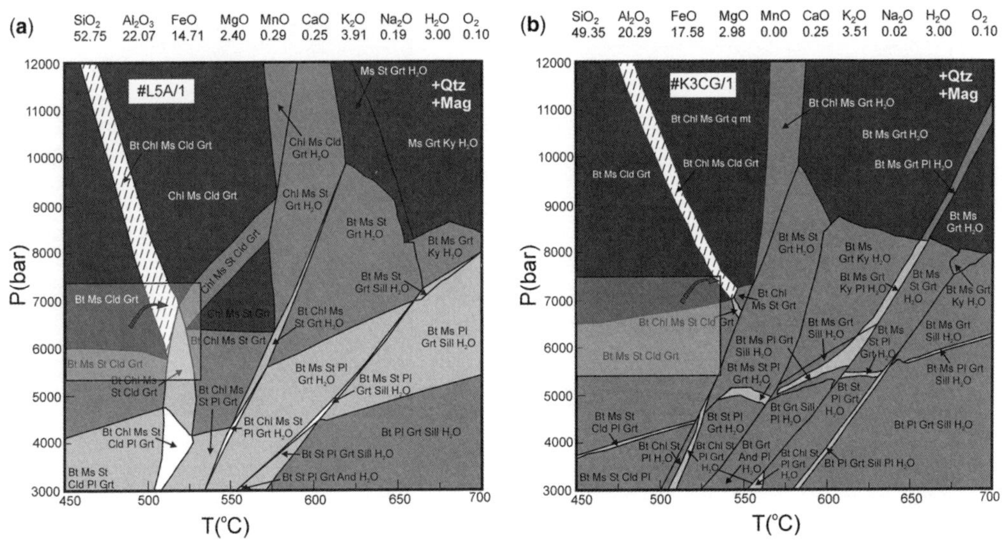

Fig. 8. P–T pseudosections constructed from bulk chemical compositions of two representative samples of chloritoid–biotite schists in the NCMnKFMASHO system. Bulk compositions are given above the figures. The transparent rectangle represents the P–T range estimated from quantitative geothermobarometry. The thick line with an arrowhead represents the likely P–T path that was followed by the chloritoid–biotite schists.

by the Mn-rich sample # L5A/1 than the Mn-impoverished sample # K3CG/1 (Table 8). The phase relation in the T–MnO diagram, therefore, explains the stabilization of the critical four-phase assemblage in #K3CG/1 within the range of estimated peak P–T conditions of 6.3 ± 1 kbar and 490 ± 40 °C.

In the μ_{H_2O}–μ_{O_2} diagram, constructed at 6 kbar and 475 °C, the assemblage garnet–chlorite–chloritoid–biotite–magnetite has a very narrow wedge-shaped field (Fig. 9b). At lower and higher μ_{O_2} values, the five-phase assemblage gives way to garnet–chloritoid–biotite–magnetite and garnet–chloritoid–chlorite–magnetite, respectively (Fig. 9b). The assemblage garnet–chloritoid–biotite–chlorite ± magnetite becomes unstable at lower μ_{H_2O} irrespective of the μ_{O_2} of the metamorphic fluids (Fig. 9b). The stability field of the four- and five-phase assemblages (garnet–chloritoid–biotite–chlorite ± magnetite) shrinks at higher temperatures (Fig. 9c) and completely disappears at 550 °C (Fig. 9d).

It is evident from the above analyses that, barring pressure and temperature, several factors such as Mn in the bulk chemical composition of the rock, and the and the chemical potentials of H_2O and O_2 (μ_{H_2O} and μ_{O_2}, respectively) of the metamorphic fluid, exert a strong control on the stabilization of the critical assemblages garnet–chlorite–chloritoid–biotite in pelites. This study also demonstrates that the effects of μ_{H_2O} and μ_{O_2} can be so strong that they may obliterate the effects of other physico-chemical parameters. The net influence of these entire parameters make this assemblage a rare occurrence in medium- grade metapelites.

Discussion

Metamorphic P–T *path and their geodynamic significance*

When interpreted in the reaction topology of the KFMASH system and in the pseudosections in the system NCMnKFMASHO, textural relationships deduced for the chloritoid–biotite schist define a clockwise P–T path of evolution, with P_{max} preceding or reached at the same time as T_{max} (Figs 7 & 8a, b). The results of the quantitative geothermobarometry indicate that T_{max} was close to 490 ± 40 °C at a pressure of 6.3 ± 1 kbar. In view of the geometry of the P–T path it is, however, not possible to confirm whether the estimated pressure was equal or lower than the P_{max} of metamorphism of the studied areas. Metamorphism along a clockwise P–T path is generally explained in term of a subduction–obduction tectonic model commonly observed in the fold belts formed at the periphery of two or more cratonic blocks (England & Thompson 1984; Ernst 1988; Brown 1995; Goscombe *et al.* 2003).

Detailed structural analyses coupled with limited petrological studies have demonstrated that the

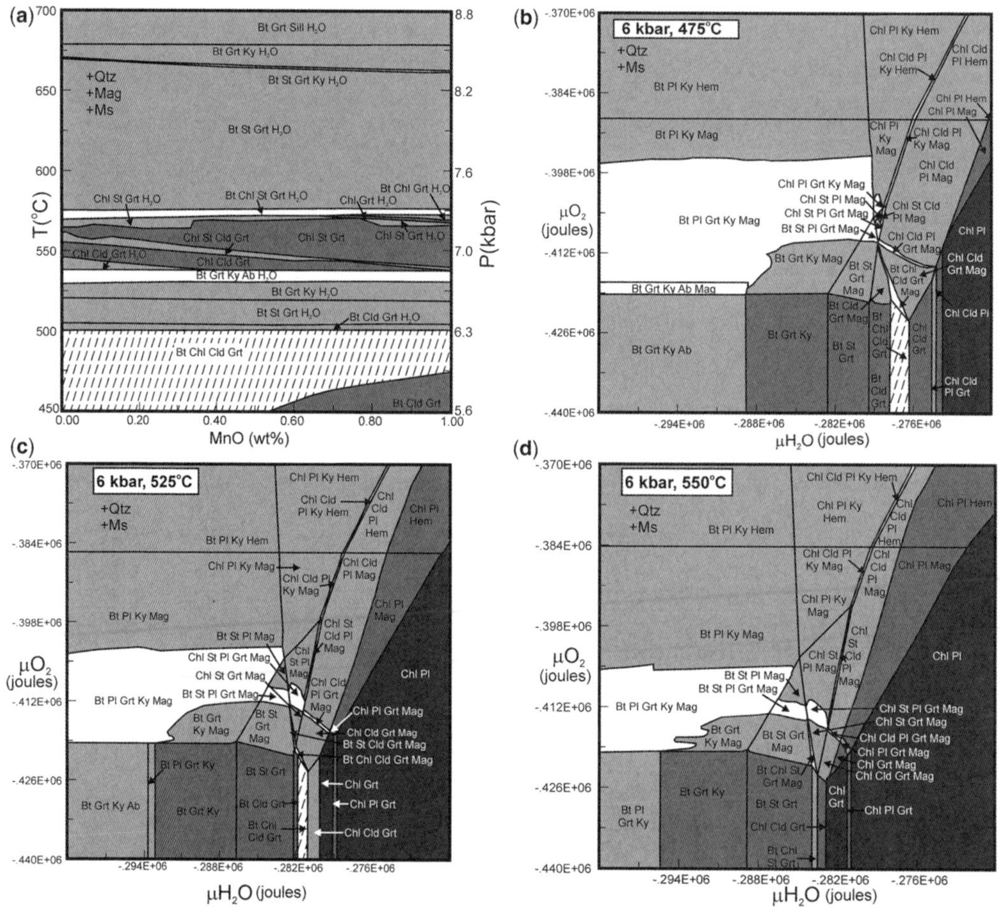

Fig. 9. (**a**) Effects of the variation of Mn and (**b**)–(**d**) μ_{H_2O} and μ_{O_2} on the stability of the chloritoid–biotite–garnet–chlorite assemblage. The bulk chemical composition of #K3CG/1 was used to construct all of the pseudosections in this figure. Note that: (a) the stability of the four-phase assemblage, which is stable between 450 and 500 °C, shrinks and is replaced by garnet–chloritoid–biotite in Mn-rich bulk conditions; and (b) the assemblages garnet–chlorite–chloritoid–biotite ± magnetite have a narrow stability field in the isothermal–isobaric $\mu_{H_2O}-\mu_{O_2}$ diagram; (c) which is further reduced at higher temperatures; and (d) eventually vanishes at a temperature close to 550 °C.

rocks in the SSZ and in the adjoining parts of the NSFB share a common structural and metamorphic history, with the localization of strain in the SSZ occurring in response to a strong compression across the interface between the NSFB and the Singhbhum Craton (Naha 1965; Sarkar 1984; Ghosh & Sengupta 1990; Mukhopadhyay & Deb 1995; Sengupta *et al.* 2005; Ghosh *et al.* 2006). Furthermore, the geometry of the kinematic markers in the SSZ confirms a thrust movement towards the south and SW as the rocks of the NSFB override the Singhbhum Craton (Fig. 1). The SSZ, which represents the lowermost part of the NSFB and is characterized by strong crystalloblastic deformation, marks the main dislocation zone between the NSFB and the Singhbhum Craton (Ghosh & Sengupta 1990). These structural attributes together with the growth of porphyroblasts during and subsequent to the development of pervasive planar fabric in the SSZ are, thus, consistent with a clockwise *P–T* trajectory inferred from the petrological studies.

The estimated T_{max} of 490 ± 40 °C is much higher than the temperature expected from the Precambrian geothermal gradient in the shield areas at the estimated pressure of 6.3 ± 1 kbar (Fig. 10), which, by implication, points to thermal perturbation during metamorphism (Ganguly *et al.* 1995). Studies have shown that regional metamorphism in the collisional setting is caused

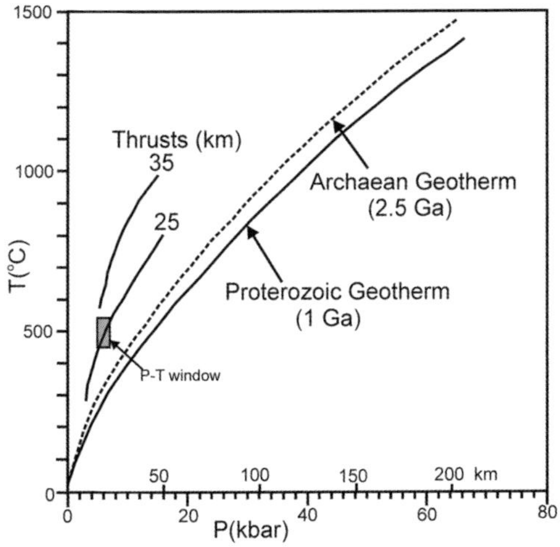

Fig. 10. Proterozoic and Archaean geotherms of the southern Indian Shield area and the perturbation of the latter due to the overthrusting of 35 and 25 km crustal blocks, as calculated by Ganguly *et al.* (1995), assuming a constant basal heat flux of 21.3 mW m^{-2} and a constant uplift rate of 0.1 mm $year^{-1}$. Note that the P–T box estimated in this study falls on the transient geothermal gradient calculated for the 25 km-thick thrust slice.

by either or both of the following heating mechanisms:

- magmatism as a consequence of the delamination of the subcrustal lithosphere (Cook *et al.* 1998);
- heat energy released by the decay of an enhanced concentration of radionuclides in the overthickened continental crust (England & Thompson 1984; Senthil Kumar *et al.* 2006).

In the absence of any synmetamorphic magmatic rocks in the studied areas, the first possibility cannot be invoked to explain the required thermal perturbation. Ganguly *et al.* (1995) computed a number of transient geothermal gradients that may develop in a collisional tectonic setting due to the decay of radionuclides for a set of thicknesses, uplift rates and internal heat generation values for the thickened crust (Fig. 10). The estimated P–T values that are constrained for the peak metamorphism of the studied area fall on the transient geothermal gradient calculated for a region that is overridden by 25 km-thick continental crust (Fig. 10). This observation, therefore, suggests that the decay of radioactive elements in 20–25 km-thick (corresponding to the pressure of 6.3 $\pm$ 1 kbar) slabs of the NSFB were sufficient to raise the temperature of metamorphism of the rocks in the SSZ to 490 $\pm$ 40 °C.

Limited geochronological data suggest that the protoliths of the metasedimentary rocks of the NSFB were deposited after 2.2 Ga, and major phases of deformation and metamorphism in the SSZ occurred in the time span of 1.5–1.8 Ga, with the main phase of shearing occurring at around 1.6 Ga (Sarkar 2000; Sengupta *et al.* 2005; Mahato *et al.* 2008). Integrating the geochronological data with the petrological information of this communication, an approximate 1.6 Ga age is tentatively assumed to be the timing of the inferred clockwise P–T path along which the Palaeoproterozoic supracrustal rocks of the NSFB and SSZ were metamorphosed. Precise geochronological work in the SSZ and in the adjoining parts of the NSFB is warranted to fix an accurate time frame of tectonothermal events in the region.

The author is indebted to Prof. D. Mukhopadhyay for his guidance in the field and during preparation of the manuscript. I thank Prof. S. C. Sarkar, Dr A. Gupta and Prof. P. Sengupta of the Department of Geological Sciences, Jadavpur University and Dr A. Matin, Department of Geology, University of Calcutta for their encouragement and many erudite discussions in the field. I am grateful to Prof. M. M. Raith and Prof. C. Ballhaus, director of the institute for allowing me to generate analytical data used for this work at the Mineralogisch Petrologisches Institut, Bonn University. Funding for fieldwork was sourced from DST Women Scientist Project 2006 awarded to the author. I convey my sincere thanks to Profs R. Mazumder and D. Saha for inviting me to write this manuscript. This work has been greatly benefited by constructive reviews from Prof. J. Ganguly and Dr I. Braun.

References

Ahn, K. S. & Nakamura, Y. 2000. The natural reaction muscovite + chlorite + chloritoid = andalusite + biotite + quartz + H_2O and a new petrogenetic grid. *Geosciences Journal*, **4**, 25–39.

Albee, A. L. 1965. A petrogenetic grid for the Fe–Mg silicates of pelitic schists. *American Journal of Sciences*, **263**, 512–536.

Albee, A. L. 1972. Metamorphism of pelitic schists: reaction relations of chloritoid and staurolite. *Geological Society of America Bulletin*, **83**, 3249–3268.

Andriessen, P. A. M., Boelrijk, N. A. I. M., Hebeda, E. H., Priem, H. N. A., Verdurmen, E. A. Th. & Verschure, R. H. 1979. Dating the events of metamorphism and granitic magmatism in the Alpine Orogen of Naxos (Cyclades, Greece). *Contributions to Mineralogy and Petrology*, **69**, 215–225.

Babůrek, J. 1998. Chloritoid Biotite Assemblage as a witness of intermediate MP/LP metamorphism in the Královský Hovzd Unit, Moldanubicum, Bohemian massif. *Geolines*, **6**, 8–9.

Baltatzis, E. & Wood, B. J. 1977. The occurrence of paragonite in chloritoid schists from Stonehaven, Scotland. *Mineralogical Magazine*, **41**, 211–216.

Bandyopadhyay, N. 2003. *Metamorphic History of the Rocks in the Southeastern Sector of the Proterozoic*

Singhbhum Shear Zone and its Environs. Unpublished PhD thesis, University of Calcutta.

BICKLE, M. J. & ARCHIBALD, N. J. 1984. Chloritoid and staurolite stability: implications for metamorphism in the Archaean Yilgarn Block, western Australia. *Journal of Metamorphic Geology*, **2**, 179–203.

BROWN, M. 1995. P–T–t evolution of orogenic belts and the causes of regional metamorphism. *In*: LE BAS, M. J. (ed.) *Milestones in Geology*. Geological Society of London, London, Memoirs, **16**, 67–81.

CHATTERJEE, N., BANERJEE, M., BHATTACHARYA, A. & MAJI, A. K. 2010. Monazite chronology, metamorphism–anatexis and tectonic relevance of the mid-Neoproterozoic Eastern Indian Tectonic Zone. *Precambrian Research*, **179**, 99–120.

CHOPIN, C., SEIDEL, E., THEYE, T., FERRARIS, G., IVALDI, G. & CATTI, M. 1992. Magnesiochloritoid, and the Fe–Mg series in the chloritoid group. *European Journal of Mineralogy*, **4**, 67–76.

CONNOLLY, J. A. D. 2005. Computation of phase equilibria by linear programming: a tool for geodynamic modeling and its application to subduction zone decarbonation. *Earth and Planetary Science Letters*, **236**, 524–541.

COOK, F. A., VAN DER VELDEN, A. J., HALL, K. W. & ROBERTS, B. J. 1998. Tectonic delamination and subcrustal imbrication of the Precambrian lithosphere in northwestern Canada mapped by LITHOPROBE. *Geology*, **26**, 839–842.

DASGUPTA, S., CHAKRABORTY, S. & NEOGI, S. 2009. Petrology of an inverted Barrovian sequence of metapelites in Sikkim Himalaya, India: constraints on the tectonics of inversion. *American Journal of Science*, **309**, 43–84.

DICKENSON, M. P. 1988. Local and regional differences in the chemical potential of water in amphibolite facies pelitic schists. *Journal of Metamorphic Geology*, **6**, 365–381.

DICKENSON, M. P. & HEWITT, D. A. 1986. A garnet–chlorite geothermometer. *Geological Society of America, Abstracts with Programs*, **18**, 584.

DROOP, G. T. R. 1987. A general equation for estimating Fe^{3+} concentrations in ferromagnesian silicates and oxides from microprobe analyses, using stoichiometric criteria. *Mineralogical Magazine*, **51**, 431–435.

DROOP, G. T. R. & HARTE, B. 1995. The effect of Mn on the phase relations of medium grade pelites: constraints from natural assemblages on petrogenetic grid topology. *Journal of Petrology*, **36**, 1549–1578.

ENGLAND, P. C. & THOMPSON, A. B. 1984. Pressure–temperature–time paths of regional metamorphism: I. Heat transfer during the evolution of regions of thickened continental crust. *Journal of Petrology*, **25**, 894–928.

ERNST, W. G. 1988. Tectonic history of subduction zones inferred from retrograde blueschist P–T paths. *Geology*, **16**, 1081–1084.

GANGULY, J. 1968. Analysis of the stabilities of chloritoid and staurolite and some equilibria in the system $FeO–A1_2O_3–SiO_2–H_2O–O_2$. *American Journal of Sciences*, **266**, 277–298.

GANGULY, J. 1972. Staurolite stability and related parageneses: theory, experiments and applications. *Journal of Petrology*, **13**, 335–365.

GANGULY, J. & SAXENA, S. K. 1984. Mixing properties of aluminosilicate garnets: constraints from natural and experimental data, and application to geothermobarometry. *American Mineralogist*, **69**, 88–97.

GANGULY, J., CHENG, W. & TIRONE, M. 1996. Thermodynamics of aluminosilicate garnet solid solution: new experimental data, an optimized model, and thermometric applications. *Contributions to Mineralogy and Petrology*, **126**, 137–151.

GANGULY, J., DASGUPTA, S., CHENG, W. & NEOGI, S. 2000. Exhumation history of a section of the Sikkim Himalayas, India: records in the metamorphic mineral equilibria and compositional zoning in garnet. *Earth and Planetary Science Letters*, **183**, 471–486.

GANGULY, J., SINGH, R. N. & RAMANA, D. V. 1995. Thermal perturbation during charnockitization and granulite facies metamorphism in southern India. *Journal of Metamorphic Geology*, **13**, 419–430.

GHENT, E. D., STOUT, M. V., BLACK, P. M. & BROTHERS, R. N. 1987. Chloritoid bearing rocks associated with blueschists and eclogites, northern New Caledonia. *Journal of Metamorphic Geology*, **5**, 239–254.

GHOSH, S. K. & SENGUPTA, S. 1987. Progressive development of structures in a ductile shear zone. *Journal of Structural Geology*, **9**, 277–287.

GHOSH, S. K. & SENGUPTA, S. 1990. Singhbhum shear zone: structural transition and kinematic model. *Proceedings of the Indian Academy of Sciences (Earth and Planetary Sciences)*, **99**, 229–247.

GHOSH, M., MUKHOPADHYAY, D. & SENGUPTA, P. 2006. Pressure–temperature–deformation history for a part of the Mesoproterozoic fold belt in North Singhbhum, Eastern India. *Journal of Asian Earth Sciences*, **26**, 555–574.

GOSCOMBE, B., HAND, M., GRAY, D. & MAWBY, J. 2003. The metamorphic architecture of a transpressional Orogen: the Kaoko Belt, Namibia. *Journal of Petrology*, **44**, 679–711.

GRAMBLING, J. A. 1983. Reversals in Fe–Mg partitioning between chloritoid and staurolite. *American Mineralogist*, **68**, 373–388.

GRAMBLING, J. A. 1990. Internally consistent geothermometry and H_2O barometry in metamorphic rocks: the example garnet–chlorite–quartz. *Contributions to Mineralogy and Petrology*, **105**, 617–628.

HARTE, B. & HUDSON, N. F. C. 1979. Pelitic facies series and the temperatures and pressures of Dalradian metamorphism in eastern Scotland. *In*: HARRIS, A. L., HOLLAND, C. H. & LEAKE, B. E. (eds) *The Celadonides of the British Isles: Reviewed*. Geological Society, London, Special Publications, **8**, 323–337.

HIROI, Y. 1983. Progressive metamorphism of the Unazuki pelitic schists in the Hida terrane, central Japan. *Contributions to Mineralogy and Petrology*, **82**, 334–350.

HOLDAWAY, M. J. 2000. Application of new experimental and garnet Margules data to the garnet–biotite geothermometer. *American Mineralogist*, **85**, 881–892.

HOLDAWAY, M. J., DUTROW, B. L. & HINTON, R. W. 1988. Devonian and Carboniferous metamorphism in west-central Maine: the muscovite–almandine geobarometer and the staurolite problem revisited. *American Mineralogist*, **73**, 20–47.

HOLLAND, T. J. B. & POWELL, R. 1998. An internally consistent thermodynamic data set for phases of petrological interest. *Journal of Metamorphic Geology*, **16**, 309–343.

HOLLAND, T., BAKER, J. & POWELL, R. 1998. Mixing properties and activity–composition relationships of chlorites in the system $MgO-FeO-Al_2O_3-SiO_2-H_2O$. *European Journal of Mineralogy*, **10**, 395–406.

HYNES, A. & FOREST, R. C. 1988. Empirical garnet–muscovite geothermometry in low-grade metapelites, Selwyn Range (Canadian Rockies). *Journal of Metamorphic Geology*, **6**, 297–309.

JERABEK, P., JANAK, M. K., FARYAD, S. W., FINGER, F. & KONECNY, P. 2008. Polymetamorphic evolution of pelitic schists and evidence for Permian low-pressure metamorphism in the Vepor Unit, West Carpathians. *Journal of Metamorphic Geology*, **26**, 465–485.

KALT, A., ALTHERR, R. & LUDWIG, T. 1998. Contact metamorphism in pelitic rocks on the island of Kos (Greece, Eastern Aegean Sea): a test for the Na-in cordierite thermometer. *Journal of Petrology*, **39**, 663–668.

KRETZ, R. 1983. Symbols for rock forming minerals. *American Mineralogist*, **68**, 277–279.

LABOTKA, T. C. 1980. Petrology of a medium pressure regional metamorphic terrain, Funeral Mountains, California. *American Mineralogist*, **65**, 670–689.

LABOTKA, T. C. 1981. Petrology of an andalusite type regional metamorphic terrain, Paramint Mountains, California. *Journal of Petrology*, **22**, 261–296.

LIKHANOV, I. I., REVERDATTO, V. V., SHEPLEV, V. S., VERSCHININ, A. E. & KOZLOV, P. S. 2001. Contact metamorphism of the Fe and Al-rich metapelites in the Transangarian region of the Yenisei Ridge, eastern Siberia, Russia. *Lithos*, **58**, 55–80.

LOOMIS, T. P. 1986. Metamorphism of metapelites: calculations of equilibrium assemblages and numerical simulations of the crystallization of garnet. *Journal of Metamorphic Geology*, **4**, 201–229.

MAHATO, S., GOON, S., BHATTACHARYA, A., MISHRA, B. & BERNHARDT, H. J. 2008. Thermo-tectonic evolution of the North Singhbhum Mobile Belt (eastern India): a view from the western part of the belt. *Precambrian Research*, **162**, 102–127.

MPOSKOS, E. 1989. High-Pressure metamorphism in gneisses and pelitic schists in the East Rhodope Zone (N. Greece). *Mineralogy and Petrology*, **41**, 25–39.

MUKHOPADHYAY, D. & DEB, G. 1995. Structural and textural development in Singhbhum shear zone, eastern India. *Proceedings of the Indian Academy of Sciences (Earth and Planetary Sciences)*, **104**, 385–405.

NAHA, K. 1965. Metamorphism in relation to stratigraphy, structure and movements in part of East Singhbhum, Eastern India. *Quarterly Journal of the Geological, Mineralogical and Metallurgical Society of India*, **37**, 41–88.

NEWTON, R. C., CHARLU, T. V. & KLEPPA, O. J. 1980. Thermochemistry of the high structural state plagioclases. *Geochimica et Cosmochimica Acta*, **44**, 933–941.

PERCHUK, L. L. & LAVRENTIEVA, I. V. 1983. Experimental investigation of exchange equilibria in the system cordierite–garnet–biotite. *In*: SAXENA, S. K. (ed.) *Kinetics and Equilibrium in Mineral Reactions*. Advances in Physical Geochemistry, **3**, 199–239.

PITRA, P. & GUIRAUD, M. 1996. Probable anticlockwise P–T evolution in extending crust: Hlinsko region, Bohemian Massif. *Journal of Metamorphic Geology*, **14**, 49–60.

POUCHOU, L. & PICHOIR, F. 1984. A new model for quantitative X-ray microanalysis, Part-I: Application to the analysis of homogeneous samples. *La Recherche Aérospatiale*, **3**, 167–192.

POWELL, R. & HOLLAND, T. 1990. Calculated mineral equilibria in the pelite system, KFMASH ($K_2O-FeO-MgO-Al_2O_3-SiO_2-H_2O$). *American Mineralogist*, **75**, 367–380.

POWELL, R. & HOLLAND, T. 1999. Relating formulations of the thermodynamics of mineral solid solutions: activity modeling of pyroxenes, amphiboles, and micas. *American Mineralogist*, **84**, 1–14.

RAO, N. K. & RAO, G. V. U. 1983. Uranium mineralization in Singhbhum shear zone, Bihar: IV. Origin and geological time frame. *Journal of the Geological Society of India*, **24**, 615–627.

ROY, A., SARKAR, A., JEYAKUMAR, S., AGGRAWAL, S. K. & EBIHARA, M. 2002. Sm–Nd age and mantle source characteristics of the Dhanjori volcanicrocks, Eastern India. *Geochemical Journal*, **36**, 503–518.

SAHA, A. K. 1994. *Crustal Evolution of Singhbhum–North, Orissa, Eastern India*. Geological Society of India, Memoirs, **27**.

SARKAR, S. C. 1984. *Geology and Ore Mineralisation Along the Singhbhum Copper–Uranium Belt, Eastern India*. Jadavpur University, Calcutta.

SARKAR, S. C. 2000. Crustal evolution and metallogeny in the eastern Indian craton. *In*: *National Seminar Commemorating Dr M. S. Krishnan's Birth centenary (Proceedings Volume)*. Geological Survey of India, Special Publications, **55**, 169–194.

SENGUPTA, N., MUKHOPADHYAY, D., SENGUPTA, P. & HOFFBAUER, R. 2005. Tourmaline-bearing rocks in the Singhbhum shear zone, eastern India: evidence of boron infiltration during regional metamorphism. *American Mineralogist*, **90**, 1241–1255.

SENTHIL KUMAR, P., MENON, R. & REDDY, G. K. 2006. The role of radiogenic heat production in the thermal evolution of a Proterozoic granulite–facies orogenic belt: Eastern Ghats, Indian Shield. *Earth and Planetary Science Letters*, **254**, 39–54.

SHARMA, M., BASU, A. R. & RAY, S. L. 1994. Sm–Nd isotopic and geochemical study of the Archaean tonalite–amphibolite association from the eastern India craton. *Contributions to Mineralogy and Petrology*, **117**, 45–55.

SHARMA, R. S. 2010. *Cratons and Fold Belts of India*. Lecture Notes in Earth Sciences, **127**, doi: 10.1007/978-3-642-01459-8_2.

SPEAR, F. S. 1993. *Metamorphic Phase Equilibria and Pressure–Temperature–Time Paths*. Mineralogical Society of America, Washington, DC.

SPEAR, F. S. & CHENEY, J. T. 1989. A petrogenetic grid for pelitic schists in the system $SiO_2-Al_2O_3-FeO-MgO-K_2O-H_2O$. *Contributions to Mineralogy and Petrology*, **101**, 149–164.

TRIBOULET, C. 1983. Uni- and divariant equilibria between staurolite, chloritoid, garnet, chlorite, biotite in medium pressure meta-acidites from Lorient-Concarneau area (South Brittany, France). *Contributions to Mineralogy and Petrology*, **82**, 195–204.

WANG, P. & SPEAR, F. S. 1991. A field and theoretical analysis of garnet + chlorite + chloritoid + biotite assemblages from the tri-state (MA, CT, NY) area, USA. *Contributions to Mineralogy and Petrology*, **106**, 217–235.

WATERS, D. J. & LOVEGROVE, D. P. 2002. Assessing the extent of disequilibrium and overstepping of prograde metamorphic reactions in metapelites from the Bushveld Complex aureole, South Africa. *Journal of Metamorphic Geology*, **20**, 135–149.

WEI, C. J. & POWELL, R. 2004. Calculated phase relations in high-pressure metapelites in the system NKFMASH (Na_2O–K_2O–FeO–MgO–Al_2O_3–SiO_2–H_2O) with application to natural rocks. *Journal of Petrology*, **44**, 183–202.

WHITNEY, D. L., MECHUM, T. A., KUEHNER, S. M. & DILEK, Y. R. 1996. Progressive metamorphism of pelitic rocks from protolith to granulite facies, Dutchess County, New York, USA: constraints on the timing of fluid infiltration during regional metamorphism. *Journal of Metamorphic Geology*, **14**, 163–181.

WU, C. M. & ZHAO, G. 2006. Recalibration of the Garnet–Muscovite (GM) geothermometer and the Garnet–Muscovite–Plagioclase–Quartz (GMPQ) geobarometer for metapelitic assemblages. *Journal of Petrology*, **47**, 2357–2368.

WU, C. M., ZHANG, J. & REN, L. D. 2004. Empirical Garnet–Biotite–Plagioclase–Quartz (GBPQ) geobarometry in medium- to high-grade metapelites. *Journal of Petrology*, **45**, 1907–1921.

ZANE, A., SASSI, R. & GUIDOTTI, C. V. 1998. New data on metamorphic chlorite as a petrogenetic indicator mineral, with special regard to greenschist-facies rocks. *The Canadian Mineralogist*, **36**, 713–726.

Metamorphic evolution of the Chotanagpur Granite Gneiss Complex of the East Indian Shield: current status

S. SANYAL* & P. SENGUPTA

Department of Geological Sciences, Jadavpur University, Kolkata 700032, India

**Corresponding author (e-mail: sanyal.sn@gmail.com)*

Abstract: Geological information on the Chotanagpur Granite Gneiss Complex (CGGC) has been reviewed with a view to: (a) identifying the different metamorphic episodes; (b) developing an event stratigraphy in the high-grade blocks; and (c) correlating the different metamorphic episodes with the globally extensive orogenic processes. Integrating the existing geological information, the geological evolution of the high-grade block of the CGGC has been divided into four stages associated with four distinct metamorphic events ($M_I - M_{IV}$). The earliest metamorphic event (M_I) that is recorded in granulite enclaves in the regionally extensive felsic gneisses culminated in ultrahigh temperatures (>900 °C at *c.* 5–8 kbar) at around 1.87 Ga. In the second stage, voluminous felsic magmas were intruded – the M_I granulites – and were metamorphosed to form migmatitic felsic gneisses (M_{II}) within about 1.66–1.55 Ga. The third stage witnessed intrusions of a suite of anorthosite and porphyritic granitoids (*c.* 1.55–1.51 Ga), followed by high-grade metamorphism (700 ± 50 °C, 6.5 ± 1 kbar, M_{III}) during approximately 1.2–0.93 Ga. The fourth stage (M_{IV}) is marked by the intrusion of a suite of mafic dykes, followed by infiltration-driven metamorphism (600–750 °C at 7 ± 1) during 0.87–0.78 Ga. The proposed metamorphic events have been correlated with the supercontinental cycles in the Proterozoic time.

Studies in the past two decades have revealed repeated the growth and demise of supercontinents throughout the Precambrian time (reviewed in Li *et al.* 2008). Impressions of these mega-events that eventually controlled the evolution of the lithosphere, hydrosphere, atmosphere and biosphere of the Earth are ciphered in the extinct orogenic belts of Precambrian age (Condie 2004; Reddy & Evans 2009). In the reconstructed Gondwanaland – the Neoproterozoic supercontinent – India is juxtaposed against Antarctica, in which the Eastern Ghats Belt (EGB) and the Chotanagpur Granite Gneiss Complex (CGGC) of Peninsular India are placed against the Rayner Complex and eastern part of the Vestfold Hills of East Antarctica, respectively (reviewed in Dasgupta & Sengupta 2003; Chatterjee *et al.* 2010). While studies from the EGB and the Rayner Complex have confirmed the continuity of the two crustal blocks at least during Meso- and Neoproterozoic time (reviewed in Dasgupta & Sengupta 2003), the lack of quality geological and geochronological data from the CGGC until 2005 was the major hindrance in comparing the geological history between the CGGC and the eastern part of the Archaean Vestfold Hills. In the last 5 years, some detailed studies in the light of modern petrology and geochronology have been carried out in a few areas of the CGGC (Sanyal *et al.* 2007; Maji *et al.* 2008; Singh & Krishna 2009; Chatterjee *et al.* 2010). However, no attempt has been made to synthesize the results of these studies to decipher the different tectonic pulses that shaped the rocks of the CGGC. This exercise is expected to impose some constraints for comparing geological events between the CGGC and its supposed counterpart of East Antarctica. In this paper, an attempt has been made to synthesize the extant petrological, structural and geochronological information from the CGGC with a view to: (1) identify the gaps in knowledge and to characterize different metamorphic events that shaped the crystalline basement of the CGGC on the basis of existing information; (2) link the tectonic processes of the CGGC with the well-established supercontinental cycles that are presumed to have occurred during Proterozoic time; and (3) test the hypothesis that the CGGC was once linked to the crustal block lying east of the Vestfold Hills of East Antarctica.

Extent and boundary relationship

The Proterozoic Chotanagpur Granite Gneiss Complex (CGGC) is an east–west-trending subarcuate belt of the East Indian Shield that covers approximately 100 000 km^2 (*c.* 500 km long in the east–west direction and 200 km in width) across the states of Chattisgarh, Jharkhand, Bihar and West Bengal (reviewed in Mahadevan 2002; Acharyya 2003; Sharma 2009). The southern margin of the CGGC is marked by an east–west- to ENE–WSW-trending crustal-scale brittle–ductile shear zone, which is variously termed as the Tamar–Porapahar–Khatra

From: Mazumder, R. & Saha, D. (eds) 2012. *Palaeoproterozoic of India*. Geological Society, London, Special Publications, **365**, 117–145. http://dx.doi.org/10.1144/SP365.7

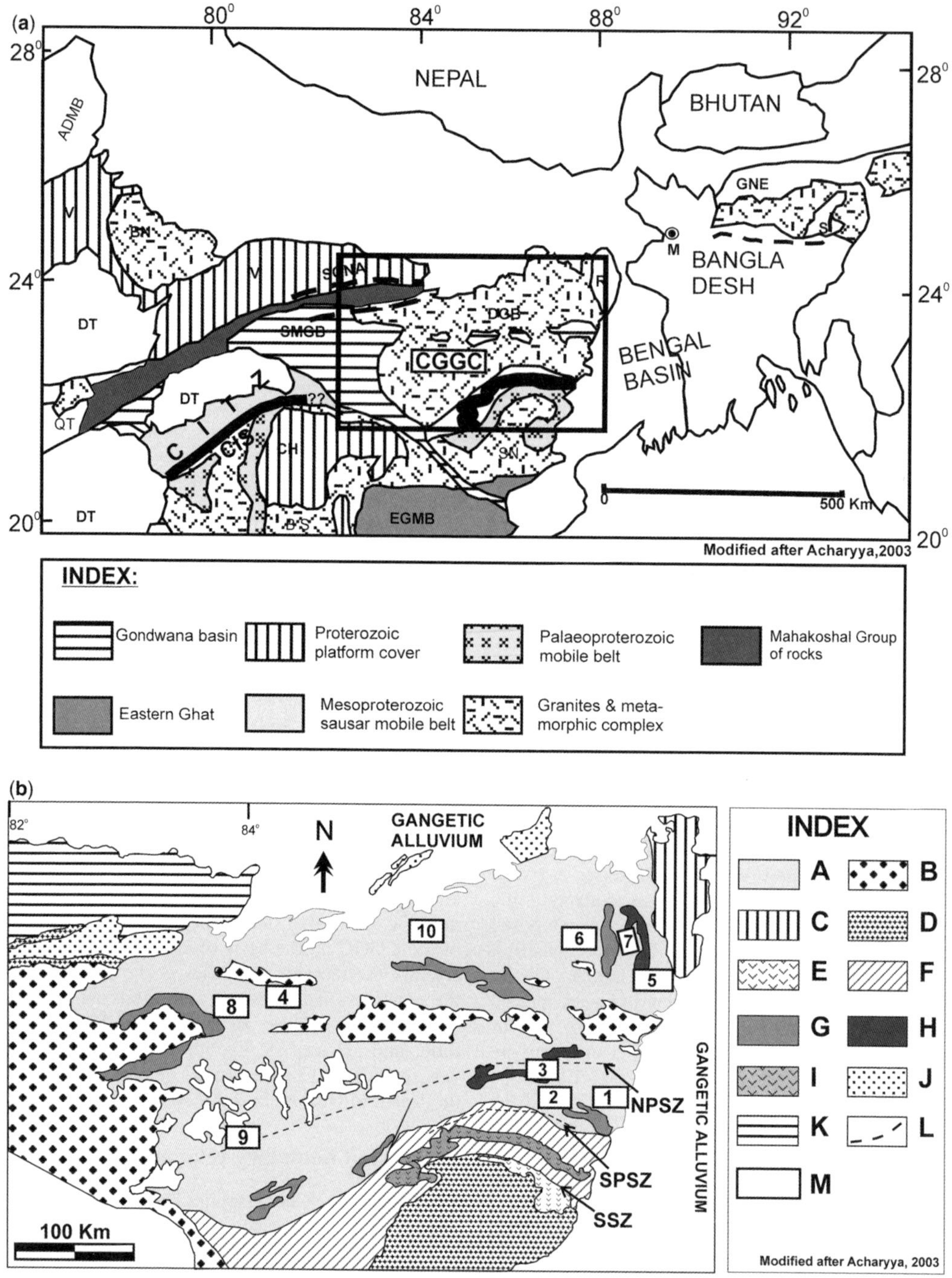

Fig. 1. (**a**) General geological map of central and east India showing different crustal blocks. The dark rectangular block marks the CGGC and the rocks of adjoining areas. ADMB, Aravalli–Delhi Mobile Belt; BN, Bundelkhand massif; BS, Bastar Craton; CH, Chattisgarh Basin; CIS, Central Indian Shear; CITZ, Central Indian Tectonic Zone; DGB, Damodar Gondwana Basin; DT, Decan Trap; EGMB, Eastern Ghats Mobile Belt; GNE, Gneissic Complex of NE India; M, Maddhapara; S, Shillong Plateau; SMGB, Son–Mahanadi Gondwana Basin; SN, Singhbhum Craton; SONA, Son–Narmada lineament; V, Vindhyan Basin. (**b**) Geological map of the Chotanagpur Granite Gneiss Complex (CGGC) and part of the Singhbhum Mobile Belt (SMB). A, CGGC gneissic rocks; B, Gondwana rocks; C, Rajmahal Basalt; D,

Fault Zone (TPKF: Mahadevan 2002), the Northern Shear Zone (NSZ: Kumar *et al.* 1978) or the South Purulia Shear Zone (SPSZ: Mazumdar 1988). This shear/fault zone separates the CGGC from the Palaeo- to Mesoproterozoic North Singhbhum Fold Belt (NSFB), girdling the Archaean nucleus of the Singhbhum Craton (Fig. 1a). Extant geological information shows that the northerly convex arcuate SPSZ hosts an ensemble of lithologies including a suite of nepheline-bearing syenite, alkaline ultramafic rocks, alkali granite and carbonatite, with an economically viable apatite deposit, quartzite, tuffaceous rocks and Fe-ore (Banerji 1985; Basu 1993; Acharyya *et al.* 2006). Kinematic markers in the shear zone indicate the thrusting of the CGGC over the NSFB along a steep northerly dipping dislocation zone (Mazumdar & Chatterjee 1990; Basu 1993; Acharyya *et al.* 2006). A concordant U–Pb SHRIMP (Sensitive High Resolution Ion Microprobe) date of zircon in nepheline-bearing syenite fixes the crystallization age of alkaline magma at 0.92 Ga (Reddy *et al.* 2008). This provides the maximum age of shearing. The CGGC is covered in the north by the Gangetic alluvium, and in the east by the Mesozoic Rajmahal trap basalt. In the west the CGGC is covered by Gondwana deposits of the Son–Mahanadi Valley, and supracrustal rocks and granitoids of the Palaeoproterozoic Mahakoshal Group of the Central Indian Tectonic Zone (CITZ, Fig. 1b). Because of the younger Gondwana cover sediments, the relationship between the Mahakoshal Group and the CGGC cannot be determined. Recent studies have revealed that the rocks of the CGGC have extended further east beneath the thick cover of the Gangetic alluvium in the Maddhapara of Bangladesh (Fig. 1b, Hossain *et al.* 2007). This observation points to the fact that the original extent of the CGGC is much larger than the area occupied by surface exposures, and continues beneath the younger unmetamorphosed cover sediments and extrusive rocks.

Classification

Several attempts have been made to classify the rocks of the CGGC in the past two decades. In the following sections we briefly present the rationale behind the different classification schemes hitherto proposed for the CGGC. Ghose (1983) and Banerji (1991) classified the litho ensemble of the CGGC into three distinct lithostratigraphic units – viz.: crystalline basement; older metasediments; and late intrusives – and considered that these units are the products of three major orogenic phases, namely the Chotanagpur Orogeny (1.6–1.5 Ga), the Satpura Orogeny (0.90–0.85 Ga) and the Munger Orogeny (0.42–0.35 Ga). Singh (1998), on the other hand, opined that the rocks of the CGGC can only be classified on the basis of lithodemic groups, rather than lithostratigraphic subdivision, as the rocks do not in general obey the 'law of superposition'. He proposed a three-tier classification where the CGGC lithodemic group appears as the oldest member followed by the Kodarma Group and the Rajgir Group, respectively.

The CGGC is further subdivided into the following three main suites: (a) the enclave suite comprising the granulite pockets of the Dumka, Purulia and Palamau areas, as well as the low- to medium-grade schist, BIF, marble and mafites of Chakrabanda, Sua and many other areas; (b) the Chatra Gneiss suite comprising several granite and tonalitic gneisses and migmatites; and (c) the intrusive suite of rocks comprising haplogranite, pegmatite and quartz veins. Mahadevan (2002) classified the CGGC rocks into five east–west subdivisions, taking the Chotanagpur Plateau and the Domodar as well as the Koel River Gondwana basins as reference frames (Fig. 2). The subdivisions are:

(i) the South Palamau–Gumla–Ranchi–Purulia belt south of the Damadar Valley basins;
(ii) the Daltangunj (North Palamau)–Hazaribag–Dumka belt between the Damador and Koel river basins;
(iii) the North Garhwa–Chatra–Giridi– Deoghar belt north of the Koel River Basin;
(iv) the Bihar Mica Belt (BMB) north of the Chotanagpur plateau;
(v) the Rajgir–Kharagpur belt north of the BMB.

Abundant granulite-facies rocks that are in different stages of retrogression are found in subdivisions I–III (Mahadevan 2002).

In the following sections we shall argue that geological information from the CGGC is too meager to attempt any meaningful classification.

Regional structural trend

In most parts, the pervasive gneissic foliation of the CGGC has a general trend that varies from an east–west to a ENE–WSW direction with moderate to steep dip towards the north (Fig. 2). The regional

Fig. 1. (*Continued*) Singhbhum and other granites; E, Dhanjori volcanics; F, SMB metamorphites; G, metasediments in CGGC; H, granulite domains; I, Dalma volcanics; J, Mahakoshal Group of rocks; K, Vindhyan Supergroup of rocks; L, major lineaments/shear zones; M, Quaternary sediments. (1) BSS sector; (2) MPR sector; (3) JBN sector; (4) GB sector; (5) MBR sector; (6) DJR sector; (7) DJG sector; (8) RT sector; (9) RK sector; (10) BMB sector.

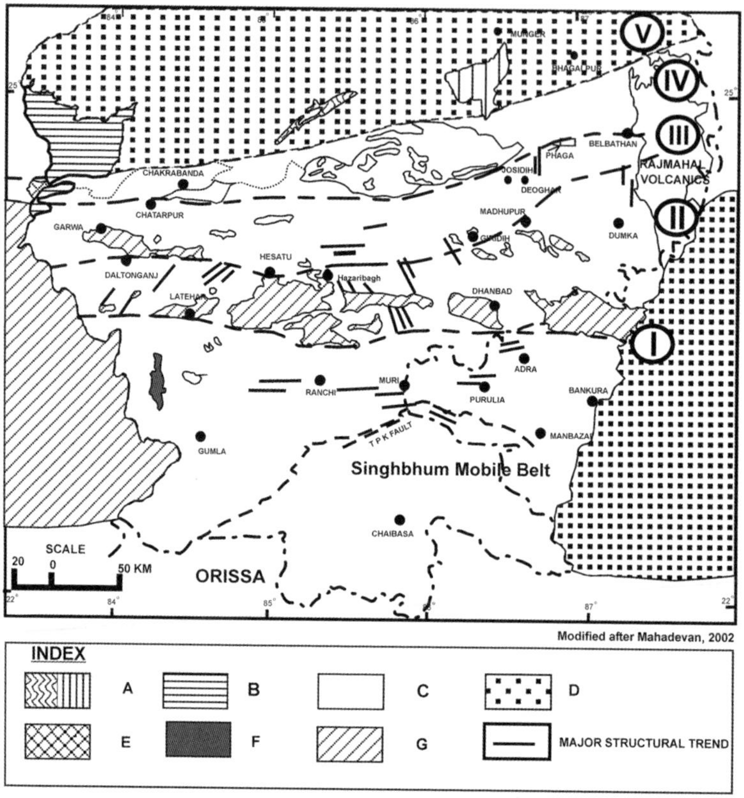

Fig. 2. Geological map of the Chotanagpur Granite Gneiss Complex (CGGC) and the adjoining areas showing major subdivisions (modified after Mahadevan 2002). A, Rajgir Group/Kharagpur Group of rocks; B, Vindhyan Supergroup of rocks; C, CGGC rocks; D, alluvium; E, Mahakoshal Group of rocks; F, laterite; G, Gondwana rocks. (I) Ranchi–Purulia subdivision; (II) Hazaribagh–Dumka subdivision; (III) Giridih–Deoghar subdivision; (IV) Bihar Mica Belt subdivision; (V) Rajgir–Kharagpur subdivision.

east- to ENE-trending foliation veers to the NE (in areas lying NW of Gomoh, west of Hazaribagh and SW of Daltongunj) and to the NW (areas lying SE of Hazaribagh and west of Giridih), possibly owing to the superposition of late open warps on east- to ENE-trending regional foliation during east–west compression (Fig. 2). In areas around Josidih–Deoghar and Jamua–Dumka towards the eastern margin of the CGGC, the intensity of east–west compression is very strong. This compression folded the regional foliation into isoclinal folds with the development of a prominent axial-planar foliation, with north to NNE strike and a moderate dip towards the east (Bhattacharya 1976; Sanyal *et al.* 2007; Chatterjee *et al.* 2010) (Fig. 2).

Review of petrological, structural and geochronological studies of the CGGC

The geological investigation of the rocks of the CGGC, although started some years ago (Dunn & Dey 1942), are restricted in small and scattered areas However, there are a few areas within the CGGC from where some detail petrological, structural and geochronological information are available. For the convenience of description and in view of the concentration of geological information, the CGGC is divided under the following sectors (shown in Fig. 2):

(1) the Bero–Saltora–Santuri (BSS) sector near Bankura, West Bengal;
(2) the Murguma–Purulia–Raghunathpur (MPR) sector, West Bengal;
(3) the Jabarban–Belamu–Nawahatu (JBN) sector, West Bengal;
(4) the Garu–Baresanr (GB) sector, Jharkhand;
(5) the Massanjor–Baglan–Rangaliya (MBR) sector, Jharkhand;
(6) the Deoghar–Josidihi–Rohini (DJR) sector, Jharkhand;
(7) the Dumka–Jamua–Ghormara (DJG) sector, Jharkhand;

(8) the Ramanujganj–Tatapani (RT) sector, Chattisgarh;
(9) the Raikera – Kunkuri (RK) sector, Chattisgarh;
(10) the Bihar Mica Belt (BMB), Bihar.

Sectors 1, 2, 3 and 9 fall within subdivision I of Mahadevan (2002) that lies south of the Damodar Valley Gondwana basins; sectors 4, 5 and 8 occur within subdivision II between the Damodar and Koel River Valley Gondwana deposits; while sectors 6 and 7 fall within subdivision III, north of the Koel Valley (Mahadevan 2002). The BMB (sector 10) forms subdivision IV of Mahadevan (2002).

In the following sections the existing geological and geochronological information of each of these sectors are summarized.

Sector 1: the Bero–Saltora–Santuri (BSS) sector

'This is one of the most well-studied sectors in the SE part of the CGGC and has been extensively studied by a number of workers since the early 1970s (e.g. Manna & Sen 1974; Roy 1977; Bhattacharyya & Mukherjee 1987; Sen & Bhattacharya 1993; Mukherjee *et al.* 2005; Ghose & Chatterjee 2008; Maji *et al.* 2008). Migmatitic quartzofeldspathic gneiss (quartz + K-feldspar + plagioclase + garnet + biotite + hornblende ± orthopyroxene) with enclaves of metapelitic gneiss (almandine-rich garnet + sillimanite + K-feldspar + quartz + hercynite + ilmenite + biotite) form the country rocks that are intruded successively by massif-type anorthosite (plagioclase + orthopyroxene ± clinopyroxene + ilmenite + hornblende + biotite ± chlorite + quartz + garnet) and porphyritic granitoids (quartz + K-feldspar + plagioclase + hornblende + garnet + biotite). The latter magmatic rocks contain distended lenses of mafic granulite(orthopyroxene + clinopyroxene + plagioclase + hornblende + garnet + quartz) and calc-silicate gneiss (clinopyroxene + scapolite ± wollastonite + plagioclase + K-feldspar + calcite + quartz). The most pervasive planar structure of the country rocks is a gneissic banding that is defined by stromatic to laterally discontinuous leucosomes with or without garnet (reviewed in Maji *et al.* 2008). A recent study by Maji *et al.* (2008) has identified three phases of deformation (D_1–D_3) and four phases of metamorphism (M_1–M_4). The S_1 fabric that is defined by granulite-facies mineralogy (garnet + orthopyroxene in quartzofeldspathic rock and garnet + sillimanite + spinel in metapelite) developed along the axial planes of D_1 folds seen in calc-silicate rocks and rarely in metapelites. Dehydration melting in quartzofeldspathic gneisses and metapelites produced leucosomes and melanosomes of granulite-facies assemblages during culmination of the M_1 metamorphism (Maji *et al.* 2008). S_1 planes, along with the lithological interfaces, were coaxially deformed by D_2 – D_3 folds that have east–west-trending and northerly dipping axial planes. In most places, a prominent planar fabric (S_2) is developed parallel to the axial planes of the D_2 folds. D_3 folds locally have a sheath-like geometry. Structural features indicate that the massif-type anorthosite (commonly known as the Bengal anorthosite) was emplaced during the interlude period of the D_1 and D_2 deformations (Bhattacharya & Mukherjee 1987; Maji *et al.* 2008), whereas the interlude period of the D_2 and D_3 deformations was marked by the intrusion of the protolith of a suite of granitoids to fayalite-bearing syanitic gneisses (Maji *et al.* 2008). Both D_2 and D_3 folds are accompanied by amphibolite-facies ensembles (M_2–M_3) that replace the granulite-facies minerals of M_1 (Maji *et al.* 2008). M_4 metamorphism is marked with the development of coronal garnet, commonly occuring as a symplectic intergrowth with ilmenite and/or quartz, in practically all of the lithologies including anorthosite and granites.

On the basis of the results of quantitative geothermobarometry on the M_1 assemblages, the maximum pressure–temperature (P–T) condition during the M_1 event has been bracketed within 5–6 kbar and 750–850 °C (Sen & Bhattacharya 1993; Maji *et al.* 2008). The estimated temperatures are distinctly lower to induce extensive melting in metapelites and quartzofeldspathic rocks that consumed practically all of the primary hydrous minerals and to stabilize wollastonite + scapolite (eqAn > 80) in calc-silicate rocks (eqAn in scapolite (X) = 100 × (Al – 3)/3 where Al is the number of Al atoms per formula unit of scapolite; Harley & Buick 1992; Sengupta *et al.* 1999). Studies on Fe and Mg diffusion between garnet and pyroxene have demonstrated that geothermometers involving garnet and pyroxene are unlikely to record temperatures in excess of 850 °C unless the cooling rates are very high (Chakraborty & Ganguly 1991; Fitzsimons & Harley 1994). In view of this observation, the estimated temperature should be treated as the minimum temperature values for the M_1 metamorphism. The Rb–Sr whole-rock isochrons of the high-grade gneisses and the chemical date of a few monazite inclusions in M_1 garnet have yielded an age of approximately 1.70 Ga (Table 1). This has been considered to be the timing of the M_1 metamorphism (reviewed in Chatterjee *et al.* 2010).

The recent U–Pb date of zircon in anorthosite fixes the age of crystallization of anorthosite at around 1.55 Ga (Chatterjee *et al.* 2008) (Table 1), which by implication marks the maximum age of the D_2 – D_3 deformational events. The margin of the anorthosite is, however, marked by a

Table 1. *Geochronology from different sectors of CGGC*

Sector	Dating method	Rock type/mineral	Age (Ma)
(1) BSS	U–Th–Pb monazite age dating (Maji *et al.* 2008)	Garnetiferous metapelitic gneiss	1176 ± 9: Mnz hosted in Gt(P) 1082 ± 14: Mnz core in matrix 1041 ± 20: Mnz core in Gt(P) 981 ± 26: Mnz in recrystallized matrix 825 ± 26: Mnz rim in Gt(P) 818 ± 11: Mnz rim in matrix
		Foliated granite	1021 ± 26: Mnz hosted in Gt(P) 992 ± 11: Mnz in recrystallized matrix 977 ± 15: Mnz hosted in Gt(P) 967 ± 11: Mnz in recrystallized matrix
	U–Th–Pb monazite age dating (near the southern CGC margin) (Chatterjee *et al.* 2010)	Granite gneiss	1442–1305:Mnz core 939–967: Mnz outer core 859 ± 87: Mnz rim
		Foliated porphyritic granite	1870–1691: Mnz core 965 ± 51: Mnz rim 778–860: Mnz core-rim
		Metapelitic granulite	937 ± 30
	U–Pb zircon age dating in ID-TIMS method (Bengal Anorthosite) (Chatterjee *et al.* 2008)	Gabbroic anorthosite	1550 ± 12 947 ± 27
(2) MPR	Rb–Sr whole-rock isochron age dating (Ray Barman *et al.* 1994)	Migmatitic granite gneiss	1178 ± 61
		Porphyritic Hy-granite gneiss	1071 ± 64
(3) JBN	K–Ar whole-rock age dating (Baidya *et al.* 1987)	Biotite granite	870 ± 40
		Leuco granite	810 ± 40
(4) GB	No age data available		
(5) MBR	U–Pb zircon age dating using the ICP-MS technique (Ray Barman *et al.* 1994)	Hy-granite gneiss	1624 ± 5, 1000
		Massive charnockite	1515 ± 5, 1000
		Basic granulite	1515
	Rb–Sr whole-rock isochron age dating (Mallik *et al.* 1991)	Migmatitic granite gneiss	1599–1522
(6) DJR	U–Th–Pb monazite age dating (Sanyal *et al.* 2007) (east of Trikut Pahar near Deoghar, Jharkhand)	Quartzofeldspathic granulite and khondalite	1628: Mnz core in recryst. matrix 1518: Mnz core in recryst. matrix 1118: Mnz outer core in matrix 1088: Mnz outer core in matrix 979: Mnz rim in quartzose matrix 942: Mnz rim in quartzose matrix 862: Mnz included in biotite 842: Mnz included in biotite 788: Mnz rim in quartzose matrix
(7) DJG	U–Th–Pb monazite age dating (NW of Dumka) (Chatterjee *et al.* 2008)	Metapelitic granulite	1190 ± 26: Mnz rim Gt hosted 995 ± 24:Mnz rim Gt hosted 950 ± 20: Mnz in matrix
	U–Th–Pb monazite age dating (NE of Dumka near the northern CGC margin) (Chatterjee *et al.* 2010)	Migmatitic quartzofeldspathic gneiss	1824–1659:Mnz core Gt hosted 984–930: Mnz core in Gt and matrix 876–828; Mnz core in Gt and matrix
		Porphyritic granite	1272 ± 35: Mnz core 954 ± 32: Mnz rim
		Metapelitic granulite	1720 ± 31: Mnz core 1298–1070: Mnz outer core 940–924: Mnz rim

(*Continued*)

Table 1. *Continued*

Sector	Dating method	Rock type/mineral	Age (Ma)
	Rb–Sr whole-rock isochron age dating (near Jarmundi) (Ray Barman *et al.* 1994)	Quartz syenite Charnockite	1457 ± 63 1331 ± 125
	K–Ar whole-rock age dating (NE part of the CGGC)(Sarkar 1980)	Migmatitic granite gneiss	1416–1246
(8) RT	No available age data		
(9) RK	Rb–Sr whole-rock isochron age dating (Singh & Krishna 2009)	Grey granite gneiss Pink granite	1005 ± 51 815 ± 47
	K–Ar whole-rock age dating (Sarkar 1980)	Foliated granite gneiss near Ranchi–Muri	1086–850
(10) BMB	Pb–Pb mineral age dating (Singh *et al.* 2001)	Galena from sulphide-bearing metasediments from the Hesatu–Belbathan area	1700–1650
	Rb–Sr whole-rock isochron age dating (Pandey *et al.* 1986*a*, *b*)	Migmatitic granite gneiss	1300–1110
	Fission-track age dating (Lal *et al.* 1976)	Garnet in mica pegmatite Muscovite in mica pegmatite Biotite in mica pegmatite Apatite in mica pegmatite	830 760 595 590

metamorphic fabric (S_2) defined by recrystallized aggregates of magmatic mineralogy that are variably replaced by amphibole and biotite (Roy 1977; Maji *et al.* 2008). Chemical dates of monazite suggest that the hydration of M_1 assemblages (M_2-M_3) and the development of the S_2 fabric occurred between 1.3 and 1.1 Ga (Maji *et al.* 2008). The origin of the coronal garnet in the mafic rock of the Saltora area has been a subject of debate. Manna & Sen (1974) and Sen & Bhattacharya (1993) opined that the garnet-forming reaction was triggered during retrograde metamorphism as a consequence of near-isobaric cooling. Whereas a prograde growth of garnet (at the expense of hornblende + plagioclase + ilmenite ± quartz) has been advocated by Bhattacharya & Mukherjee (1987) and Maji *et al.* (2008). Quantitative geothermobarometry in conjunction with a petrogenetic grid in parts of the NCFMASH system tightly constrain the $P-T$ conditions of the M_4 metamorphism (650 ± 50 °C, 4–5 kbar: Maji *et al.* 2008). It is, however, not clear whether the estimated temperature of 650 ± 50 °C is the maximum temperature or represents a 'snap-shot' on the cooling path. The chemical dates for monazite indicate that the M_4 garnet corona was formed between 1.0 and 0.95 Ga (Maji *et al.* 2008) (Table 1). Some of the Grenvillian (1.0–0.95 Ga) monazite shows 0.83–0.68 Ga old overgrowths (Maji *et al.* 2008) (Table 1). The Neoproterozoic dates are also confirmed by Chatterjee *et al.* (2010) (EPMA dates of monazite 0.86–0.77 Ga: Table 1). These younger dates are presumed to reflect another amphibolite- to granulite-facies metamorphism in the CGGC (Chatterjee *et al.* 2010: see the subsection on 'Sector 7: the Dumka – Jamua – Ghormara (DJG) sector' later in this section).

Sector 2: the Murguma – Purulia – Raghunathpur (MPR) sector

This sector, which represents the southernmost part of the CGGC and adjoins the Palaeoproterozoic Singhbhum Mobile Belt (Fig. 2), has been studied by a number of workers (Dunn 1929; Sen 1956, 1959; Baidya *et al.* 1987, 1989; Ray Barman *et al.* 1994). The most dominant rock type of the area is migmatitic quartzofeldspathic gneiss with stromatic leucosomes (locally with garnet) and well-foliated mesosome consisting of biotite and hornblende. Distended bands of calc-silicate rocks are scattered throughout the migmatitic gneiss of the area. The gneissosity of the quartzo-feldspathic rock has an east–west trend that is axial planar with a set of fold defined by calc-silicate bands. The gneissic banding is folded by two sets of coaxial folds (D_2-D_3). D_3 represents a broad warp on the east–west-trending gneissosity. Field features demonstrate that the orthopyroxene-bearing porphyritic granite (porphyritic charnockite) intruded between the D_1 and D_2 folding events (Ray Barman *et al.* 1994).

The mineral assemblage of the porphyritic charnockite is plagioclase + microcline + quartz + hornblende + biotite + orthopyroxene + garnet. The textural feature that orthopyroxene was replaced by hornblende represents the amphibolite-facies retrogression of porphyritic charnockite. The absence of precision geochemical data does not allow us to support or negate the view that the porphyritic charnockite represents charnockitic magmatism (Kilpatrick & Ellis 1992). Detailed petrological information is lacking in this sector. Nevertheless, the formation of orthopyroxene + plagioclase symplectite between the garnet and hornblende in porphyritic charnockite (Ray Barman *et al.* 1994) suggests that hydration of the orthopyroxene-bearing assemblage at high pressure was followed by decompression (Pattison 2003).

Rb–Sr whole-rock dating of porphyritic charnockite and migmatitic gneiss yielded 1.07 and 1.17 Ga respectively (Ray Barman *et al.* 1994) (Table 1). These ages are interpreted as the ages of the protolith formation of these rock types (Ray Barman *et al.* 1994). This is just one option for Rb–Sr whole-rock dating, however; the partial resetting of the Rb–Sr clock during intense metamorphism and deformation is another option. In this second case, the numbers have no geological significance.

Sector 3: the Jabarban–Belamu–Nawahatu (JBN) sector

This sector occurs in the NE part of Puruliya, West Bengal. Migmatitic to non-migmatitic granite gneisses are the host rock of this area, which contains enclaves of metapelite and calc-gneisses. Mafic rock (amphibolite) intrudes the protoliths of composite gneiss, and biotite-bearing porphyritic granite (locally pegmatite) is the latest intrusive in the area (Baidya *et al.* 1989). The east–west-trending gneissic banding (S_1) is axial planar to a set of folds defined by mafic bands (D_1). S_1 is folded by D_2 folds with an east–west trend of the axial plane. D_3 folds are superposed on D_1–D_2 folds, and are north–south-closing open warps. Field features suggest that porphyritic granite was emplaced during the interlude period of the D_1 and D_2 folding (Baidya *et al.* 1989). No detailed petrological study has been undertaken in this sector. A few K–Ar dates from the composite gneiss and biotite-bearing porphyritic granites have been found to cluster between 1.08 and 0.80 Ga (Sarkar 1980; Baidya *et al.* 1987) (Table 1). It is not clear whether these dates represent any geological event or are partially reset ages of no geological significance.

Sector 4: the Garu–Baresanr (GB) sector

This sector falls north of the Damodar Valley Gondwana basins in the central part of the CGGC (Fig. 2). Majumdar (1988) gave a brief account of rock types exposed in this part of CGGC. Migmatitic to non-migmatitic composite gneisses are the main rock types of this area, which is intruded by porphyritic granite (Mazumder 1988). Graphite-bearing khondalite and calc-silicate rocks occur as mappable bands and lenses within the host gneiss where, as dismembered bodies of amphibolite, they represent the mafic intrusive/extrusive. Ultramafic rock-hosted base–metal deposits have been reported from areas lying south of Daltonganj (Ghose 1992). No detail petrological or geochronological information is available from this sector.

Sector 5: the Massanjor–Baglan–Rangalia (MBR) sector

This sector occurs near the eastern extremity of the CGGC, north of the Damodar Valley Gondwana Basin, and forms a part of what is known as the Mor Valley Granulite of eastern India (Mahadevan 2002; Sharma 2009) (Fig. 2). In the following sections we present a synopsis of our ongoing study on the rocks in and around Massanjor.

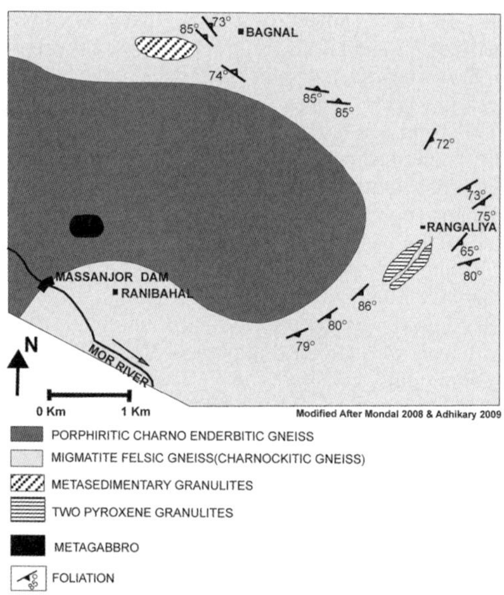

Fig. 3. Geological map of the Massanjor–Bagnal–Rangaliya (MBR) sector, Jharkhand.

The dominant rock type of this sector (Fig. 3) is migmatitic charnockite (orthopyroxene + K-feldspar + plagioclase ± garnet + ilmenite) that, in most places, is retrogressed to hornblende ± biotite-bearing quartzofeldspathic gneiss. The migmatitic charnockite and the quartzofeldspathic gneiss contain bands of khondalite (quartz + K-feldspar + plagioclase + sillimanite + garnet + ilmenite), Mg–Al granulite (orthopyroxene + sillimanite + quartz + perthite + plagioclase ± garnet + ilmenite + biotite + cordierite), calc–silicate (clinopyroxene + quartz + sphene + K-feldspar + plagioclase) and mafic granulite (plagioclase + clinopyroxene + garnet + hornblende + ilmenite + titanite) (Fig. 4a, b). The migmatitic charnockite was intruded by a large body of porphyritic charnockite (quartz + K-feldspar + plagioclase + orthopyroxene + clinopyroxene + garnet + hornblende + ilmenite + biotite + magnetite) that contains enclaves of the rocks it intruded (Fig. 4c). Structural elements in the rocks support three episodes of deformation after the formation of the protolith of migmatitic charnockite (D_1–D_3). East–west-trending migmatitic banding in charnockite formed during D_1. Khondalite, Mg–Al granulite and calc-silicate rock developed a gneissic banding that predates D_1. The migmatitic banding of D_1 were folded by coaxial D_2 and D_3 folds. Porphyritic charnockite was emplaced between D_1 and D_2 (Fig. 4d). Porphyritic charnockite developed a crude foliation parallel to the axial plane of D_2. A few mafic dykes (now amphibolite with the mineralogy: plagioclase + amphibole + clinopyroxene + chlorite + epidote + calcite + quartz + ilmenite) cut across this foliation but was folded by open and north–south-closing D_3 folds.

In the Mg–Al granulite the orthopyroxene and sillimanite form intergrowths that rim and include magnetite locally intergrown with hercynite (Fig. 5a, b). Cordierite and biotite are retrograde phases that separate sillimanite and orthopyroxene from quartz (Fig. 5c). Cordierite also surrounds garnet

Fig. 4. Field features around the Massanjor–Baglan–Rangaliya (MBR) sector of the CGGC. (**a**) Metapelitic granulite with garnetiferous leucosomes forming an irregular network. The rock occurs as enclave within the charnockite gneiss. (**b**) Mafic granulite enclave showing the development of megacrystic garnet clots. (**c**) Lensoid enclave of mafic granulite (dark green in colour) within porphyritic charno-enderbitic rock with invading tongues of later rock within the former rock. (**d**) Emplacement of porphyritic charnockite along the F_2 fold core of the migmatitic charnockite gneiss.

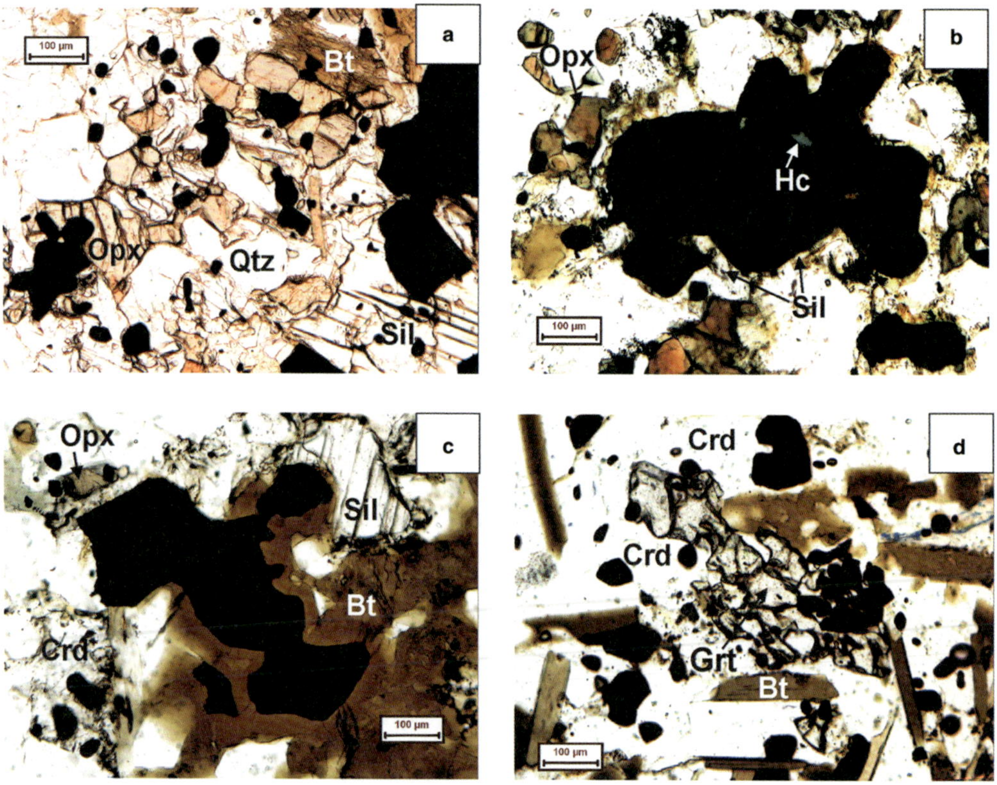

Fig. 5. Microtextural features in Mg–Al granulite of the MBR sector of the CGGC. (**a**) Orthopyroxene (Opx)–sillimanite (Sil) intergrowths forming a corona around magnetite (dark) against the direct contact with the adjoining quartz (Qtz). (**b**) Hercynite (Hc)–magnetite $(dark)_{ss}$ grain with a spatially organized corona of orthopyroxene (Opx) and sillimanite (Sil). Biotite (Bt) grains replacing all of the ferromagnesian phases. (**c**) Cordierite (Crd)–biotite (Bt) mass developing inbetween orthopyroxene (Opx) and sillimanite (Sil), and also replacing magnetite (dark) along the margin. (**d**) Development of a cordierite (Crd)–biotite (Bt) groundmass around garnet (Grt).

(Fig. 5d). Representative compositions of minerals present in the rock are presented in Table 2. The core of orthopyroxene has more than 7 wt% Al_2O_3 and garnet shows a significant spessertine component ($X^{Spss} = 0.26$, where X is Mn/(Fe + Mg + Ca + Mn) in garnet; Table 2). The #mg number (Mg/Mg + Fe) of the ferromagnesian minerals changes in the order: orthopyroxene ($X_{Mg} = 0.73$) < biotite ($X_{Mg} = 0.89$) < cordierite ($X_{Mg} = 0.90$, Table 2). Textural features and compositions of the phases suggest stabilization of the assemblages $(spinel–magnetite)_{ss}$ + quartz + melt at the peak metamorphic conditions, with intergrowths of orthopyroxene + sillimanite ± garnet formed during cooling. In recent communications, Harley & Thompson (2004) and Kelsey (2008) opined that in the $K_2O–FeO–MgO–Al_2O_3–SiO_2–H_2O$ (KFMASH) system the assemblage aluminous orthopyroxene (>7 wt% Al_2O_3) + sillimanite + quartz + garnet + melt is a good indicator of ultrahigh-temperature metamorphism (>900 °C) at a pressure in excess of 8 kbar (Fig. 6a, b). Pressure of metamorphism that stabilized the assemblage orthopyroxene + sillimanite + quartz + melt can go below 8 kbar depending on the bulk rock compositions (Kelsey 2008). The topology presented in Figure 6b predicts that the assemblage orthopyroxene + sillimanite + garnet + quartz would be shifted to a higher temperature if MnO is added to the KFMASH system. This follows that the Mg–Al granulite (with Mn-rich garnet) and other enclaves in the migmatitic charnockite must have cooled through ultrahigh-temperature metamorphism at a pressure of around 8 kbar. Geothermobarometry of the mafic granulite enclaves (representative mineral compositions are given in Table 3) yield 8–9 kbar at distinctly lower temperatures 825–850 °C (present study). We interpret that the lower-temperature values are a consequence of the resetting of the Fe–Mg exchange thermometer (Harley & Thompson 2004; Baldwin *et al.* 2007; Kelsey 2008). Representative mineral compositions of

Table 2. *Representative mineral compositions from Mg–Al granulite of the MBR sector, CGGC. Sample No. SM45*

Analysis No.	54/1	56/1	62/1	64/1	86/1	65/1	66/1	69/1	93/1	94/1
Phase	opx (core)	opx (core)	opx (core)	grt (core)	grt (core)	grt (rim)	bt	bt	crd	crd
SiO_2	49.79	49.86	49.62	39.68	39.33	38.98	39.08	38.7	50.85	50.34
TiO_2	0	0.05	0	0	0	0	2.68	2.16	0.03	0
Al_2O_3	7.32	7.13	7.37	22.33	22.28	22.09	17.24	17.87	34.43	33.86
Cr_2O_3	0	0	0.01	0	0	0	0	0.02	0	0
FeO	16.49	15.89	15.31	18.84	19.85	17.3	6.12	4.88	2.43	2.49
MnO	2.08	2.46	2.06	8.04	7.4	11.69	0.02	0.16	0.4	0.35
MgO	24.07	23.65	23.36	9.81	10.07	8.09	20.96	21.54	12.41	11.98
CaO	0.05	0.04	0.03	1.31	1.33	1.37	0	0	0.04	0
ZnO	0	0.12	0.14	0.01	0	0	0	0	0.11	0
NiO	0.07	0.02	0.11	0	0.12	0.06	0.11	0.03	0.14	0
Na_2O	0.02	0.03	0.04	0.04	0	0	0.1	0.08	0.1	0.03
K_2O	0	0.02	0.02	0	0	0	9.91	9.59	0.03	0.02
Total	99.9	99.28	98.09	100.05	100.39	99.59	96.22	95.04	100.97	99.08
Oxygen	6	6	6	12	12	12	22	22	18	18
Si	1.81	1.83	1.84	3.04	3.00	3.03	5.53	5.50	4.99	5.01
Ti	0.00	0.00	0.00	0.00	0.00	0.00	0.29	0.23	0.00	0.00
Al	0.31	0.31	0.32	2.01	2.00	2.02	2.87	2.99	3.98	3.97
Cr	0.00	0.00	0.00	0.00	0.00	0.00	0.00	0.00	0.00	0.00
Fe	0.44	0.46	0.48	1.21	1.27	1.12	0.72	0.58	0.20	0.21
Mn	0.06	0.08	0.06	0.52	0.48	0.77	0.00	0.02	0.03	0.03
Mg	1.31	1.29	1.29	1.12	1.14	0.94	4.42	4.56	1.81	1.78
Ca	0.00	0.00	0.00	0.11	0.11	0.11	0.00	0.00	0.00	0.00
Zn	0.00	0.01	0.02	0.00	0.00	0.00	0.00	0.00	0.01	0.00
Ni	0.01	0.01	0.01	0.00	0.01	0.01	0.01	0.00	0.01	0.00
K	0.00	0.00	0.00	0.00	0.00	0.00	1.79	1.74	0.00	0.00
#mg	0.75	0.74	0.73	0.48	0.47	0.45	0.86	0.89	0.90	0.90
X^{Spss}				0.18	0.16	0.26				

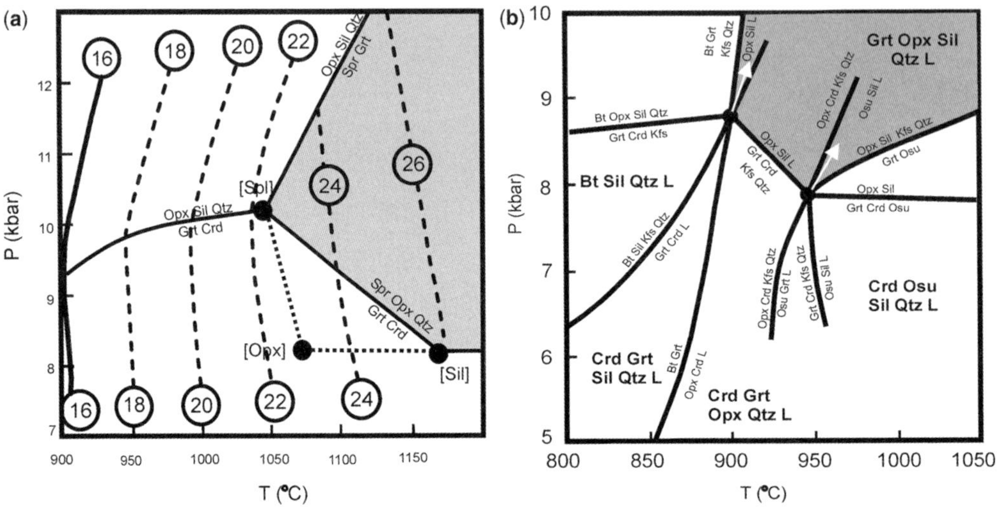

Fig. 6. (**a**) Partial petrogenetic grid in the FeO–MgO–Al_2O_3–SiO_2 (FMAS) system. Dashed lines with numbers in the open circles represent Tschermak's molecules (in per cent) in orthopyroxene coexisting with garnet, sillimanite, cordierite and quartz (numbers within the circle – after fig. 4 of Harley 2004). The bold line indicates the per cent of Tschermak's molecule of orthopyroxene in the Massanjor area. Filled circles represent invariant points. See the text. Abbreviations for minerals are after Kretz (1983). (**b**) Partial petrogenetic grid in the system K_2O–FeO–MgO–Al_2O_3–SiO_2–H_2O (KFMASH) at low fO_2 (after fig. 2a of Harley 2004). The stability field of the ultrahigh-temperature assemblage garnet + orthopyroxene + sillimanite + quartz + liquid is shaded. White lines with arrowheads indicate the displacement of KFMASH in variant points (filled circles) due to the addition of MnO in the system. See the text. All abbreviations are after Kretz (1983).

the porphyritic charnockite are given in Table 4. Geothermobarometry (Lee & Ganguly 1988; Bhattacharya *et al.* 1991) involving garnet, pyroxene and plagioclase yielded results of 700 ± 50 °C, 6.5 ± 1 kbar (Table 5). These P–T values are assigned to the metamorphism accompanying D_2 deformation. The mineral assemblage in the mafic dykes suggests that the D_3 deformation occurred in amphibolite-facies conditions.

No published geochronological data are available in this sector. However, citing unpublished from Ray Barman on zircon U–Pb data, Acharyya (2003) fixed the protolith ages of migmatitic and porphyritic charnockite at 1.62 ± 0.05 and 1.51 ± 0.05 Ga, respectively (Table 1). The lower intercept of the discordant zircon (*c.* 1000 Ma) is stated to be the age of metamorphism for the porphyritic charnockite (Acharyya 2003) (Table 1).

Sector 6: the Deoghar–Josidihi–Rohini (DJR) sector

Migmatitic quartzofeldspathic gneiss (hornblende + biotite + quartz + K-feldspar + plagioclase + ilmenite + sphene ± clinopyroxene ± garnet) and augen gneiss (K-feldspar + quartz + biotite + plagioclase ± apatite ± sericite ± zoisite) are the country rocks of this sector, which are invaded by a swarm of mafic dykes (hornblende + biotite + plagioclase + titanite ± quartz ± apatite: Ghosh & Sengupta 1999; Ray *et al.* 2011). Several generations of pegmatite veins cut across the gneissic basement rocks (Ghosh & Sengupta 1999; Mahmoud *et al.* 2008; Ray *et al.* 2011). Different generations of mafic dykes and pegmatite veins are also reported from other sectors of the CGGC, and these intrusives are considered to be the consequence of regional extension under brittle–ductile regimes (Ghosh *et al.* 2008; Mahmoud *et al.* 2008). Field studies in and around Josidihi and Deoghar (viz. the Ratanpahar and Nandanpahar areas) have shown that the protolith of the augen gneiss, presumably porphyritic granitoids, intruded the quartzofeldspathic gneiss (Fig. 7a, b) (Ghosh & Sengupta 1999; Ray *et al.* 2011). The gneissic banding in the quartzofeldspathic rock is defined by centimetre-thick, laterally continuous to discontinuous leucosomes with rare garnet. The volume of the leucosome varies between 20 and 30 vol% of the gneissic rock. Rafts of quartzofeldspathic gneiss with its gneissic banding truncated by the foliation of the host augen gneiss indicate that the protolith of the augen gneiss intruded after the felsic host rocks developed gneissic banding (Ray *et al.* 2011). Strong east–west compression, together with a large rheological contrast between the mafic dykes and their host rocks, produced spectacular folding and

Table 3. *Representative mineral compositions from mafic granulite of the MBR sector, CGGC. Sample No. SM17*

Analysis No.	27/1	25/1	32/1	22/1	23/1	37/1	38/1	24/1	33/1	31/1
Phase	grt	grt	grt	cpx	cpx	pl	pl	ill	amp	amp
SiO_2	37.38	37.02	37.97	51.28	52.60	49.04	46.15	0.00	41.58	41.70
TiO_2	0.05	0.00	0.05	0.24	0.29	0.01	0.07	50.94	2.03	2.15
Al_2O_3	20.92	21.21	21.36	1.26	1.39	32.85	34.31	0.07	13.21	12.55
Cr_2O_3	0.05	0.01	0.04	0.06	0.13	0.10	0.00	0.00	0.09	0.02
FeO	26.30	28.45	28.96	10.86	10.15	0.20	0.21	47.03	17.38	17.35
MgO	3.94	3.88	4.10	12.93	12.41	0.02	0.00	0.92	9.74	9.47
MnO	1.43	1.48	1.25	0.22	0.24	0.04	0.00	0.39	0.15	0.06
CaO	10.67	8.21	8.03	23.57	23.50	16.24	17.91	0.25	11.65	11.74
Na_2O	0.04	0.00	0.00	0.31	0.20	2.73	1.59	0.00	2.14	2.06
K_2O	0.02	0.00	0.01	0.00	0.00	0.03	0.01	0.00	0.50	0.45
Total	100.91	100.33	101.81	96.22	95.04	101.33	100.44	99.60	98.57	97.64
Oxygen	12.00	12.00	12.00	6.00	6.00	8.00	8.00	3.00	23.00	23.00
Si	2.91	2.91	2.94	1.93	1.96	2.22	2.12	0.00	6.22	6.29
Ti	0.00	0.00	0.00	0.01	0.01	0.00	0.00	0.97	0.23	0.24
Al	1.92	1.97	1.95	0.05	0.06	1.75	1.86	0.00	2.33	2.23
Cr	0.00	0.00	0.00	0.00	0.00	0.00	0.00	0.00	0.01	0.00
Fe^{2+}	1.46	1.67	1.51	0.34	0.32	0.01	0.01	1.00	2.17	2.19
Fe^{3+}	0.26	0.20	0.13	0.00	0.00	0.00	0.00	0.00	0.00	0.00
Mg	0.46	0.45	0.43	0.72	0.69	0.00	0.00	0.03	2.17	2.13
Mn	0.09	0.10	0.09	0.01	0.01	0.00	0.00	0.01	0.02	0.01
Ca	0.89	0.69	0.91	0.95	0.94	0.79	0.88	0.01	1.87	1.90
Na	0.01	0.00	0.00	0.02	0.01	0.24	0.14	0.00	0.62	0.60
K	0.00	0.00	0.00	0.00	0.00	0.00	0.00	0.00	0.10	0.09
X^{alm}	0.50	0.57	0.51							
X^{gr}	0.31	0.24	0.31							
#mg	0.24	0.21	0.22	0.68	0.69			0.03	0.50	0.49
X^{an}						0.77	0.86			

$X^{alm} = Fe/(Fe + Mg + Ca + Mn)$ in garnet; $X^{gr} = Ca/(Fe + Mg + Ca + Mn)$ in garnet; and $X^{an} = Ca/(Ca + Na + K)$ in plagioclase.

boudinage structures of the mafic dykes (Fig. 8a, b). Geometrical analyses of the preserved structural elements identified three sets of folds in the mafic dykes (D_4-D_6) that also involved and succeeded the development of the gneissic banding in the host felsic gneiss (pre-D_4) (Sanyal *et al.* 2007; Ray *et al.* 2011). D_4 and D_5 are coaxial folds with north–south- to NNE–SSW-trending, moderately dipping regional foliations that are axial planer to the D_4 (and rarely the D_5) folds (Fig. 8c). Intense layer-parallel stretching during both of these deformations produced boudin structures of different sizes and shapes in the competent mafic dykes, whereas comparatively less competent surrounding felsic and augen gneisses show bending folds along the boudin necks (Sengupta *et al.* 1998; Ghosh & Sengupta 1999; Ray *et al.* 2011). Ovoid to irregular quartzofeldspathic segregates, presumed to be quenched melt fractions, were found along the boudin necks (Fig. 8d). The changes in boudin morphology from rectangular to barrel-shaped to fishhead-shaped has been explained by the rheological inversion of mafic dykes and the host quartzofeldspathic and augen gneiss during D_5 folding (Ghosh & Sengupta 1999; Ray *et al.* 2010). In augen gneiss, biotite and the long axes of the feldspar grains define the pervasive north–south-trending foliation. The augen gneiss and the quartzofeldspathic gneiss develop patches of leucosomes with biotite selvages (Fig. 9a). These leucosomes are developed along the axial panes of D_4-D_5 folds (Fig. 9b). The third-generation folds (D_6) are east–west-verging open warps (Fig. 9c). Homophanous biotite granite and several mappable bodies of biotite ± muscovite-bearing pegmatite were emplaced during different stages of the D_4–D_6 folding events (Fig. 9d).

Studies have shown that the rheological inversion of mafic dykes was caused by successive replacement of clinopyroxene by amphibole and biotite (Ray *et al.* 2011). On the basis of structure and mineralogy, two events of metamorphism, M_1 and M_2, were recognized. The M_1 event produced the stromatic leucosomes in the quartzofeldspathic gneiss. No $P-T$ estimates are available for the M_1 event. Nevertheless, the presence of abundant leucosome (20–30 vol%) suggests that the migmatitic structure was formed under granulite-facies

Table 4. *Representative mineral compositions from porphyritic charnockite gneiss of the MBR sector, CGGC. Sample No. SM72A*

Analysis No. Phase	91/1 grt (core)	90/1 grt (rim)	76/1 grt (rim)	84/1 opx (rim)	67/1 opx (core)	66/1 opx (rim)	74/1 cpx (core)	73/1 cpx (rim)	65/1 pl (core)	64/1 pl (rim)	78/1 amp	80/1 ill
SiO_2	37.77	37.28	37.65	49.12	49.11	49.22	49.97	51.1	60.5	60.47	40.75	0
TiO_2	0.03	0	0.03	0.05	0.08	0.05	0.15	0.14	0.01	0.02	2.14	50.01
Al_2O_3	20	20.39	20.29	0.54	0.51	0.41	1.27	1.19	24.12	24.75	10.7	0.01
Cr_2O_3	0	0.06	0.12	0	0.03	0.02	0	0	0	0	0	0
MgO	1.65	1.34	1.63	9.91	10.21	10.23	7.81	8.02	0.15	0.01	5.7	0.35
FeO	30.25	29.59	30.91	37.81	38.14	37.88	17.86	17.48	0.82	0.36	23.44	47.21
CaO	7.67	7.88	7.38	0.69	0.69	0.53	21.03	21.38	6.64	6.76	11.44	0.02
MnO	1.72	1.66	1.95	0.62	0.65	0.65	0.3	0.27	0.05	0.01	0.17	0.42
Na_2O	0	0.01	0.01	0.07	0	0	0.43	0.4	7.78	7.69	1.41	0
K_2O	0.01	0	0	0.01	0.02	0.02	0.01	0.01	0.31	0.3	1.71	0.04
Total	99.1	98.21	99.97	98.82	99.44	99.01	98.83	99.99	100.38	100.37	97.46	98.06
Oxygen basis	12	12	12	6	6	6	6	6	8	8	23	3
Si	3.05	3.04	3.03	2.01	2.00	2.01	1.97	1.98	2.70	2.688	6.39	0
Ti	0.00	0.00	0.00	0.00	0.00	0.00	0.00	0.00	0.00	0.001	0.25	0.98
Al	1.91	1.96	1.92	0.03	0.02	0.02	0.06	0.05	1.27	1.297	1.98	0
Cr	0.00	0.00	0.01	0.00	0.00	0.00	0.00	0.00	0.00	0	0.00	0
Fe^{2+}	2.05	2.02	2.05	1.29	1.30	1.29	0.59	0.57	0.00	0.002	3.08	0.91
Mg	0.20	0.16	0.20	0.60	0.62	0.62	0.46	0.46	0.01	0.001	1.33	0.01
Ca	0.67	0.69	0.64	0.03	0.03	0.02	0.89	0.89	0.32	0.322	1.92	0.00
Mn	0.12	0.12	0.13	0.02	0.02	0.02	0.01	0.01	0.00	0	0.02	0.01
Na	0.00	0.00	0.00	0.01	0.00	0.00	0.03	0.03	0.67	0.663	0.43	0.00
K	0.00	0.00	0.00	0.00	0.00	0.00	0.00	0	0.02	0.017	0.34	0.00
X^{alm}	0.68	0.68	0.68									
X^{py}	0.07	0.05	0.06									
X^{gr}	0.22	0.23	0.21									
X^{an}									0.31	0.32		

$X^{alm} = Fe/(Fe + Mg + Ca + Mn)$ in garnet; $X^{py} = Mg/(Fe + Mg + Ca + Mn)$ in garnet; $X^{gr} = Ca/(Fe + Mg + Ca + Mn)$ in garnet; and $X^{an} = Ca/(Ca + Na + K)$ in plagioclase.

Table 5. *Representative thermobarometric calculations from the MBR sector of the CGGC*

Rock type	Barometry	Method	Association	P (kbar)	Thermometry	Method	*T* (°C)
Mafic granulite	GCPS	Eckert *et al.* (1991)	$gt^{27}-cpx^{23}-pl^{37}$ (core)	8.57	$gt^{27}-cpx^{23}$(core)	Ganguly (1979)	824
			$gt^{25}-cpx^{22}-pl^{38}$ (rim)	4.88	$gt^{25}-cpx^{22}$ (rim)	Ganguly (1979)	625
Mg–Al granulite	GRAIL	Bohlen *et al.* (1983)	gt^{72}–rt–ilm–sil–q(core)	8.7	$gt^{64}-opx^{54}$ (core)	Lee & Ganguly (1988)	807
					$gt^{86}-opx^{56}$ (core)	Lee & Ganguly (1988)	823
					$gt^{64}-bt^{66}$	Dasgupta *et al.* (1991)	509
Porphyritic charnockite	GOPS	Bhattacharyya *et al.* (1991)	$gt^{91}-opx^{67}-pl^{65}$(core)	6.86	$gt^{91}-opx^{67}$	Lee & Ganguly (1988)	690
		Newton & Perkins (1982)	$gt^{91}-opx^{67}-pl^{65}$(core)	6.86	$gt^{76}-opx^{84}$	Lee & Ganguly (1988)	698
		Bhattacharyya *et al.* (1991)	$gt^{90}-opx^{66}-pl^{64}$ (rim)	5.73	$gt^{91}-cpx^{74}$	Ganguly (1979)	764
					$gt^{90}-cpx^{73}$	Ellis & Green (1979)	664

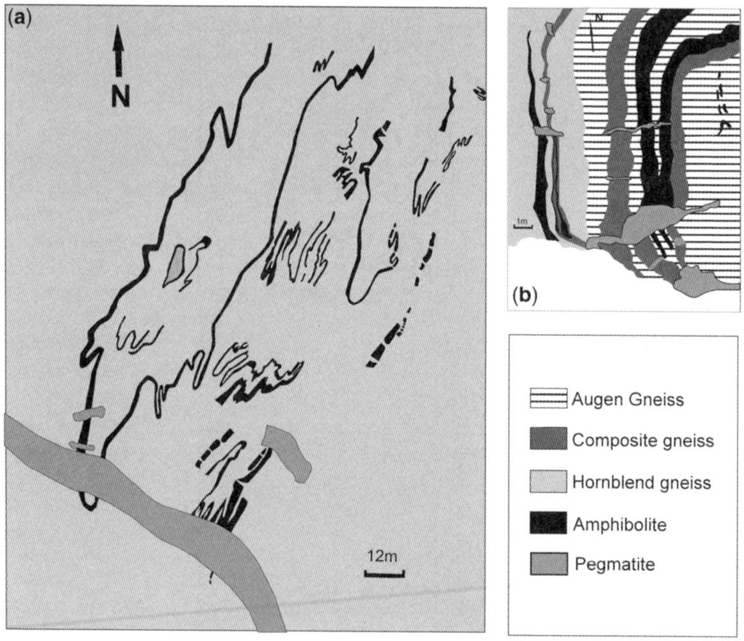

Modified after Mondal, 2006

Fig. 7. Large-scale geological mapping in the Deoghar–Josidihi–Rohini (DJR) sector in the CGGC. (**a**) Outcrop-scale map in the Ratanpahar area near Josidihi. (**b**) Outcrop-scale map in the Nandanpahar area around Deoghar.

conditions. Mineralogy of the mafic dykes and the augen gneiss are consistent with amphibolite-facies metamorphism (M_2) that was associated with the D_4–D_6 folding events. The partial to complete replacement of the magmatic mineralogy of mafic dykes by hornblende-bearing assemblages suggest infiltration-driven metamorphism during the M_2 event. Quantitative geothermobarometry and numerical analyses of appropriate bulk compositions indicate that the M_2 metamorphism culminated in the P–T range of 7 ± 1 kbar and 600–750 °C (Ray *et al.* 2007). The estimated temperatures are in good agreement with vapour-present melting that produced second-generation leucosomes with biotite-rich selvages in quartzofeldspathic gneiss and augen gneiss. No geochronological data are available from this sector.

Sector 7: the Dumka–Jamua–Ghormara (DJG) sector

This sector, which represents the NE part of the CGGC, is sandwiched between the DJR sector to the west and the Mesozoic Rajmahal volcanics to the east (Fig. 2). Dunn (1941) provided some preliminary geological information about this sector, and Bhattacharya (1976) published a detailed map showing the general lithological distribution and broad structural pattern (Fig. 10). Like all other sectors, migmatitic quartzofeldspathic gneiss is the dominant lithology that contains mappable enclaves of khondalite (quartz + K-feldspar + plagioclase + sillimanite + garnet + illmenite), mafic rocks (hornblende + plagioclase + clinopyroxene + illmenite ± orthopyroxene ± garnet) and calc-silicate rocks (plagioclase + garnet + clinopyroxene + sphene + quartz + hornblende). A number of mafic dykes (hornblende + plagioclase + sphene + quartz + biotite) that cut across the migmatitic structure of the quartzofeldspathic rock are present in the western part of this sector. The migmatitic structure of the country rock is defined by abundant (>20 vol%) leucosomes with or without garnet, whereas amphibole + biotite + feldspar + garnet forms the melanosomes. Locally, the migmatitic quartzofeldspathic rock contains corroded patches of greasy coloured charnockite showing a migmatitic structure (orthopyroxene + biotite + hornblende + quartz + K-feldspar ± garnet: Fig. 11a). The migmatitic layers in charnockite resemble and laterally grade into the migmatitic structure in the host quartzofeldspathic gneiss. This structural feature, together with the preservation of relict orthopyroxene within the amphibole in the quartzofeldspathic gneiss, is consistent with the view that the quartzofeldspathic rocks in the wider area represent retrograded charnockite. Khondalite is also

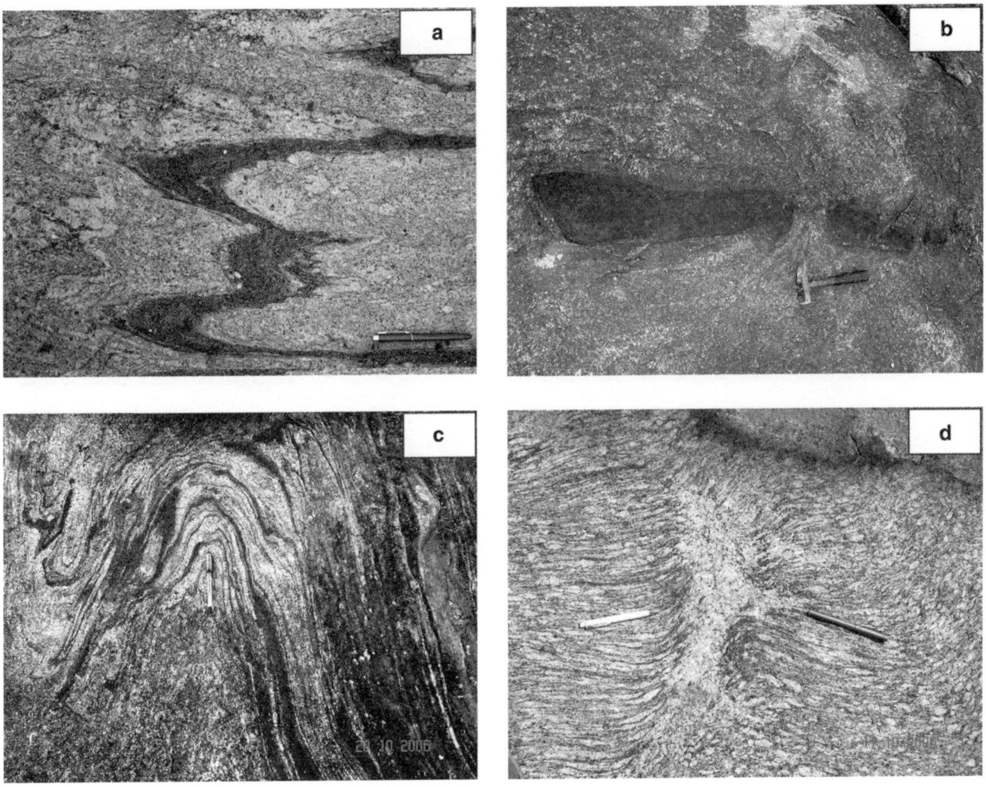

Fig. 8. Field features around the Deoghar–Josidihi–Rohini (DJR) sector in the CGGC. (**a**) Amphibolite bands (dark) within migmatitic charnockite gneiss showing a folded structure with thicked hinge and drawn-out limbs. (**b**) Development of a rectangular boudin structure in mafic bands (amphibolite) within the felsic gneiss. (**c**) Coaxial D_3–D_4 folds in amphibolite and migmatitic charnockite gneiss. (**d**) Development of irregular quenched melt pockets along the boudin necks in felsic gneiss.

migmatitic with garnet + sillimanite-bearing melanosomes that are alternated with garnet-bearing leucosomes (Fig. 11b). The migmatitic structure of Khondalite is transposed by the gneissic (migmatitic) layering of the host quartzofeldspathic rock (Fig. 11c). This feature supports polyphase migmatizations in the area.

On the basis of structural features, this sector can be divided into eastern and western domains. In the eastern domain (near Dumka), the gneissic foliation of the quartzofeldspathic rocks (S_1) are folded into coaxial D_2–D_3 folding, with roughly east–west-trending axial planes. Field studies demonstrate that elliptical bodies of syenite and anorthosite were emplaced between the D_1 (S_1) and D_2 deformation events (Bhattacharya 1976; Sanyal *et al.* 2007). In contrast to the eastern domain, the regional trend of the western domain is north–south, which is axial planar to the folds defined by the mafic dykes (Fig. 11d). The geometry and the generations of folding of the mafic dykes are similar to those observed in the adjoining DJR sector. North–south-closing folds and axial-planar fabric have been reported from one place from the eastern domain near Dumka (Chatterjee *et al.* 2010). The development of north–south-closing folds and accompanied axial-planar foliation indicate strong east–west compression.

Field and petrological studies identify two phases of high-grade metamorphism that produced migmatitic bandings in khondalite and charnockite, and more than one phases of amphibolite-facies metamorphism. During the first phase, charnockite was retrogressed to quartzofeldspathic gneiss (Sanyal *et al.* 2007, our unpublished data). Mineralogy of the mafic dykes is consistent with another amphibolite-facies metamorphism that accompanied the youngest structural event.

Sanyal *et al.* (2007) demonstrated that the migmatitic structure in khondalite was formed under ultrahigh-temperature conditions in the middle crust (5–6 kbar and 930–950 °C). This event consumed all of the primary biotite, and produced extensive fluid-undersaturated melt. No detailed P–T estimates

Fig. 9. (**a**) Dark coloured biotite selvages around leucosomal segregates in the felsic gneiss. (**b**) Development of leucosomes of different generations in the migmatitic felsic gneiss. Thick leucosomes become folded with the amphibolite bands, while thin leucosomes develop axial planer to it. (**c**) Development of east–west-verging open to moderately closing folds in amphibolite. (**d**) Development of a pegmatite vein in the felsic gneiss–amphibolite intercalated unit paralleling the D_5 axial planes.

are available for the second high-grade event (that produced migmatitic charnockite). Nevertheless, extant experimental data suggest 700–800 °C at 5–6 kbar of pressure for the development of orthopyroxene in quartzofeldspathic lithologies (Pattison *et al.* 2003). Similar P–T values (4–6 kbar, 750–780 °C) were also obtained from the adjoining syenitic gneisses (Roy Barman *et al.* 1994). Chatterjee *et al.* (2010) obtained high-pressure (9–12 kbar) metamorphism at moderate temperatures (730–800 °C) that accompanied north–south-closing fold development in the Dumka area.

Mallik *et al.* (1991) obtained Rb–Sr whole-rock ages of 1.52–1.59 Ga from the felsic gneisses from Mor Valley. The charnockite gneiss of Jamua–Jarmundi, however, record Rb–Sr whole-rock ages of 1.45 ± 0.63 Ga, with an initial $^{87}Sr/^{86}Sr$ ratio of 0.70776 ± 0.00425 (Roy Barman *et al.* 1994). The same method records lower ages (1.33 ± 0.12 Ga) from the adjoining syenite, with a high initial $^{87}Sr/^{86}Sr$ ratio of 0.71390 (Ray Barman *et al.* 1994). In view of the mobility of Rb–Sr during polymetamorphism, the exact meanings of these ages are not clear.

Recent studies have shown that U–Th–Pb chemical dates of monazite from felsic gneisses and metapelite (khondalite) enclaves show the following age clusters: (1) *c.* 1.87–1.72 Ga; (2) 1.66–1.27 Ga; (3) *c.* 1.1–0.93 Ga; and (4) *c.* 0.85–0.75 Ga (Sanyal *et al.* 2007; Chatterjee *et al.* 2008, 2010). The oldest age cluster is defined by the oldest part of the monazite grains included in the cores of porphyroblastic garnet in felsic (charnockite) gneisses and khondalite (Chatterjee *et al.* 2010). In view of the geological features of these rocks, approximately 1.87 Ga is presumed to be the age of ultrahigh-temperature metamorphism in the khondalite enclaves. The second age cluster is defined by monazite grains present in the rim of garnet porphyroblasts, as well as in the quartzofeldspathic matrix of the charnockite gneiss and khondalite (Sanyal *et al.* 2007; Chatterjee *et al.* 2010).

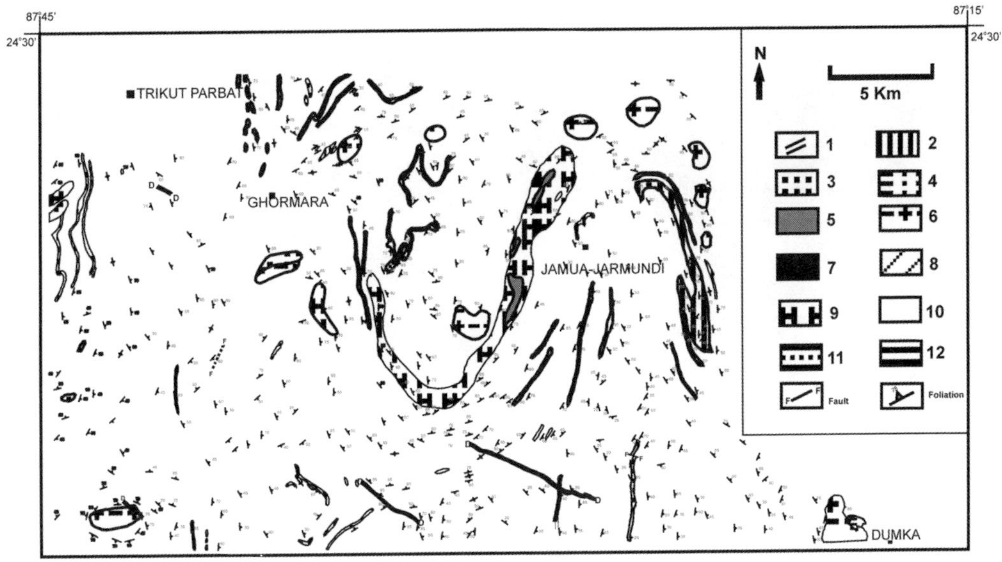

Fig.10. Geological map of the Dumka–Jamua–Ghormara (DJG) sector in the CGGC. Modified after Bhattacharyya (1976). 1, dolerite dyke; 2, para-amphibolite; 3, khondalite; 4, anorthosite; 5, charnockite gneiss; 6, augen gneiss; 7, feldspathic quartzite; 8, pegmatite; 9, hornblende gneiss; 10, migmatitic biotite gneiss; 11, metanorite (hybridised); 12, orthoamphibolite.

It is likely that the maximum age of this cluster (*c.* 1.66 Ga) dates the second high-grade metamorphism that produced gneissic banding in charnockite. Monazite showing ages of the third and fourth clusters grew over the monazite portions that show Palaeoproterozoic ages (Chatterjee *et al.* 2010). Overgrowths of Neoproterozoic monazite (fourth cluster) over the Grenvillian monazite (third cluster) is a common feature (Chatterjee *et al.* 2007, 2010; Sanyal *et al.* 2007). Petrological features and the monazite dates, therefore, support repeated high-grade events. This may explain, at least partially, the large spread of ages of monazite in clusters (1) and (2). Chatterjee *et al.* (2010) opined that 0.88–0.83 Ga dated the high-pressure event that accompanied regional east–west compression.

Sector 8: the Ramanujganj–Tatapani (RT) sector

This sector constitutes the westernmost part of the CGGC and adjoins the Mesozoic Gondwana rocks of the Mahanadi Basin. Palaeoproterozoic rocks of the Mahakoshal Group occur immediately west of the Gondwana Basin. Limited geological information available in Patel (2007) and Patel *et al.* (2007) shows the presence of a NW–SE-trending domal structure in and around Ramanujganj that dominantly constituted an alternate sequence of metapelites and psammopelites. Metapelites locally contain layers of quartzite with abundant manganiferous (>24 wt% MnO) garnet, and rare ferroactinolite and grunerite and metacarbonate. In the SE part of the domal structure, metapelites are represented by megacrystic andalusite and porphyroblastic garnet and biotite that are set in a fine-grained matrix of quartz, muscovite and biotite. The grade of metamorphism increases towards the NW direction where metapelites become distinctly coarse grained, with andalusite replaced by sillimanite. In the western and northern part of the domal structure, the metasedimentary rocks (metapelites, psammopelites and quartzite) are intruded by granitic to granodioritic rocks containing partly digested xenoliths of the supracrustals. Three sets of folds, viz. D_1-D_3, have affected the metasedimentary rocks. D_1 and D_2 folds are coaxial with D_2, with a NW–SE-trending axial-planer schistosity. Interference of D_2 and D_3 produced dome and basin structures on the outcrop scale, and is presumed to be responsible for the regional domal structure. Structural features suggest that the protoliths of the granitoids have intruded the metasedimentary rocks prior to or during D_1 deformation. Post-D_3 pegmatite veins and mafic dykes are common in the area. In the Tatapani area, occurring south of Ramanujganj, calc-silicate rocks develop the rare assemblage vesuvianite + grossular garnet + diopside + wollastonite + quartz. Interpreting the $P-T$ stability field of the vesuvianite-bearing assemblage, together with the $P-T$ stability of andalusite, Patel

Fig. 11. Field features in the Dumka–Jamua–Ghormara (DJG) sector in the CGGC. (**a**) Development of a greasy coloured charnockite patch within the migmatitic quartzofeldspathic gneiss. (**b**) Thick gneissic banding in khondalite showing alternation of pink garnet-rich layers and white sillimanite–quartz–feldspar-rich layers. (**c**) Gneissic bands in the metapelitic rock are transposed by the migmatitic layering in the host felsic gneiss. (**d**) Development of north–south-oriented foliation axial planer to folds in the amphibolite bands.

(2007) calculated pressures of less than 4 kbar at metamorphic temperatures in the range of 590–650 °C. Fluid during metamorphism was presumed to be H_2O-rich, with X_{CO_2} ($=CO_2/CO_{2+}H_2O$) less than 0.15. No geochronological information is available from the area.

Sector 9: the Raikera–Kunkuri (RK) sector

The Raikera–Kunkuri region, that lies approximately 200 km south of the RT sector stated earlier in the western part of the CGGC, is characterized by an interbanded sequence of pelitic schist (with chlorite, biotite and hornblende schist), cross-bedded pebbly quartzite and dolerite dykes/sills, now amphibolite(reviewed in Singh & Krishna 2009). The litho-packet is intruded by voluminous suites of biotite- and muscovite + biotite-bearing granitoids. Older granitoids (grey coloured) yield an 11-point Rb–Sr whole-rock isochron age of approximately 1.0 ± 0.05 Ga, with an initial $^{87}Sr/^{86}Sr$ ratio of 0.7047 ± 0.0065. The same dating method yielded a nine-point whole-rock age of around 0.81 ± 0.05 Ga (initial $^{87}Sr/^{86}Sr = 0.7539 \pm 0.0066$) for the younger pink granitoids (Singh & Krishna 2009). Compared to the younger pink granitoids, the grey granitoids, the dominant variety here, contain higher modal plagioclase and biotite, and lesser modal K-feldspar. Rounded inclusions of quartz and plagioclase were observed in the K-feldspar grains of pink granitoids. On the basis of this non-specific petrographical observation and the initial $^{87}Sr/^{86}Sr$ values, Singh & Krishna (2009) proposed a metasomatic origin for pink granitoids that is also the source of Y-mineralization in the area. Structural geometry of the Raikera–Kunkuri area are similar to that of Ramanujganj, with grey granitoids intruded syntectonically with the D_3 folds (Singh & Krishna 2009). The granitoid intrusives of this area have been correlated with heavy-metal-bearing granitoids to pegmatoids of the Bihar Mica Belt (Singh & Krishna 2009).

Sector 10: the Bihar Mica Belt (BMB) sector

The BMB, which produces one of the best industrial-grade mica deposits of the world, is situated north of the Hazaribagh Plateau at the northern extremity of the CGGC and is separated from the gneissic terrain by the peneplained Barakar River Valley. The northern extremity of this belt is obscured owing to the cover of thick Gangatic alluvium (reviewed in Mahadevan 2002). The geological formations in this belt include an ensemble of metamorphosed arenaceous and argillaceous rocks interbanded with hornblende schist, orthoamphibolite and calc-silicate rocks. A conglomerate horizon that occurs at the base of the BMB contains clasts of high-grade rocks akin to the southern part of the CGGC (Ghose & Chatterjee 2008). This feature suggests that the sediments of the BMB were sourced from the high-grade block of the CGGC. Large bodies of granitoid rocks, dolerite dykes, gabbro anorthosite and mica pegmatites intrude this schistose litho-package. Mica pegmatites are also the source of rare earth elements (REEs) and rare metal deposits (Mahadevan 2002; Singh & Krishna 2009). At the Hesatu–Belbathan Belt, economic-grade base-metal mineralization is found associated with carbonate-bearing metasediments (Ghose 1992).

The metasedimentary rocks of the BMB have been affected by three sets of folding (D_1-D_3). Granitoids were emplaced syn- to post-D_2 folding, whereas different generations of mica-bearing granite–pegmatites were emplaced pre-D_2 to post-D_3 folding (Mahadevan 2002). No detailed petrological information is available from the BMB. On the basis of mineralogy, Mahadevan (2002) postulated amphibolite-facies metamorphism for the metasedimentary rocks and the embedded mafic rocks (now amphibolite). The Pb/Pb age of 0.96 ± 0.05 Ga of Vinogradov *et al.* (1964) provided the age of emplacement of the oldest REE-rare metal-bearing mica pegmatites of the BMB. Within the uncertainties, this age is in good agreement with the approximately 0.91 ± 0.02 Ga U–Pb and Pb/Pb ages of the Columbite–Tantalite minerals from pegmatite (Krishna *et al.* 2003). Whole-rock and Rb–Sr mineral isochron ages were obtained from granitic rocks of the BMB. The oldest age for the granite that intruded the sediments of the BMB have been dated to be about 1.6 Ga, while most of the younger granite intrusive clusters are within 1.3–1.1 Ga (Pandey *et al.* 1986*a*, *b*). K–Ar dating of granitoids yields lower ages (1.08–0.85 Ga) (Sarkar 1980). Notwithstanding the susceptibility of Rb–Sr and K–Ar isotopic systems to metamorphism, metasomatism and deformation, the reported ages of the younger granite overlap with the U–Pb and Pb/Pb ages recorded from pegmatite. The Pb/Pb age of galena from base-metal deposits constrains that sedimentation in the BMB occurred in the time span of 1.70–1.65 Ga (Singh *et al.* 2001). This is in good agreement with the Rb–Sr whole-rock age of older intrusive granite (Pandey *et al.* 1986*a*, *b*). Lal *et al.* (1976) presented fission-track ages of garnet (0.83 Ga), muscovite (0.76 Ga), biotite (0.59 Ga) and apatite (0.59 Ga) from mica pegmatites. Closure temperatures of fission track for these minerals are functions of several parameters including the cooling rate (Faure 2001). Nevertheless, the existing information suggests that the closure temperature of garnet (assumed to be as high as that of zircon) and apatite should lie between approximately 250 and 100 °C (higher and lower values for garnet, and apatite respectively, Wagner & Van der Haute 1992). This follows that the amphibolite-facies rocks of the BMB were exhumed at or close to the surface before the beginning of the Phanerozoic Eon. The closure temperature of apatite (*c.* 100 °C) further suggests that the rocks of this area had not received any significant thermal pulse, which by implication means cratonization, since 0.59 Ga (Lal *et al.* 1976).

Discussion

The CGGC is constituted of a high-grade gneissic block girdled by amphibolite-facies BMB, RT and RK sectors. Boundary relationships between the amphibolite- and granulite-facies blocks have not yet been studied. Lithological association and the inferred grade of metamorphism of the RT sector and the adjoining Mahakoshal Group show a striking resemblance (both the regions show andalusite-type metamorphism: Roy *et al.* 2002; Patel *et al.* 2007; Chakraborty *et al.* 2011). This, together with the marked contrast of geological features between the RT sector and the high-grade gneissic block of the CGGC, raises the possibility that the RT could be a part of the Mahakoshal Group. This possibility needs to be corroborated by field mapping and geochronological data to unveil the stratigraphical status of the RK and RT sectors. Geological features suggest that a basin was formed and filled by detritus from the gneissic block of the CGGC during Palaeoproterozoic time (*c.* 1.70–1.65 Ga: Singh *et al.* 2001; Ghose & Chatterjee 2008). Precision U–Pb dates of zircon from the metasediments and amphibolite from the BMB are required to put a tighter constrain on the timing of sedimentary basin formation and their metamorphism. Nevertheless, published U–Pb and Pb/Pb ages indicate that the supracrustal rocks of the BMB were infiltrated by voluminous mica pegmatite between about 1.0 and 0.90 Ga, which is coeval with a phase of high-grade metamorphism in the gneissic block in the south.

Further detailed geochemical study may help in understanding whether or not these two events are genetically linked. Intensive studies in the light of recent advancement are also necessary to understand whether the apparent amphibolite-facies metamorphism is the product of a single metamorphic episode or is superimposed on an earlier granulite-facies event.

Metamorphic evolution of the high-grade gneissic block

A review of the published literature that is presented in the earlier sections has revealed the fact that practically no meaningful geological and geochronological information is available from a vast expanse of the high-grade gneissic block of the CGGC. Existing studies are few in number and are strongly clustered in the eastern part of the high-grade block. The published information too rarely meets the standard of modern petrology. Of particular concern is the paucity of quality geochronological information. Rb–Sr and K–Ar radio-clocks are susceptible to reset under high-temperature conditions and deformation (Faure 2001). Therefore, the interpretation of Rb–Sr and K–Ar dates in polydeformed high-grade terrains is not straightforward. The generation of detailed petrological information for different metamorphic episodes and fixing the time of these events through robust geochronological methods (e.g. U–Pb dates of zircon through LA-ICP-MS (laser ablation-inductively coupled plasma-mass spectrometry) and SHRIMP) from wider areas of the CGGC are desperately needed in order to constrain the tectonic pulses in the CGGC. Without this information no meaningful classification of the CGGC can be made.

Notwithstanding the uncertainties mentioned above, extant geological and geochronological information demonstrate that the CGGC has a protracted history that spans for more than one billion years (10^9 years) (Table 6). Integrating this information, an attempt has been made in the following section to develop an event stratigraphy for the high-grade block of the CGGC.

The geological evolution of the CGGC has been divided into four stages (Table 6). Information about the earliest stage is gathered from the enclaves/boudins hosted in the migmatitic charnockite gneisses (Stage I, Table 6). An ultrahigh-temperature metamorphism (>900 °C) has been documented at 5–8 kbar, interpreting the mineral assemblages of metasedimentary rocks (Khondalite and quartzofeldspathic rock in the DJG sector: Sanyal *et al.* 2007) and Mg–Al granulites (MBR sector, this study). This metamorphism is designated as M_I. On the basis of the result of geothermobarometry, Maji *et al.* (2008) obtained distinctly lower temperatures (750–850 °C) at 5–6 kbar for M_1 episode from the BSS sector. Studies have shown that, owing to the faster diffusion of ions through minerals at temperatures above 850 °C, ion-exchange geothermometers record 60–100 °C lower temperatures than the temperature at the peak metamorphism in many ultrahigh-temperature belts (reviewed in Kelsey 2008). This observation, together with the presence of highly calcic scapolite in calc-silicates, tends to support an ultrahigh-temperature metamorphism in the BSS sector as well. The CGGC, therefore, joins to a select band of granulites in the world where the temperature of metamorphism exceeded 900 °C (reviewed in Kelsey 2008). Published U–Th–Pb chemical dates of earliest monazite preserved at the cores of garnet porphyroblasts constrains that the minimum age of the ultrahigh-temperature metamorphism (M_I) is approximately 1.87 Ga (Table 6, reviewed in Chatterjee *et al.* 2010). Studies have shown that monazite may lose Pb during high-grade metamorphism (Hawkins & Bowring 1997). A more robust technique (U–Pb dating of zircon with SHRIMP or LA-ICP-MS) is needed for the precise dating of the ultrahigh-temperature metamorphism in the CGGC. Studies have identified two fundamental types of orogenic systems – accretionary and collisional – as potential sites for ultrahigh-temperature metamorphism (reviewed in Harley 1998; Kelsey 2008). Thermal perturbation by advective heat transport in accretionary tectonic settings seems to be a plausible mechanism to explain the ultrahigh-temperature metamorphism at mid-crustal depth in the CGGC (Harley 1998; Kelsey 2008). The idea that the large enclaves of mafic granulite bodies are a source of heat for the ultrahigh-temperature metamorphism is, indeed, a viable option but requires confirmation by rigorous petrological and geochronological studies. More studies on the metapelite and calcareous enclaves will surely identify distinctive assemblages for ultrahigh-temperature metamorphism and help to pinpoint the tectonic setting of the ultrahigh-temperature metamorphism.

The second stage of evolution (Stage II, Table 6) is marked by intrusions of the protolith of the regionally dominant granitoids followed by their deformation and migmatization to produce charnockite gneiss (M_{II}). Studies in high-grade terrains show that voluminous felsic magma from which charnockite (*sensu lato*) are formed develop in arc settings (reviewed in Frost & Frost 2008). Detailed geochemical characterization, including isotope compositions of the migmatitic charnockite gneiss, is required to test the hypothesis that the regionally extensive felsic magmas in the CGGC were also formed in an arc setting. Experimental data on

Table 6. *Deformational and metamorphic events in CGGC*

Time	Metamorphic and igneous events	Deformational events	Age (Ga)
Neoproterozoic (Stage IV)	Final cooling of the complex		*c*. 0.80–0.60
	Emplacement of syn- to post-D_5 leucogranite. Pegmatite emplacement parallel to the axial planes of D_4 and D_5. Bihar Mica Belt pegmatites syn- to post-D_6	D_6 cross warps	*c*. 0.85–0.80
	Upper amphibolite-facies metamorphism (M_{IV}) to produce amphibolite, foliated granite and augen gneiss; incipient migmatization and pegmatite emplacement along the $D_4 - D_5$ axial plane: $T = 750–600$ °C at 7 ± 1 kbar (DJR sector); $T = 800–730$ °C at 9–12 kbar (DJG sector)	Development of the D_4–D_5 fabric (north–south); transposition of early granulite fabrics; coaxial D_4 and D_5 folding event	*c*. 0.85–0.78 (DJR); *c*. 0.87–0.78 (DJG)
	Intrusion of mafic dyke swarms and porphyritic granite within polymetamorphosed high-grade rocks.	Post-M_{III}/D_3 intrusion during the early Neoproterozoic	*c*. 0.93
Meso- to Neoproterozoic (Stage III)	Granulite to upper amphibolite-facies metamorphism (M_{III}). $T = 700 \pm 50$ °C at 6.5 ± 1 kbar	Development of thin gneissic banding (S_2) with accompanied D_2 and D_3 deformations. Fabric development is east–west	*c*. 1.2–0.93
	Intrusion of porphyritic granitoid, anorthosite and syenite within migmatitic charnockite gneiss with M_I granulite enclaves	Post-M_{II}/D_1 magmatism during early Mesoproterozoic time	*c*. 1.55–1.51
Palaeo- to Mesoproterozoic (Stage II)	High-grade metamorphism (M_{II}) and migmatization of granitoid protoliths and the development of migmatitic charnockite gneiss; $T = 700–800$ °C at 5–7 kbar	Formation of stromatic gneissic banding in the charnockite gneiss: S_1/D_1(granulite) fabric	*c*. 1.66–1.55
	Intrusion of the protolith (granitoids) of migmatitic charnockite gneiss within M_I granulites		*c*. 1.66
Palaeoproterozoic (Stage I)	Ultrahigh-temperature metamorphism (M_I) in khondalite, quartzofeldspathic gneiss and Mg–Al granulite enclaves within migmatitic charnockite gneiss; $T =$ 900–950 °C at 5–8 kbar	Formation of gneissic banding within granulite enclaves with ultrahigh-temperature minerals	*c*. 1.87

partial melting of quartzofeldspathic rocks suggest that the orthopyroxene-bearing leucosomes in the charnockite gneiss formed at temperature above 800 °C (Skjerlie & Johnston 1996). Estimated P–T values from conventional geothermobarometry record lower temperatures (700–800 °C at 5–7 kbar), which may be related to the resetting of mineral compositions at lower temperature (Kelsey 2008). The dearth of quality petrological data does not allow us to fix the tectonic condition in which M_{II} metamorphism occurred. The published U–Th–Pb age of the monazite (in felsic gneiss and khondalite) and the approximately 1.55 U–Pb zircon date (in anorthosite) bracketed the age of the M_{II} metamorphism and migmatization within approximately 1.66–1.55 Ga. The oldest age of 1.66 Ga provides the minimum age of widespread felsic magmatism in the CGGC (Table 1).

After M_{II} migmatization, several parts of the CGGC were intruded by anorthosite, porphyritic granitoids and charnockite (Stage III, Table 6). U–Pb dates of zircon in anorthosite (in the BSS sector: Chatterjee *et al.* 2008, 2010) and in porphyritic charnockite (in the MBR sector, Acharyya 2003) bracketed the age of these magmatism approximately within 1.55–1.51 Ga (Acharyya 2003; Chatterjee *et al.* 2008) (Tables 1 & 6). This early Mesoproterozoic magmatic rocks were subjected to another high-grade event (M_{III}) that was accompanied by regional deformation (D_2). This event produced a crude gneissic fabric (S_2) and developed coronal garnet $\pm$ amphibole in pyroxene–plagioclase-bearing rocks in the $P–T$ range of 6.5 ± 1 kbar and 700 ± 50 °C (Maji *et al.* 2008, this study). The U–Th–Pb chemical dates of monazites and the lower intercept of the U–Pb discordia of the zircon dates indicated that this regional high-grade event and the S_2 fabric development occurred within 1.2–0.93 Ga (Acharyya 2003; Sanyal *et al.* 2007; Chatterjee *et al.* 2010) (Table 6).

The next major geological event in the CGGC is marked by emplacement of a set (or swarm) of mafic dykes that cut across the $S_1–S_2$ fabrics (Stage IV, Table 6). The dykes along with the $S_1–S_2$ foliations were repeatedly folded ($D_4–D_5$) by a strong east–west compression, producing a pervasive north–south-trending foliation cutting across the $S_1–S_2$ fabric and causing a marked swing in the regional trend (Fig. 2) The deformation also produced boudinage of the competent mafic dykes within the host of relatively incompetent felsic gneiss augmented by infiltration-driven metamorphism (M_{IV}) in the $P–T$ range of 7 ± 1 kbar and 600–750 °C in the DJR sector (Ray *et al.* 2011), and at higher pressure (9–12 kbar) and higher temperature (730–800 °C) in the DJG sector (Chatterjee *et al.* 2010). The U–Th–Pb chemical age of monazite indicates that the metamorphism that accompanied the regional north–south fabric development culminated within 0.85–0.78 Ga in the DJG sector (Sanyal *et al.* 2007) and 0.87–0.78 Ga in the DJR sector (Chatterjee *et al.* 2010). Existing geological information is not adequate to provide explanations for: (a) the restriction of north–south fabrics in a few areas in the CGGC; (b) the variation in the $P–T$ conditions of metamorphism that accompanied the north–south fabric; and (c) the variation in timing (more than 100 Ma) of the east–west compression. More studies are warranted to address these critical issues.

CGGC in the context of Proterozoic supercontinental cycles

Supercontinents that contained several continental fragments are presumed to have formed and broken apart at least three times during Proterozoic time (reviewed in Reddy & Evens 2009). Several studies have documented that Columbia (Nuna), the earliest Proterozoic supercontinent, was formed through the integration of older continental fragments during approximately 1.9–1.8 Ga (Zhao *et al.* 2004; Reddy & Evans 2009; Rogers & Santosh 2009). After its final assembly at about 1.8 Ga, Columbia underwent subduction-related growth via accretion at the margin of the supercontinent between about 1.8 and 1.3 Ga (Zhao *et al.* 2004). Exact configuration of Columbia is not constrained through precise palaeomegnetic data (Reddy & Evans 2009) that opens up speculation about the position of India and north China in the proposed configuration of Columbia (reviewed in Rogers & Santosh 2009). The CGGC records approximately 1.87–1.72 Ga felsic magmatism and high-grade metamorphism. Felsic magmatism, deformation and metamorphism of similar ages have been reported from the Aravalli–Delhi Fold Belt (ADFB: Mukhopadhyay *et al.* 2000; Kaur *et al.* 2007; Bhowmik *et al.* 2010), the Central Indian Tectonic Zone (CITZ: Roy & Prasad 2001; Bhowmik *et al.* 2005, 2010; Chakraborty *et al.* 2011); the and Eastern Ghats Belt (Simmat & Raith 2008). These metamorphic belts also witnessed approximately 1.7–1.3 Ga magmato-metamorphic events (reviewed in Bhowmik 2006;

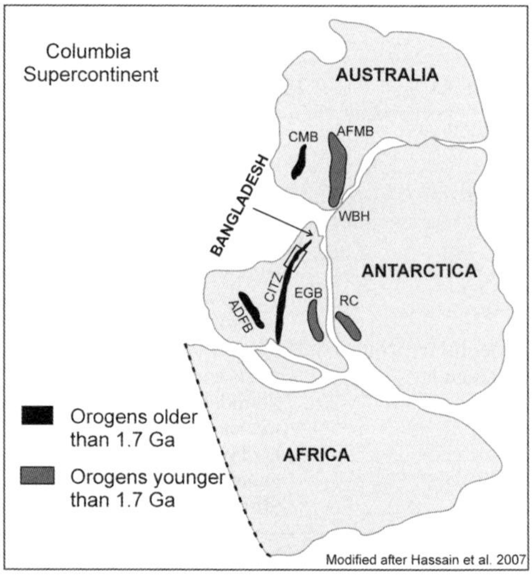

Fig. 12. Reconstructed Columbia supercontinent and its remnants in Bangladesh. The area blocked by the bold line represents the Chotanagpur Granite Gneissic Complex (CGGC) and the North Singhbhum Fold Belt (NSFB). AFMB, Albany–Fraser Mobile Belt; ADFB, Aravalli– Fold Belt; CMB, Capricorn Mobile Belt; CITZ, Central Indian Tectonic Zone; EGB, Eastern Ghats Belt; RC, Rayner Complex; WBH, Windmill Islands–Bunger Hills.

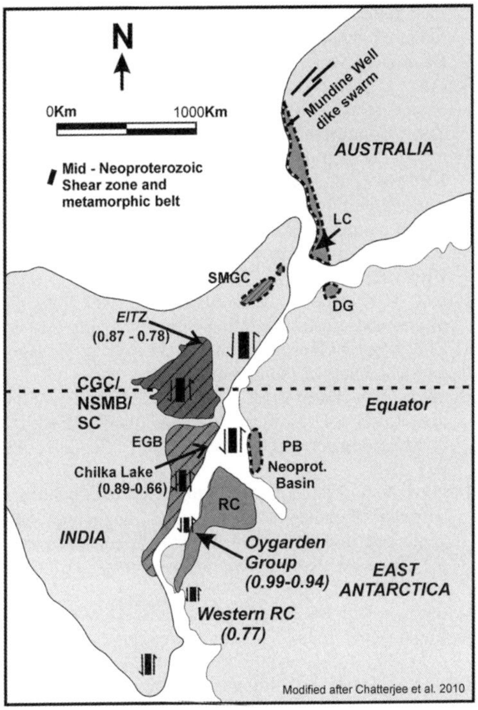

Fig. 13. Reconstructed Rodinia supercontinent with the relative positions of India, East Antarctica and Australia at 900 Ma after Li *et al.* (2008). The regions affected by the mid-Neoproterozoic orogeny are marked with filled rectangles. The bold dotted line marks the East Indian Tectonic Zone of Chatterjee *et al.* (2010). CGC, Chotanagpur Gneissic Complex; DG, Denman Glacier; LC, Leeuwin Complex; NSMB, North Singhbhum Mobile Belt; PB, Prydtz Belt; RC, Rayner Complex; SC, Singhbhum Craton; SMGC, Shillong–Meghalaya Gneiss Complex; EGB, Eastern Ghats Belt; EITZ, East Indian Tectonic Zone.

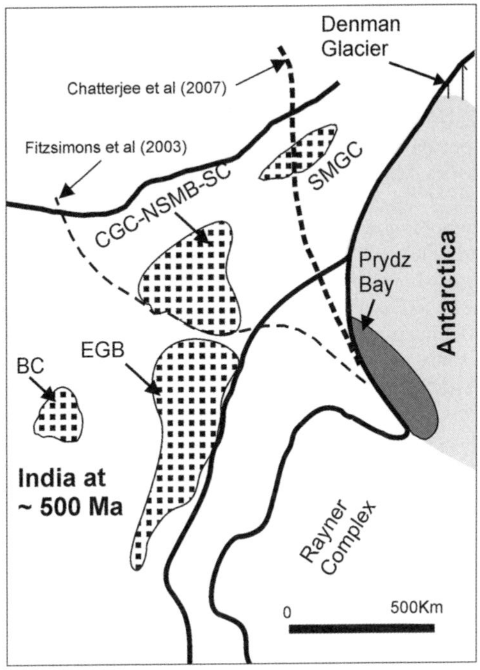

Fig. 14. Probable extent of the Pan-African suture across the eastern part of India and East Antarctica (map modified after Chatterjee *et al.* 2007). BC, Baster Craton; CGC, Chotanagpur Gneissic Complex; EGB, Eastern Ghats Belt; NSMB, North Singhbhum Mobile Belt; SC, Singhbhum Craton; SMGC, Shillong–Meghalaya Gneiss Complex.

Simmat & Raith 2008; Bhowmik *et al.* 2010; Chatterjee *et al.* 2010). Contemporaneous tectonic activities were also recorded from the North Singhbhum Mobile Belt (NSMB: Nelson *et al.* 2007; Mahato *et al.* 2008) and the gneissic belts of Shillong–Meghalay (Ghosh *et al.* 1994; Chatterjee *et al.* 2007). This information corroborated the view that Archaean crustal blocks of the north Indian Shield was stitched together by long Palaeoproterozoic orogens comprising the ADFB, CITZ, NSFB, CGGC and Meghalaya (Fig. 12). This belt can be extended to approximately 1.7–1.6 Ga old metamorphic belts of Albany–Fraser (Australia) and the Windmill Islands and Bunger Hills of Antarctica (Harris 1995) (Fig. 12). This correlation lends credence to the configuration of Columbia (Fig. 12) as proposed by Rogers & Santosh (2004). Studies have indicated that Columbia was completely fragmented after 1.3 Ga, and reassembled again at around 1.0 Ga to form another supercontinent, Rodinia (Li *et al.* 2008). A strong impression of approximately 1.0–0.93 Ga old orogenic processes have been documented in the CGGC (Maji *et al.* 2008; Chatterjee *et al.* 2010; this study), the CITZ (Lippolt & Hautmann 1994), the ADFB (Deb *et al.* 2001; Buick *et al.* 2006; Bhowmik *et al.* 2010) and Meghalaya (Chatterjee *et al.* 2007). This observation suggested that the Palaeoproterozoic sutures in the north Indian craton were rejuvenated during the formation of the Rodinia supercontinent. The east–west compression, which deformed the mafic dykes and produced the north–south fabric in the eastern part of the CGGC, is presumed to have continued further south, maintaining almost the same trend of the tectonic grain (Chatterjee *et al.* 2010). In a recent study, Chatterjee *et al.* (2010) identified an approximately 0.87 Ga orogenic pulse from the eastern part of the CGGC. This event, which produced a north–south-trending tectonic grain superposed on the east–west-trending regional fabric, can be traced down to the North Singhbhum Fold Belt, the Singhbhum Craton and the Eastern Ghats

Belt, and is termed designated as the Eastern Indian Tectonic Zone (EITZ, Fig. 13) (Chatterjee *et al.* 2010). On the basis of the U–Th–Pb chemical dating of monazite, it is postulated that deformation and metamorphism in the EITZ occurred during approximately 0.87–0.78 Ga (Chatterjee *et al.* 2010). Mafic dykes and their host rocks in the DJR sector also bear the evidence of regional east–west compression and approximately 0.87 Ga metamorphism. This Neoproterozoic tectonic event, that is also traced into the in Rayner Complex of East Antarctica (Fig. 13), is considered as an indication of the southward movement of India after its approximately 1.0 Ga collision with East Antarctica (Chatterjee *et al.* 2010). Extant geological and geochronological information from a limited area, although supporting the postulation of transcontinental EITZ, requires more studies in wider areas in order to establish the existence, extent and mechanism of orogenesis in the Indo-Antarctic landmass during about 0.78–0.67 Ga period. Existing geochronological data, however, support the view that the rifting between India and Antarctica did not start before about 0.78 Ga. Fitzsimons (2003) reconstructed an Early Cambrian configuration of Gondwanaland in which the Pan-African suture of Prydz Bay in Antarctica is suspected to have continued into the Indian Shield through the CGGC (Fig. 14). The lack of Pan-African events in the CGGC led Maji *et al.* (2008) to infer that the Pan-African suture of Prydz Bay should lie to the east of the CGGC (Fig. 14). Although this is a distinct possibility, more studies in the CGGC are warranted to testify to the inferences of Chatterjee *et al.* (2007) and Maji *et al.* (2008).

We acknowledge the financial assistance from the DST, Government of India through Research Project Grant (No. ESS/16/205/2004) and facilities under CAS, Geology, Jadavpur University. S. Sanyal carried out the monazite dating at the University of Bonn, Germany and duly acknowledges the support extended by B. Spiering during the analyses. The authors also thank R. Goswami, S. Dasgupta, S. Chaudhury, K. Mondal, S. Roy, S. Mondal, S. Ray, D. Adhikary and A. Mondal for their support at many stages of the work during the last couple of years. This manuscript has greatly benefited from the erudite comments of Prof. N. Chatterjee and Prof. S. Dasgupta. However, the authors are responsible for the views expressed in this article. We express our sincere thanks to Prof. R. Majumder for inviting us to write this lengthy review and for bearing all of the inconveniences of a delayed submission with utter patience until the end. The work is a contribution to IGCP 509, India.

References

Acharyya, A., Roy, S., Chaudhuri, B. K., Basu, S. K., Bhaduri, S. K. & Sanyal, S. K. 2006. Proterozoic rock suites along South Purulia Shear Zone, Eastern India: evidence for rift related setting. *Journal of the Geological Society of India*, **68**, 1069–1086.

Acharyya, S. K. 2003. A plate tectonic model for Proterozoic crustal evolution of Central Indian Tectonic Zone. *Gondwana Research*, **7**, 9–31.

Baidya, T. K., Chakravorty, P. S., Drubetskoy, E. & Khiltova, V. J. 1987. New geochronologic data on some granitic phases of the Chotanagpur granite gneiss complex in the northwestern Purulia dist., West Bengal. *Indian Journal of Earth Sciences*, **14**, 136–141.

Baidya, T. K., Maity, N. & Biswas, P. 1989. Tectonic phases and crustal evolution in a part of the Eastern Chotanagpur Gneissic Complex. *Journal of the Geological Society of India*, **34**, 318–324.

Baldwin, J. A., Brown, M. & Schmitz, M. D. 2007. First application of titanium-in-zircon thermometry to ultrahigh-temperature metamorphism. *Geology*, **35**, 295–298.

Banerji, A. K. 1985. On the nature of a part of boundary between Chotanagpur plateau and Singhbhum orogenic belt and its role in mineralization. *Bulletin of the Geology, Mining and Metallurgical Society of India*, **53**, 171–180.

Banerji, A. K. 1991. Presidential address. Geology of the Chotanagpur region. *Indian Journal of Geology*, **63**, 275–282.

Basu, S. K. 1993. Alkaline-carbonatite complex in Precambrian of South Purulia Shear Zone, Eastern India: its characteristics and mineral potentialities. *Indian Minerals*, **47**, 179–194.

Bhattacharya, A., Krishnakumar, K. R., Raith, M. & Sen, S. K. 1991. An improved set of a–x parameters for Fe–Mg–Ca garnets and refinements of the orthopyroxene–garnet thermometer and the orthopyroxene–garnet–plagioclase–quartz barometer. *Journal of Petrology*, **32**, 629–656.

Bhattacharya, B. P. 1976. Metamorphism of the Precambrian rocks of the central part of Santhal Parganas district, Bihar. *Quarternary Journal of the Geology, Mining and Metallurgical Society of India*, **48**, 183–196.

Bhattacharyya, P. K. & Mukherjee, S. 1987. Granulites in and around the Bengal anorthosite, eastern India: genesis of coronal garnet and evolution of the granulite–anorthosite complex. *Geological Magazine*, **124**, 21–32.

Bhowmik, S. K. 2006. Ultra high-temperature metamorphism and its significance in the Central Indian Tectonic Zone. *In*: Sajeev, K. & Santosh, M. (eds) Extreme Crustal Metamorphism and Related Crust Mantle Process. *Lithos*, **92**, 484–505.

Bhowmik, S. K., Basu Sarbadhikari, A., Spiering, B. & Raith, M. M. 2005. Mesoproterozoic reworking of Palaeoproterozoic ultrahigh-temperature granulites in the Central Indian Tectonic Zone and its implications. *Journal of Petrology*, **46**, 1085–1119.

Bhowmik, S. K., Bernhardt, H. J. & Dasgupta, S. 2010. Grenvillian age high-pressure upper amphibolites–granulite metamorphism in the Aravalli-Delhi Mobile Belt, Northwestern India: New evidence from monazite chemical age and its implication. *Precambrian Research*, **178**, 168–184.

Bohlen, S. R., Wall, V. J. & Boettcher, A. L. 1983. Geobarometry in granulites. *In*: Saxena, S. K. (ed.) *Kinetics and Equilibrium in Mineral Reactions*. Advances in Physical Geochemistry, **3**, 141–172.

Buick, I. C., Allen, C., Pandit, M., Rubatto, D. & Hermann, J. 2006. The Proterozoic magmatic and metamorphic history of the Banded Gneissic Complex, central Rajasthan, India: A-ICP-MS U–Pb zircon constraints. *Precambrian Research*, **151**, 119–142.

Chakraborty, K., Sanyal, S. & Sengupta, P. 2011. Fluid induced metamorphism in a suite of mafic dykes from Palaeo Proterozoic Mahakoshal Group, Central India. *In*: Srivastava, R. K. (ed.) *Dyke Swarms: Keys for Geodynamic Interpretation*. Springer, Berlin, 219–240.

Chakraborty, S. & Ganguly, J. 1991. Compositional zoning and cation diffusion in aluminosilicate garnets. *In*: Ganguly, J. (ed.) *Diffusion, Atomic Ordering and Mass Transfer*. Advances in Physical Geochemistry, **8**, 120–175.

Chatterjee, N., Banerjee, M., Bhattacharya, A. & Maji, A. K. 2010. Monazite chronology, metamorphism-anatexis and tectonic relevance of the mid-Neoproterozoic Eastern Indian Tectonic Zone. *Precambrian Research*, **179**, 99–120.

Chatterjee, N., Crowley, J. I. & Ghose, N. C. 2008. Geochronology of the 1.55 Ga Bengal anorthosite and Grenvillian metamorphism in the Chotanagpur Gneissic Complex, eastern India. *Precambrian Research*, **161**, 303–316.

Chatterjee, N., Mazumdar, A. C., Bhattacharya, A. & Saikia, R. R. 2007. Mesoproterozoic granulites of the Shillong–Meghalaya Plateau: Evidence of westward continuation of the Prydz Bay Pan-African suture into Northeastern India. *Precambrian Research*, **152**, 1–26.

Condie, K. C. 2004. Supercontinents and superplume events: distinguishing signals in the geologic record. *Physics of the Earth and Planetary Interiors*, **146**, 319–332.

Dasgupta, S. & Sengupta, P. 2003. Indo-Antarctic correlation: a metamorphic perspective from the Eastern Ghats Belt, India. *In*: Yoshida, M. & Dasgupta, S. (eds) *Proterozoic East Gondwana: Supercontinent Assembly and Breakup*. Geological Society, London, Special Publications, **206**, 131–143.

Dasgupta, S., Sengupta, P., Guha, D. & Fukuoka, M. 1991. A refined garnet–biotite Fe–Mg exchange geothermometer and its application in amphibolites and granulites. *Contributions to Mineralogy and Petrology*, **109**, 130–137.

Deb, M., Thorpe, R. I., Krstic, D., Corfu, F. & Davis, D. W. 2001. Zircon U–Pb and galena Pb isotope evidence for an approximate 1.0 Ga terrane constituting the western margin of the Aravalli–Delhi orogenic belt, north western India. *Precambrian Research*, **108**, 195–213.

Dunn, J. A. 1929. The geology of north Singhbhum including parts of Ranchi and Singhbhum districts. *Geological Survey of India Memoirs*, **54**, 1–280.

Dunn, J. A. 1941. *The Economic Geology of Mineral Resources of Bihar Province*. Geological Society of India, Memoir, **LXXVIII**.

Dunn, J. A. & Dey, A. K. 1942. Geology of Eastern Singhbhum. Geological Society of India, Memoir, **69**, 452.

Eckert, J. O. Jr., Newton, R. C. & Kleppa, O. J. 1991. ?H of reaction and calibration of garnet–quartz geothermometers in CMAS system by solution calorimetry of stoichiometric mineral mixes. *American Mineralogist*, **76**, 148–160.

Ellis, D. J. & Green, D. H. 1979. An experimental study of the effect of Ca upon garnet–clinopyroxene Fe–Mg exchange equilibria. *Contributions to Mineralogy and Petrology*, **71**, 13–22.

Faure, G. 2001. *Origin of Igneous Rocks: the Isotopic Evidence*. Springer, Berlin.

Fitzsimons, I. C. W. 2003. Proterozoic basement provinces of southern and southwestern Australia, and their correlation with Anterctica. *In*: Yoshida, M., Windley, B. F. & Dasgupta, S. (eds) *Proterozoic East Gondwana: Supercontinent Assembly and Break Up*. Geological Society, London, Special Publications, **206**, 93–130.

Fitzsimons, I. C. W. & Harley, S. L. 1994. The influence of retrograde cation exchange on granulite P–T estimates and a convergence technique for the recovery of peak metamorphic conditions. *Journal of Petrology*, **35**, 543–576.

Frost, B. R. & Frost, C. D. 2008. On charnockites. *Gondwana Research*, **13**, 30–44.

Ganguly, J. 1979. Garnet and clinopyroxene solid-solutions and geothermometry based on Fe–Mg distribution coefficient. *Geochimica et Cosmochimica Acta*, **43**, 1021–1029.

Ghose, N. C. 1983. Geology,tectonics and evolution of the Chotanagpur granite–gneiss complex, Eastern India. *Recent Reserches in Geology*, **10**, 211–247.

Ghose, N. C. 1992. Chotanagpur Gneiss–Granulite Complex, Eastern India: present status and future prospect. *Indian Journal of Geology*, **64**, 100–121.

Ghose, N. C. & Chatterjee, N. 2008. Petrology, tectonic setting and source of dykes and related magmatic bodies in the Chotanagpur Gneissic Complex, Eastern India. *In*: Srivastava, R. K., Sivaji, C. & Chalapathi Rao, N. V. (eds) *Indian Dykes: Geochemistry, Geophysics and Geochronology*. Narosa, New Delhi, 471–493.

Ghose, N. C., Chatterjee, N., Mukherjee, D., Kent, R. W. & Saunders, A. D. 2008. Mineralogy and geochemistry of the Bengal Anorthosite Massif in the Chotanagpur Gneissic Complex at the Eastern Indian Shield Margin. *Journal of the Geological Society of India*, **72**, 263–277.

Ghosh, S., Chakraborty, S., Paul, D. K., Bhalla, J. K., Bishui, P. K. & Gupta, S. N. 1994. New Rb–Sr isotopic ages and geochemistry of granitoids from Meghalaya and their significance in middle to late Proterozoic crustal evolution. *Indian Minerals*, **48**, 33–44.

Ghosh, S. K. & Sengupta, S. 1999. Boudinage and composite boudinage in superposed deformations and syntectonic migmatization. *Journal of Structural Geology*, **21**, 97–110.

Harley, S. L. 1998. On the occurrence and characterization of ultrahigh temperature. *In*: Treloar, P. J. & O'Brien, P. J. (eds) *What Drives Metamorphism and Metamorphic Reactions?* Geological Society, London, Special Publications, **138**, 81–107.

Harley, S. L. 2004. Extending our understanding of ultrahigh temperature crustal metamorphism. *Journal of*

Mineralogical and Petrological Sciences, **99**, 140–158.

HARLEY, S. L. & BUICK, I. S. 1992. Wollastonite–scapolite assemblages as indicators of granulite pressure temperature fluid histories – the Rauer Group, East Antarctica. *Journal of Petrology*, **33**, 693–728.

HARLEY, S. L. & THOMPSON, P. 2004. The influence of Cordierite on melting and mineral-melt equilibria in ultra-high-temperature metamorphism. *Transactions of the Royal Society of Edinburgh, Earth Sciences*, **95**, 87–98.

HARRIS, L. B. 1995. Correlation between the Albany, Fraser and Darling Mobile Belts of Western Australia and Mirnyy to Windmill Islands in the East Antarctic shield: Implications for Proterozoic Gondwanaland reconstructions. *In*: YOSHIDA, M. & SANTOSH, M. (eds) *India and Antarctica During the Precambrian*. Geological Society of India, Memoir, **34**, 47–71.

HAWKINS, D. P. & BOWRING, S. A. 1997. U–Pb systematics of monazite and xenotime: case studies from the Paleoproterozoic of the Grand Canyon, Arizona. *Contributions to Mineralogy and Petrology*, **127**, 87–103.

HOSSAIN, I., TSEUNAGE, T., RAJESH, M. H., CHEN, B. & ARAKAWA, Y. 2007. Palaeoproterozoic U–Pb SHRIMP zircon age from basement rocks in Bangladesh: a possible remnant of the Columbia supercontinent. *Comptes Rendus Geoscience*, **339**, 979–986.

KAUR, P., CHAUDHRI, N., RACZEG, I., KRONER, A. & HOFFMAN, A. W. 2007. Geochemistry, zircon ages and whole-rock Nd isotopic systematic for Palaeoproterozoic A-type granitoids in the northern part of the Delhi, belt, Rajasthan, NW India: implications for late Palaeoproterozoic crustal evolution of the Aravalli craton. *Geological Magazine*, **144**, 361–378.

KELSEY, D. E. 2008. On ultrahigh temperature metamorphism. *Gondwana Research*, **13**, 1–29.

KILPATRICK, J. A. & ELLIS, D. H. 1992. C-type magmas: igneous charnockites and their extrusive equivalents. *Transactions of the Royal Society of Edinburgh, Earth Sciences*, **83**, 155–164.

KRETZ, R. 1983. Symbols for rockforming minerals. *American Mineralogist*, **68**, 277–279.

KRISHNA, V., SASTRY, D. V. L. N., PANDEY, B. K. & SINHA, R. P. 2003. U–Pb and Pb–Pb ages on columbite–tantalite minerals from pegmatites of Bihar Mica Belt, Jharkhand, India. *In*: *Indian Society of Mass Spectrometry Silver Jubilee Symposium*. National Institute of Oceanography, Goa, 650–653.

KUMAR, M. N., DAS, N. & DASGUPTA, S. 1978. Geology and mineralization along the Northern Shear Zone, Purulia District, West Bengal- an up-to-date appraisal. *Records of the Geological Survey of India*, **133**, 25–31.

LAL, N., SAINI, H. S., NAGPAUL, K. K. & SHARMA, K. K. 1976. Tectonic and cooling history of the Bihar Mica Belt, India, as revealed by fission-track analysis. *Tectonophysics*, **34**, 163–180.

LEE, H. Y. & GANGULY, J. 1988. Equilibrium compositions of co-existing garnet and orthopyroxene: experimental determinations in the system Feo–Mgo–Al_2O_3–SiO_2 and applications. *Journal of Petrology*, **29**, 93–113.

LI, Z. X., BOGDANOVA, S. V. *ET AL.* 2008. Assembly, configuration and break-up history of Rodinia: a synthesis. *Precambrian Research*, **160**, 179–210.

LIPPOLT, H. J. & HAUTMANN, S. 1994. $^{40}Ar/^{39}Ar$ ages of Precambrian manganese ore minerals from Sweden, India and Morocco. *Mineralium Deposita*, **18**, 195–215.

MAHADEVAN, T. M. 2002. *Geology of Bihar & Jharkhand*. Geological Society of India, Bangalore.

MAHMOUD, M. Y., MITRA, A. K., DHAR, R., SARKAR, S. & MANDAL, N. 2008. Repeated emplacement of syntectonic pegmatites in Precambrian granite gneisses: indication of pulsating brittle–ductile rheological transitions. *In*: SRIVASTAVA, R. K. (ed.) *Indian Dykes: Geochemistry, Geophysics, and Geochronology*. Narosa, New Delhi.

MAHATO, S., GOON, S., BHATTACHARYA, A., MISHRA, B. & BERNHARDT, H. J. 2008. Thermo-tectonic evolution of the North Singhbhum Mobile Belt: a view from the western part of the belt. *Precambrian Research*, **162**, 102–107.

MAJI, A. K., GOON, S., BHATTACHARYA, A., MISHRA, B., MAHATO, S. & BERNHARDT, H. J. 2008. Proterozoic polyphase metamorphism in the Chotanagpur Gneiss Complex (India), and implications for transcontinental Gondwana correlation. *Precambrian Research*, **162**, 385–402.

MALLIK, A. K., GUPTA, S. N. & RAY BARMAN, T. 1991. Dating of early Precambrian granite–greenstone complex of the Eastern Indian Precambrian shield with special reference to the Chotanagpur granite gneiss complex. *Records of the Geological Survey of India*, **124**, 20–21.

MANNA, S. S. & SEN, S. K. 1974. Origin of garnet in basic granulites around Saltora, West Bengal, India. *Contributions to Mineralogy and Petrology*, **44**, 195–218.

MAZUMDAR, S. K. 1988. Crustal evolution of the Chotanagpur gneissic complex and the Mica Belt of Bihar. *In*: MUKHOPADHYAY, D. (ed.) *Precambrian of the Eastern Indian Shield*. Geological Society of India, Memoir, **8**, 49–84.

MAZUMDAR, S. K. & CHATTERJEE, S. 1990. A note on the geotectonic status of the Tamar–Porapahar fracture zone near the southern edge of the CGGC. Seminar on Tectono-magmatic evolution of CGGC. *Bulletin of the Geology, Mining and Metallurgical Society of India*, **56**, 12.

MUKHERJEE, D., GHOSE, N. C. & CHATTERJEE, N. 2005. Crystallization history of a massif anorthosite in the eastern Indian shield margin based on borehole lithology. *Journal of Asian Earth Sciences*, **25**, 77–94.

MUKHOPADHYAY, D., BHATTACHARYA, T., CHATTOPADHYAY, N., LOPEZ, R. & TOBISCH, O. T. 2000. Anasagar gneiss: A folded granitoid pluton in the Proterozoic South Delhi Fold Belt, central Rajasthan. *Proceedings of the Indian Academy of Science (Earth and Planetary Science)*, **109**, 21–37.

NELSON, D. R., BHATTACHARYA, H. N., MISRA, S., DASGUPTA, N. & ALTERMANN, W. 2007. New SHRIMP U–Pb zircon dates from the Singhbhum craton, Jharkhand–Orissa region, India. *In*: BANERJEE, S. (ed.), *Abstract Volume of the International Conference on Precambrian sedimentation and tectonics and Second GPSS Meeting*. Indian Institute of Technology, Bombay, 47.

NEWTON, R. C. & PERKINS, D. III. 1982. Thermodynamic calibration of geobarometers based on the assemblages garnet – plagioclase – orthopyroxene – (clinopyroxene) – quartz. *American Mineralogist*, **67**, 203–222.

PANDEY, B. K., GUPTA, J. N. & LALL, Y. 1986*a*. Whole rock and mineral isochron ages for the granites from Bihar mica belt of Hazaribagh, Bihar, India. *Indian Journal of Earth Sciences*, **13**, 157–162.

PANDEY, B. K., UPADHYAY, D. L. & SINHA, K. K. 1986*b*. Geochronology of Jajawal–Binda–Nagnaha granitoids in relation to Uranium mineralization. *Indian Journal of Earth Sciences*, **13**, 163–168.

PATEL, S. C. 2007. Vesuvianite–wollastonite–grossular-bearing calc-silicate rock near Tatapani, Surguja district, Chhattisgarh. *Journal of Earth System Science*, **116**, 143–147.

PATEL, S. C., SUNDARARAMAN, S., DEY, R., THAKUR, S. S. & KUMAR, M. 2007. Deformation pattern in a Proterozoic low pressure metamorphic belt near Ramanujganj, western Chotanagpur terrane. *Journal of the Geological Society of India*, **70**, 207–216.

PATTISON, D. R. M. 2003. Petrogenetic significance of orthopyroxene-free garnet + clinopyroxene + plagioclase ± quartz-bearing metabasites with respect to the amphibolites and granulite facies. *Journal of Metamorphic Geology*, **21**, 21–34.

RAY BARMAN, T., BISHUI, P. K. & MUKHOPADHYAY RAY, J. N. 1994. Rb–Sr geochronology of the high-grade rocks from Purulia, West Bengal and Jamua–Dumka sector, Bihar. *Indian Minerals*, **48**, 45–60.

RAY, S., SANYAL, S. & SENGUPTA, P. 2011. Mineralogical control on rheological inversion on a suite of deformed mafic dykes from parts of the Chotanagpur Granite Gneiss Complex of east India. *In*: SRIVASTAVA, R. K. (ed.) *Dyke swarms: Keys for Geodynamic Interpretation*. Springer, Berlin, 550.

REDDY, S., CLARK, C. & MAJUMDAR, R. 2008. Was there a pre-Nunan India iV Antarctica connection? Evidence from SHRIMP U–Pb zircon data. *In*: *33rd International Geological Congress, Oslo, 6–14 August 2008*.

REDDY, S. M. & EVANS, D. A. D. 2009. Palaeoproterozoic supercontinents and global evolution: correlations from core to atmosphere. *In*: REDDY, S. M., MAZUMDER, R., EVANS, D. A. D. & COLLINS, A. (eds) *Palaeoproterozoic Supercontinents and Global Evolution*. Geological Society, London, Special Publications, **323**, 1–26.

ROGERS, J. J. W. & SANTOSH, M. 2004. *Continents and Supercontinents*. Oxford University Press, Oxford.

ROGERS, J. J. W. & SANTOSH, M. 2009. Tectonics and surface effects of the supercontinent Columbia. *Gondwana Research*, **15**, 373–380.

ROY, A. K. 1977. Structure and metamorphic evolution of the Bengal anorthosite and associated rocks. *Journal of the Geological Society of India*, **18**, 203–223.

ROY, A. & PRASAD, H. M. 2001. *Precambrian of Central India: A possible tectonic model*. Geological Survey of India, Special Publications, **64**, 177–197.

ROY, A., DEVARAJAN, M. K. & PRASAD, H. 2002. Ductile shearing and syntectonic granite emplacement along the southern margin of the Palaeoproterozoic Mahakoshal supracrustal belt: evidence from Singrauli Area, Madhya Pradesh. *Journal of the Geological Society of India*, **59**, 9–21.

SANYAL, S., SENGUPTA, P. & GOSWAMI, R. 2007. Evidence of Mesoproterozoic ultra-high temperature metamorphism from parts of CGGC, Jharkhand, India. *In*: *Abstract Volume of the International Conference on Precambrian sedimentation and tectonics and Second GPSS Meeting. Indian Institute of Technology*, Bombay, 62–63.

SARKAR, S. N. 1980. Precambrian stratigraphy and geochronology of Peninsular India: a review. *Indian Journal of Earth Sciences*, **7**, 12–26.

SEN, S. 1956. Structures of porphyritic granite and associated metamorphic rocks of East Manbhum, Bihar, India. *Geological Society of America Bulletin*, **67**, 647–670.

SEN, S. 1959. Mineral genetic trends in the evolution of metamorphic rocks and origin of granites of East Manbhum, India. *Proceedings of the National Institute of Science of India*, **25A**, 118–138.

SEN, S. K. & BHATTACHARYA, A. 1993. Post-peak pressure–temperature–fluid history of the granulites around Saltora, West Bengal. *Proceedings of the National Academy of Sciences of India*, **63(A)**, 282–306.

SENGUPTA, P., SEN, J., DASGUPTA, S., RAITH, M. M., BHUI, U. K. & EHL, J. 1999. Ultrahigh temperature metamorphism of meta-pelitic granulites from Kondapalle, Eastern Ghats Belt: Implications for the Indo-Antarctic correlation. *Journal of Petrology*, **46**, 1068–1087.

SENGUPTA, S., GHOSH, S. K. & NANDI, N. 1998. Polyphase deformation and migmatization in the Chotanagpur gneissic complex around Jasidih and Deoghar, Bihar, India. *In*: PALIWAL, B. S. (ed.) *The Indian Precambrian*. Scientific Publishers (India), Jodhpur, 9–17.

SHARMA, R. 2009. *Cratons and Fold Belts of India*. Springer, Berlin, 304.

SIMMAT, R. & RAITH, M. M. 2008. U–Th–Pb monazite geochronometry of the Eastern Ghats Belt, India: timing and spatial disposition of poly-metamorphism. *Precambrian Research*, **162**, 16–39.

SINGH, R. N., THORPE, R. & KRISTIC, D. 2001. Galena Pb isotope date of base metal occurrences in the Hesatu–Belbathan belt, eastern Precambrian shield, Bihar. *Journal of the Geological Society of India*, **57**, 535–538.

SINGH, S. P. 1998. Precambrian stratigraphy of Bihar – an overview. *In*: PALIWAL, B. S. (ed.) *The Indian Precambrian*. Scientific Publishers (India), Jodhpur, 376–408.

SINGH, Y. & KRISHNA, V. 2009. Rb–Sr geochronology and petrogenesis of granitoids from the Chotanagpur Granite Gneiss Complex of Raikera–Kunkuri region, Central India. *Journal of the Geological Society of India*, **74**, 200–208.

SKJERLIE, K. P. & JOHNSTON, A. D. 1996. Vapour-absent melting from 10 to 20 kbar of crustal rocks that contain multiple hydrous phases: implications of anatexis in the deep to very deep continental crust and active continental margins. *Journal of Petrology*, **37**, 661–691.

VINOGRADOV, A. P., TUGARINOV, A. I., ZHYKOV, C., STAPNIKOVA, N., BIBIKOVA, E. & KHORRE, K. 1964. *Geology of the Indian Precambrian*. Report of 22nd International Geological Congress, **10**, 553–567.

WAGNER, G. A. & VAN DER HAUTE, P. 1992. *Fission-Track Dating*. Kluwer Academic, Dordrecht.

ZHAO, G. C., SUN, M., WILDE, S. A. & LI, S. Z. 2004. A Palaeo-Mesoproterozoic supercontinent: assembly, growth and break up. *Earth-Science Reviews*, **67**, 91–123.

Original nature and source of khondalites in the Eastern Ghats Province, India

S. BHATTACHARYA[1]*, A. K. CHAUDHARY[2] & M. BASEI[3]

[1]*Geological Studies Unit, Indian Statistical Institute, Kolkata 700108, India*

[2]*Indian Institute of Technology, Roorkee 247667, India*

[3]*Institute of Geosciences,São Paulo University, São Paulo 05508-080,Brazil*

**Corresponding author (e-mail: profsamar.bhatta@gmail.com)*

Abstract: The khondalites of the Eastern Ghats Province, India represent a sedimentary sequence of shale–greywacke, but the relatively more ferruginous composition can be attributed to some mafic rocks in the provenance. While a high La/Sc ratio and an insignificant Cr-enrichment over V in the majority of samples indicate dominantly granitic source rocks; a low La/Sc ratio and high Cr/V ratio in some samples indicate some mafic source rocks as well. High La/Sc ratio, LREE (light rare earth element)-enrichment, negative Eu anomaly and negative ε values, along with evidence of significant weathering and recycling in the provenance, indicate Old Upper Continental Crust-type provenance. Nd-model dates indicate Palaeoproterozoic provenance, which could be the Eastern Dharwar Craton and/or the Napier Complex of East Antarctica. $(La/Yb)_N$ ratios are variable and their co-variance with both $(La/Sm)_N$ and $(Gd/Yb)_N$ are noticeable, and could be attributed to the complexity of possible provenance composition.

Khondalites, garnet–sillimanite–K-feldspar gneisses, form a significant component in the regional granulite terrain of the Eastern Ghats, India. Erstwhile studies have revealed their restitic nature and the granulite–granite connection via biotite-dehydration melting under granulite-facies conditions (Sen & Bhattacharya 1997, 2000; Bhattacharya & Kar 2002). Although Dash *et al.* (1987) suggested a palaeosol (deeply weathered soil) origin for the khondalite precursor of the Eastern Ghats, the common presence of centimetre- to metre-scale quartzite layers are difficult to explain in this model, as pointed out by Nanda & Pati (1991). On geochemical grounds, Condie *et al.* (1992) also argued against a palaeosol origin for Archaean and early Proterozoic khondalites. In addition, Sen & Bhattacharya (1997) argued against the palaeosol origin from the viewpoint of the demonstrably granulite–granite connection. Thus, a sedimentary parentage is evident for the khondalite suite in the Eastern Ghats Belt, which also includes calc-silicate rocks in some parts of this granulite belt (Sen *et al.* 1995; Bhowmik 1997; Sen & Bhattacharya 2000).

Although Sen & Bhattacharya (1997) and Bhattacharya & Kar (2002) described different types of pelitic precursors, the common and dominant variety of khondalite is characterized by ferruginous composition, both in bulk composition and in garnet composition, and was described as original shale. Some early work by Divakara Rao (1983) suggested a mixed source of Na-rich trondhjemites and early Archaean mafic rocks.

However, more ferruginous composition relative to the average shale needs some explanation. In this paper, we address this question with additional trace element (including rare earth element (REE)) data to reveal the original nature of the sediments and their source rocks. Nd–Sr isotopic data on some selected samples from the Eastern Ghats Province, together with the geochemical data, were utilized to constrain the depositional age of the sediments and the possible source terrane (provenance).

Geological setting

The Eastern Ghats granulite belt, along the east coast of India, is bounded by granite–greenstone terrains of the Singhbhum Craton to the north, and the Bastar and East Dharwar cratons to the west. The lithological make-up can be broadly grouped into the following: (a) metapelitic granulite including khondalites, quartzite and calc-granulite; (b) charnockitic gneisses and associated mafic granulites; (c) peraluminous granite gneisses and migmatites; (d) anorthosites; and (e) alkaline complexes (Fig. 1).

Granulite lithologies record a polyphase deformation and complex metamorphism. Isoclinal and rootless early folds with a NE–SW-trending steep axial-plane foliation and common structural repetition suggest a regional NW–SE-directed compression

From: MAZUMDER, R. & SAHA, D. (eds) 2012. *Palaeoproterozoic of India*. Geological Society, London, Special Publications, **365**, 147–159. http://dx.doi.org/10.1144/SP365.8

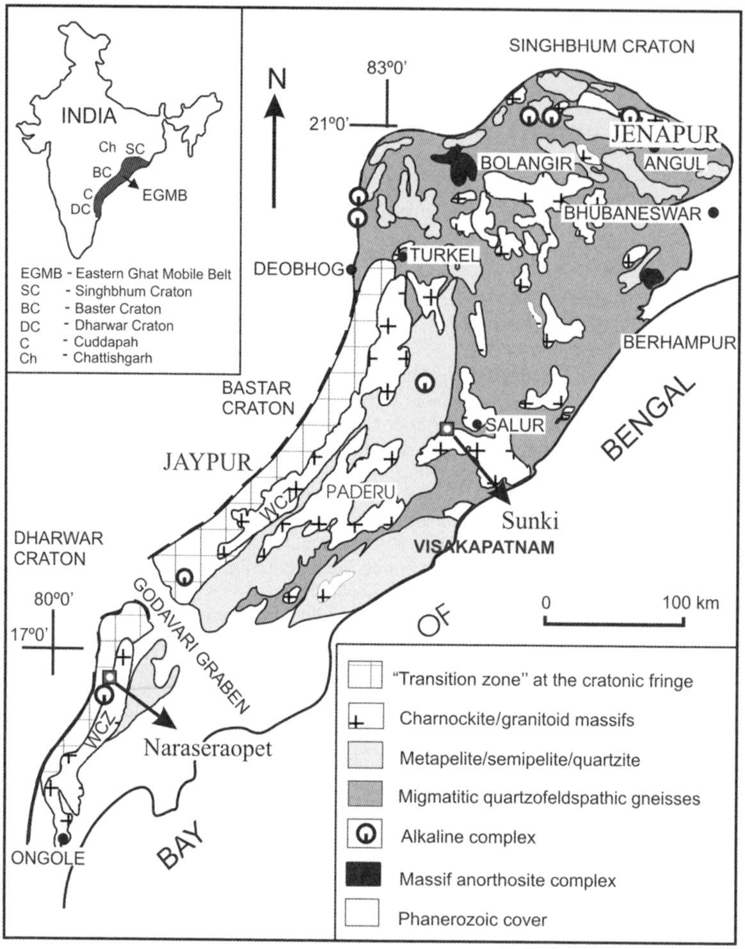

Fig. 1. Generalized geological map of the Eastern Ghats Belt, India.

during the development of first-generation folds (Bhattacharya 1997). Significant crustal thickening during the first deformation is also evident from the record of 8–9 kbar of pressure in the exhumed granulites, particularly in the metapelitic granulites (reviewed in Bhattacharya 1996).

Complex tectonothermal records include: ultrahigh temperature in metapelitic migmatites (Dasgupta *et al.* 1994; Sen *et al.* 1995; Mohan *et al.* 1997; Sengupta *et al.* 1999; Bhattacharya & Kar 2002); dehydration melting in different crustal protoliths (Dasgupta *et al.* 1992; Sen & Bhattacharya 1997, 2000; Bhattacharya & Kar 2002; Kar *et al.* 2003); and complex and possibly distinct pressure–temperature ($P–T$) paths in different sectors of this regional granulite terrain (Sengupta *et al.* 1990; Sen *et al.* 1995; Mohan *et al.* 1997; Mukhopadhyay & Bhattacharya 1997; Bhattacharya *et al.* 2003).

Based on Nd-mapping, different crustal domains with unconnected pre-metamorphic history have been identified in the Eastern Ghats granulite belt (Bhattacharya *et al.* 2001; Rickers *et al.* 2001; Dobmeier & Raith 2003).

In this paper, we have selected for analysis typical khondalite samples – garnet–sillimanite–quartz–K-feldspar–Fe–Ti oxides – from the three suites of Chilka, Rayagada and Paderu in the Eastern Ghats Province (northern Eastern Ghats Belt). All of the three suites belong to crustal domains 2 and 3, for which Nd-model dates of khondalites have been reported as between 2.0 and 2.4 Ga (Rickers *et al.* 2001). Although multiple granulite-facies events have been recorded, the earliest granulite-facies event in the Eastern Ghats Province is reported as approximately 1.2 Ga (Simmat & Raith 2008) and this could provide an important constraint for the sediment deposition.

Metamorphism vis-à-vis chemistry

Most studies of the effects of granulite-facies metamorphism on chemical composition have concluded that, although certain large-ion lithophile elements (LILE) (U, Th, Rb, Cs, Pb) can be strongly depleted, many elements, including REE and high field strength elements (HFSE) are relatively immobile, if no metasomatism accompanies metamorphism (Condie *et al.* 1992 and references therein). In terms of the major elements, Sen & Bhattacharya (1997) considered the Eastern Ghats Province khondalites of the Chilka suite and concluded that these khondalites closely follow the advanced weathering profile of Nesbitt & Young (1984). It may also be emphasized here that no evidence of metasomatism was recorded from these Eastern Ghats Province khondalites (Sen & Bhattacharya 1997, 2000; Bhattacharya & Kar 2002).

Fe-enrichment

Earlier publications have described the precursors of these khondalites of the Eastern Ghats Province as shale. However, the relatively ferruginous bulk composition remains unexplained so far. Here we attempt to explain the Fe-enriched bulk composition of 25 khondalite samples from three different suites of Chilka, Rayagada and Paderu from published literature (Sen & Bhattacharya 1997, 2000; Bhattacharya & Kar 2002).

Alumina-enrichment is expected in the restitic khondalites along with depletion of potash and iron. Hence, Fe-enrichment against Al could be appropriate in deciphering precursor characteristics. Following Li & Schoonmaker (2004), we have computed the enrichment factor, E_{Al}^{Fe}, for 25 khondalites and compared them with those found in average shale and pelagic clay (Table 1).

The pelagic clay could have been a possible candidate for the ferruginous composition of the khondalites, as indicated by a comparative Fe-enrichment factor of 0.724 in the pelagic clay and by a khondalite average of 0.732 (SD = 0.036). However, some other characteristics of pelagic clays, such as Mn- and Na-enrichment factors, 0.0798 and 0.333, respectively, do not match with those of the khondalite averages of 0.0121 (SD = 0.0014) and

Table 1. *Fe-enrichment relative to alumina in khondalites in the Eastern Ghats Belt Province, India**

Sample	Al (%)	Fe (%)	Mn (%)	Na (%)	F_{Al}^{Fe}	E_{Al}^{Mn}	E_{Al}^{Na}
C-TA215	10.180	7.420	0.132	0.112	0.730	0.013	0.011
C-13/93	13.480	7.860	0.155	0.170	0.580	0.011	0.013
C-K-5	13.740	8.360	0.232	0.156	0.610	0.017	0.011
C-SD50	6.620	6.560	0.077	0.060	0.990	0.012	0.009
C-3/93	6.620	5.820	0.070	0.178	0.880	0.011	0.027
C-10/93	7.760	7.080	0.054	0.148	0.910	0.007	0.019
C-SL-4	11.920	11.300	0.364	0.074	0.950	0.031	0.006
C-J-6	12.940	7.640	0.008	0.022	0.590	0.001	0.002
C-4/1	10.600	11.520	0.217	0.008	1.090	0.020	0.001
C-T-10	7.800	6.480	0.008	0.014	0.830	0.001	0.002
R-95/14A	9.520	6.480	0.132	0.074	0.718	0.014	0.008
R-95/17B	7.200	4.040	0.062	0.550	0.561	0.008	0.076
R-95/18	10.880	4.680	0.116	0.104	0.430	0.011	0.009
R-95/33	8.700	7.200	0.163	0.074	0.828	0.019	0.009
R-97/11	7.000	4.520	0.062	1.280	0.646	0.009	0.182
R-97/20	6.760	4.080	0.093	0.990	0.604	0.014	0.147
R-97/23A	7.380	2.820	0.116	2.230	0.518	0.015	0.302
R-95/3	3.800	3.720	0.023	0.074	0.979	0.006	0.019
R-95/17A	4.480	3.880	0.030	0.074	0.866	0.006	0.017
R-97/5A	4.260	4.240	0.108	0.882	0.995	0.026	0.209
P-53	6.800	4.360	0.031	0.740	0.641	0.005	0.109
P-20A	9.860	6.820	0.161	1.570	0.691	0.016	0.159
P-88	7.700	3.680	0.070	2.580	0.478	0.009	0.335
P-18	9.760	5.840	0.108	0.149	0.598	0.011	0.153
P-60A	9.720	5.800	0.085	1.340	0.597	0.009	0.138
Mean	8.620	6.090	0.107	0.546	0.732	0.012	0.079
SD	0.530	0.440	0.015	0.145	0.036	0.001	0.023
Shale	8.800	4.720	0.085	0.590	0.536	0.010	0.067
Pelagic clay	8.400	6.500	0.670	2.800	0.774	0.080	0.333

*C, Chilka suite; R, Rayagada suite; P, Paderu suite.

0.0789 (SD = 0.0226), respectively. These values in khondalites are comparable to those of average shale, 0.0097 and 0.067, respectively. The relatively high Na in some samples could represent greywacke-like precursors; incidentally, plagioclase feldspars have been observed in them (cf. Bhattacharya & Kar 2002).

It is important to note that pelagic clays accumulate at abyssal depths, below compensation depths for carbonate minerals (Li & Schoonmaker 2004), but the khondalites of the Eastern Ghats Province are associated with calc-silicate rocks and this precludes pelagic clays as the khondalite precursors.

The association khondalite, quartzite, calc-silicate rocks and the evidence of some greywacke-like precursors in the khondalite are more akin to a shale–greywacke sequence, commonly described as marine turbidites. Khondalites of the Lapland granulite belt was described as marine turbidites by Barbey *et al.* (1982).

The above considerations lead us to view the khondalite precursors of the Eastern Ghats Province as a shale–greywacke sequence; the ferruginous composition relative to average shale is more likely to be a feature of the source rocks in the provenance. In other words, the Fe-enriched composition of these khondalites could be explained by the presence of some mafic source rocks in the provenance.

A ternary plot of A–CN–K, designed to illustrate weathering of typical rock types and their products, may be used to constrain the source rocks of the khondalite precursors. In this diagram some samples plot close to the A–CN join, these are the likely products of weathering of mafic rocks (Fig. 2).

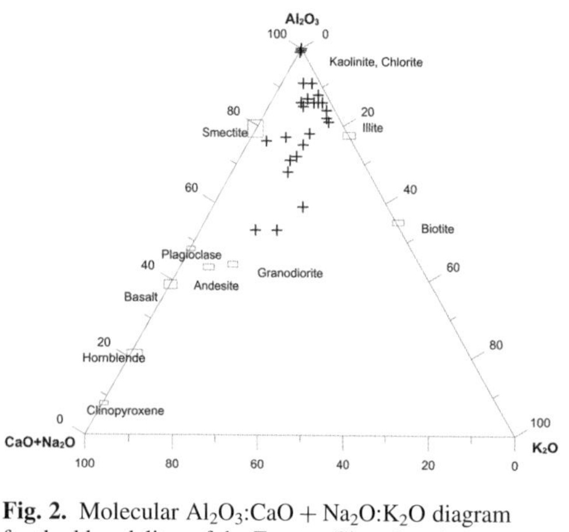

Fig. 2. Molecular Al_2O_3:CaO + Na_2O:K_2O diagram for the khondalites of the Eastern Ghats Province.

Trace element constraints

Analytical procedure

For the determination of concentrations of trace and rare earth elements, about 20 mg of each of the samples were weighed in a screw-capped Teflon (Savilex) vial. The samples were digested in the conventional way by adding an HF and HNO_3 mixture to them and leaving the tightly closed vials on a hot plate at approximately 100 °C for 2 days. The digested samples were evaporated to dryness, fluxed with concentrated HNO_3 and evaporated to dryness again. The dry sample cakes were dissolved and diluted to 100 g of the sample solution in 2N HNO_3. Ultra-pure acids were used for the digestion. All measurements were performed on a Perkin-Elmer Sciex ELAN DRC 6000 inductively coupled plasma-mass spectrometry (ICP-MS) at the Institute Instrumentation Center of the Indian Institute of Technology, Roorkee. For calibrating the instrument two USGS rock standards, GSP-2 and AGVO-2 were used. The standards were digested in an identical manner to the samples. The calibration curve was a straight line through zero. The drift of the instrument was monitored by measuring one of these standard solutions after every 10 samples. The instrument demonstrated good stability and no correction was necessary for drift; and the instrument software had built-in empirical formulae for interference correction. All of the data was corrected for blanks. The precision of the data was expressed as the residual standard deviation (RSD) in per cent defined as (the standard deviation ÷ mean concentration) × 100 for a particular element. In general, the RSD for all of the elements was much less than 4%. Individually, they are less than 2.5% for alkali and alkaline earth metals (Rb 2.23%; Sr 1.38%; Ba 1.55%), less than 2% for the transition elements, except for Sc and Hf (Zr 0.9%; Nb 1.07%; Ni 1.8%; Cr 1.4%; V 1.43%; Co 1.28%; Sc 10%; Hf 3.18%; Cu 1.6%), within 3% for LREE (La 1.2%; Ce 1.0%; Nd 1.6%; Eu 2.5%; Sm 3.0%; Pr 2%) and within 4% for HREE (Dy 2.2%; Tb 2.2%; Gd 2.4%; Ho 2.8%; Er 3.2%; Yb 3.4%; Tm 3.4; Lu 3.9%).

Results

Trace element data for 24 khondalite samples are presented in Table 2. The distribution of certain trace elements in terrigenous sedimentary rocks is especially sensitive to the type of the provenance because during most types of sedimentary processes elements such as Cr, V, Ni, Th, U, Sc, La, Zr, Hf, Y and rare earth elements (REEs) are strongly retained in particulate material (McLennan *et al.* 2003).

Table 2. *Trace element composition in ppm of khondalites, Eastern Ghats Belt Province, India*

Sample Elements	C-10/93	C-T-10	C-Ta215	C-SL-4	C-3-93	C13-93	C4/1	C-J6	C-K-5	C-SD-50	P-97/88	P-97/20A	P-9760A	P-18
Cr	817.8	102.8	667.8	121.4	1035.5	559.4	142.8	153.7	116.4	686.5	21.7	119.5	88.2	82.3
Ni	22.1	22.8	27.4	40.3	20.4	70.7	18.9	16.6	37.7	18.6	7.8	17.7	8.4	10.0
Co	14.8	78.8	23.5	121.8	16.9	19.3	556.5	23.9	84.0	23.9	119.7	150.8	112.4	183.4
Sc	8.3	9.4	17.0	27.2	10.7	20.4	9.4	11.1	10.1	10.3	9.6	9.6	6.8	10.0
V	76.8	62.9	78.1	139.1	50.9	96.4	118.9	108.1	87.8	49.8	3.1	87.6	61.5	63.8
Rb	124.8	1.3	92.7	53.8	131.4	126.7	1.2	1.4	152.3	57.6	111.0	53.3	102.9	112.9
Ba	516.0	22.9	495.7	642.9	547.2	664.1	304.7	59.7	680.9	283.5	638.8	479.4	594.7	1058.6
Sr	56.4	7.5	70.3	77.5	31.5	92.6	16.3	4.7	96.5	23.5	164.8	53.8	61.4	116.8
Nb	31.0	20.9	23.7	23.5	25.2	25.1	18.1	23.9	26.7	25.4	14.4	7.2	19.3	5.7
Zr	532.5	454.0	552.1	378.6	478.3	358.3	135.2	175.0	494.1	483.4	294.6	512.6	711.5	217.8
Hf	10.3	9.4	9.4	6.3	7.6	6.0	3.8	3.9	8.3	8.9	6.3	9.4	11.5	4.4
Y	41.9	10.9	62.4	43.7	40.5	40.2	13.1	4.6	48.3	48.9	140.9	72.5	48.5	68.2
Th	31.4	16.5	17.6	8.6	10.7	13.8	18.5	17.1	23.9	11.2	108.1	17.7	47.7	59.2
U	1.6	1.6	1.6	0.6	0.8	2.4	1.1	0.7	2.2	1.2	3.5	0.7	1.5	1.4
La	83.5	68.9	18.5	10.2	10.2	12.9	61.1	51.6	19.8	48.2	83.9	51.6	35.9	66.2
Cr/V	10.65	1.63	8.55	0.87	20.34	5.80	1.20	1.42	1.33	13.79	7.00	1.36	1.43	1.29
V/Ni	3.48	2.76	2.85	3.45	2.50	1.36	6.29	6.51	2.33	2.68	0.40	4.95	7.32	6.38
Ni/Co	1.49	0.29	1.17	0.33	1.21	3.66	0.03	0.69	0.45	0.78	0.07	0.12	0.07	0.05
Zr/Hf	51.70	48.30	58.73	60.10	62.93	59.72	35.58	44.87	59.53	54.31	46.76	54.53	61.87	49.50
Th/U	20.26	10.16	10.96	15.19	13.38	5.75	16.82	23.75	10.86	9.33	30.89	25.29	31.80	42.90
Th/Sc	3.78	1.76	1.04	0.32	1.00	0.68	1.97	1.54	2.37	1.09	11.26	1.84	7.01	5.92
Zr/Sc	64.16	48.30	32.48	13.92	44.70	17.56	14.38	15.77	48.92	46.93	30.69	53.40	104.63	21.78
Rb/Sr	2.21	0.17	1.32	0.69	4.17	1.37	0.07	0.30	1.58	2.45	0.67	0.99	1.68	0.97
La/Sc	10.06	7.33	1.09	0.38	0.95	0.63	6.50	4.65	1.96	4.68	8.74	5.38	5.28	6.62
La/Th	2.66	4.18	1.05	1.19	0.95	0.93	3.30	3.02	0.83	4.30	0.78	2.92	0.75	1.12
Y/Ni	1.90	0.48	2.28	1.08	1.99	0.57	0.69	0.28	1.28	2.63	18.06	4.10	5.77	6.82

(Continued)

Table 2. *Continued*

Sample Elements	R-95/3	R-95-14A	R95-17A	R95-17B	R-95/33	R-97-5A	R-97-11	R-97-20	R-97-23A	R-95-18
Cr	48.2	111.7	48.9	8.2	133.5	41.4	105.9	54.7	14.5	50.8
Ni	15.6	24.7	2.4	20.4	74.8	12.2	51.5	5.0	2.9	23.1
Co	125.4	109.0	105.4	108.2	168.8	153.7	175.1	169.8	137.1	95.4
Sc	3.1	8.8	1.8	6.6	10.6	4.8	5.6	5.4	3.2	25.4
V	13.1	77.7	21.2	10.5	98.4	17.8	68.6	26.6	30.6	47.9
Rb	20.9	45.4	23.8	160.1	42.3	17.6	71.4	76.3	113.5	165.6
Ba	224.6	250.8	318.9	1266.5	425.8	176.4	730.6	748.7	1045.9	1368.0
Sr	16.1	28.1	8.5	141.9	44.8	8.6	79.9	75.8	202.8	88.6
Nb	13.5	23.8	10.9	10.9	20.4	14.2	21.6	17.1	14.9	31.4
Zr	1241.3	696.9	308.6	171.3	188.5	837.4	229.5	647.1	196.4	1008.4
Hf	19.3	11.2	6.6	3.5	3.6	14.9	4.2	10.4	3.3	17.8
Y	19.8	57.3	20.8	15.8	78.5	34.9	17.4	54.5	36.4	62.7
Th	22.1	16.3	10.4	12.3	30.9	20.6	1.9	28.4	26.8	11.3
U	1.6	1.2	0.4	0.7	0.8	1.8	0.2	1.2	1.4	2.1
La	38.5	18.3	55.2	31.7	31.9	61.9	18.6	22.0	26.7	35.4
Cr/V	3.68	1.44	2.31	0.78	1.36	2.33	1.54	2.06	0.47	1.06
V/Ni	0.84	3.15	8.83	0.51	1.32	1.46	1.33	5.32	10.55	2.07
Ni/Co	0.12	0.23	0.02	0.19	0.44	0.08	0.29	0.03	0.02	0.24
Zr/Hf	64.32	62.22	46.76	48.94	52.36	56.20	54.64	62.22	59.52	56.65
Th/U	13.63	13.74	24.76	17.18	37.41	11.44	8.80	23.67	18.72	5.38
Th/Sc	7.13	1.85	5.78	1.86	2.92	4.29	0.34	5.26	8.38	0.44
Zr/Sc	400.42	79.19	171.44	25.95	17.78	174.46	40.98	119.83	61.38	39.70
Rb/Sr	1.30	1.62	2.79	1.13	0.94	2.05	0.89	1.01	0.56	1.87
La/Sc	12.42	2.08	30.67	4.80	3.01	12.90	3.32	4.07	8.34	1.39
La/Th	1.74	1.12	5.31	2.58	1.03	3.00	9.79	0.77	1.00	3.13
Y/Ni	1.27	2.32	8.67	0.77	1.05	2.86	0.34	10.90	12.55	2.71

However, as pointed out in a previous section, granulite-grade metamorphism may remobilize/deplete some trace elements and, hence, caution should be exercised while interpreting trace element distribution in the khondalites.

HFSE and Y

Zr and Hf content is variable: between 135 and 1241 ppm and between 3.3 and 19.3 ppm, respectively. However, Hf correlates well with Zr (Zr/Hf = 55) and this indicates zircon's control on both of these elements. PAAS (Post-Archaean Australian Shale)-normalized trace element distribution patterns for the Eastern Ghats Province khondalites show a remarkable similarity to those of the shales (Fig. 3a).

Transition metals

These elements are normally enriched (compatible) in the restitic granulites and, hence, their absolute concentrations cannot be used to decipher source rock compositions. However, certain ratios of similarly compatible elements may reflect the source rock type. Particularly sensitive is the Cr/V ratio. While most of the samples show only a marginal enrichment of Cr over V (average Cr/V = 1.28; $n = 15$), indicating granitic source rocks (Fig. 3a), some khondalite samples from all of the three suites show a significant Cr-enrichment (average Cr/V: 7.8; $n = 9$), and this suggests the involvement of some mafic source rocks (Fig. 3b). Samples with significant Cr-enrichment are also Y-enriched (average 4.5; $n = 9$), indicating no ophiolite or ultra-mafic source. A Cr/V v. Y/Ni plot (Fig. 4) further confirms the dominant granitic source rocks in the provenance.

Other important ratios

The Th/Sc and La/Sc ratios, representing the ratios of a highly incompatible to relatively compatible element, are useful in identifying source components in the provenance (McLennan *et al.* 1993). High La/Sc ratios in most samples, average 8.16 ($n = 19$), indicate a relatively felsic average provenance composition (McLennan *et al.* 1993). Low values of La/Sc in a few samples, average 1.2 ($n = 5$), indicate some mafic component in the

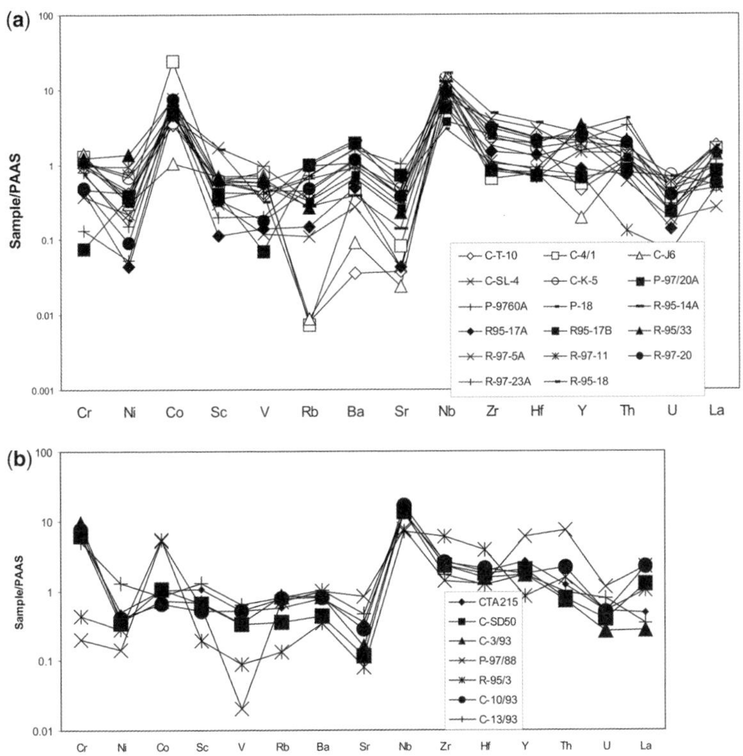

Fig. 3. PAAS-normalized spider plot of the khondalites of the Eastern Ghats Province. (**a**) For the dominant group, with marginal Cr-enrichment over V. (**b**) For the other group with significant Cr-enrichment over V. PAAS-normalizing values are from Taylor & McLennan (1985).

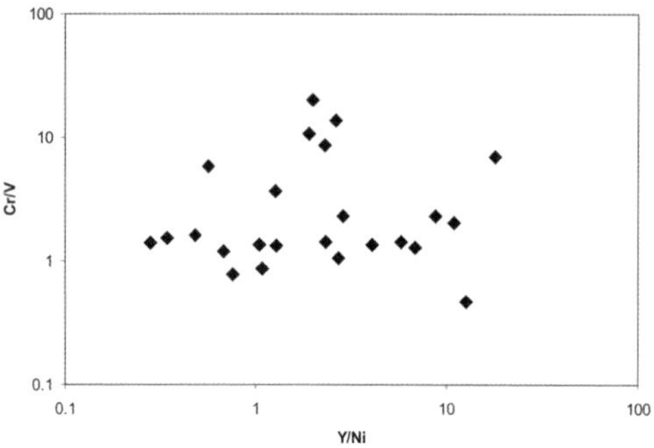

Fig. 4. Cr/V v. Y/Ni diagram for the khondalites of the Eastern Ghats Province.

provenance. La/Th ratios are variable between 0.75 and 9.79 but the average of 2.39 is similar to the PAAS value of 2.62. A few samples with a high La/Th > 4.0 may relate to mafic source rocks (Taylor & McLennan 1985). These variations in the khondalites of the Eastern Ghats Province relative to the PAAS are depicted in Figure 3a, b.

Rb/Sr and Th/U ratios are high in many samples, average 2.04 ($n = 12$) and 20 ($n = 21$), respectively, and may indicate significant weathering and recycling in the provenance. Highly variable Th/Sc and Zr/Sc ratios are recorded in the khondalites and two groups can be identified, namely: high Th/Sc (average 6.5; $n = 9$) and Zr/Sc (average 100; $n = 15$); and low Th/Sc (average 1.38; $n = 15$) and Zr/Sc (average 21; $n = 9$), respectively. Th/Sc v. Zr/Sc plot (Fig. 5) shows a substantial increase in Zr/Sc but less of an increase in Th/Sc, and this could have resulted from zircon concentration (sorting) during sedimentary recycling. Relatively low Zr/Sc values in many samples may relate to mafic source composition with a low zircon content.

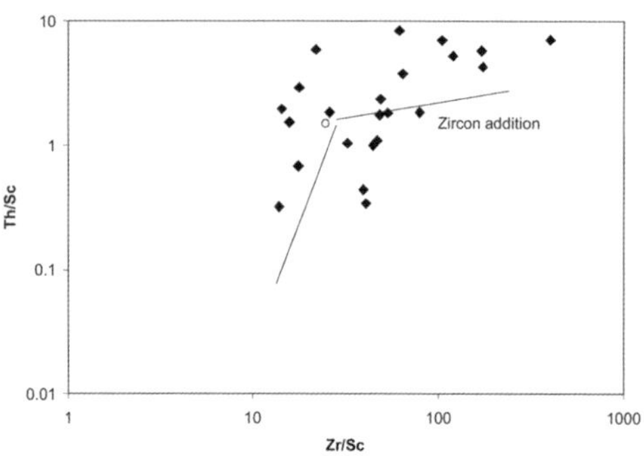

Fig. 5. Th/Sc v. Zr/Sc diagram for the khondalites of the Eastern Ghats Province. Open circle for the Upper Continental Crust and trend lines after McLennan *et al.* (2003).

REE

The distribution of rare earth elements (REEs) in sedimentary rocks most faithfully reflects the character of the provenance (Taylor & McLennan 1985) (Table 3). The khondalites are variably light rare earth element (LREE)-enriched, $(La/Sm)_N$ between 3 and 24, indicating an LILE-enriched source. $(La/Yb)_N$ ratios are variable between 1.76 and 24.86, and co-variance with both $(La/Sm)_N$ and $(Gd/Yb)_N$ are noticeable and could be attributed to the complexity of possible provenance composition (Fig. 6). Eu depletion (negative anomaly) reflects the intracrustal differentiation of the igneous source rocks. Two groups can be identified, namely: the significant negative Eu anomaly (0.11–0.33); and the moderate negative anomaly (average 0.61; $n = 16$). The group with the moderate negative anomaly comparable to upper-crustal rocks is also enriched in heavy rare earth elements (HREEs), with a very low $(Gd/Yb)_N$ average 0.37 ($n = 18$), and this indicates a considerable concentration of xenotime during sorting in the provenance. This group also has an elevated Zr content of between 454 and 1241 ppm, and could reflect the zircon control. The other group with a significant negative Eu anomaly (average 0.22; $n = 8$) is are relatively less enriched in HREE (average $(Gd/Yb)_N$ 0.71; $n = 6$) and has a nearly flat HREE pattern, commonly observed in upper-crustal rocks (Fig. 6). No HREE depletion, with low $(Gd/Yb)_N$ ratios, would imply that garnet fractionation has been unimportant in the history of the igneous precursors, unlike those in the Archaean. A Eu/Eu* v. $(Gd/Yb)_N$ plot may be used to monitor the source compositions. While the majority of samples have relatively low $(Gd/Yb)_N$ values (less than 0.50) and higher Eu/Eu* values (greater than 0.4), comparable to those of granitic rocks, there are also some samples with higher $(Gd/Yb)_N$ and lower Eu/Eu* values comparable to those in basaltic rocks (Fig. 7). While greater enrichment of LREE may be ascribed to significant monazite, HREE-enrichment may be ascribed to xenotime components (Fig. 6). It is important to note that while monazite is commonly observed in the khondalites of the Eastern Ghats Province (Simmat & Raith 2008), a significant xenotime component has also been recorded from khondalites in Eastern Ghats Province (Dobmeier & Raith 2003).

Isotopic constraints

Analytical procedure

The isotopic analyses were performed at the Centre of Research in Geochronology, São Paulo University, using a VG-354 multi-collector mass spectrometer.

Table 3. *Rare earth element composition of khondalites, Eastern Ghats Belt Province, India*

Sample Elements	C-10/93	C-T-10	C-Ta215	C-SL-4	C-3-93	C13-93	C4/1	C-J6	C-K-5	C-SD-50	P-97/88	P-97/20A	P-9760A	P-18
La	83.60	68.90	18.50	10.20	10.20	12.92	61.14	51.58	19.79	48.22	83.92	51.57	35.99	66.20
Ce	165.00	121.30	45.90	25.60	28.40	42.82	121.65	81.64	62.65	86.73	217.61	99.05	107.92	160.03
Nd	84.20	73.00	11.50	8.90	5.90	8.67	65.50	46.77	13.60	51.59	82.93	57.78	24.36	66.17
Sm	14.50	13.10	0.80	0.90	0.47	0.54	11.40	8.41	0.83	9.84	11.29	10.53	1.44	9.90
Eu	2.80	2.60	0.06	0.07	0.04	0.07	2.25	1.66	0.08	1.92	1.24	1.78	0.06	1.18
Gd	12.00	10.60	0.90	1.00	0.61	0.76	9.36	6.76	1.02	8.21	7.07	7.92	1.43	5.90
Tb	1.40	0.80	1.20	1.30	0.83	0.99	1.06	0.51	1.25	1.29	2.98	1.90	1.37	1.90
Dy	21.60	19.20	1.90	1.80	1.21	1.48	16.84	12.17	1.78	15.24	12.41	14.53	1.68	10.00
Ho	1.14	0.43	1.60	1.40	1.00	1.09	0.66	0.23	1.32	1.28	3.54	1.99	1.27	1.76
Er	15.50	13.90	1.10	0.90	0.66	0.73	12.12	8.86	0.85	10.93	8.59	10.23	0.82	6.88
Yb	27.20	24.40	1.90	1.50	1.12	1.17	21.00	15.43	1.30	18.98	14.71	17.50	1.22	11.71
Lu	5.80	5.40	0.09	0.05	0.04	0.06	4.56	3.38	0.07	3.97	2.52	3.51	0.06	2.28
∑REE	434.7	353.6	85.5	53.6	50.5	71.3	327.5	237.4	104.5	258.2	448.8	278.3	177.6	343.9
(La/Yb)N	2.09	1.92	6.61	4.62	6.19	7.50	1.98	2.27	10.34	1.73	3.88	2.00	20.04	3.84
(La/Sm)N	3.60	3.28	14.44	7.08	13.55	23.93	3.35	3.83	14.89	3.06	4.64	3.06	15.61	4.18
(Gd/Yb)N	0.36	0.35	0.38	0.54	0.44	0.53	0.36	0.35	0.63	0.35	0.39	0.37	0.95	0.41
Eu/Eu*	0.65	0.67	0.22	0.22	0.23	0.33	0.66	0.67	0.27	0.65	0.42	0.59	0.11	0.47

Sample Elements	R-95/3	R-95-14A	R95-17A	R95-17B	R-95/33	R-97-5A	R-97-11	R-97-20	R-97-23A	R-95-18	Chondrite
La	38.5	18.3	55.2	31.7	31.9	62.0	18.6	22.0	26.7	35.4	0.2370
Ce	87.8	47.2	97.4	70.1	81.1	123.1	30.6	67.0	78.7	74.0	0.6130
Nd	36.9	14.8	61.2	31.0	25.9	65.4	18.4	14.9	17.3	33.1	0.4570
Sm	5.5	1.8	11.4	4.8	3.0	11.6	3.6	0.9	0.9	5.6	0.1480
Eu	1.0	0.3	2.3	0.9	0.5	2.3	0.8	0.1	0.1	1.1	0.0563
Gd	4.6	1.8	9.4	4.0	2.9	9.7	3.0	1.0	0.9	4.9	0.1990
Tb	0.8	1.1	0.8	0.7	1.6	1.1	0.3	1.2	0.9	1.1	0.0361
Dy	8.0	3.5	17.1	6.9	5.6	17.4	5.6	1.7	1.2	9.2	0.2460
Ho	0.5	1.5	0.6	0.4	2.0	0.9	0.5	1.4	0.9	1.6	0.0546
Er	5.6	2.2	12.4	4.8	3.6	12.7	4.1	0.9	0.6	6.9	0.1600
Yb	9.7	3.7	21.3	8.2	6.0	22.2	7.1	1.4	0.7	12.6	0.1610
Lu	2.1	0.6	4.6	1.8	1.0	4.7	1.5	0.1	0.0	2.3	0.0246
∑ REE	201.0	96.8	293.6	165.3	165.1	333.0	93.8	112.4	128.9	187.8	
(La/Yb)N	2.71	3.33	1.76	2.63	3.63	1.90	1.79	10.61	24.86	1.90	
(La/Sm)N	4.35	6.35	3.01	4.13	6.60	3.35	3.26	16.18	18.53	3.99	
(Gd/Yb)N	0.39	0.39	0.36	0.40	0.39	0.35	0.34	0.57	1.01	0.31	
Eu/Eu*	0.62	0.46	0.68	0.65	0.50	0.67	0.71	0.23	0.20	0.66	

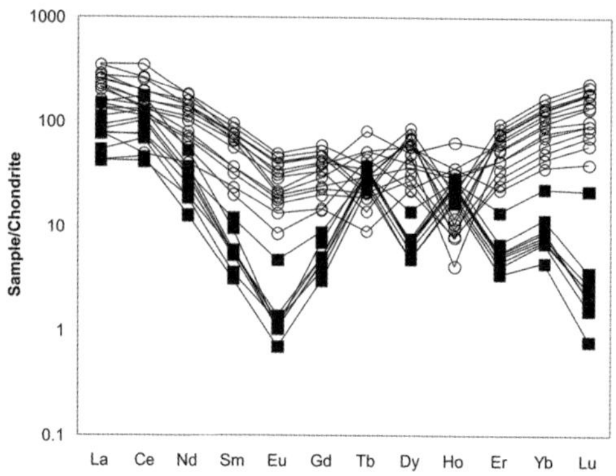

Fig. 6. Chondrite-normalized REE plot of the khondalites of the Eastern Ghats Belt. Normalizing values are from McDonough & Sun (1995).

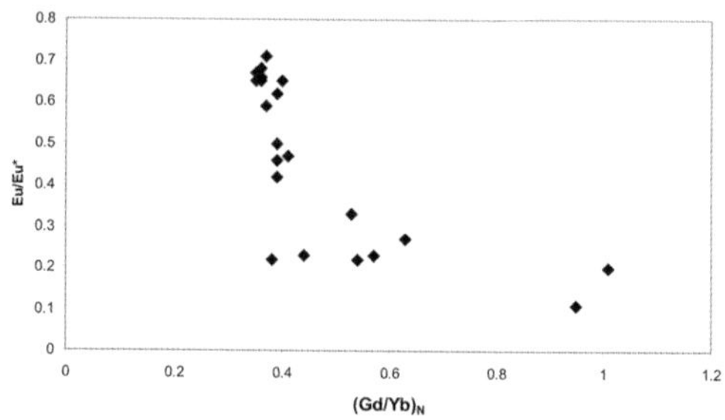

Fig. 7. Eu/Eu* v. $(Gd/Yb)_N$ plot of the Eastern Ghats Province khondalites.

For the Rb–Sr analytical work, the measured ratio of $^{87}Sr/^{86}Sr$ obtained for the NBS 987 standard was 0.710254 $\pm$ 22 (2σ) and the laboratory blank for the chemical procedure during the period of analysis yielded a maximum value of 4 ng for Sr.

For the Sm–Nd analytical work, the measured ratio of $^{143}Nd/^{144}Nd$ obtained for the La Jolla standard was 0.511857 $\pm$ 0.000046 (2σ), and the laboratory blanks for the chemical procedure during the period of analysis yielded maximum values of 0.4 ng for Nd and 0.7 ng for Sm.

Rb–Sr results

Rb–Sr isotopic data of the four samples may provide some constraints on the age of sedimentation. The Sr-model dates (Table 4) may indicate deposition at around 1.6 Ga.

Sm–Nd results

Sm–Nd isotopic data of selected khondalite samples from different segments provide age constraints for the provenance. The Nd-model dates indicate a provenance of the age clusters between 2.0 and 2.5 Ga (Table 5). Also, negative ε values of between -1 and -7, calculated at 1.6 Ga, indicates older continental crust in the provenance.

Discussion

Original nature of the sediments

The bulk composition is consistent with average shale, but the Fe-enriched composition could be related to some mafic source rocks in the provenance. The presence of smectite-type (presently plagioclase) minerals in some samples would indicate a shale–greywacke sedimentary sequence, which corresponds to marine turbidite deposits.

Provenance features

Only marginal Cr-enrichment over V, high La/Sc ratios, and a high Rb and Ba content in the majority of samples indicate dominantly granitic or felsic source rocks. However, significant Cr-enrichment over V, low values of La/Sc, high values of La/Th and relatively high values of $(Gd/Yb)_N$, but relatively low Eu/Eu* values, in some samples are strong indications of the involvement of mafic/basaltic source rocks in the provenance of the khondalite protoliths. On similar ground, Condie *et al.* (1992) concluded that sediment protoliths of SGG (sillimanite–garnet–gneiss) and QGG (quartz–garnet–gneiss) of the Datong khondalite suite may have been derived from sources in which granite and smaller amounts of basalt dominated.

Table 4. *Rb–Sr isotopic data for khondalites of the Eastern Ghats Belt Province, India*

Location	Sample	Rb (ppm)	Sr (ppm)	$^{87}Rb/^{86}Sr$	Error, 2σ	$^{87}Sr/^{86}Sr$	Error, 2σ	T (Ma), $\lambda = 1.42$	Sr^i
Paikmal	4 KH	117	50.9	6.755	0.078	0.864593	0.000117	1644	0.70511
Taptapani	7 KH	77.48	33.06	6.891	0.0542	0.8715	0.0001	1665	0.70667
Sunki	8 KH	2.51	5.54	1.3178	0.0106	0.73536	0.00002	1616	0.70477
Gunpur	14 KH	74.52	35.36	6.1851	0.0493	0.85117	0.00006	1644	0.70508

λ, decay constant; Sr^i, $^{87}Sr/^{86}Sr$ initial.

Table 5. *Sm–Nd isotopic data for khondalites of the Eastern Ghats Belt Province, India*

Location	Sample	Sm (ppm)	Nd (ppm)	$^{147}Sm/^{144}Nd$	Error, 2σ	$^{143}Nd/^{144}Nd$	Error, 2σ	$f_{Sm/Nd}$	T_{DM} (Ga)	ε^0	ε at 1.6 Ga
Paikmal	4 KH	3.67	19.7	0.1125	0.0007	0.511381	0.000008	−0.43	2.5	−24.5	−7.34
Taptapani	7 KH	9.47	62.14	0.0921	0.0005	0.511481	0.000012	−0.53	2	−22.5	−1.22
Sunki	8 KH	14.61	87.09	0.1015	0.0006	0.511578	0.000012	−0.48	2	−20.7	−1.25
Gunupur	14 KH	5.02	22.85	0.1327	0.0008	0.511801	0.000013	−0.33	2.3	−16.3	−3.27

$f_{Sm/Nd}$, enrichment of $^{147}Sm/^{144}Nd$ with respect to CHUR; ε, $^{143}Nd/^{144}Nd$ deviation from that in CHUR; ^{0}, at present.

Significant LREE-enrichment indicates LILE-enriched (granitic) source rocks; however, a large variation in the $(La/Yb)_N$ ratios (1.73–24.86), and co-variance with both $(La/Sm)_N$ (3.06–18.53) and $(Gd/Yb)_N$ ratios (0.35–0.95), would suggest complexity of possible provenance compositions (McLennan *et al.* 1990).

High Rb/Sr and Th/U ratios in many samples indicate significant weathering and recycling in the provenance. Substantial increases in the Zr/Sc ratios, but lesser increases in the Th/Sc ratios, might have resulted from zircon concentration during sedimentary recycling (sorting). A more LREE-enriched composition, coupled with HREE-enrichment in many samples, probably reflects monazite and xenotime concentrations during sedimentary recycling.

A negative Eu anomaly, approximately 0.61 in the majority of samples, and LREE-enrichment are similar to those observed in Phanerozoic turbidites (Bhatia 1985). However, unlike the Phanerozoic turbidites, HREE-enrichment in the majority of khondalite samples reflects the monazite concentration during sorting. Interestingly significant monazite is commonly observed in the khondalites of the Eastern Ghats Province.

The khondalites of the Eastern Ghats Province with LREE-enrichment, high La/Sc ratios, a significant negative Eu anomaly, resulting from plagioclase separation during intracrustal fractionation/differentiation processes, and low $^{143}Nd/^{144}Nd$ (non-radiogenic Nd isotope) ratio, as indicated by negative ε values, display typical features of Old Upper Continental Crust. Such terranes are also characterized by severe weathering and sedimentary recycling, as indicated by high Rb/Sr and Th/U ratios.

Palaeoproterozoic provenance

In a recent study of detrital zircon grains/domains from the Eastern Ghats Province metapelites, Upadhyay *et al.* (2009) indicated that they were sourced from Late Archaean/early Proterozoic–Mesoproterozoic provenances. These authors described Hf depleted mantle model ages (T_{DM}) of between 2.7 and 1.8 Ga, and referred to the whole-rock Nd model dates (T_{DM}) of between 2.5 and 1.9 Ga given in Rickers *et al.* (2001). Our whole-rock Nd model dates of between 2.5 and 2.1 Ga for the khondalites of the Eastern Ghats Province are consistent with the earlier reports mentioned above. Thus, Eastern Ghats Province sediment provenance appears to be dominantly Palaeoproterozoic. Upadhyay *et al.* (2009) also identified several later detrital zircon populations, defined by $^{207}Pb/^{206}Pb$ age clusters, namely, 1.8, 1.64–1.6 and 1.5–1.4 Ga, respectively, and discussed their different possible provenances. For the Palaeoproterozoic provenance, two possible terranes have been suggested by these authors, namely the Eastern Dharwar Craton and the Napier Complex. Continental reconstruction models suggest that proto-India and East Antarctica were in close proximity in Mesoproterozoic times (Yoshida *et al.* 1992). The Eastern Dharwar Craton with juvenile magmas of approximately 2.5 Ga (Jayananda *et al.* 2000) also have some basaltic intrusions dated at around 2.1 Ga (Sinha *et al.* 1998). In the Napier Complex magmatism around 3.0 Ga, and metamorphism at approximately 2.8 and 2.5 Ga, have been reported by Kelly & Harley (2005); while 2450–2470 Ma zircon populations have been reported by Carson *et al.* (2002). However, Palaeoproterozoic basaltic magmas have not yet been reported from the Napier Complex. In view of the geochemical evidence of some basaltic source rocks in the provenance of Eastern Ghats Province sediments, the Eastern Dharwar Craton is our favoured Palaeoproterozoic provenance for the Eastern Ghats Province sediments.

Onset of sedimentation

It was suggested that the onset of sedimentation in the Eastern Ghats Province coincided with the alkaline magmatism along the western margin of the Eastern Ghats Belt and, hence, related to the same rifting event (Upadhyay 2008). Upadhyay *et al.* (2009) used the two constraints for the Eastern Ghats Province sedimentation, namely, alkaline magmatism (1.42 Ga) and oldest metamorphic zircons (1.2 Ga) in the Eastern Ghats Province metasediments. However, different intrusion ages between 1.5 and 1.25 Ga have been reported for the alkaline plutons, and this led Vijay Kumar & Leelanandam (2008) to argue that a simple model of continental rifting alone may not be viable for the genesis of all the alkaline plutons of the Eastern Ghats Belt. Again, Srivastava & Singh (2004) described mafic dykes in the Bastar Craton and suggested a continental rift environment for their emplacement. Recently, French *et al.* (2008) provided age information for two sets, BD1 and BD2 mafic dykes described from the Bastar Craton: 1891 and 1883 Ma, respectively.

Although, there appears to be a general consensus on the riftogenic nature of the sedimentary basin of the Eastern Ghats Province, the exact timing of the sedimentation in the Eastern Ghats Province has not been unequivocally established. The Sr-model date for the Eastern Ghats Province khondalites may be viewed in this context as another tentative age constraint only.

Conclusions

- The relatively ferruginous composition of the khondalites (relative to average shale) can be a feature of the provenance; namely, the presence of some mafic rocks in the provenance.
- Trace element features indicate dominantly granitic/felsic provenance composition; but some mafic components in the provenance are also indicated by, for example, high Cr/V, low La/Sc ratios, and low Rb and Ba content in some samples.
- High La/Sc, LREE-enrichment and negative Eu anomalies, together with negative ε values of the khondalites of the Eastern Ghats Province, indicate Old Upper Continental Crust-type provenance.
- The Palaeoproterozoic provenance (2–2.5 Ga) of the Eastern Ghats Province khondalites could be the Eastern Dharwar Craton and or the Napier Complex of East Antarctica.
- Geochemical signatures attesting to the complexity of the provenance composition could relate to the later detrital zircon populations recently described from the Eastern Ghats Province metasediments.

The Indian Statistical Institute, Kolkata, provided the infrastructural facilities.

References

Barbey, P., Capdevila, R. & Hameurt, J. 1982. Major and transition trace element abundances in the Khondalite suite of the granulite belt of Lapland (Fennoscandia): evidence for an early Proterozoic flysch belt. *Precambrian Research*, **16**, 273–290.

Bhatia, M. R. 1985. Composition and classification of Paleozoic flysch mudrocks of eastern Australia: implications in provenance and tectonic setting interpretation. *Sedimentary Geology*, **41**, 249–268.

Bhattacharya, S. 1996. Eastern Ghats granulite terrain of India: an overview. *Journal of Southeast Asian Earth Science*, **14**, 165–174.

Bhattacharya, S. 1997. Evolution of Eastern Ghats granulite belt of India in a compressional tectonic regime and juxtaposition against iron ore craton of Singhbhum by oblique collision–transpression. *Proceedings of the Indian Academy of Sciences (Earth & Planetary Science)*, **106**, 65–75.

Bhattacharya, S. & Kar, R. 2002. High-temperature dehydration melting and decompressive P–T path in a granulite complex from the Eastern Ghats, India. *Contributions to Mineralogy and Petrology*, **143**, 175–191.

Bhattacharya, S., Kar, R., Misra, S. & Teixeira, W. 2001. Early Archaean continental crust in the Eastern Ghats granulite belt, India: isotopic evidence from a charnockite suite. *Geological Magazine*, **138**, 609–618.

Bhattacharya, S., Kar, R., Teixeira, W. & Basei, M. 2003. High-temperature crustal anatexis in a clockwise P–T path: isotopic evidence from a granulite–granitoid suite in the Eastern Ghats belt, India.*Journal of Geological Society, London*, **160**, 39–46.

Bhowmik, S. K. 1997. Multiple episodes of tectonothermal processes in the Eastern Ghats granulite belt, India. *Proceedings of the Indian Academy of Sciences (Earth & Planetary Science)*, **106**, 131–146.

Carson, C. J., Ague, J. J., Grove, M., Coath, C. D. & Harrison, T. M. 2002. U–Pb isotopic behavior of zircon during upper-amphibolite–facies fluid infiltration in the Napier Complex, East Antarctica. *Earth and Planetary Science Letters*, **199**, 287–310.

Condie, K. C., Boryta, M. D., Liu, J. & Qian, X. 1992. The origin of khondalites: geochemical evidence from Archaean to early Proterozoic granulite belt in the North China Craton. *Precambrian Research*, **59**, 207–223.

Dasgupta, S., Sanyal, S., Sengupta, P. & Fukuoka, M. 1994. Petrology of granulites from Anakapalle evidence for Proterozoic decompression in the Eastern Ghats, India. *Journal of Petrology*, **35**, 433–459.

Dasgupta, S., Sengupta, P., Fukuoka, M. & Chakraborti, S. 1992. Dehydration melting, fluid buffering and decompressional P–T path in a granulite complex from the Eastern Ghats, India. *Journal of Metamorphic Geology*, **10**, 777–788.

Dash, B., Sahoo, K. N. & Bowes, D. R. 1987. Geochemistry and original nature of Precambrian khondalites in the Eastern Ghats, Orissa, India. *Transactions of the Royal Society of Edinburgh, Earth Science*, **78**, 115–127.

Divakara Rao, V. 1983. Khondalites from the Eastern Ghats granulite belt: geochemistry and origin. *Geophysical Research Bulletin*, **21**, 233–242.

Dobmeier, C. J. & Raith, M. M. 2003. Crustal architecture and evolution of the Eastern Ghats Belt and adjacent regions of India. *In*: Yoshida, M. & Dasgupta, S.

(eds) *Proterozoic East Gondwana: Supercontinent Assembly and Breakup*. Geological Society, London, Special Publications, **206**, 145–168.

French, J. E., Heaman, L. M., Chako, T. & Srivastava, R. K. 2008. 1891–1883 Ma Southern Bastar-Cuddapah mafic igneous events, India; a newly recognized large igneous province. *Precambrian Research*, **160**, 308–322.

Jayananda, M., Moyen, J. F., Martin, H., Peucat, J. J., Auvray, B. & Mahabaleswar, B. 2000. Late Archaean (2550–2520 Ma) juvenile magmatism in the Eastern Dharwar craton, southern India: constraints from geochronology, Nd–Sr isotopes and whole rock geochemistry. *Precambrian Research*, **99**, 225–254.

Kar, R., Bhattacharya, S. & Sheraton, J. W. 2003. Hornblende dehydration melting in mafic rocks and the link between massif-type charnockite and associated granulites, Eastern Ghats granulite belt, India. *Contributions to Mineralogy and Petrology*, **145**, 707–729.

Kelly, N. M. & Harley, S. L. 2005. An integrated microtextural and chemical approach to zircon geochronology: refining the Archaean history of the Napier Complex, East Antarctica. *Contributions to Mineralogy and Petrology*, **149**, 57–84.

Li, Y. H. & Schoonmaker, J. E. 2004. Chemical composition and mineralogy of marine sediments. *Treatise on Geochemistry*, **7**, 1–36.

McDonough, W. F. & Sun, S. S. 1995. The composition of the Earth. *Chemical Geology*, **120**, 223–253.

McLennan, S. M., Bock, B., Hemming, S. R., Hurowitz, J. A., Lev, S. M. & McDaniel, D. K. 2003. The roles of provenance and sedimentary processes in the geochemistry of sedimentary rocks. *Geological Association of Canada GEOtext*, **5**, 1–31.

McLennan, S. M., Hemming, S., McDaniel, D. K. & Hanson, G. N. 1993. Geochemical approaches to sedimentation, provenance and tectonics. *Geological Society of America, Special Paper*, **284**, 21–40.

McLennan, S. M., Taylor, S. R., McCulloch, M. T. & Maynard, J. B. 1990. Geochemical and Nd–Sr isotopic composition of deep-sea turbidites: crustal evolution and plate tectonic associations. *Geochimica et Cosmochimica Acta*, **54**, 2015–2050.

Mohan, A., Tripathy, P. & Motoyoshi, Y. 1997. Reaction history of sapphirine granulites and a decompressional P–T path in a granulite complex from the Eastern Ghats. *Proceedings of the Indian Academy of Sciences (Earth & Planetary Science)*, **106**, 115–129.

Mukhopadhyay, A. K. & Bhattacharya, A. 1997. Tectonothermal evolution of the gneiss complex at Salur in the Eastern Ghats granulite belt of India. *Journal of Metamorphic Geology*, **15**, 719–734.

Nanda, J. K. & Pati, U. C. 1991. Geochemistry and original nature of Precambrian khondalites in the Eastern Ghats, Orissa, India. A discussion. *Transactions of the Royal Society of Edinburgh, Earth Science*, **82**, 87–88.

Nesbitt, H. W. & Young, G. M. 1984. Prediction of some weathering trends of plutonic and volcanic rocks based on thermodynamic and kinetic considerations. *Geochimica et Cosmochimica Acta*, **48**, 1523–1534.

Rickers, K., Mezger, K. & Raith, M. M. 2001. Evolution of the continental crust in the Proterozoic Eastern Ghats Belt, India and new constraints for Rodinia reconstruction: implications from Sm–Nd, Rb–Sr and Pb–Pb isotopes. *Precambrian Research*, **112**, 183–210.

Sen, S. K. & Bhattacharya, S. 1997. Dehydration melting of micas in the Chilka Lake Khondalites: the link between the metapelites and granitoids. *Proceedings of the Indian Academy of Sciences (Earth & Planetary Science)*, **106**, 277–297.

Sen, S. K. & Bhattacharya, S. 2000. Diverse signatures of deformation, pressure–temperature and anatexis in the Rayagada sector of the Eastern Ghats granulite terrane. *Proceedings of the Indian Academy of Sciences (Earth & Planetary Science)*, **109**, 347–369.

Sen, S. K., Bhattacharya, S. & Acharyya, A. 1995. A multi-stage pressure–temperature record in the Chilka Lake granulites: the epitome of the metamorphic evolution of Eastern Ghats, India? *Journal of Metamorphic Geology*, **14**, 287–298.

Sengupta, P., Dasgupta, S., Bhattacharya, P. K., Fukuoka, M., Chakraborti, S. & Bhowmik, S. 1990. Petrotectonic imprints in the sapphirine granulites from Anantagiri, Eastern Ghats Mobile Belt, India. *Journal of Petrology*, **31**, 971–996.

Sengupta, P., Sen, J., Dasgupta, S., Raith, M. M., Bhui, U. K. & Ehl, J. 1999. Ultrahigh temperature metamorphism of metapelitic granulites from Kondapalle, Eastern Ghats Belt: implications for the Indo-Antarctic correlation. *Journal of Petrology*, **40**, 1065–1087.

Simmat, R. & Raith, M. M. 2008. U–Th–Pb monazite geochronometry of the Eastern Ghats Belt, India: timing and spatial disposition of poly-metamorphism. *Precambrian Research*, **162**, 16–39.

Sinha, D. K., Rai, A. K., Parihar, P. S. & Hansoti, S. K. 1998. Khairagarh basin, Central India: its geology, geochemistry, geochronology and uranium metallogeny. *International Seminar: UNESCO-IUGS-IGCP 368*. Bhubaneswar, 135–137.

Srivastava, R. K. & Singh, R. K. 2004. Trace element geochemistry and genesis of Precambrian sub-alkaline mafic dykes from the central Indian craton: evidence for mantle metasomatism. *Journal of Asian Earth Science*, **23**, 373–389.

Taylor, S. R. & McLennan, S. M. 1985. *The Continental Crust: Its Composition and Evolution*. Blackwell, Oxford.

Upadhyay, D. 2008. Mesoproterozoic alkaline magmatism at the southeastern cratonic margin: constraints on craton-Eastern Ghats Belt suturing and implications for crustal evolution in Peninsular India. *Precambrian Research*, **162**, 59–69.

Upadhyay, D., Gerdes, A. & Raith, M. M. 2009. Unraveling sedimentary provenance and techtonothermal history of high-temperature metapelites, using zircon and monazite chemistry: a case study from the Eastern Ghats Belt, India. *Journal of Geology*, **117**, 665–683.

Vijaya Kumar, K. & Leelanandam, C. 2008. Evolution of the Eastern Ghats Belt, India: a plate tectonic perspective. *Journal of the Geological Society of India*, **72**, 720–749.

Yoshida, M., Funaki, M. & vitanage, P. W. 1992. Proterozoic to Mesozoic East Gondwana: the juxtaposition of India, Sri Lanka and Antarctica. *Tectonics*, **11**, 381–391.

Palaeoproterozoic sedimentation in the Cuddapah Basin, south India and regional tectonics: a review

DILIP SAHA[1]* & VIKASH TRIPATHY[1]

[1]*Geological Studies Unit, Indian Statistical Institute, 203 B.T Road, Kolkata 700108, India*

**Corresponding author (e-mail: sahad.geol@gmail.com)*

Abstract: The Proterozoic Cuddapah Basin of south India hosts a number of unconformity-bound sequences deposited over continental crust. The Palaeoproterozoic Papaghani Group with basal polymict conglomerate unconformably overlies the Archaean Peninsular Gneiss and set the stage for Palaeoproterozoic sedimentation following a long hiatus. Two cycles of fluvial–shallow-marine psammite–carbonate–pelite sedimentation in the western part of the basin (Papaghni sub-basin) is punctuated at different levels by mafic flows, sills and dykes, and less common acidic tuffs. Each of the sedimentation cycles fosters extensive carbonate platforms with dolomitic stromatolites and algal laminites. Emplacement of shallow crustal intrusives at the end of the first cycle was possibly linked to thermal anomalies in the mantle and associated widespread crustal extension in south India. Brittle fault-slip analysis from the western Cuddapah indicates repetitive extensional to strike-slip regimes. This paper proposes a likely late Palaeoproterozoic shallow-marine sedimentary sequence of the Nallamalai Fold Belt as allochthonous, thrusted over the successions of the Papaghni sub-basin and the Kurnool Group in the western part of the basin. Inversion of the Papaghni sub-basin and the development of regional erosional unconformities is discussed in the context of Palaeoproterozoic and early Mesoproterozoic orogenic events at the SE margin of India.

The Indian peninsula hosts a number of Proterozoic intracratonic basins that are traditionally referred to as the Purana basins (Holland 1909; Radhakrishna 1987; recent review by Chakraborti *et al.* 2010). Of these, the Cuddapah Basin, occurring along the eastern margin of the Dharwar Craton, is the largest of the Proterozoic basins in south India (Fig. 1). The complex history of the Cuddapah Basin dates back to the Palaeoproterozoic. It records tectonic events that shaped the SE margin of proto-India from when it became part of Columbia (Saha 2002; Rogers & Santosh 2004; Santosh 2010). Mafic igneous activity in the lower part of the basin and in surrounding parts of the adjacent cratons has been linked to an approximately 1900 Ma Large Igneous Province (LIP) involving the East Dharwar and Bastar cratons (French *et al.* 2008; Heaman 2008). Cyclic sedimentation and interaction with mafic magmatism within the Cuddapah Basin over a prolonged period of time starting from the Palaeoproterozoic has generated considerable interest (e.g. Bhattacharji & Singh 1984; Meijerink *et al.* 1984; Nagaraja Rao *et al.* 1987; Bhaskar Rao *et al.* 1995; Chaudhuri *et al.* 2002; Anand *et al.* 2003; Dasgupta & Biswas 2006; Saha *et al.* 2006). The inverted Cuddapah Basin provides a unique opportunity to examine the controls of extensive Palaeoproterozoic intracratonic sedimentation around the Dharwar Craton. Globally, the Palaeoproterozoic saw the onset of a plate tectonics related regime comparable to that in the Phanerozoic (Condie 1997; Brown 2008, 2010; cf. Condie *et al.* 2009). The Eastern Ghats Belt (EGB) in India is considered to have correlatives in the high-grade belts of East Anatarctica, as part of reconstructions of the growth of India during the Mesoproterozoic and Neoproterozoic that lead to its amalgamation into the supercontinent Gondwana (Dasgupta & Sengupta 2003; Meert 2003; Yoshida *et al.* 2003; Rogers & Santosh 2004; Collins & Pisarevsky 2005).

The geographical proximity of the Cuddapah Basin, the Nellore Schist Belt (NSB) and the EGB, all of which have Palaeoproterozoic components, also provides an opportunity to assess the tectonic influence of plate margin processes on the continental interior, particularly on the inversion of marginal to intracratonic basins in the Proterozoic (e.g. Glen *et al.* 2005; Saha *et al.* 2006). Although four unconformity bound sedimentary sequences are recognized within the basin, the present contribution focuses largely on the stratigraphy and sedimentation of the two older rock groups, namely the Papaghni Group and the Chitravati Group (in older literature referred to as the Lower Cuddapah Group: Holland 1909), and their regional tectonic set-up. The major outcrops of these sequences are in the SW part of the basin, marked by a positive gravity anomaly. This is interpreted as due to a large igneous body at depth

From: Mazumder, R. & Saha, D. (eds) 2012. *Palaeoproterozoic of India*. Geological Society, London, Special Publications, **365**, 161–184. http://dx.doi.org/10.1144/SP365.9

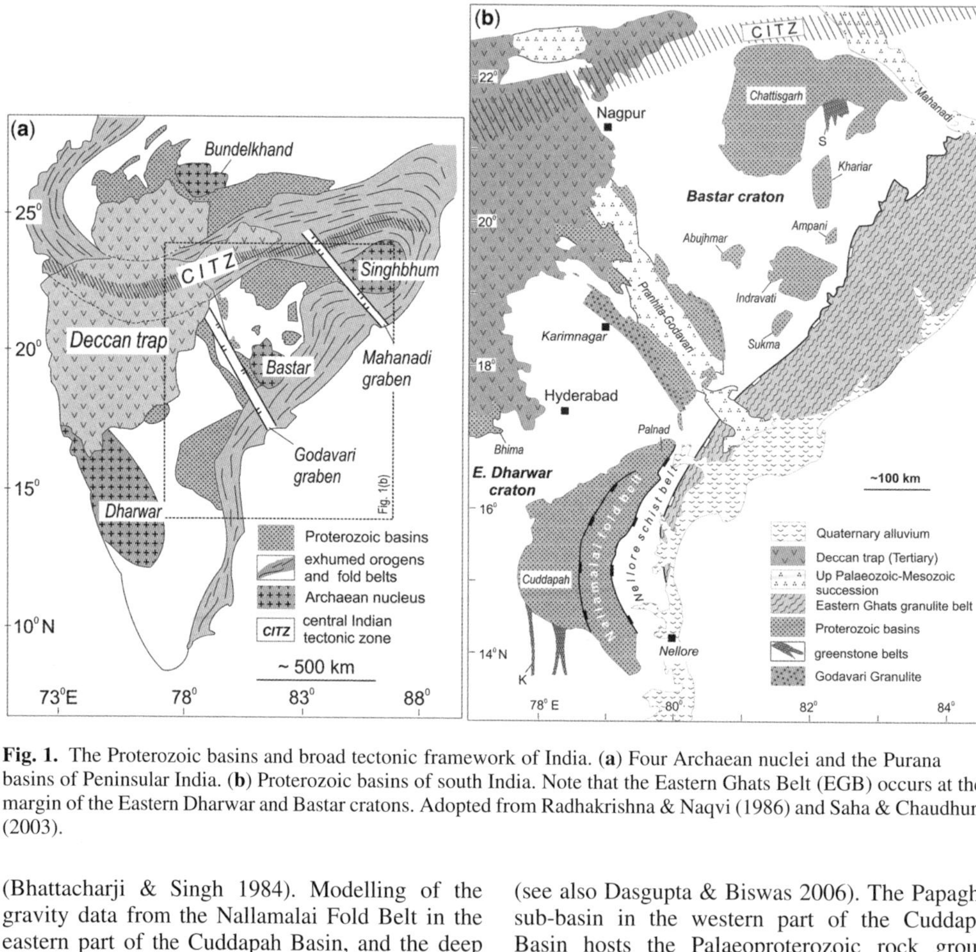

Fig. 1. The Proterozoic basins and broad tectonic framework of India. (**a**) Four Archaean nuclei and the Purana basins of Peninsular India. (**b**) Proterozoic basins of south India. Note that the Eastern Ghats Belt (EGB) occurs at the margin of the Eastern Dharwar and Bastar cratons. Adopted from Radhakrishna & Naqvi (1986) and Saha & Chaudhuri (2003).

(Bhattacharji & Singh 1984). Modelling of the gravity data from the Nallamalai Fold Belt in the eastern part of the Cuddapah Basin, and the deep seismic sounding (DSS) profiles (Kaila & Tiwari 1985), suggest that basement to the Nallamalai Fold Belt (NFB) is comparable to that below the western part of the Cuddapah Basin. The Peninsular Gneiss on the western margin of the Cuddapah Basin represents the surface expression of the dominantly granitoid basement, with slivers of Archaean greenstone belts. The deformation history of the NFB is also briefly reviewed in order to elucidate the tectonic framework of the basin inversion in an intracratonic setting.

An overview of the physical stratigraphy, Cuddapah Basin

On the basis of sedimentation pattern, spatial distribution and thickness variation in the constituent rock groups, Nagaraja Rao *et al.* (1987) recognized four sub-basins within the Cuddapah Basin, namely: the Papaghni sub-basin; the Kurnool sub-basin; the Srisailam sub-basin; and the Palnad sub-basin (Fig. 2) (see also Dasgupta & Biswas 2006). The Papaghni sub-basin in the western part of the Cuddapah Basin hosts the Palaeoproterozoic rock groups whose age of sedimentation has been well constrained (Bhaskar Rao *et al.* 1995; Anand *et al.* 2003). The Nallamalai Fold Belt in the eastern part of the basin hosts a thick shallow-marine succession (the Nallamalai Group) that is intensely deformed. The available geochronological data from the Nallamalai Group suggest the Nallamalai Group to be at least older than about 1400 Ma (age of the Chelima lamproite intruding the Nallamalai rocks: Chalapathi Rao *et al.* 1999). However, the maximum age of sedimentation is not well constrained. A major intracontinental thrust, the Maidukuru Thrust (cf. Rudravaram line: Saha *et al.* 2010), in the western part of the NFB suggests that the Nallamalai succession may be allochthonous. Possible late Mesoproterozoic to younger rock groups separated by angular unconformities from the older rock groups of the Papaghni sub-basin or the Archaean basement occur in the Srisailam, Palnad and Kurnool sub-basins (Fig. 3). The Papaghni sub-basin has an arcuate western boundary where arkosic–feldspathic arenites and conglomerates of the Gulcheru

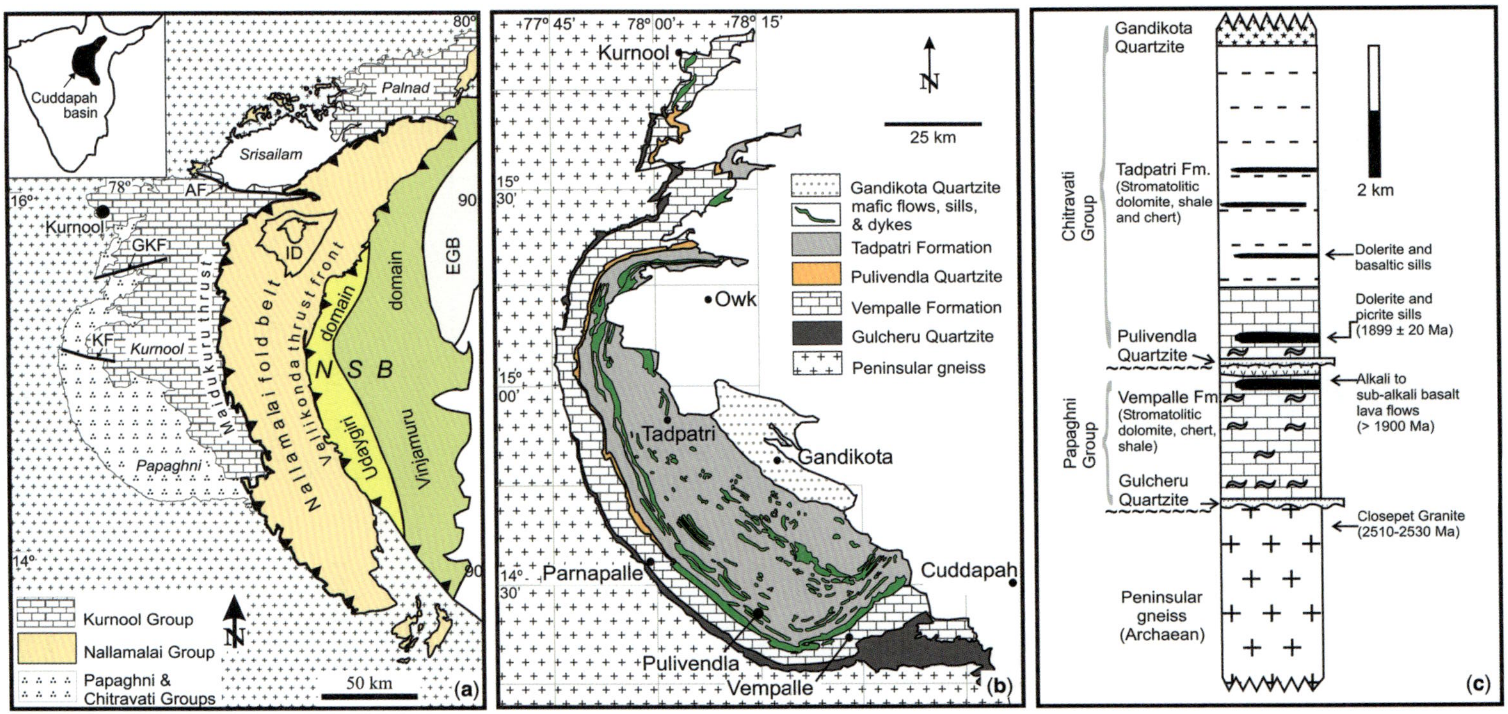

Fig. 2. Geology of the Cuddapah Basin and its lithostratigraphy. (**a**) Sketch geological map of the Cudddaph Basin showing the sub-basins, and the boundary thrusts of NFB and NSB. Udaigiri and Vinjamuru represent two distinct domains within the NSB (after Ramam & Murty 1997). (**b**) Geological map of western Cuddapah showing the lower Cuddapah rock groups (after Nagaraja Rao *et al.* 1987). GKF, Gani–Kalva Fault; AF, Atmakur Fault; KF, Kona Fault. (**c**) Generalized litholog for the rock groups in the Papaghni sub-basin and the main geochronological constraints (after Anand *et al.* 2003).

Supergroup	Group	Formation	Lithology
	Kurnool Group *500+ m*	Nandyal Shale	
		Koilkuntala Limestone	
		Paniam Quartzite	
		Owk Shale	
		Narji Limestone	
		Banganapalli Quartzite	
~~ Unconformity ~~~~~~~			
		Srisailam Formation	Pebbly grit, quartzite, heterolithic shale-sandstone
– x – x – x – tectonic contact – x – x – x –			
	Nallamalai Group *~1200 m*	Cumbum Formation (≈ Pullampet Shale)	Shale, dolomitic limestone, quartzite
		Bairenkonda Quartzite (≈ Nagari Quartzite)	Pebbly grit, quartzite, heterolithic shale-sandstone
– x – x – x – tectonic contact – x – x – x –			
Cuddapah Supergroup	**Chitravati Group** *4975 m*	Gandikota Quartzite	quartzite, pebble beds
		Tadpatri Formation	Shale, ash fall tuffs, quartzite, stromatolitic dolomite with mafic flows, sills and dykes
		Pulivendla Quartzite	Conglomerate and quartzite
	~~ Unconformity ~~~~~~~		
	Papaghni Group *2110 m*	Vempalle Formation	Stromatolitic dolomite, shale, basic flows and intrusive
		Gulcheru Quartzite	Conglomerate, feldspathic sandstone and quartzite
~~ Unconformity ~~~~~~~			
Peninsular Gneiss			

Fig. 3. Lithostratigraphic subdivisions, Cuddapah Basin. The Palaeoproterozoic interval is represented by the Papaghni Group and the Chitravati Group in the Papaghni sub-basin, and possibly also the largely allochthonous Nallamalai Group.

Formation unconformably overlie the Peninsular Gneiss with slivers of Archaean greenstone belts. Two cycles of fluvial to shallow-marine sedimentation in the Papaghni sub-basin are represented by the Papaghni Group and the Chitravati Group, which are separated by an unconformity (Laxminarayana *et al.* 2001; Chaudhuri *et al.* 2002).

Papaghni Group

The Papaghni Group consists of two formation rank units: the siliciclastic-dominant Gulcheru Quartzite; and the mixed siliciclastic–carbonate-bearing Vempalle Formation. The latter has many mafic flows, sills and dykes in the upper part of the formation. Although the formation takes its name from Guvvalcheruvu in the Cuddapah district, the basal conglomeratic facies is poorly developed/preserved around this locality. The present description is based on a section from south of Kurnool. Similar facies association are also observed from around Parnapalle.

Gulcheru Quartzite. The basal Gulcheru Quartzite consists of a matrix to clast-supported thick-bedded polymictic conglomerate with occasional interbeds of gritty trough cross-bedded feldspathic sandstone (Fig. 4a). Subangular–subrounded pebbles of vein quartz, pegmatite, granite, fine micaceous sandstone, black chert and grey shale/argillite suggest their derivation from the adjoining granitoid basement (Peninsular Gneiss) with greenstone patches. The coarse gritty matrix consists of quartz and pink feldspar, locally with ferruginous patches. The basal conglomeratic unit around Chinna Tandrapadu, SE of Kurnool is about 15 m thick. The common occurrence of trough cross-bedding in the gritty interbeds, channel lags, outsized clasts, lateral thinning out of the stack of conglomeratic beds and a generally fining-upwards facies suggest an alluvial-fan setting for the conglomerate–gritty feldspathic sandstone facies.

The basal facies grades upward to massive to trough cross-bedded gritty–pebbly feldspathic sandstone (7 m), which possibly represents sheetflood deposits. The trough azimuths show a unimodal distribution with wide dispersion and with modal direction towards the ESE (discussed in a later section on palaeocurrents). Higher up in the section trough cross-stratified units are overlain by medium- to coarse-grained, rippled to cross-stratified glauconitic sandstone with occasional pebbly interbeds. This unit consists of well-sorted subrounded quartz sand with rare feldspar grains. The topmost unit crops out around Tandrapadu and consists of heterolithic dark brown micaceous shale–fine sandstone (14 m) with bipolar trough cross–strata, mudcracks and occasional lag pebbles.

Fig. 4. Facies and sedimentary structures in the Papaghni Group. (**a**) Basal conglomerate and feldspathic grit, Gulcheru Quartzite, off Chinna Tandrapadu. (**b**) Heterolithic calcareous sandstone and dolomitic limestone, lower Vempalle Formation, Maddimadugu. Note the rippled to cross-stratified units. (**c**) Herringbone cross-stratification, calcareous sandstone, lower Vempalle Formation, Maddimadugu.

The association of straight-crested ripples, interference ripples, flat-crested ripples, shallow troughs with bipolar trough-axis azimuths, mudcracks in the heterolithic facies and occasional lag pebbles suggest the uppermost part of the Gulcheru Quartzite to be of tidal-flat origin.

In some sections (e.g. around Maddimadugu in Cuddapah district) the heterolithic unit grades upwards to the stromatolite-bearing Vempalle Formation. The Gulcheru Quartzite is also known to host putative trace fossils (Saha 2006), and organic-mat-induced sedimentary structures (Chakraborti & Shome 2010) representing the interaction of microbial communities and siliciclastic sediments in one of the oldest Palaeoproterozoic intracratonic basins in India.

Vempalle Formation. North of Maddimadugu the heterolithic facies of the the Gulcheru Formation grades upwards to a mixed siliciclastic–carbonate unit with silicified stromatolitic limestone–dolomite beds, calcareous cross-bedded to rippled sandstone and laminated grey-green–brown shale. The lower part of the formation consists of thin-bedded ripple-laminated calcareous sandstone with the local occurrence of herringbone cross-beds, thin gritty units with clasts of lime mud (intraformational flat pebbles) and desiccation cracks in silty units (Fig. 4b, c). Further up-plane laminated to rippled impure dolomites are intercalated with thin calcareous rippled sandstone beds. Both symmetric and asymmetric ripples are present, and muddy interlaminae show common desiccation crack fills. Biosedimentary structures include algal laminites–stromatolites with isolated stacked hemispheroids to laterally linked hemispheroidal forms (SH/LLH: Logan *et al.* 1964) in the thinly laminated dolomitic units. The upper part of the formation consists of scaly red shales (25 m) with very thin, fine sandy/silty laminae. The lithological association, and the sedimentary and common biosedimentary structures suggest an intertidal–subtidal origin for the carbonates of the Vempalle Formation, with an approximate thickness of 110 m in the Maddimadugu section. The laminated shales at the top suggest shoreface–inner-shelf deposition, possibly during the maximum rise of sea level in the first cycle of sedimentation.

The middle–upper part of the formation is marked by the common occurrence of up to 1 m-thick sills and thinner dykes of basalt and/or dolerite with chilled margins. Outcrops of thick basaltic sills are common around Vempalle (*Geological Survey of India Map 1980*, quoted in Nagaraja Rao *et al.* 1987) and basaltic flows are common in outcrops SW of Tadpatri.

Chitravati Group

The Chitravati Group consists of three formation rank units: the Pulivendla Quartzite; the Tadpatri Formation; and the Gandikota Quartzite. The transition from the Papaghni Group to the Chitravati Group is marked by the intermittent occurrence of mafic flows and associated shallow-crustal intrusives in the form of sills and dykes in the underlying Vempalle Formation. Broadly coeval with the mafic igneous activity, we also see the deposition of a siliciclastic system – that is, the Pulivendla Quartzite – that grades upwards into a mixed siliciclastic–carbonate system of the Tadpatri Formation. The latter is also marked by the extensive development of mafic sills and dykes (Fig. 2).

Pulivendla Quartzite. Compared to the Gulcheru Quartzite, the Pulivendla Quartzite has more restricted strike continuity across the Papaghni sub-basin. The formation consists of medium- to thick-bedded well-sorted quartz arenite with sparse pebble beds in the basal part. The lower part of the formation NW of Yagantipalli consists of large (1–2 m across) trough cross-bedded, medium-grained well-sorted sandstone (quartzite) with rare lag pebbles. Locally, deformed cross-strata are also seen. Trough axis azimuth varies between north and west (Dasgupta & Biswas 2006). This grades upwards into a coarser sandstone with trough cross-bed and plane-parallel strata, normally graded mass-flow beds and low-angle cross-strata. Shallow-channel forms with lag pebbles and well-rounded and well-sorted quartz sands, and a wide dispersal of the trough axis suggest a subtidal bar environment, shallowing upwards into an intertidal environment, locally with signs of intermittent exposure. The Pulivendla Quartzite, with a thickness of about 90 m, grades upwards into the Tadpatri Formation (Fig. 5a).

Tadpatri Formation. The Pulivendla Quartzite grades upwards into a heterolithic shale–fine calcareous sandstone unit with 30–40 cm-thick dolomitic limestone interbeds (Fig. 5b, c). The sandy units contain symmetric to asymmetric ripples. The dolomitic units show common algal laminites–stromatolitic mounds. In some sections (near Yagantipalle) the dolomitic limestone is intercalated with carbonaceous shale, which grades upwards into plane-laminated splintery grey-green sand-free shale (40–50 m). In the Yagantipalli section, or in Panem–Betamcherla road sections, the dolomitic limestone is commonly intruded by thick doleritic sills (individual sills up to 4 m thick, and laterally continue for several hundred of metres; Fig. 5d). These usually show chilled margins and/or contact metamorphic effect in the host carbonates. In other sections, such as around Tamarjupalle, Kalvabugga and south of Yerraguntla, rhyolitic ignimbrites are intercalated with silicified dolomites in the uppermost part of the Tadpatri Formation. Stromatolites in the Tadpatri Formation usually have low-amplitude hemispheroidal (SH) forms, some attaining a diameter of 1 m. Smaller laterally linked hemispheroids (LLH) are also common.

The lithofacies association of algal laminites, stromatolites and rippled fine calcareous sandstone suggests an intertidal–subtidal regime. Local asymmetry of the internal laminae of stromatolite forms

Fig. 5. Facies and sedimentary structures in the Chitravati Group. (**a**) Pulivendla Quartzite grading up into the heterolithic facies of the Tadpatri Formation, Muddulettiswami Temple Road, off Krishnagiri. (**b**) Heterolithic thin-bedded sandstone–shale intercalated with stromatolitic dolostone, Tadpatri Formation, north of Krishnagiri. (**c**) Mafic sills and stromatolitic limestone, Tadpatri Formation, Yadiki. (**d**) Plane-laminated green shale with thick dolerite sill (bar scale), upper Tadpatri Formation, south of Yagantipalle.

suggests the influence of tidal currents. The development of plane-laminated shales in the upper part of the formation suggests a maximum sea-level rise, suppressing the carbonate factory. The sedimentation is punctuated by mafic and minor acid volcanism.

Gandikota Quartzite. In the east-central part of the Papaghni sub-basin (Fig. 2), the Tadpatri Formation is overlain by another formation rank unit, the Gandikota Quartzite. The contact is gradational with thinly bedded sand–shale intercalations in the transitional zone giving way to amalgamated quartzite beds with sheet geometry (Fig. 6a). The bulk of the formation consists of well-rounded well-sorted medium- to coarse-grained quartz arenite with large (up to 0.7 m) planar tabular to large trough cross-stratification. Cross-stratified units are intercalated with plane-parallel units and those with straight crested ripples (Fig. 6b, c) Deformed cross-beds are common in the upper part. Nagaraja Rao *et al.* (1987) reported herringbone cross-beds. On the whole, Gandikota Quartzite represents a bar–interbar facies with a tidal influence.

Upper boundary of the Chitravati Group. In the western part of the Cuddapah Basin the deposition of the Chitravati Group is followed by a major hiatus, marked by an angular unconformity that extends over the entire Papaghni sub-basin. The basal polymict conglomerate horizon in the younger Kurnool Group overlies the Papaghni and Chitravati groups, and onlaps the basement Peninsular Gneiss in the NW part of the Cuddapah Basin (Saha *et al.* 2006).

Nallamalai Group

While the relatively undeformed strata of the Papaghni Group and the Chitravatim Group (Fig. 3) crop out in the western part of the Cuddapah Basin, the Nallamalai Fold Belt (NFB) in the eastern part has folded and faulted metasedimentary rocks constituting the Nallamalai Group. The contact

Fig. 6. Facies and sedimentary structures in the Gandikota Quartzite. (**a**) Sharp boundary (arrowheads) between shales of the Tadpatri Formation (Fm) and the Gandikota Quazrtizte (Qzt), Pennuru Gorge off Jammalamadugu. (**b**) Parallel stratification to large planar tabular cross-stratification. (**c**) Straight crested to bifurcating ripples. (b) and (c) are from the outcrop west of Gandi Fort.

between the NFB and the western part of the Cuddapah Basin has been interpreted as a thrust (Fig. 2) (Chakraborti & Saha 2006; Saha *et al.* 2010). The Nallamalai Group consists of the lower sandstone-dominated Bairenkonda Quartzite and the upper Cumbum Formation, consisting mainly of shales with sandstone and dolomite intercalations (Fig. 7). In the Balrajupalle–Sanipai (Kottapeta) section, the basal part of the Nagari Quartzite, considered by some authors to be equivalent of the Bairenkonda Quartzite (Nagaraja Rao *et al.* 1987), shows coarse to pebbly trough cross-stratified and rippled sandstones overlain by a hummocky cross-stratified sandstone–siltstone. Straight crested ripples are common in coarse- to medium-grained sandstones with desiccation features. This then grades into plane-laminated shale interstratified with glauconitic sandstone. Further up, a shaly unit with intercalations of dolomite and ferruginous quartzite mark the transition to the Pullampet Formation (probable equivalent of the Cumbum Formation) in the Bagidipalli area. An oolitic ironstone facies occurs within the ferruginous quartzite. The exposed top is marked by plane-laminated shale with occasional dolomitic interbeds. Further east in the Naranrajupalli–Rajampet area, massive to graded, gritty to coarse arenites with common carbonate clasts are intercalated with rippled calcareous sandstone and plane-laminated shale (Mukherjee 2000; Saha 2004). Dasgupta & Biswas (2006) reported volcaniclastics within the shales that also contain common slump horizons. The latter grades upwards through intercalations of thin fine massive sandstone beds to a cross-bedded to rippled quartzite. Saha (2006) reported *Rhysonetron* and other suspected trace fossils (cf. *Planolites*, *Cochlichnus*) from the Nagari Quartzite.

In the east-central part of the NFB, the Nallamalai Group consists of quartzite, quartz phyllite, slate (shale) and minor dolomitic limestone, with a cumulative thickness of 1200 m, as estimated from an exposed section of the Vellikonda range east of Porumamilla. The quartzite-dominant part with minor quartz phyllites (protoliths of micaceous sandstone and shale) in the lower part is regarded as the Bairenkonda Subgroup, and the upper phyllitic and quartzite intercalation is regarded as the Cumbum Subgroup (Meijerink *et al.* 1984). In the Porumamilla area, the constituent members are, in ascending order, the Tekurapeta Quartz–Phyllite Member, the Porumamilla Quartzite Member, the Kavalakuntla Quartzite Member, the Chintalapalle Slate Member, the Soman Cheruvu Phyllite Member and the Pandurangpuram Quartzite Member (Tripathy & Saha 2010) (Fig. 7). The original sedimentary features in these units are often overprinted by tectonic cleavages and fabric. However, strain heterogeneity in the NFB allows the local preservation of sedimentary structures, even in the east-central NFB. The lower part of the succession includes thin wavy bedded/rippled fine-grained quartzite (glauconitic sandstone) with intercalations of laminated slate (shale), which passes upwards to a thin bedded massive to graded-stratified fine sandstone–siltstone. Further up in the succession, the moderately thick-bedded, well-sorted, medium-grained Porumamilla Quartzite

	Nagaraja Rao et al. (1987)		after Tripathy and Saha (2010)	
Nallamalai Group	Cumbum Formation	slate, quartzite, dolomite	**Cumbum Formation**	Pandurangapuram Quartzite member (175-200m)
				Soman Cheruvu Phyllite Member (Green phyllite; 200-225m)
				Chintalapalle Slate Member (brown-black slate, > 100m)
	Bairenkonda Quartzite	quartzite, shale	**Bairenkonda Quartzite**	Kalavakuntla Member (heterolithic quartzite-slate, ferruginous; 200-225m)
				Porumamilla Quartzite member (thick bedded, coarse quartz arenite; 325-350m)
				Tekurupeta Quartz-phyllite Member (~100m)

Fig. 7. Lithostratigraphic subdivisions of the Nallamalai Group in the east-central part of the Nallamalai Fold Belt.

Member shows trough cross-strata, low-angle strata and plane-parallel strata alternating with thin rippled units (Fig. 8). Local overturning of cross-strata, and ball and pillow structures, are common in the upper part.

The lower part of the Cumbum equivalent succession at Chintalapalle consists of grey-green slate (shale) with centimetre-thick fine-grained sandstone interbeds, and, locally, with massive dolomitic interbeds. A thick green phyllite with thin quartzite intercalation at Soman Cheruvu grades upwards to a medium- to fine-grained quartzite with plane-parallel strata, ripples and large planar tabular cross-strata with tidal bundles (Fig. 9). South and west of the Iswarakuppam dome, thick shale–siltstone (slates of the Cumbum Formation) show repeated cycles of massive to normally graded siltstone, followed upwards by plane-parallel units and, finally, by plane-laminated shale (Fig. 9d).

While the lower part of the Nagari Quartzite shows features of a fan delta, it grades upwards to a pertidal succession. Further up, subtidal mixed siliciclastic–carbonates in the transition zone to the Pullampet Formation give way to oolitic ironstone facies, indicating inner-shelf deposition. The transition possibly represents a major transgressive event. The presence of ash beds, mass-flow conglomerates with clasts of intrabasinal carbonates, slumped beds, etc., suggest synsedimentary faulting and volcanism during the deposition of the Pullampet–Cumbum Formation. The shale–siltstone-dominant middle part of the Cumbum Formation may represent turbidites (see also Dasgupta & Biswas 2006). The quartzite-dominant upper Cumbum Formation represents subtidal bars–peritidal deposits.

Unconformities and sequence boundaries

The Papaghni and Chitravati groups in the western part of the Cuddapah Basin represent Palaeoproterozoic sedimentation in an intracratonic basin. Unconformity-bound sequences with a major basin-wide erosional unconformity below the Papaghni Group and above the Chitravati Group have been recognized. Subaerial mafic flows on top of the Vempalli Formation (Papaghni Group) and the fluvial facies at the base of the Pulivendla Quartzite in contrast to plane-laminated shales (shelfal mud) of the Vempalli Formation suggest another major unconformity between the Papaghni and Chitaravati groups (see also Lakshminaraya *et al.* 2001). The distribution of broad sedimentary facies, interpreted sequence boundaries and maximum flooding

Fig. 8. Facies and sedimentary structures in the Bairenkonda Quartzite. (**a**) Alternate thick (amalgamated beds) and thin bedded (rippled) quartzite, lower Bairenkonda Quartzite. (**b**) Unit with symmetrical ripples alternating with those with plane-parallel or low-angle stratification. (**c**) Trough cross-bedded quartzite; note the herringbone strata. Outcrops around Porumamilla.

surfaces are shown in a schematic section across the Papaghni sub-basin (Fig. 10). The alluvial fans of the Gulcheru Quartzite with polymict, matrix-supported to clast-supported thick conglomerate beds lie directly over the Peninsular Gneiss of the Eastern Dharwar Craton. The western boundary of the Gulcheru Quartzite truncates the NNW-trending greenstone belts within the Peninsular Gneiss (Fig. 2) (Saha & Chaudhuri 2003). A major angular unconformity (nonconformity) at the base thus bounds the Papaghni Group. The transition from the Gulcheru Quartzite to the Vempalle Formation shows the gradual establishment of a carbonate platform as the fluvial system gives way to supratidal–intertidal systems within a shallow embayment. Grey-brown shales with occasional thin carbonate intercalations in the upper part of the Vempalle Formation represent a highstand systems tract. A sequence boundary (van Wagoner *et al.* 1990) at the top of the Vempalle Formation (the paraconformity of Nagaraja Rao *et al.* 1987) is indicated by the siliciclastic-dominant Pulivendla Quartzite resting on top of the Vempalle Formation. While at some sections mafic–ultramafic flows occur immediately below the Pulivendla Quartzite, sills and dykes are very common in the upper part of the Vempalle Formation and in the Tadpatri Formation, indicating a major tectonic disturbance in the Papaghni sub-basin. As discussed earlier, the igneous intrusions were probably linked to a Large Igneous Province (LIP) that affected the southern Bastar and Eastern Dharwar cratons (French *et al.* 2008; Heamann 2008). However, plume control of this proposed LIP is still hypothetical as coeval mafic dykes or associated igneous rocks are yet to be reported from other continental fragments.

Taking into account the much wider and extensive outcrop belt of stromatolitic limestones/dolostones in the Tadpatri Formation compared to the Vempalle Formation (Fig. 2), the former appears to represent a stable carbonate platform, punctuated by the emplacement of mafic sills and dykes, and the eruption of acid tuffs that were partially reworked to be deposited as laminated shales in the upper part with very little input from terrigenous coarser clastics. The changeover from calcareous shales with sparse silicified limestone/dolostone into thick quartzite beds of the Gandikota Quartzite suggests another sequence boundary at the top of the Chitravati Group. While the boundary between the Tadpatri Formation and the Gandikota Quartzite is gradational and conformable at the Penneru River gorge around Gandikota in the central part of the basin (Fig. 6a), it is possible that the Tadpatri Formation was partly subaerially exposed further up-dip. Notably, the younger Banganapalle Formation (Kurnool Group) rests on a regional erosional surface (angular unconformity) that straddles over the Gandikota Quartzite and older formations including the Tadpatri Formation and the Pulivendla Quartzite.

There is a major hiatus in deposition at the site of the Papaghni sub-basin after the deposition of the Gandikota Quartzite. The latter may represent delta progradation, with tidal reworking in the

Fig. 9. Facies and sedimentary structures in the Cumbum Formation. (**a**) Plane-laminated shale with fine sandstone interbeds, Chintalapalle slate; note the slaty cleavage overprinting the stratification. (**b**) Plane-parallel to cross-stratified quartzite in the upper Cumbum Formation, north of Soman Cheruvu. (**c**) Tidal bundles in the quartzites, upper Cumbum; note the thin muddy layer above the large cross-stratified quartzite layer, Rapur. (**d**) Graded beds in the middle part of the Cumbum Formation, Atmakur road section.

distal part. The carbonate platform of the Tadpatri Formation faced a natural demise as there was a major retreat of the sea, indicated by the progradational Gandikota Quartzite. East of the Cuddapah Basin, the Nellore Schist Belt (NSB) is known to be affected by Palaeoproterozoic subduction as indicated by the Kandra ophiolites, which are interpreted to be suprasubduction zone ophiolites (Saha 2010; Vijaya Kumar *et al.* 2010). The NSB probably includes a collage of accreted terranes, as shown by

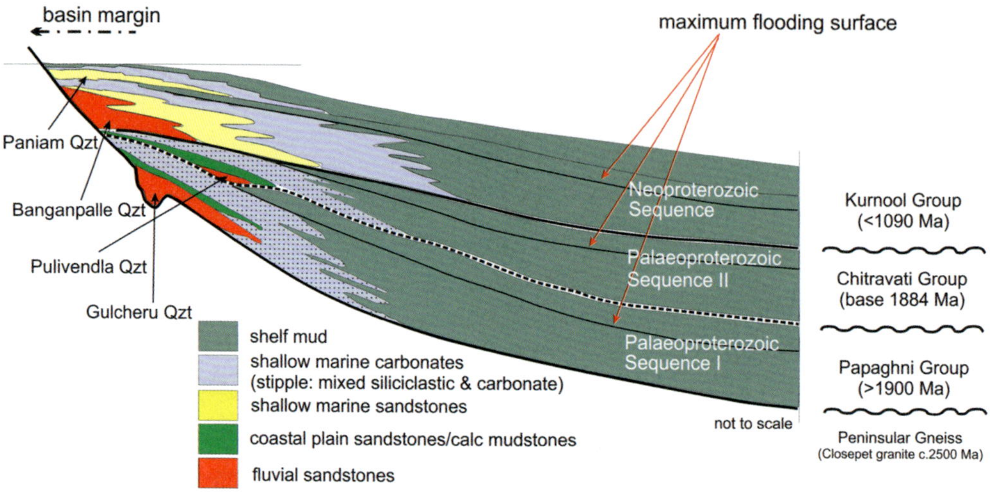

Fig. 10. Schematic section across the western part of the Cuddapah Basin showing the major sequence boundaries. Three unconformity-bound sequences are indicated.

the still-younger ophiolitic mélange of Kanigiri (Dharma Rao *et al.* 2011). A Palaeoproterozoic convergence east of the Dharwar Craton margin probably led to the reorganization of the surrounding oceans and sea lanes. Thus, the progradational Gandikota Quartzite over the carbonate platform of the Tadpatri Formation may represent a major retreat of the sea, probably related to the approximately 1850 Ma plate subduction (Vijaya Kumar *et al.* 2010) in the NSB leading to the partial Palaeoproterozoic amalgamation outboard of the East Dharwar Craton margin. On the basis of comparable D_1 deformation in the Nellore Schist Belt (Vinjamuru domain) and in the NE part of the NFB, Saha (2002, 2004) suggested possible late Palaeoproterozoic docking of the Vinjamuru domain against the NFB.

Status of the Nallamalai Group

Although the Cuddapah Supergroup is conventionally regarded as including the Nallamalai Group (Nagaraja Rao *et al.* 1987), the map relations and vertical sections do not show any vertical stacking of the Nallamalai over the lower Cuddapah sequence. The discordant relationship between the Nagari Quartzite in the SW part of the NFB and the Papaghni Group south of Maddimadugu is generally put forward as indicating an unconformable relationship between the Nallamalai Group and the Chitravati Group of the lower Cuddapah sequence (Nagaraja Rao *et al.* 1987). The discordant relationship may as well be explained as being due to a thrust contact. The Maidukuru Thrust on the western margin the NFB extends right from Atmakur in the north to Cuddapah town in the south (Chakraborti & Saha 2006). In this sector the contact is largely between the Kurnool Group and the Nallamalai Group. Signatures of a thrust contact are also present in the stretch between Cuddapah and Sanipai (D. Saha unpublished work).

The Nagari Quartzite and the Bairenkonda Quartzite are considered to be equivalent, although the published maps show these two quartzite-dominant successions as separated by a wide expanse of the Cumbum Formation (e.g. *Geological Survey of India Map 1981* adapted in Nagaraja Rao *et al.* 1987). The Nagari Quartzite has direct contact with the Peninsular Gneiss in the extreme SW part of the NFB, and there are divergent views on its lithostratigraphic status (e.g. King 1872; Narayanswami 1966; Lakshminaryana *et al.* 2001; Ramakrishnan & Vaidyanathan 2008). The lithostratigraphic association within the Bairenkonda Quartzite in the east-central part of NFB (Tripathy & Saha 2010) is somewhat different from that of the typical Nagari Quartzite in the SW NFB. However, there are certain commonalities in lithological character, and the variation may be within the broad range of a formation rank unit.

However, recognition of a major thrust (fig. 2 of Chakraborti & Saha 2006) at the western margin of the NFB suggests that this belt is allochthonous (Fig. 11). Development of carbonate mylonites derived from the Narji Limestone belonging to the Kurnool Group in the footwall of this thrust suggests that the thrusting took place after the cessation of sedimentation in the Kurnool sub-basin. However, as mentioned earlier, the status of the Nagari

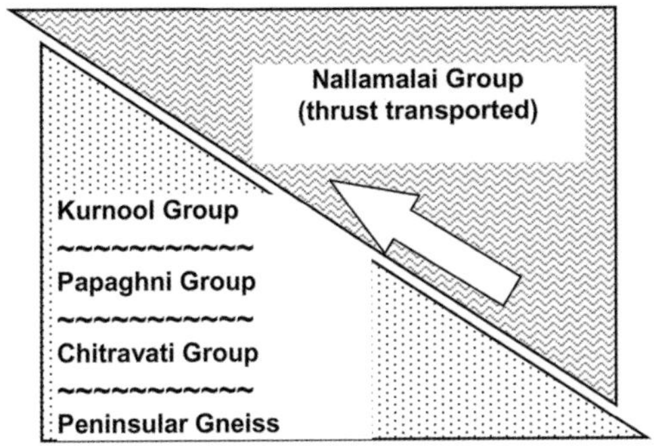

Fig. 11. Schematic diagram showing the relationship between the major rock groups in the Cuddapah Basin and the NFB.

Quartzite – as far as the correlation with the Bairenkonda Formation and the contact relationship with the Peninsular Gneiss is concerned – are crucial issues that can be addressed properly with more detailed investigation in the SW corner of the NFB and with, perhaps, further input from detrital zircon geochronological work in the extreme SW part of the NFB. On the balance of evidence, the Nallamalai Group is largely allochthonous, and is thrusted over the Kurnool Group and the Papaghni Group.

Status of the Cuddapah Supergroup

In view of the discussion above, we suggest a revision of the Cuddapah Supergroup as far as putting the constituent group rank units into a supergroup rank unit is concerned. While the first two group rank lithostratigrpahic units, namely the Papaghni Group and the Chitravati Group, are restricted to the Papaghni sub-basin in western Cuddapah (Fig. 2), and apparently represent two cycles of sedimentation with similar lithologies, the Nallamalai Group has a thrust contact with the rock groups in the west. Moreover, the basaltic flows and sills interstratified with the mixed siliciclastic and carbonate horizons in the Vempalle Formation or the Tadpatri Formation are absent in the Nallamalai Group. However, the Pullampet Formation (Nallamalai Group) is known to host volcanogenic-bedded barite deposits and associated acid volcanism (Ramam & Murty 1997), which is rare in the Papghni sub-basin. Thus, following the International Stratigraphic Guide (World Wide Web Address: http://www.stratigraphy.org/upload/bak/litho.htm) we propose that the name Cuddapah Supergroup be retained to include the Papaghni Group and the Chitravati Group only, while the Nallamalai Group should be regarded as an independent group rank lithostratigraphic unit (Fig. 3).

Palaeoproterozoic sedimentation in the Cuddapah Basin and tectonic controls

Given the granitic nature of the basement of the lower Cuddapah sequences with overall fluvial–shallow-marine cyclic sedimentation punctuated by basin-wide unconformities, the Cuddapah Supergroup represents one of the finest examples of intracratonic basinal sequence dating back to the Palaeoproterozoic. Although there exists still some uncertainties on the exact age of sedimentation of the Nallamalai Group, syn- to late-D_2 intrusive granite along the eastern margin of NFB are as old as 1584 Ma (Crawford & Compston 1973; Saha & Chakraborti 2007). These raise the possibility of the minimum age of Nallamalai Group to be late Palaeoproterozoic–early Mesoproterozoic. Recent work in the Nellore Schist Belt (NSB) and the Eastern Ghats Belt (EGB) suggest approximately 1850 Ma old suprasubduction zone ophiolites (Saha 2010; Vijaya Kumar *et al.* 2010), which constrain the minimum age of associated metasediments as Palaeoproterozoic.

The NSB in south India consists of metamorphosed and migmatized volcano-sedimentary rocks with intrusive granites and pegmatites, ranging in age from Neoarchean to late Proterozoic (Ramam & Murty 1997; Babu 1998; Dobmeier & Raith 2003; Dobmeier *et al.* 2006; Ravikant 2010; Vijaya Kumar *et al.* 2010). The NSB is sandwiched between the high-grade EGB and the NFB (Fig. 1). The belt is accreted to the SE margin of the Eastern Dharwar Craton along the Vellikonda thrust front (Venkatakrishnan & Dotiwala 1987), which has been interpreted as a major intracontinental thrust (Saha 2004; Saha *et al.* 2010). The NSB is marked by multiple deformation and metamorphism, with variation in metamorphic grade from low-greenschist facies to upper-amphibolite facies across a 35–70 km-wide belt (Ramam & Murty 1997; Saha 2002, 2004). The northern NSB also hosts deformed alkaline complexes (Prakasam alkaline province: Leelanandam *et al.* 2006). Two lithotectonic domains, namely the eastern Vinjamuru domain and the western Udaigiri domain, have been recognized within the NSB (Fig. 2b) (Dobmeier & Raith 2003). Ramam & Murty (1997) recognized to groups within the NSB, namely the older Vinjamauru Group and the younger Udaigiri Group. The Vinjamuru domain (group) consists of a heterogeneous assemblage of pillowed to vesicular metabasaltic flows and sills, minor amounts of acid–intermediate metavolcanics, quartzites, thin BHQ and thin metapelites. The rocks of the Vinjamuru domain have undergone epidote–amphibolite- to upper-amphibolite-facies metamorphism (Moeen 1998), and in the eastern part of the domain the metabasalts and metapelites are intensely migmatized.

The Udaigiri domain (group) consists of gritty feldspathic sandstones, phyllites, quartzites interstratified with vesicular basalt, and acid-tuff and minor rhyolites. The metasediments in the Udaigiri domain have suffered low-greenschist-facies metamorphism. In addition, a distinct association of sheeted dolerite dykes and hornblende-gabrros occurring in the extreme south of the NSB constitute the Kandra Igneous Complex (also known as Kandra Ophiolite Complex).

The close proximity and similarity of sedimentary sequences in the NSB (Udaigiri Group) and the Nallamalai Group suggest a possible relationship between sedimentation and/or deformation in the two domains. The contact between the Vinjamuru domain and the Udaigiri domain has earlier been suspected to be a fault/thrust (Ramam & Murty 1997; Saha 2004). Work in progress by the present authors suggests a strong left-lateral strike-lip displacement along a steep shear zone straddling the contact between the two domains.

Basin opening

As described earlier, the basal Gulcheru Quartzite is marked by the extensive development of alluvial fans with provenance to the south and west of the present western margin of the Cuddapah Basin. The presence of deep-seated faults close to the margin attests to early passive rifting of the upper granitic crust that led to the creation of initial accommodation space in the Papaghni sub-basin. The siliciclastic regime in a fluvial to supratidal–intertidal system give way to a mixed siliciclastic–carbonate regime, as indicated by the peritidal Vempalle Formation with its periodic influx of reworked siliciclastics. The sedimentation was largely in the photic zone, as evidenced by extensive development of algal laminites and LLH/SH stromatolites (Logan *et al.* 1964). The shaly units in the upper part of the Vempalle Formation possibly represent the establishment of a highstand system tract over an embayment in the Eastern Dharwar Craton. The minimum age of the basin opening (Papaghni sub-basin) is constrained by the age of mafic flows/sills within the Vempalle Formation at approximately 1900 Ma (e.g. Anand *et al.* 2003).

Outcrop belts of the Papaghni Group and the Chitravati Group, including the mafic dykes and sills, are displaced by a number of transverse faults occurring in the western part of the Cuddapah Basin (Fig. 2). Fault-slip analysis from the Ramallakota sector of the Gani–Kalva Fault indicates early extension across the Gani–Kalva or Kona faults (Tripathy & Saha 2009). It is difficult to date the actual fault slips associated with the Gani–Kalva Fault or the Kona Fault. However, by segregating data from the older stratigraphic units (Chitravati Group) from the younger stratigraphic units (Kurnool Group) we see a clear change in the type of stress regime (compare the shaded rows in Table 1 with those without any shading). The extensional stress regime is apparent only from the data obtained from the stratigraphic units belonging to the Papaghni sub-basin. The data from the younger stratigraphic units belonging to the Kurnool sub-basin suggest reactivation under strike-slip and compressional regime. Based on the above fault-slip analysis (Tripathy 2011), we suggest that intermittent extensional tectonic activity occurred during the progressive evolution of the Papaghni sub-basin).

Large Igneous Province, crustal extension jand basin subsidence

Drury (1984) dealt at length with the dyke swarms surrounding the Cuddapah Basin and their implications for the thermal evolution of south India. French *et al.* (2008) recognized an approximately 1890 Ma Large Igneous Province (LIP) involving the Palaeoproterozoic dyke swarms in the East Dharwar and the southern Bastar cratons. One set of east–west dykes in the East Dharwar and the contemporaneous mafic flows, dykes and sills in the transition zone between the Vempalle Formation and the Tadpatri Formation are manifestations of this LIP. As the LIPs are known to be associated with hot mantle upwellings, or plumes in intraplate setting (e.g. Bryan & Ernst 2008), the early extension in the Papaghni sub-basin may be linked to this intraplate tectonic activity. If this event represents a precursor to a major break-up event in the Palaeoproterozoic proto-India and its still uncertain correlative continental fragments, the event heralds the generation of new continental margins and extensive shallow seas. The thick carbonate platform development evidenced by the common stromatolite-bearing Tadpatri Formation suggests a slow and stable subsidence of the Papaghni sub-basin. The succeeding Gandikota Quartzite probably represents a prograding alluvial fan.

Palaeocurrent patterns in the Gulcheru Quartzite and the Nagari Quartzite

In the middle–upper part of the alluvial-fan complex of the Gulcheru Quartzite, the trough cross-beds in feldspathic–subarkosic sandstones have a distinct unimodal palaeocurrent pattern with large dispersion. Across the arcuate outcrop of the Gulcheru Quartzite in the SW part of the basin, the modal palaeoflow is generally towards the interior of the Papghni sub-basin, except at Parnapalle where the modal direction varies between the NW and west (Dasgupta & Biswas 2006). In the

Table 1. *Estimated palaeostress orientation and stress regime from the Gani–Kalva Fault and the Kona Fault*

Site/set name	Rock unit	n	σ_1	σ_2	σ_3	R	Type of stress tensor
Gani–Kalva Fault							
g3	Paniam Quartzite	26	14/110	12/015	68/242	0.09	Compressional
g4	Paniam Quartzite	22	12/065	12/333	73/197	0.34	Compressional
g5	Paniam Quartzite	49	10/106	80/281	01/197	0.63	Strike-slip
g5_1	Paniam Quartzite	32	10/027	09/120	77/254	0.22	Compressional
g6	Paniam Quartzite	42	10/149	80/299	05/059	0.58	Strike-slip
g6a	Paniam Quartzite	25	18/300	70/152	10/034	0.36	Strike-slip
g7	Paniam Quartzite	24	02/242	81/140	09/332	0.49	Strike-slip
g11	Paniam Quartzite	46	15/008	03/101	75/201	0.70	Compressional
g2	Banganapalli Quartzite	35	19/327	01/058	71/147	0.57	Compressional
g8	Banganapalli Quartzite	31	16/301	03/211	73/112	0.46	Compressional
g9	Pulivendla Quartzite	49	19/211	69/058	09/305	0.68	Strike-slip
g10	Tadpatri dyke	69	88/168	02/030	02/300	0.85	Extensional
g10_1	Tadpatri dyke	33	07/202	82/045	02/293	0.24	Strike-slip
g1	Vempalli dyke	51	68/347	06/096	18/190	0.67	Extensional
g1_1	Vempalli dyke	25	63/245	23/037	10/132	0.63	Extensional
Kona Fault							
k6	Paniam Quartzite	43	11/042	10/310	75/168	0.64	Compressional
k7	Banganapalli Quartzite	93	03/278	86/110	00/188	0.94	Strike-slip
k8	Banganapalli Quartzite	93	04/213	04/303	85/090	0.75	Compressional
k8_1	Banganapalli Quartzite	46	13/044	68/167	18/313	0.15	Strike-slip
k9	Banganapalli Quartzite	58	00/355	79/080	11/265	0.24	Strike-slip
k1	Pulivendla Quartzite	69	72/208	17/002	08/093	0.54	Extensional
k2	Pulivendla Quartzite	34	72/031	07/275	15/182	0.93	Extensional
k3	Tadpatri Formation	102	86/192	00/094	02/005	0.05	Radial extension
k4	Tadpatri Formation	51	72/077	14/297	11/203	0.46	Extensional
k5	Tadpatri Formation	79	74/051	15/202	07/292	0.67	Extensional
k5_1	Tadpatri Formation	48	07/156	81/301	06/066	0.63	Strike–slip

Shaded rows correspond to fault-slip data from the Palaeoproterozoic rock units in the western Cuddapah Basin. n, the number of fault sets; σ_1, σ_2, σ_3, principal stresses; σ_1, maximum compression; R is a measure of the shape of stress ellipsoid (after Tripathy 2011).

Maddimadugu area, and further north around Lingareddipalle, the palaeocurrents are toward the north or NNE; that is, away from the southern margin of the Papaghni sub-basin. Similarly, in areas south of Kurnool the dominant palaeocurrents are toward the east and ESE, again directed away from the western margin of the basin (Fig. 12). Thus, coalescing alluvial-fan complexes in the lower part of the Gulcheru Quartzite were possibly controlled by boundary faults representing the SW margin of the early stages of Papaghni sub-basin. As described earlier, the trough cross-bedded coarse-gritty sandstone (quartzite) facies gives way to supratidal–intertidal heterolithic sandstone–shale facies heralding the transition to the stromatolitic dolostones of the Vempalle Formation.

In contrast, the basal gritty–pebbly units of the Nagari Quartzite show a westerly palaeocurrent mode with an occasional easterly palaeocurrent. Trough cross-bedded units in Tummakona vagu (between Sanipaya and Balarajupalle) show a dominant easterly unimodal palaeocurrent (Fig. 13). Thus, the difference in lithofacies association between the Papaghni Group and the Nallamalai Group is accompanied by their difference in palaeocurrent patterns, as shown by the available data from the Gulcheru Quartzite and the Nagari Quartzite.

NFB precursor and links with the NSB

It has been proposed earlier that the possible Palaeoproterozoic rock succesions in the NFB (Nallamalai Group) are thrust transported (see also Chakraborti & Saha 2006; Saha *et al.* 2010) and, therefore, allochthonous. The suprasubduction zone ophiolites in Kandra (southern NSB) have recently been dated at 1847 ± 14 Ma (Vijaya Kumar *et al.* 2010) indicating an ancient plate convergence along the SE margin of India. The early deformation (D_1: Saha 2002) in the NFB could be related to this convergence. Ophiolitic remnants in the southern NSB indicate an open-ocean system outboard of the Palaeoproterozoic margin of the East Dharwar Craton. The sedimentary successions in the NFB show a tidal and storm influence, indicating connections with the open ocean (Chaudhuri *et al.* 2002;

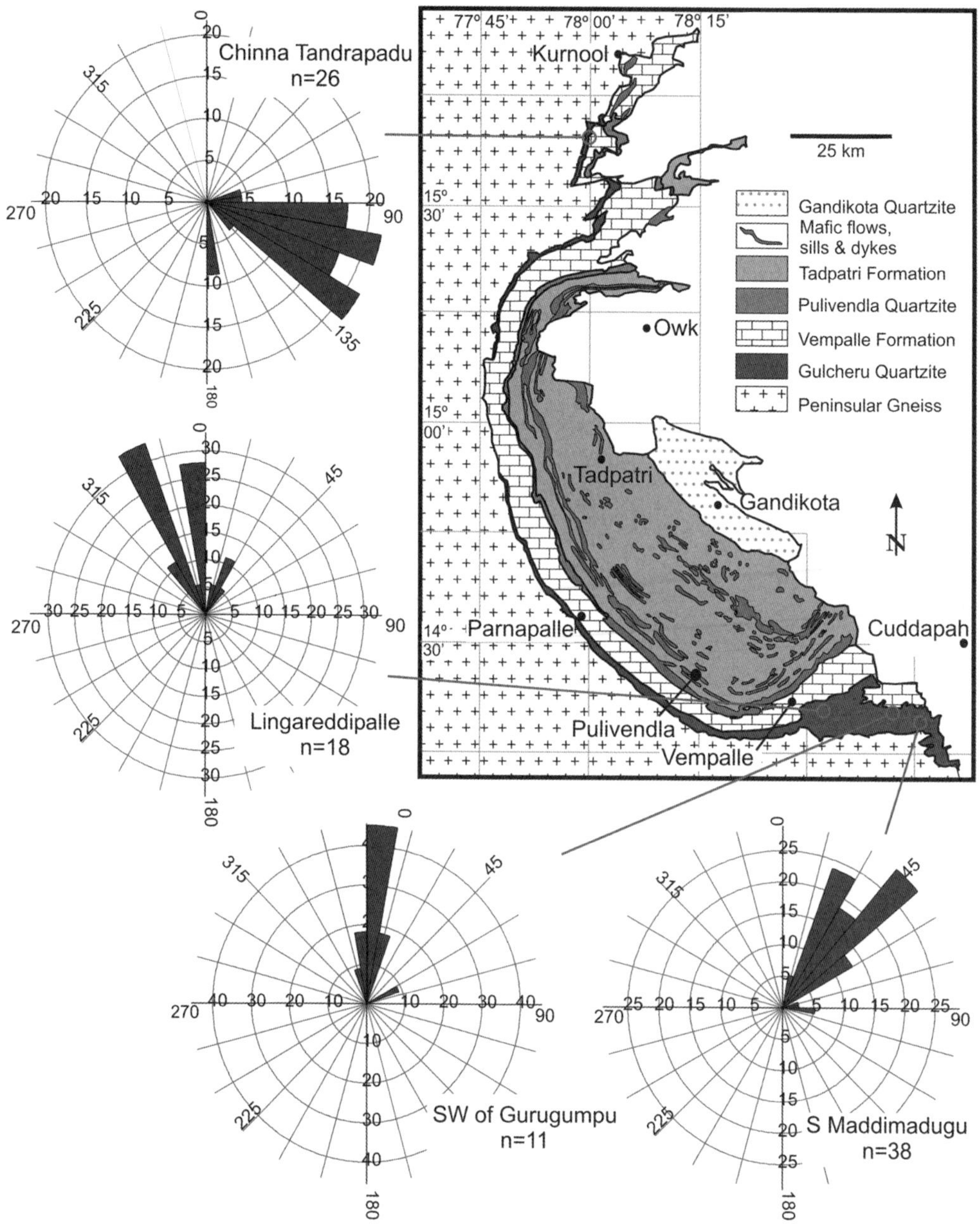

Fig. 12. Unimodal palaeocurrent roses, trough cross-beds, Gulcheru Quartzite. Note that the modal directions in the four measured sites point away from the western and southern margin of the of the Papaghni sub-basin, and except for South Maddimadugu, towards the interior of the sub-basin.

Saha *et al.* 2006). The Nallamalai Group is likely to have been deposited in a marginal basin connected to the Palaeoproterozoic ocean basin.

The overall eastward dip of the Moho reflector (Kaila & Tiwari 1985) has been interpreted by Singh & Mishra (2002) as indicating the eastward subduction of the East Dharwar Craton. However, the southern NSB ophiolites, in conjunction with the Kondapalli gabbro–anorthosite complexes (magmatic arc), in the western margin of the EGB

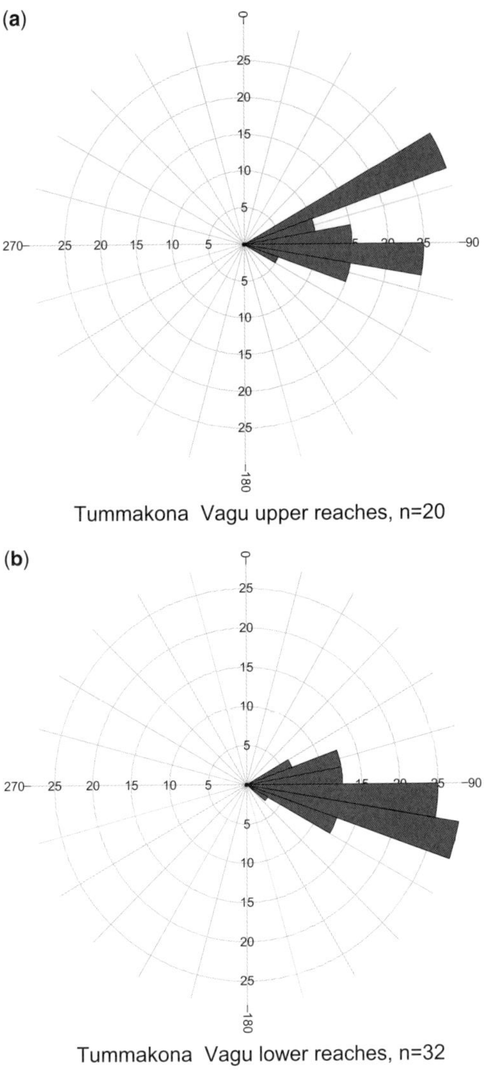

Fig. 13. Palaeocurrent roses from the Nagari Quartzite, Tummakona vagu between Sanipaya and Balrajupalli. Note the strong easterly directed modal orientation obtained from trough cross-stratified units.

have been fitted into a westward subduction model in the Palaeoproteozoic (Leelanandam *et al.* 2006; Vijaya Kumar & Leelanandam 2008; Vijaya Kumar *et al.* 2010), which probably led to the amalgamation of Columbia. Based on geochronology and geochemical studies, Ravikant (2010); suggested that the NSB probably had the Archaean precursor (cf. greenstone belts of East Dharwar: Pascoe 1973). According to Ravikant (2010), emplacement of enriched Mid Ocean Ridge Basalt (MORB)-type basalts and gabbros in the NSB occurred at approximately 1911 Ma. Ravikant (2010) proposed that Palaeoproterozoic magmatism formed in response to a major extensional event along the Eastern Dharwar Craton margin that probably initiated break-up from the SE margin of the North China Craton. It may be noted that while Ravikant (2010) considered an extensional event – that is, the break-up of India and North China – for the origin of Kandra and related gabbros and basalts, Vijaya Kumar *et al.* (2010) contemplated a convergent setting for Kandra ophiolites (see also Saha 2010).

Sedimentary successions across the north–south-tending NFB indicate an overall palaeoslope towards the east (present-day geographical co-ordinates). Although bi-directional palaeocurrents are not uncommon in the peritidal succession in the lower part of the Nagari Quartzite, trough cross-bedded units in the same formation show dominant unimodal easterly palaeocurrents (Fig. 13). Herringbone cross-stratification indicating bipolar palaeocurrents is also present in the upper part of Bairenkonda Quartzite around Porumamilla (Fig. 8c). Even in the eastern margin of the NFB around Rapur, the tidal bundles within planar tabular cross-beds in the Cumbum Formation quartzites show dominant easterly palaeoflow (Fig. 9c). The linear outcrop belt, large thickness, coarse to fine siliciclastic sediments with minor carbonates, occasional turbidites, and common tidal and storm influence in the Nallamalai Group suggest deposition in an elongated trough, the relatively deeper part of which was fed by an intermittent turbidite system. The soft-sediment deformation structures in the laminated shale–siltstones or calcareous/dolomitic mudstones of the Cumbum Formation have been interpreted as the result of seismic disturbances (Dasgupta & Biswas 2006) originating from volcanic tremors, which may have also triggered the turbidite systems. The siliciclastic systems were probably fed by fluviodeltaic systems originating in the craton margin.

Crustal structure in East Dharwar Craton and the SE margin of India

The present state of knowledge on the crustal structure across the Cuddapah Basin and its basement is based primarily on gravity data (Qureshy *et al.* 1968; Kailasam 1976; Venkatakrishnan & Dotiwala 1987; Krishna Brahmam 1989; Verma & Satyanarayana 1990; Ramakrishnan 1991; Ram Babu 1993; Subba Rao 2002), aeromagnetic data (Babu Rao *et al.* 1987; Mishra *et al.* 1987; Babu Rao 1991; Ramakrishnan 1991; Anand & Rajaram 2002) and available deep seismic sounding profiles (Kaila *et al.* 1979, 1987; Kaila & Tiwari 1985; Singh & Mishra 2002; Reddy *et al.* 2004). Published geological maps (e.g. Meijerink *et al.* 1984; Nagaraja Rao *et al.* 1987; Chakraborti & Saha 2006; Saha *et al.* 2010) showing major faults, including boundary thrusts, and other geological details from the Cuddapah Basin and adjoining region only corroborate the first-order interpretations based on the geophysical data.

Gravity data interpretation and geological controls

In comparison to the Peninsular Gneiss, which constitutes the bulk of the upper crust in the East Dharwar, the western part of the Cuddapah Basin is marked by an elliptical positive gravity anomaly feature coincident with the NW–SE trend of the Papaghni Basin, which hosts the Papaghni and Chitravati groups. The common occurrence of thick mafic sills and lava flows associated with the Tadpatri Formation and the upper part of the Vempalle Formation is also restricted to the SW part of the Cuddapah Basin. The positive gravity anomaly feature is interpreted in terms of a lower-crustal lensoid mafic body (Bhattacharji & Singh 1984). The maximum thickness of sedimentary succession in the Papaghni sub-basin is estimated to be around 6 km. In contrast, the NFB is marked as a north–south elongated negative gravity anomaly feature interpreted as being due to greater thickness (*c.* 12 km) of sedimentary strata. The apparent larger thickness in the NFB may be due to tectonic thickening and duplication in the multiply deformed fold-and-thrust belt (Saha 2002; Saha & Chakraborti 2007; Saha *et al.* 2010). The transition from the eastern margin of the NFB to the Eastern Ghats Granulite Belt or the Nellore Schist Belt (NSB) is, again, marked by a steep gradient in Bouger anomaly pattern, with strong positive anomalies in the EGB and the NSB. While high-density granulites in the EGB compared to thick, low-grade metasedimentary rocks in the NFB account for the steep gradient in the transition zone, the NSB outcrops are marked by the common occurrence of mafic rocks, particularly in the Vinjamuru domain. The NSB is thrusted over the NFB along a major intracontinental thrust (Saha *et al.* 2010; the Vellikonda thrust front: Venkatakrishna & Dotiwala 1987).

Deep seismic sounding profile and deep structure of the Cuddapah Basin

The Deep Seismic Sounding (DSS) studies in south India were carried out along the Kavali–Udipi and Koniki–Alampur traverses passing through the Cuddapah Basin (Kaila *et al.* 1979, 1987; Kaila & Tiwari 1985). The east–west Kavali–Udipi section runs from Kavali in the east coast of India through the NSB, the NFB and SW part of the Cuddapah Basin, and further west across the Dharwar Craton. The Koniki–Alampur section, around 16°N latitude, crosses the EGB, the NSB, the NFB and the western part of the Cuddapah Basin (Fig. 14). Across the Cuddapah Basin, the shallowest depth to the crystalline basement is around 200 m near Parnapalle (the western boundary of the Cuddapah Basin), while in the eastern part beneath the NFB it is about 10 km (Reddy *et al.* 2004). A number of deep faults have been inferred from the Kavali–Parnapalle sector of DSS profiles (Kaila *et al.* 1979). In the Cuddapah Basin, all of the reflectors from shallow depths to the Moho boundary show a consistent down-dip towards the east starting from the Parnapalle right up to the eastern boundary. The Moho depth is highest (*c.* 50 km) below the NSB and the shallowest (*c.* 35 km) below the western boundary of the Cuddapah Basin near Parnapalle (Kaila *et al.* 1979; Kaila & Tiwari 1985; Kaila & Krishna 1992).

Some of the interpreted faults apparently offset the Moho. A low-angle thrust (fault No. 2 in DSS profile, Fig. 14b) in the upper crust juxtaposes the upthrusted NSB against the NFB along the eastern margin of the Cuddapah Basin, coinciding with the Vellikonda thrust front. Similarly, the easterly dipping Maidukuru Thrust (Rudravaram line of Meijerink *et al.* 1984), coincides with a steep fault (No. 7) defining the boundary between the NFB and the western Cuddapah (Papaghni and Kurnool sub-basins). Near Parnapalle, in the SW margin of the Cuddapah Basin, apparent normal displacement (along fault No. 9) offsets the Moho at 30–40 km depth. The initial phases of opening of the Cuddapah Basin (Papaghni sub-basin) could have been controlled by this deep-seated fault (Kaila *et al.* 1979, 1987; Verma & Satyanarayana 1990).

The east–west Koniki–Alampur profile passes through the northern part of the Cuddapah Basin crossing the Iswarakuppam dome. Kaila *et al.* (1987) observed that the Iswarakuppam dome is bound by normal faults on either side, with a sediment thickness of 5 km above the granitic basement. The deep structures and gravity anomalies favour the tectonic juxtaposition of the NFB against both the Papaghni and Kurnool sub-basins.

Nallamalai Fold Belt (NFB), craton margin deformation and basin inversion

The NFB is a north–south-trending fold-and-thrust belt with two major intracontinental thrusts along its eastern and western boundaries (Saha 1994; Saha *et al.* 2010). The eastern boundary thrust, the Vellikonda thrust front, separates the Nellore Schist Belt with supposed Archaean greenstone belt remnants

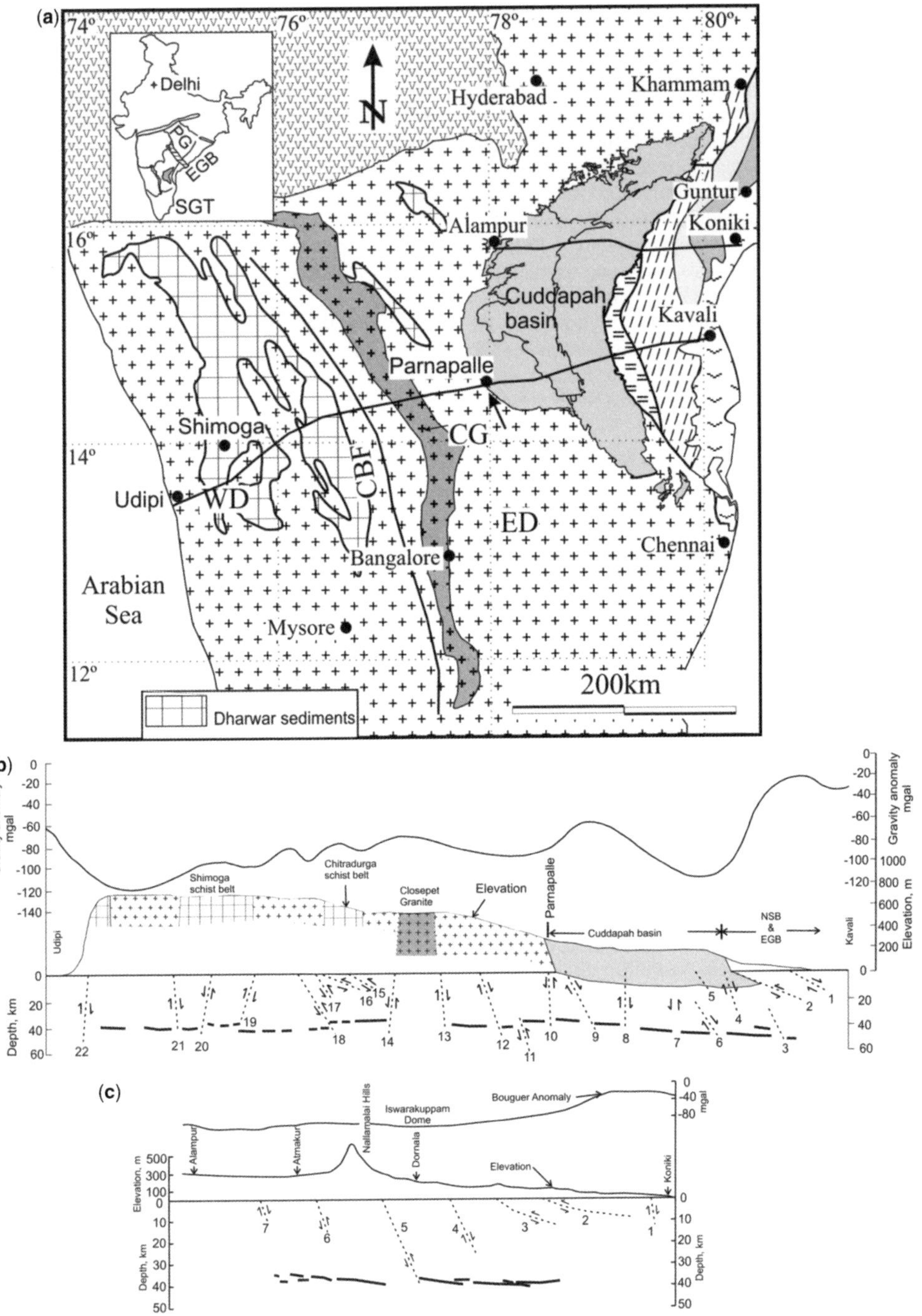

Fig. 14. Deep structure across the Dharwar Craton and the Cuddapah Basin (after Kaila *et al.* 1979, 1987; Kaila & Tiwari 1985). (**a**) Map showing the broad regional geology around the Cuddapah Basin and the locations of the two deep seismic sounding traverses. CG, Closepet Granite; CBF, Chitradurga Boundary Fault; ED, Eastern Dharwar Craton; WD, Western Dharwar Craton. Dharwar sediments refer to Archaean sedimentary successions in greenstone belts. Inset: EGB, Eastern Ghats Belt; PG, Pranhita–Godavari Valley Basin; SGT, Southern Granulite Terrain. (**b**) Udipi–Kavali profile. Note that the major boundary faults apparently continue up to Moho depth (e.g. fault Nos 8 and 9 with net normal displacement). (**c**) The Alampur–Koniki profile through the Iswarakuppam dome.

and Palaeoproterozoic ophiolites (Narayanswami 1966; Meijerink *et al.* 1984; Nagaraja Rao *et al.* 1987; Ramam & Murty 1997; Saha 2002, 2004; Ravikant 2010; Saha *et al.* 2010; Vijaya Kumar *et al.* 2010). Recent work suggests a major thrust along the western boundary of the NFB, along which folded Nallamalai rocks are juxtaposed against the generally flat-lying Kurnool Group of rocks (Chakraborti & Saha 2006; Saha *et al.* 2006, 2010). Recognition of the approximately 1850 Ma old Kandra nappe complex as a suprasubduction zone ophiolite (Dilek 2003) provides the basis for linking Palaeoproterozoic deformation in the Nellore Schist Belt with a west-directed plate subduction (Saha 2010, 2011; Vijaya Kumar *et al.* 2010). The high-grade Eastern Ghats Belt represents an old arc complex in this model (Vijaya Kumar & Leelanandam 2008).

In the NE NFB, early deformation ($D_1 - D_2$ deformation phases) is recorded in superposed folds (F_1 and F_2 fold systems) and overprinting cleavages (Saha 2002). The common occurrence of west-vergent F_1 folds, easterly dipping slaty cleavage and shear zones/thrusts with mylonites suggest an oblique transpressional deformation along the craton margin (Saha *et al.* 2010; Tripathy & Saha 2010). There is a regional variation in the trend of the F_2 folds from NNW in the south, to north–south in the central part to NE–SW in the northern part, consistent with the arcuate form of the NFB (Fig. 2). F_2 folds are associated with a steep crenulation cleavage. The emplacement of syn- to late-D_2 granites in the northern part of the belt is broadly contemporaneous with deformed granites/granite gneisses in the NSB (Saha & Chakraborti 2007). While the emplacement age of Vinukonda granite in the NSB is about 1589 Ma (Gupta *et al.* 1984; Dobmeier *et al.* 2006), the Vellaturu Granite in the NFB is dated at 1584 Ma (Crawford & Compston 1973). The Chelima lamproite emplaced in the folded Nallamalai rocks is dated at 1400 Ma (Chalapathi Rao *et al.* 1999). Late deformation (D_3) in the NFB is recorded in east–west fault zones, such as in Vami Konda SE of Nekarikallu (Saha 2002) and in the eastern extension of the Atmakur Fault north of the Iswarakuppam dome (Tripathy & Saha 2010; Tripathy 2011). Matin & Guha (1996) also recorded late east–west structures in the extreme southern part of the NFB. Local development of east–west structures in association with dominant west-vergent folds and east-dipping cleavages has been reported from the western part of the southern NFB (Mukherjee 2001). The Palnad Nappe in the NE corner of the Cuddapah Basin and the associated footwall deformation in the Kurnool Group in Palnad (Saha & Chakraborty 2003), and footwall deformation along the Maidukuru Thrust along the western margin of the NFB, represent late contractional deformation (Chakraborti & Saha 2006). These are postulated to be related to late Neoproterozoic events as Gondwana amalgamated.

For this present review, the critical issue is a Palaeoproterozoic convergence along the SE margin of India, which possibly has imprinted in the D_1 deformation of the NFB and NSB. The western part of the Cuddapah Basin (Papaghni sub-basin) does not record any early contractional deformation, strengthening the view that the NFB precursor was spatially separated from the Papaghni sub-basin in the Palaeoproterozic. The transverse faults, viz. the Gani–Kalva and Kona faults, record early shallow-crustal deformation under an extensional/strike-slip regime (Table 1) (Tripathy & Saha 2009; Tripathy 2011) possibly linked to basin opening in the western part. Although these transverse faults were reactivated during late contractional deformation, as evident from results of fault-slip analysis (Tripathy 2011), the tectonic juxtaposition of the NFB against the western Cuddapah Basin probably did not happen until late Neoproterozic times (Saha *et al.* 2006). The final juxtaposition of the EGB against the Bhandara and the Bastar cratons, north of the Pranhita–Godavari rift, is directly or indirectly dated at around 500 Ma (Bhadra *et al.* 2004; Dobmeier *et al.* 2006; Biswal *et al.* 2007; cf. Okudaira *et al.* 2001). While the Papaghni Group is separated from the Tadpatri Group (Fig. 3) by a paraconformity (Nagaraja Rao *et al.* 1987), a major depositional hiatus is indicated by the erosional unconformity above the Tadpatri Group. A positive inversion of the Papaghni sub-basin could have been triggered by the late Palaeoproterozoic–early Mesoproterozoic crustal convergence along the Eastern Dharwar Craton margin, evidenced by the D_2 deformation in the NFB and granite emplacements in the NE NFB and adjoining NSB.

Concluding remarks

The lower Cuddapah succession unconformably lying over the Peninsular Gneiss in the East Dharwar Craton represents one of the finest examples of intracratonic basinal sequence dating back to the Palaeoproterozoic in India. Although there still exists some uncertainties on the exact age of sedimentation of the Nallamalai Group, syn- to late-D_2 intrusive granite along the eastern margin of the NFB are as old as 1584 Ma (Crawford & Compston 1973; Saha & Chakraborti 2007) rendering the Nallamalai Group to be late Palaeoproterozoic–early Mesoproterozoic. As the NFB is thrust transported, the Nallamalai Group is possibly allochthonous. Overall, fluvial–shallow-marine cyclic sedimentation in the lower Cuddapah (Papaghni sub-basin) is episodically interrupted by basin-wide unconformities. The earliest cycle of sedimentation began with extensive alluvial fans along the SW margin of the basin, with sediments derived from exposed granitic terrain of the East Dharwar Craton to the west. The subsequent mixed siliciclastic–carbonate sedimentation in an extensive shallow sea occurred over a stable platform, which gave way to shelfal mud deposition in the upper part of the Vempalle Formation indicating a maximum flooding event in the first cycle of sedimentation. The emplacement of dykes, sills and subaerial mafic flows fixed the initial setting for the second cycle of sedimentation recorded in the Pulivendla Quartzite. Coastal–shallow-marine carbonate sedimentation of the Tadpatri Formation was accompanied by mafic igneous activity. Relative sea-level rise in the second cycle is, again, marked by a transition to offshore shale free from any terrigenous coarser clastics. Gradual shallowing led to the development of stromatolitic bioherms in the upper part of the Tadpatri Formation, which also contains rhyolitic ash beds. The second cycle is terminated with the deposition of the progradational Gandikota Quartzite.

Deposition of the Nallamalai Group occurred in a basin further east and independent of the Papaghni sub-basin. The bulk of the arenaceous Bairenkonda Quartzite was deposited in shallow seas above storm wave base. Shelf mud of the Cumbum Formation ultimately gave way to peritidal quartzites in the uppermost part. Inversion of the Nallamalai sub-basin is possibly linked to contractional deformation of the Nellore Schist Belt, which reportedly contains suprasubduction zone ophiolites. Given the limited geochronological data, early sedimentation (first cycle) in the Papaghni sub-basin occurred prior to Columbia amalgamation (*c.* 1800 Ma). Mafic intrusion and second cycle of Papaghni sedimentation (Chitravati Group) is broadly contemporaneous with the suprasubduction-zone ophiolites in the NSB. But the plate subduction did not have any direct influence on the Papaghni sedimentation as the intracratonic basin was spatially separated. A major hiatus in deposition followed the approximately 1589 Ma granite emplacement in the NSB and NFB, and the amalgamation of these two belts.

This work is supported by the Indian Statistical Institute, Kolkata in the form of several research grants to D. Saha during the past decade. Discussions with A. Chaudhuri and S. Patranabis-Deb helped in crystallizing some ideas in stratigraphic analysis. The critical and constructive review comments from T. van Loon, G. Ghosh and A. Collins helped in improving the paper. However, the authors are responsible for any remaining shortcomings. An initial version of this paper was presented in the IGCP-509 India conference. V. Tripathy acknowledges a senior research fellowship granted by ISI. We also acknowledge a DST grant (SR/S4/ES-307/2007) and a DST-AISRF grant (INT/AUS/P-28/2009ii), which partly supported this work.

Appendix

Table A.1. *Abbreviations explained*

Abbreviation	Explanation
LIP	Large Igneous Province
EGB	Eastern Ghats Belt
NSB	Nellore Schist Belt
NFB	Nallamalai Fold Belt
SH/LLH	Stacked hemispheroids/laterally linked hemispheroids
KF	Kona Fault
GKF	Gani–Kalva Fault
AF	Atmakur Fault
ED	Eastern Dharwar Craton
CBF	Chitradurga Boundary Fault
WD	Western Dharwar Craton

Table A.2. *The latitude/longitude of places referred to in the text, but not shown on maps*

Location	Latitude/longitude
Atmakur	N15°52′42″/E78°35′18″
Bagidipalli	N14°09′21″/E79°00′40″
Betamcherla	N15°26′56″/E78°08′46″
Balrajupalle	N14°07′41″/E79°01′13″
Chinna Tandrapadu	N15°45′31″/E78°03′31″
Chintalapalle tank	N15°06′08″/E79°00′36″
Jammalamadugu	N14°50′21″/E78°22′49″
Kavalakuntla Reserve Forest	N15°00′00″/E79°07′10″
Maddimadugu	N14°19′45″/E78°46′29″
Naranrajupalle	N14°12′11″/E79°05′05″
South of Yagantipalle	N15°18′55″/E78°11′40″
South of Yerraguntla	N14°34′25″/E78°33′36″
Mudduletiswami Temple Road	N15°20′00″/E78°05′13″
North of Krishnagiri	N15°20′51″/E78°05′07″
Pandurangapuram	N14°59′34″/E79°07′41″
Rajampet	N14°11′44″/E79°09′35″
Rapur	N14°12′04″/E79°28′32″
Sanipai (Kottapeta)	N14°06′29″/E78°57′00″
Soman cheruvu	N15°00′00″/E79°07′01″
Tekurupeta	N15°00′00″/E79°02′25″
Tummukona vagu, upper reaches	N14°06′50″/E78°59′04″
Tummukona vagu, lower reaches	N14°07′21″/E79°00′05″
Yadiki	N15°02′47″/E77°52′15″

References

Anand, M., Gibson, S. A., Subbarao, K. V., Kelley, S. P. & Dickin, A. P. 2003. Early Proterozoic melt generation processes beneath the intra-cratonic Cuddapah basin, Southern India. *Journal of Petrology*, **44**, 2139–2171.

ANAND, S. P. & RAJARAM, M. 2002. Aeromagnetic data to probe the Dharwar craton. *Current Science*, **83**, 162–167.

BABU, V. R. R. M. 1998. The Nellore schist belt: an Archean greenstone belt, Andhra Pradesh, India. *In*: RAO, A. T., DIVI, S. R. & YOSHIDA, M. (eds) *Precambrian Crustal Processes in East Coast Granulites–Greenstone Regions of India and Antarctica within East Gondwana*. Gondwana Research Group, Memoir, **4**, 97–136.

BABU RAO, V. 1991. Geological significance of magnetic depth estimates over western part of Cuddapah basin, south India. *Journal of the Geological Society of India*, **38**, 605–614.

BABU RAO, V., ATCHUTA RAO, D., RAM RAO, Ch., SARMA, B. S. P., BHASKARA RAO, D. S., VEERASWAMY, K. & SARMA, M. R. L. 1987. Some salient results of interpretation of aeromagnetic data over Cuddapah basin and adjoining terrain, South India. *In*: RADHAKRISHNA, B. P. (ed.) *Purana Basins of Peninsular India*. Geological Society of India, Memoir, **6**, 295–312.

BHADRA, S., GUPTA, S. & BANERJEE, M. 2004. Structural evolution across the Eastern Ghats Mobile Belt–Bastar craton boundary: hot over cold thrusting in an ancient collision zone. *Journal of Structural Geology*, **26**, 233–245.

BHASKAR RAO, Y. J., PANTULU, G. V. C., DAMODAR REDDY, V. & GOPALAN, K. 1995. Time of early sedimentation and volcanism in the Proterozoic Cuddapah basin, South India: evidence from Rb–Sr age of Pulivendla mafic sill. *Geological Society of India, Memoir*, **33**, 329–338.

BHATTACHARJI, S. & SINGH, R. N. 1984. Thermo-mechanical structure of the southern part of the Indian Shield and its relevance to Precambrian basin evolution. *Tectonophysics*, **105**, 103–120.

BISWAL, T. K., DE WAELE, B. & AHUJA, H. 2007. Timing and dynamics of the juxtaposition of the Eastern Ghats Mobile Belt against the Bhandara Craton, India: a structural and zircon U–Pb SHRIMP study of the fold–thrust belt and associated nepheline syenite plutons. *Tectonics*, **26**, TC4006, doi: 10.1029/2006TC002005.

BROWN, M. 2008. Characteristic thermal regimes of plate tectonics and their metamorphic imprint throughout Earth history: when did Earth first adopt a plate tectonics mode of behavior? *In*: CONDIE, K. & PEASE, V. (eds) *When did Plate Tectonics Begin?* Geological Society of America, Special Paper, **440**, 97–128.

BROWN, M. 2010. Geodynamic regimes and tectonic settings for metamorphism: relation to supercontinent cycles. *Indian Journal of Geology*, **80**, 3–21.

BRYAN, S. E. & ERNST, R. E. 2008. Revised definition of Large Igneous Provinces (LIPs). *Earth Science Reviews*, **86**, 175–202.

CHAKRABARTI, G. & SHOME, D. 2010. Interaction of microbial communities with clastic sedimentation during Palaeoproterozoic time – An example from basal Gulcheru Quartzite, *Sedimentary Geology*, **226**, 22–28, doi: 10.1016/j.sedgeo.2010.02.003.

CHAKRABORTI, S. & SAHA, D. 2006. Tectonic stresses and thin-skinned tectonics in a Proterozoic fold-and-thrust belt read from calcite mylonites in the Cuddapah basin, south India. *Indian Journal of Geology*, **78**, 37–54.

CHAKRABORTI, P. P., DEY, S. & MOHANTY, S. 2010. Proterozoic platform sequences of Peninsular India: Implications towards basin evolution and supercontinent assembly. *Journal of Asian Earth Sciences*, **39**, 589–607, doi: 10.1016/j.jseaes.2010.04.030.

CHALAPATHI RAO, N. V., MILLER, J. A., GIBSON, S. A., PYLE, D. M. & MADHAVAN, V. 1999. Precise $^{40}Ar/^{39}Ar$ age determinations of the Kotakonda kimberlite and Chelima lamproite, India: implication for the mafic dyke swarm emplacement in the Eastern Dharwar craton. *Journal of Geological Society of India*, **53**, 425–432.

CHAUDHURI, A. K., SAHA, D., DEB, G. K., PATRANABIS-DEB, S., MUKHERJEE, M. K. & GHOSH, G. 2002. The Purana basins of southern cratonic province of India – a case study for Mesoproterozoic fossil rifts. *Gondwana Research*, **5**, 23–33.

COLLINS, A. S. & PISAREVSKY, S. A. 2005. Amalgamating eastern Gondwana: the evolution of the Circum-Indian Orogens. *Earth Science Reviews*, **71**, 229–270.

CONDIE, K. C. 1997. *Plate Tectonics and Crustal Evolution*. Butterworth-Heinemann, Oxford.

CONDIE, K. C., O'NEIL, C. & ASTER, R. C. 2009. Evidence and implications for a widespread magmatic shutdown for 250 My on Earth. *Earth and Planetary Science Letters*, **282**, 294–298.

CRAWFORD, A. R. & COMPSTON, W. 1973. The age of the Cuddapah and Kurnool systems, southern India. *Journal of the Geological Society of Australia*, **19**, 453–464.

DASGUPTA, P. K. & BISWAS, A. 2006. *Rhythms in Proterozoic Sedimentation: An Example from Peninsular India*. Satish Serial Publishing, Delhi.

DASGUPTA, S. & SENGUPTA, P. 2003. Indo-Antarctic correlatio: a perspective from the Eastern Ghats granulite belt, India. *In*: YOSHIDA, M., WINDLEY, B. F. & DASGUPTA, S. (eds) *Proterozoic East Gondwana: Supercontinent Assembly and Breakup*. Geological Society, London, Special Publications, **206**, 131–143.

DHARMA RAO, C. V., SANTOSH, M. & YUAN BAO, W. 2011. Mesoproterozoic ophiolitic melange from the SE periphery of the Indian plate: U–Pb zircon ages and tectonic implications. *Gondwana Research*, **19**, 384–401.

DILEK, Y. 2003. Ophiolite pulses, mantle plumes and orogeny. *In*: DILEK, Y. & ROBINSON, P. (eds) *Ophiolites in Earth History*. Geological Society, London, Special Publications, **218**, 9–19.

DOBMEIER, C. J. & RAITH, M. M. 2003. Crustal architecture and evolution of the Eastern Ghats Belt and adjacent regions of India. *In*: YOSHIDA, M., WINDLEY, B. F. & DASGUPTA, S. (eds) *Proterozoic East Gondwana: Supercontinent Assembly and Breakup*. Geological Society, London, Special Publications, **206**, 145–168.

DOBMEIER, C., LÜTKE, S., HAMMERSCHMIDT, H. & MEZGER, K. 2006. Emplacement and deformation of the Vinukonda granite – implications for the geological evolution of peninsular India and for Rodinia reconstructions. *Precambrian Research*, **146**, 165–178.

DRURY, S. A. 1984. A Proterozoic intracratonic basin, dyke swarms and thermal evolution in South India. *Journal of the Geological Society of India*, **25**, 437–444.

FRENCH, J. E., HEAMAN, L. M., CHACKO, T. & SRIVASTAVA, R. K. 2008. 1891–1883 Ma Southern Bastar–Cuddapah mafic igneous events, India: a newly recognized large igneous province. *Precambrian Research*, **160**, 308–322.

Glen, R. A., Hancock, P. L. & Whittaker, A. 2005. Basin inversion by distributed deformation: the southern margin of the Bristol Channel Basin, England. *Journal of Structural Geology*, **27**, 2113–2134.

Gupta, J. N., Pandey, B. K., Chabria, T., Banerjee, D. C. & Jayaram, K. M. V. 1984. Rb–Sr Geochronologic studies on the granites of Vinukonda and Kanigiri, Prakasam district, Andhra Pradesh, India. *Precambrian Research*, **26**, 105–109.

Heaman, L. M. 2008. Precambrian Large Igneous provinces: an overview of geochronology, origins and impact on earth evolution. *Journal of the Geological Society of India*, **72**, 15–34.

Holland, T. H. 1909. Indian Empire. Geology. *Imperial Gazetteer of India*, **1**, 50–103.

Kaila, K. L. & Krishna, V. G. 1992. Deep Seismic Sounding studies in India and major discoveries. *Current Science*, **62**, 117–154.

Kaila, K. L. & Tiwari, H. C. 1985. Structural trends in the Cuddapah basin from deep seismic soundings (DSS) and their tectonic implication. *Tectonophysics*, **115**, 69–86.

Kaila, K. L., Roy Chowdhury, K. et al. 1979. Crustal structure along Kavali–Udipi profile in the Indian peninsular shield from deep seismic sounding. *Journal of the Geological Society of India*, **20**, 307–333.

Kaila, K. L., Tiwari, H. C., Roy Chowdhury, K., Rao, V. K., Sridhar, A. R. & Mall, D. M. 1987. Crustal structure of the northern part of the Proterozoic Cuddapah basin of India from deep seismic soundings and gravity data. *Tectonophysics*, **140**, 1–12.

Kailasam, L. N. 1976. Geophysical studies of the major sedimentary basins of the Indian craton, their deep structural features and evolution. *Tectonophysics*, **36**, 225–245.

King, W. 1872. Kudapah and Karnul Formations in the Madras Presidency. *Geological Survey of India, Memoir*, **8**, 346.

Krishna Brahmam, N. 1989. Gravity and seismicity of the Cuddapah basin and surrounding places. *Journal of the Geological Society of India*, **34**, 373–384.

Lakshminarayana, G., Bhattacharjee, S. & Ramanaidu, K. V. 2001. Sedimentation and stratigraphic framework in the Cuddapah basin. *In*: *National Seminar Commemorating Dr M. S. Krishnan's Birth centenary (Proceedings Volume)*. Geological Survey of India, Special Publications, **55**, 31–58.

Leelanandam, C., Burke, K., Ashwal, L. D. & Webb, S. J. 2006. Proterozoic mountain building in Peninsular India: an analysis based primarily on alkaline rock distribution. *Geological Magazine*, **143**, 195–212.

Logan, B. W., Rezak, R. & Ginsburg, R. N. 1964. Classification and environmental significance of algal stromatolites. *Journal of Geology*, **72**, 68–83.

Matin, A. & Guha, J. 1996. Structural geometry of the rocks of the southern part of the Nallamalai fold belt, Cuddapah Basin, Andhra Pradesh. *Journal of the Geological Society of India*, **47**, 535–545.

Meert, J. 2003. A synopsis of events related to the assembly of eastern Gondwana. *Tectonophysics*, **362**, 1–40.

Meijerink, A. M. J., Rao, D. P. & Rupke, J. 1984. Stratigraphic and Structural development of the Precambrian Cuddapah basin, SE India. *Precambrian Research*, **26**, 57–104.

Mishra, D. C., Baburao, V., Laxman, G., Rao, M. B. S. V. & Rao, M. B. S. V. 1987. Three-dimensional structural model of Cuddapah basin and adjacent eastern part from geophysical studies. *In*: Radhakrishna, B. P. (ed.) *Purana Basins of Peninsular India (Middle to Late Proterozoic)*. Geological Society of India, Memoir, **6**, 311–329.

Moeen, S. 1998. P–T estimates from Nellore schist belt, India and evidence for superimposed metamorphic events. *Geological Journal*, **33**, 1–15.

Mukherjee, M. K. 2000. *Structures, Kinematic Framework and Shallow Crustal Deformation Mechanisms in the Southern Nallamalai Fold Belt, Cuddapah Dt, Andhra Pradesh, South India*. PhD thesis, Calcutta University.

Mukherjee, M. K. 2001. Structural pattern and kinematic framework of deformation in the southern Nallamalai fold–fault belt, Cuddapah district, Andhra Pradesh, Southern India. *Journal Asian Earth Sciences*, **19**, 1–15.

Nagaraja Rao, B. K., Rajurkar, S. T., Ramalingaswamy, G. & Ravindra Babu, B. 1987. Stratigraphy, Structure and Evolution of the Cuddapah basin. *In*: Radhakrishna, B. P. (ed.) *Purana Basins of Peninsular India (Middle to Late Proterozoic)*. Geological Society of India, Memoir, **6**, 33–86.

Narayanswami, S. 1966. Tectonics of the Cuddapah basin. *Journal of the Geological Society of India*, **7**, 33–50.

Okudaira, T., Hamamoto, T., Hari Prasad, B. & Kumar, R. 2001. Sm Nd and Rb–Sr dating of amphibolite from the Nellore–Khammam schist belt, SE India: constraints on the collision of the Eastern Ghats terrane and Dharwar–Bastar craton. *Geological Magazine*, **138**, 495–498.

Pascoe, E. H. 1973. *A Manual of the Geology India and Burma*. 3rd edn. **1**. Geological Survey of India, Calcutta.

Qureshy, M. N., Krishna Brahmam, N., Aravamadhu, P. S. & Naqvi, S. M. 1968. Role of Granitic intrusions in reducing the density of the crust, and other related problems as illustrated from a gravity study of the Cuddapah basin India. *Proceedings of the Royal Society of London. Series A, Mathematical and Physical Sciences*, **304**, 449–464.

Radhakrishna, B. P. (ed.) 1987. *Purana Basins of India*. Geological Society of India, Memoir, **6**.

Radhakrishna, B. P. & Naqvi, S. M. 1986. Precambrian continental crust of India and its evolution. *Journal of Geology*, **94**, 145–166.

Ramakrishnan, T. S. 1991. Cuddapah basin configuration – Another account based on gravity and aeromagnetics. *Journal of the Geological Society of India*, **37**, 109–118.

Ram Babu, H. V. 1993. Basement structure of the Cuddapah basin from gravity anomalies. *Tectonophysics*, **223**, 411–422.

Ramakrishnan, M. & Vaidyanadhan, R. 2008. *Geology of India Volume I*. Geological Society of India, Bangalore.

Ramam, P. K. & Murty, V. N. 1997. *Geology of Andhra Pradesh*. Geological Society of India, Bangalore.

Ravikant, V. 2010. Palaeoproterozoic (_1.9 Ga) extension and breakup along the eastern margin of the

Eastern Dharwar Craton, SE India: New Sm–Nd isochron age constraints from anorogenic mafic magmatism in the Neoarchean Nellore greenstone belt. *Journal of Asian Earth Sciences*, **37**, 67–81.

Reddy, P. R., Chandrakala, K., Prasad, A. S. S. S. R. S. & Rama Rao, Ch. 2004. Lateral and vertical crustal velocity and density variations in the Southwestern Cuddapah Basin and adjoining Eastern Dharwar Craton. *Current Science*, **87**, 1607–1614.

Rogers, J. J. W. & Santosh, M. 2004. *Continents and Supercontinents*. Oxford University Press, Oxford.

Saha, D. 1994. Fold–Fault structures of the Nallamalai range, Diguvametta–Nandi Ka Nama Pass, Prakasam district, Andhra Pradesh, South India. *Indian Journal of Geology*, **66**, 203–213.

Saha, D. 2002. Multi-stage deformation in the Nallamalai fold belt, Cuddapah basin, South India – Implications for Mesoproterozoic tectonism along southeastern margin of India. *Gondwana Research*, **5**, 701–719.

Saha, D. 2004. Structural asymmetry and Plate tectonic set-up for a Proterozoic Fold and Thrust Belt: Nallamalai fold belt and Adjoining Terrane, South India. *In*: Roy, A., Jain, S. C. & Chakrabarti, U. (eds) *Uniformitarianism Revisited: Comparison Between Ancient and Modern Orogens of India: Proceedings of the Workshop on IGCP 453*. Geological Survey of India, Special Publications, **84**, 101–119.

Saha, D. 2006. Rhysonetron and probable trace fossils in the Palaeoproterozoic to Mesoproterozoic quartzites, Cuddapah basin, south India – debate on early metazoan antiquity. *In*: *Proceedings of the National Seminar on Precambrain Life: Indian Scenario*. Durgapur Government College, Durgapur. 25–26.

Saha, D. 2010. *Palaeoproterozoic dismembered ophiolites in Kandra Nappe Complex, south India. In*: *Asia Oceania Geosciences Society (AOGS 10), 5–9 July, Hyderabad.* Asia Oceania Geosciences Society, Singapore, SE09-A011.

Saha, D. 2011. Dismembered ophiolites in Palaeoproterozoic nappe complexes of Kandra and Gurramkonda, South India. *Journal of Asian Earth Sciences*, **42**, 158–175.

Saha, D. & Chakraborty, S. 2003. Deformation pattern in the Kurnool and Nallarnalai Groups in the northeastern part (Palnad area) of the Cuddapah basin, south India and its implication on Rodinia/Gondwana tectonics. *Gondwana Research*, **6**, 573–583.

Saha, D. & Chakraborti, S. 2007. Advective heat transfer and fabric development in a shallow crustal intrusive granite – the case of Proterozoic Vellaturu granite, south India. *Journal of Earth System Science*, **116**, 433–450.

Saha, D. & Chaudhuri, A. K. 2003. Deformation of the Proterozoic successions in the Pranhita–Godavari basin, south India – a regional perspective. *Journal of Asian Earth Sciences*, **21**, 557–563.

Saha, D., Ghosh, G., Chakraborty, A. K. & Chakraborti, S. 2006. Comparable Neoproterozoic sedimentary sequences in Palnad and Kurnool subbasins and their paleogeographic and tectonic implications. *Indian Journal of Geology*, **78**, 175–192.

Saha, D., Chakraborti, S. & Tripathy, V. 2010. Intracontinental thrusts and inclined transpression along eastern margin of the East Dharwar Craton, India. *Journal of the Geological Society of India*, **75**, 323–337.

Santosh, M. 2010. A synopsis of recent conceptual models on supercontinent tectonics in relation to mantle dynamics, life evolution and surface environment. *Journal of Geodynamics*, **50**, 116–133.

Singh, A. P. & Mishra, D. C. 2002. Tectonosedimentary evolution of Cuddapah basin and Eastern Ghats mobile belt (India) as Proterozoic collision: gravity, seismic and geodynamic constraints. *Journal of Geodynamics*, **33**, 249–267.

Subba Rao, D. V. 2002. Zero-free air-based gravity anomaly (preliminary) map of South India – A refined and redefined Bouguer map. *Current Science*, **82**, 1373–1378.

Tripathy, V. 2011. *Brittle deformation in western Cuddapah Basin: Implications for Proterozoic intracontinental deformation*. Unpublished PhD thesis, Jadavpur University.

Tripathy, V. & Saha, D. 2009. *Late Structures in the Cuddapah Basin, south India and its possible relation with the Gondwana assembly. In*: *Asia Oceania Geosciences Society (AOGS 9), 11–15 August, Singapore*.Asia Oceania Geosciences Society, Singapore, SE61-A006.

Tripathy, V. & Saha, D. 2010. Structure and low grade metamorphism of the east central part of the Proterozoic Nallamalai fold belt, South India – thrust stacking and discontinuous metamorphic gradients along eastern margin of east Dharwar craton. *Indian Journal of Geology*, **80**, 173–188.

Venkatakrishnan, R. & Dotiwala, F. E. 1987. The Cuddapah salient: a tectonic model for the Cuddapah basin, India, based on Landsat image interpretation. *Tectonophysics*, **136**, 237–253.

Verma, R. K. & Satyanarayana, Y. 1990. Gravity field, deep-seismic sounding and crust–mantle structure over the Cuddapah basin and Dharwar Craton of India. *Tectonophysics*, **178**, 337–356.

Vijaya Kumar, K. & Leelanandam, C. 2008. Evolution of the Eastern Ghats belt, India: a plate tectonic perspective. *Journal of the Geological Society of India*, **72**, 720–749.

Vijaya Kumar, K., Ernst, W. G., Leelanandam, C., Wooden, J. L. & Grove, M. J. 2010. First Palaeoproterozoic ophiolite from Gondwana: Geochronologic–geochemical documentation of ancient oceanic crust from Kandra, SE India. *Tectonophysics*, **487**, 22–32.

van Wagoner, J. C., Mitchum, R. M., Campion, K. M. & Rahmanian, V. D. 1990. Siliciclastic sequence stratigraphy in well logs, cores, and outcrops. *American Association of Petroleum Geologists, Methods in Exploration Series*, **7**, 55.

Yoshida, M., Jacobs, J., Santosh, M. & Rajesh, H. M. 2003. Role of Pan-African events in the Circum-East Antarctic Orogen of East Gondwana: a critical overview. *In*: Yoshida, M., Windley, B. F. & Dasgupta, S. (eds) *Proterozoic East Gondwana: Supercontinent Assembly and Breakup*. Geological Society, London, Special Publications, **206**, 57–75.

Mesoproterozoic emplacement and enriched mantle derivation of the Racherla alkali syenite, Palaeo-Mesoproterozoic Cuddapah Basin, southern India: insights from *in situ* Sr–Nd isotopic analysis on apatite

N. V. CHALAPATHI RAO[1]*, FU-YUAN WU[2] & M. SRINIVAS[3]

[1]*Centre of Advanced Study in Geology, Banaras Hindu University, Varanasi 221005, India*

[2]*State Key Laboratory of Lithospheric Evolution, Institute of Geology and Geophysics, Chinese Academy of Sciences, Chaoyang District, Beijing 100029, China*

[3]*Department of Geology, P.G. College of Science, Saifabad, Osmania University, Hyderabad 500004, India*

**Corresponding author (e-mail: nvcr100@gmail.com)*

Abstract: We report an *in situ* Sm–Nd isochron age of 1326 $\pm$ 73 Ma, determined by LA-MC-ICP-MS (laser ablation-multiple collector-inductively coupled plasma-mass spectrometry), on crystalline apatite grains from the Racherla alkali syenite occurring in the Palaeo-Mesoproterozoic Cuddapah Basin, southern India. The obtained Mesoproterozoic age necessitates the Racherla syenite to be an intrusive into the sedimentary rocks of the Cuddapah Basin contrary to some previous suggestion that it represents an inlier of the basement pluton. Age of the dated syenite, within its error limits, is also coeval with that of the nearby Chelima lamproite (1354 $\pm$ 52 Ma) and other syenite complexes (*c.* 1352 Ma) located elsewhere within the Eastern Ghats Mobile Belt, southern India, thereby implying a common geotectonic control in all of their generation. Incompatible trace element signatures (high La/Nb, Zr/Nb and La/Yb), $^{87}Sr/^{86}Sr$ (0.70432 $\pm$ 10) and the lower Nd isotopic ratio ($\varepsilon_{Nd}(t)$ -8.9) of the apatite suggest derivation of the Racherla syenite parent magma from an enriched mantle source. Our findings provide compelling evidence for the existence of ancient (*c.* 2.6 Ga) metasomatized lithospheric mantle, at the Archaean–Palaeoproterozoic boundary, beneath the Cuddapah Basin. This enriched mantle appears to have had played a dominant role in the generation of co-spatial and coeval lamproites within and around the Cuddapah Basin and a plethora of alkaline and sub-alkaline syenites within the Eastern Ghats Mobile Belt. Implications of our study in the context of Indo-Antarctic collisional tectonics are also explored.

The crescent-shaped Cuddapah Basin (Fig. 1) is one of the largest Purãna (Meso-Palaeoproterozoic) sedimentary basins in southern Peninsular India and has a stratigraphic thickness of more than 12 km (Nagaraja Rao *et al.* 1987). It hosts a number of world-class mineral deposits such as barite, talc, steatite, asbestos and alluvial diamond, and, hence, is important from an economic point of view. Despite being studied for well over a century (e.g. King 1872; Meijerink *et al.* 1984; Nagaraja Rao *et al.* 1987; Anand *et al.* 2003; French *et al.* 2007), the origin and evolution of the Cuddapah Basin is still not clearly understood and constitute a matter of debate. A number of competing models are in vogue for the formation of the basin, and include rifting (e.g. Chaudhuri *et al.* 2002) and initiation by meteroritic impact (e.g. Krishna Brahmam & Dutt 1985), as well as thermal subsidence (Bhattacharji & Singh 1984; Drury 1984).

A wide-range of exotic Mesoproterozoic intrusives (1.4–1.1 Ga), such as kimberlites, lamproites, lamprophyres, and alkaline- and subalkaline syenites, have been emplaced in and around the Cuddapah Basin and all of them collectively constitute an igneous province termed the ‘Cuddapah Intrusive Province’ (Madhavan *et al.* 1995). The petrology, geochemistry, geochronology and genesis of these diverse rock types are a matter of continued geological interest (e.g. Madhavan *et al.* 1994; Chalapathi Rao *et al.* 2004; Paul *et al.* 2006; Chalapathi Rao 2008). Whereas there has been a recent spurt in the geochronological information on the syenites from the Eastern Ghats Mobile Belt, towards the eastern margin of the Cuddapah Basin (e.g. Upadhyay *et al.* 2006; Kumar *et al.* 2007), the three alkali–syenite occurrences (Racherla, Giddalur and Idamakallu) within the Cuddapah Basin received little attention and as yet no radiometric

From: Mazumder, R. & Saha, D. (eds) 2012. *Palaeoproterozoic of India*. Geological Society, London, Special Publications, **365**, 185–195. http://dx.doi.org/10.1144/SP365.10

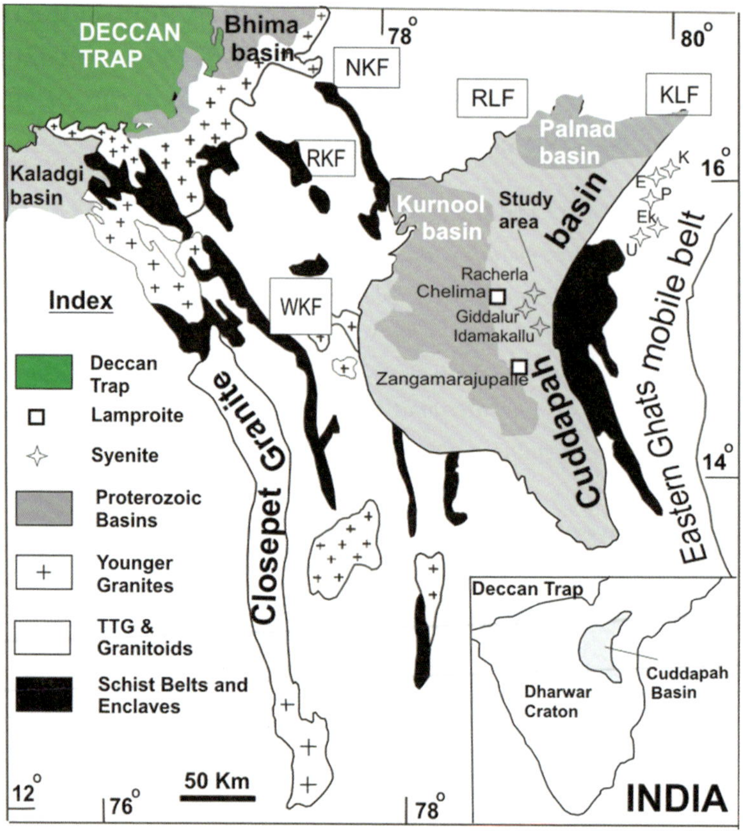

Fig. 1. A generalized geological map of the Dharwar Craton, southern India, (adapted from Naqvi 2005) showing the Palaeo-Mesoproterozoic Cuddapah Basin and surrounding areas, and the disposition of various alkaline and subalkaline syenites, kimberlite and lamproite fields. Alkaline and subalkaline syenites in the Eastern Ghats Mobile Belt, discussed in the text, are located at Elchuru (E), Kotappakonda (K), Purimetla (P), Uppalapadu (U) and Errakonda (Ek). WKF, Wajrakarur kimberlite field; RKF, Raichur kimberlite field; NKF, Narayanpet kimberlite field; RLF, Ramadugu lamproite field; KLF, Krishna lamproite field.

age is available for them. The purpose of this paper is: (i) to report the age of emplacement of one such intrabasinal alkali syenite intrusion at Racherla; (ii) to decipher the nature of the sub-Cuddapah lithosphere vis-à-vis *in situ* trace element and Sr–Nd isotopic analysis on the apatite from the Racherla syenite; and (iii) to explore the possible implications of our findings in the context of collisional tectonics involving India and Antarctica.

Background information

The occurrence of alkali syenites occurring within the Cumbum phyllites of the Palaeo-Mesoproterozoic Cuddapah Basin at Racherla (15°46′N; 78°58′E), Idamakallu (15°17′N; 79°00′E) and Giddalur (15°23′N; 78°56′E), collectively termed as the RIG syenites (Fig. 1) was first reported by King in 1872. Their petrography, geochemistry and mode of occurrence are studied by a number of previous workers (e.g. Dutt 1964; Appavadhanulu 1966; Somasekhar *et al.* 1966; Sen & Narasimha Rao 1967; Vijayam 1969; Madhavan *et al.* 1994). The syenite exposures were considered by King (1872) to either represent inliers of the basement rocks or to be related to faulting. Similar conclusions were also drawn by Vijayam (1969), who interpreted the alkali syenites to be basement plutons.

Geological set-up

The alkali syenite exposure at Racherla (Fig. 2a) stands out in the agricultural field as a conspicuous mound with a spread of about 300 m and a height of 20–25 m, and extends in an east–west direction. Its contact with the country rocks is not observed

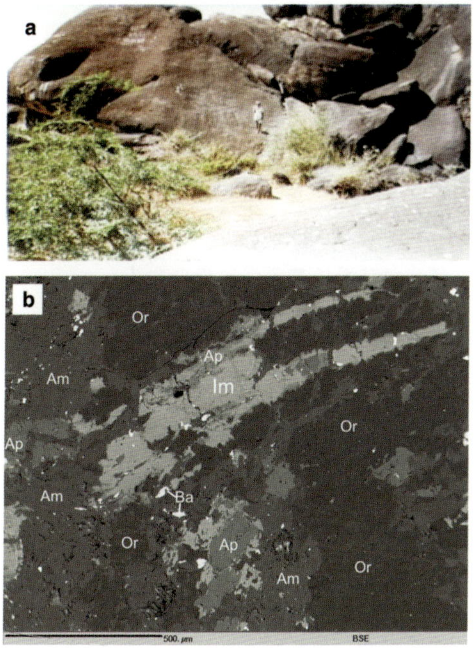

Fig. 2. (**a**) Outcrop of the Racherla alkali syenite within the Cuddapah Basin. Note the soil cover in the foreground. (**b**) Backscattered electron (BSE) image showing the overall texture of the Racherla alkali syenite. Note that apatite (Ap) is present as subhedral as well as acicular grains. Other minerals include blue amphibole riebeckite (Am), orthoclase (Or), ilmenite (Im) and a barium-rich phase (Ba).

owing to extensive soil cover. The Cumbum phyllites of the Nallamalai Group of the Cuddapah Supergroup, which occur as hillocks at some distance, are presumed to be the country rock. Megascopically, the samples of the Racherla syenite are coarse- to medium-grained, mesocratic and homogenous.

Petrology and whole-rock geochemistry

In thin section, they display equigranular and hypidiomorphic textures, and are predominantly composed of orthoclase–perthite and riebeckite (Fig. 2b). Other minerals include apatite, sphene, ilmenite, rutile, biotite and calcite present in subordinate to minor amounts. A barium-rich phase is also identified and its details will be dealt with elsewhere. Apatite is a widespread primary mineral (*c.* 5 vol%) and is present as subhedral grains, as well as in acicular laths (Fig. 2b).

The whole-rock geochemistry of the Racherla syenite is presented in Madhavan *et al.* (1994) and Anand *et al.* (2003). High TiO_2 (3.53–4.95 wt%) and P_2O_5 (1.88–2.5 wt%) content in the Racherla syenite are a reflection of the high proportion of modal rutile and apatite, respectively. The presence of normative acmite and high agpaitic index, [molar $(K_2O + Na_2O)/(Al_2O_3)$] >1, demonstrate the peralkaline nature of the parent magma.

Laser ablation-multiple collector-inductively coupled plasma-mass spectrometry (LA-MC-ICP-MS) and electron microprobe analyser (EPMA) studies on apatite

In recent years, LA-MC-ICP-MS is widely applied for the measurement of trace elements and texturally sensitive high-precision isotopic data in minerals owing to the many advantages of this technique, namely: there being no need for chemical preparation of the sample, it avoids crustal contamination and is a very rapid analysis compared to the conventional isotopic dilution techniques (e.g. Woodhead *et al.* 2005; Yang *et al.* 2008). Owing to its low Rb/Sr ratio and high Sr and Nd content, apatite constitutes an ideal choice for obtaining Sr and Nd data from the same mineral. Furthermore, its geochemical utility as a petrogenetic tracer is well understood (e.g. Chu *et al.* 2009). The major element chemistry of the individual apatite grains (Table 1) from Racherla syenite was initially measured using a CAMECA SX100 EPMA installed at the Technical University of Clausthal, Germany. The analytical details of EPMA are the same as those provided in Chalapathi Rao *et al.* (2010). About 15 apatite grains separated from the same sample of Racherla syenite were analysed

Table 1. *Major element chemistry (wt%) of apatite from Racherla syenite (R-4) by EPMA*

Oxide (wt%)	1	2	3	4	5
Na_2O	0.08	0.08	0.06	0.08	0.12
MgO	0.00	0.00	0.00	0.00	0.01
Al_2O_3	0.00	0.00	0.01	0.00	0.01
K_2O	0.00	0.00	0.00	0.00	0.02
TiO_2	0.02	0.04	0.02	0.01	0.00
FeO	0.12	0.13	0.02	0.03	0.15
CaO	55.83	56.28	55.86	55.84	55.61
MnO	0.00	0.00	0.01	0.02	0.00
SiO_2	0.02	0.02	0.01	0.01	0.00
Cr_2O_3	0.00	0.00	0.00	0.01	0.01
NiO	0.00	0.01	0.03	0.05	0.02
P_2O_5	41.14	41.42	41.31	41.49	41.14
Total	97.21	97.97	97.31	97.55	97.08

Table 2. *Trace element content (ppm) analysed by LA-MC-ICP-MS of various apatite grains in Racherla syenite (sample R-4)*

	RE 02	RE 03	RE 04	RE 05	RE 06	RE 07	RE08	RE09	RE10	RE12	RE13	RE14	RE16
Rb	0.46	0.79	0.41	0.57	0.54	0.55	0.55	0.88	0.64	2.24	0.95	0.52	2.36
Sr	12 031	10 688	12 558	10 227	23 070	13 255	9918	15 421	13 702	10 513	19 933	13 049	11 930
Y	442	839	641	673	405	496	715	560	617	675	318	604	530
Zr	9	113	206	26	108	651	9	31	24	25	5	2	17
Nb	2.7	0.9	0.4	0.0	5.3	0.8	0.4	1.2	0.1	0.2	0.1	1.1	0.1
Ba	444	260	125	127	525	143	139	152	234	73	698	235	152
Hf	0.21	1.79	2.59	0.36	1.39	8.17	0.31	0.80	1.72	0.50	0.16	0.07	0.24
Ta	0.03	0.06	0.12	0.02	0.55	0.18	0.01	0.13	0.10	0.01	0.01	0.08	0.02
Pb	5	5	8	9	7	6	6	20	14	6	18	6	5
Th	23	25	43	25	32	51	28	36	28	28	14	44	27
U	0.72	1.71	1.59	0.97	1.09	2.45	1.12	0.92	0.98	1.23	1.20	0.98	1.05
La	438	262	445	233	985	597	245	536	604	280	828	513	373
Ce	1011	804	1300	780	1987	1559	761	1365	1631	841	1684	1505	1033
Pr	152	144	191	145	246	239	137	207	241	149	206	223	171
Nd	709	759	956	768	1041	1067	752	1010	1144	786	902	1016	860
Sm	159	220	225	212	200	212	219	227	293	214	154	235	205
Eu	51	78	74	72	61	66	74	73	101	73	46	75	66
Gd	130	211	188	186	149	169	194	181	244	186	113	187	166
Tb	17	33	26	26	19	23	27	24	32	26	14	25	22
Dy	91	187	142	144	94	115	149	123	157	141	69	134	117
Ho	17	36	27	28	17	20	29	23	27	27	12	25	22
Er	41	86	64	69	39	44	73	55	59	69	29	61	52
Tm	5	10	7	8	4	4	9	6	7	8	3	7	6
Yb	24	49	36	42	21	20	45	33	32	42	17	34	30
Lu	2.3	4.6	3.4	4.2	2.0	1.8	4.4	3.3	3.0	4.1	1.7	3.2	2.9

Table 3. In situ *Sr isotopic ratio of individual apatite grains of the Racherla syenite (R-4) measured by LA-MC-ICP-MS*

Sample	$^{84}Sr/^{86}Sr$	2σ	$^{84}Sr/^{88}Sr$	2σ	$^{87}Rb/^{86}Sr$	2σ	$^{87}Sr/^{86}Sr_{(m)}$	2σ	$^{87}Sr/^{86}Sr_{(i)}$
RE 60u 01	0.0565	0.0001	0.00674	0.00001	0.00010	0.00001	0.70417	0.00003	0.70417
RE 60u 02	0.0565	0.0001	0.00674	0.00001	0.00017	0.00004	0.70423	0.00003	0.70422
RE 60u 03	0.0564	0.0001	0.00673	0.00001	0.00020	0.00002	0.70457	0.00003	0.70457
RE 30u 04	0.0565	0.0001	0.00675	0.00002	0.00001	0.00001	0.70410	0.00004	0.70410
RE 30u 06	0.0563	0.0003	0.00672	0.00004	0.00064	0.00020	0.70475	0.00009	0.70473
RE 50u 07	0.0564	0.0001	0.00674	0.00002	0.00046	0.00003	0.70496	0.00004	0.70495
RE 50u 08	0.0564	0.0002	0.00673	0.00003	0.00015	0.00002	0.70440	0.00006	0.70440
RE 50u 09	0.0565	0.0001	0.00674	0.00001	0.00005	0.00001	0.70429	0.00003	0.70429
RE 50u 11	0.0565	0.0001	0.00675	0.00001	0.00002	0.00001	0.70435	0.00004	0.70435
RE 50u 12	0.0564	0.0002	0.00674	0.00002	0.00007	0.00001	0.70445	0.00006	0.70445
RE 50u 13	0.0565	0.0002	0.00674	0.00002	0.00011	0.00001	0.70432	0.00004	0.70431
RE 50u 14	0.0566	0.0001	0.00675	0.00001	0.00024	0.00006	0.70468	0.00006	0.70467
RE 50u 15	0.0564	0.0002	0.00674	0.00002	0.00073	0.00011	0.70423	0.00005	0.70421
RE 50u 16	0.0565	0.0001	0.00674	0.00001	0.00009	0.00001	0.70436	0.00004	0.70435

$^{87}Sr/^{86}Sr_{(m)}$ and $^{87}Sr/^{86}Sr_{(i)}$, where subscript (m) is the measured ratio and (i) is the initial ratio

for their *in situ* trace element content (Table 2) and Sr and Nd isotopic ratios (Tables 3 & 4) using the Agilent 7500a ICP-MS and Neptune LA-MC-ICP-MS machines installed at the State Key Laboratory of Lithosphere Evolution, at the Chinese Academy of Sciences, Beijing. Detailed analytical protocols are provided in Yang *et al.* (2008) and Wu *et al.* (2010), only a brief introduction is given here.

The Nd isotopes were measured with the Neptune MC-ICP-MS, whereas trace elements were analysed using the Agilent 7500a. NIST610 was served as an external standard for element determination. During the Nd isotopic analysis, an in-house apatite standard (AP2) was used for external correction. AP1, another in-house apatite standard, yielded $^{147}Sm/^{144}Nd$ and $^{143}Nd/^{144}Nd$ ratios of 0.0841 ± 10 and 0.511329 ± 19 (2SD, $n = 10$), which are identical to the solution values of 0.0836 ± 14 and 0.511365 ± 32 (2SD, $n = 4$) within errors (Yang *et al.* 2008; Wu *et al.* 2010). Subsequently, Sr isotopes were measured with the Neptune MC-ICP-MS using AP1 as the external standard. Analyses of KB-3 as an unknown sample – apatite from the Khibiny syenite in Kola of Russia – gave a $^{87}Sr/^{86}Sr$ ratio of 0.70325 ± 7 (2SD, $n = 23$), which is identical to the solution value of 0.70332 ± 6 (2SD, $n = 3$) within errors (Wu *et al.* 2010). In addition, the reliabilities of our *in situ* Sr–Nd isotopic data are supported by consistent $^{84}Sr/^{86}Sr$ (0.0565 ± 1) and $^{145}Nd/^{144}Nd$ (0.348399 ± 16) readings, with the theoretical values of 0.0565 and 0.348415, respectively.

The notation of ε_{Nd} (the Nd isotopic ratio), fractionation index $f_{Sm/Nd}$, and model ages (calculated using one- and two-stage models) T_{DM1} and T_{DM2} followed in this study are given below and are adapted from Wu *et al.* (2002):

$$\varepsilon_{Nd} = [(^{143}Nd/^{144}Nd)_S/(^{143}Nd/^{144}Nd)_{CHUR} - 1] \times 10\,000$$

$$f_{Sm/Nd} = [(^{147}Sm/^{144}Nd)_S/(^{147}Sm/^{144}Nd)_{CHUR}] - 1$$

$$T_{DM1} = 1/\lambda_{Sm} \ln\{[(^{143}Nd/^{144}Nd)_S - 0.51315] / [(^{147}Sm/^{144}Nd)_S - 0.2137]\}$$

$$T_{DM2} = T_{DM1} - (T_{DM1} - t)(f_{cc} - f_s)/(f_{cc} - f_{DM})$$

where S is the sample, $(^{143}Nd/^{144}Nd)_{CHUR} = 0.512638$ and $(^{147}Sm/^{144}Nd)_{CHUR}) = 0.1967$. f_{cc}, f_s and f_{DM} are $f_{Sm/Nd}$ values of the continental crust, the sample and the depleted mantle, respectively. In this calculation, $f_{cc} = -0.4$, $f_{DM} = 0.085922$ and t is the emplacement age of the Racherla syenite; that is, 1326 Ma. The decay constant λ_{Sm} used in age calculation for ^{147}Sm is 0.00654 Ga^{-1}.

Results and discussion

The major element content of the analysed crystalline apatite shows little variation (Table 1) and overall is homogeneous. Trace element data (Table 2) reveal their Sr-rich (up to 23 070 ppm) and Rb-poor (<2.36 ppm) character, and their extreme enrichment in rare earth elements (REE) (average = $\sum$3510 ppm). High incompatible element ratios of Zr/Nb and La/Nb and their variation amongst individual apatite grains are attributed to the varying melt composition or due to the heterogeneity of the source. Likewise, the

Table 4. In situ *Nd isotopic analysis (LA-MC-ICP-MS) of the apatite grains from Racherla syenite sample (R-4)*

Sample	t(Ma)	$^{145}Nd/^{144}Nd$	2σ	$^{147}Sm/^{144}Nd$	2σ	$^{143}Nd/^{144}Nd_{(m)}$	2σ	$^{143}Nd/^{144}Nd_{(i)}$	$\varepsilon_{Nd}(0)$	$\varepsilon_{Nd}(t)$	2σ	T_{DM1}	T_{DM2}	$f_{Sm/Nd}$
RE AP 100u 8 h 01	1326	0.34844	0.00004	0.1741	0.0010	0.511999	0.000050	0.510453	−12.3	−8.44	0.97	4337	2572	−0.12
RE AP 100u 8 h 02	1326	0.34841	0.00004	0.1396	0.0004	0.511727	0.000056	0.510486	−17.6	−7.91	1.09	2889	2536	−0.29
RE AP 140u 8 h 03	1326	0.34842	0.00002	0.1790	0.0002	0.512045	0.000038	0.510455	−11.4	−8.38	0.75	4743	2565	−0.09
RE AP 140u 8 h 04	1326	0.34842	0.00002	0.1446	0.0007	0.511703	0.000031	0.510418	−18.0	−9.21	0.60	3146	2640	−0.26
RE AP 140u 8 h 05	1326	0.34841	0.00003	0.1756	0.0021	0.512019	0.000053	0.510458	−11.9	−8.33	1.04	4439	2562	−0.11
RE AP 140u 8 h 06	1326	0.34838	0.00002	0.1127	0.0005	0.511435	0.000028	0.510434	−23.3	−9.03	0.55	2559	2628	−0.43
RE AP 140u 8 h 07	1326	0.34841	0.00003	0.1257	0.0016	0.511548	0.000053	0.510432	−21.1	−9.02	1.04	2740	2627	−0.36
RE AP 140u 8 h 08	1326	0.34840	0.00002	0.1779	0.0008	0.512047	0.000035	0.510465	−11.3	−8.18	0.69	4604	2549	−0.10
RE AP 140u 8 h 09	1326	0.34840	0.00003	0.1383	0.0016	0.511615	0.000052	0.510386	−19.8	−9.87	1.01	3062	2694	−0.30
RE AP 140u 8 h 10	1326	0.34838	0.00003	0.1477	0.0008	0.511758	0.000039	0.510446	−17.0	−8.66	0.76	3166	2595	−0.25
RE AP 140u 8 h 11	1326	0.34843	0.00002	0.1062	0.0023	0.511435	0.000037	0.510491	−23.3	−7.93	0.73	2407	2540	−0.46
RE AP 140u 8 h 12	1326	0.34840	0.00003	0.1698	0.0012	0.511903	0.000036	0.510394	−14.1	−9.61	0.70	4249	2666	−0.14
RE AP 140u 8 h 13	1326	0.34837	0.00002	0.1054	0.0011	0.511363	0.000041	0.510426	−24.7	−9.20	0.80	2489	2642	−0.46
RE AP 140u 8 h 14	1326	0.34839	0.00002	0.1322	0.0007	0.511595	0.000030	0.510420	−20.1	−9.22	0.59	2871	2642	−0.33
RE AP 140u 8 h 15	1326	0.34840	0.00002	0.0751	0.0002	0.511107	0.000033	0.510440	−29.7	−9.03	0.65	2226	2630	−0.62
RE AP 140u 8 h 16	1326	0.34841	0.00002	0.1457	0.0014	0.511699	0.000035	0.510404	−18.1	−9.48	0.69	3205	2661	−0.26

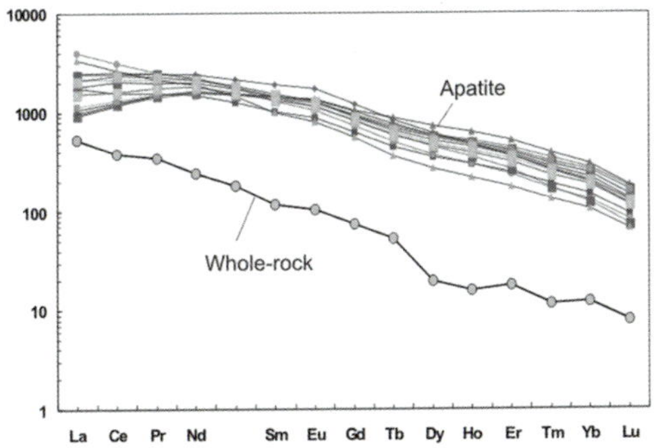

Fig. 3. Chondrite-normalized (after Evenssen *et al.* 1977) Rare earth element (REE) distribution patterns of the apatite grains from the Racherla alkali syenite (sample R-4). Whole-rock REE patterns of the same sample (from Madhavan *et al.* 1994) are also provided for comparison.

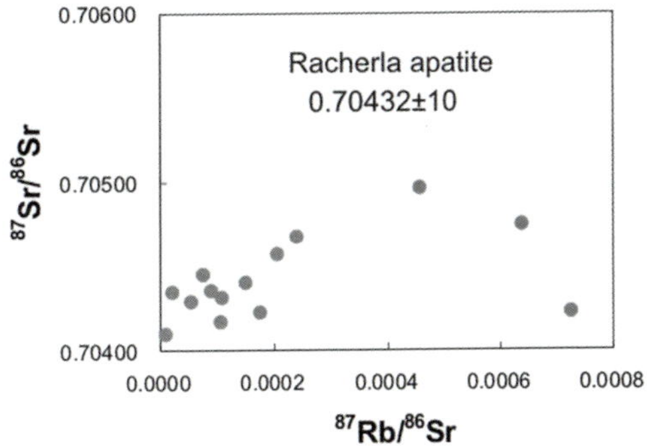

Fig. 4. $^{87}Rb/^{86}Sr$ and $^{87}Sr/^{86}Sr$ (measured by LA-MC-ICP-MS) of the apatite grains from the Racherla syenite.

chondrite-normalized REE patterns (Fig. 3) reveal that there is a considerable variation in the REE abundances in each grain. Nevertheless, relative enrichment of LREE (light REE) relative to HREE (heavy REE) is an essential feature of all of the analysed apatite implying: (i) a strong LREE fractionation into apatite; and (ii) their derivation from a mantle-derived, hydrous and LREE-enriched source. The chondrite-normalized whole-rock REE pattern of the Racherla sample (Fig. 3) is also characterized by a high LREE/HREE ratio, revealing that apatite is an important host of REE in the Racherla syenite. The absence of Eu depletion (Fig. 3) further precludes the plagioclase fractionation and the contribution of a basaltic magma in the genesis of Racherla syenite (see also Madhavan *et al.* 1994).

High incompatible trace element ratios (Zr/Nb, La/Nb and LREE/HREE) of the Racherla alkali syenite demonstrate its derivation from an enriched source in the mantle. This is further corroborated by the $^{87}Sr/^{86}Sr$ isotopic composition (average value is 0.70432 ± 10) of 14 apatite grains (Table 3). With the exception of three apatite grains, the majority of them display a restricted variation in their respective $^{87}Sr/^{86}Sr$ and $^{87}Rb/^{86}Sr$ ratios (Fig. 4). Nd isotopic analyses (Table 4) of Racherla apatite show considerable variations in $^{147}Sm/^{144}Nd$ and $^{143}Nd/^{144}Nd$ ratios, and define an isochron yielding an age of 1326 ± 73 Ma (mean square weighted deviation (MSWD) = 2.1) (Fig. 5). A Mesoproterozoic age for the Racherla syenite is also supported by its palaeomagnetic age of 1200 Ma obtained from a comparison of its chRM (characteristic remanent magnetization) direction of normal polarity with the Proterozoic APWP (*apparent polar wander path*) of the south Indian Shield (Goutham & Patil 2005).

The lower Nd isotopic ratio ($\varepsilon_{Nd}(t) = -8.9$) of the Racherla syenite places it in the 'enriched' quadrant of the Sr–Nd plot, and demonstrates its derivation from an isotopically enriched mantle source similar to that displayed by several alkaline plutons from the Eastern Ghats Mobile Belt (see Fig. 6). An extremely high proportion of crustal assimilation is required to give rise to the observed enriched ratios of the Racherla syenite, which do not find not find support from the petrographical studies. Similar conclusions were also drawn by Upadhyay (2008) for the syenite occurrences from the Eastern Ghats Mobile Belt.

The maximum age of deposition of the phyllite Cumbum Formation of the Nallamalai is constrained by the emplacement age (*c.* 1352 Ma: e.g. Chalapathi Rao *et al.* 1996, 1999; Kumar *et al.* 2001) of the Chelima lamproite, which is located about 30 km west of the Racherla syenite (Fig. 1).

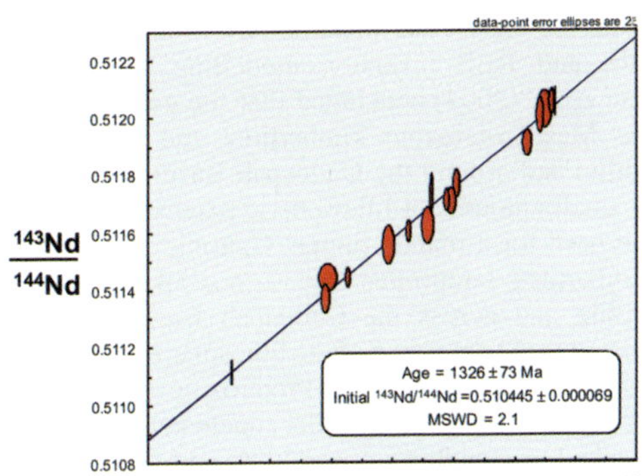

Fig. 5. Nd isotopic analyses (measured by LA-MC-ICP-MS) and isochron age of the apatite grains from the Racherla syenite.

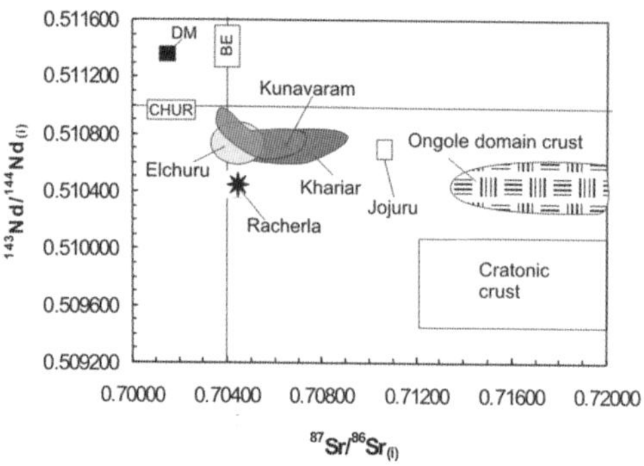

Fig. 6. Sr–Nd isotopic ratio (average of four grains) for the Racherla syenite (asterick) compared with those from other synchronous alkaline rocks (e.g. Elchuru, Khariar, Kunavaram and Jojuru) from the EGMB. The other fields are taken from Upadhyay (2008). Note that crustal material (average cratonic crust as well as that from the Ongole domain of the EGMB) shows very different isotopic signatures. An extremely high proportion of crustal assimilation is required to give rise to the observed enriched ratios, which is not supported by petrography.

A coeval, within error limits, age necessitates the Racherla syenite to be an intrusive into the sedimentary rocks of the Cuddapah Basin contrary to some previous suggestions for it to be a basement pluton (King 1872; Vijayam 1969). Ages of the Racherla syenite and the Chelima lamproite are also contemporaneous with the other alkaline and subalkaline syenite complexes emplaced at approximately 1352 Ma elsewhere within the Eastern Ghats Mobile Belt, southern India, at Uppalapadu, Errakonda, Purimetla, Elchuru and Kotappakonda (e.g. Sarkar & Paul 1998; Upadhyay *et al.* 2006; Kumar *et al.* 2007) (see Fig. 1 for locations), thereby implying a possible geotectonic control in all of their generation.

Based on the thermobarometry of mantle xenoliths and REE inversion modelling, Chalapathi Rao *et al.* (2004) concluded that the generation of the Mesoproterozoic kimberlites and lamproites within and around the Cuddapah Basin are related to small amounts of lithospheric extension without the need for a mantle plume. Cratonic zonation of kimberlites, lamproites and various alkaline rocks within and across the Cuddapah Basin is, thus, related to the extension of the lithosphere of variable thickness during the Mid-Proterozoic (Chalapathi Rao *et al.* 2004, 2010). Similar conclusions were also reached by Upadhyay *et al.* (2006) and Kumar *et al.* (2007), who attributed the distribution of various alkaline and subalkaline plutons in the Eastern Ghats Mobile Belt to continental rifting. Ravikant (2009) also provided evidence for the extension-related magmatism during the Palaeoproterozoic (1.92 Ga) in this domain and highlighted its role in break-up along the eastern margin of the Eastern Dharwar Craton. Therefore, we prefer to link the emplacement of the Racherla syenite and other RIG syenites (at Giddalur and Idamakallu) in the Cuddapah Basin to Mesoproterozoic extension at 1352 Ma in southern India.

Nd isotopic studies of the lamproites from: (i) the Cuddapah Basin (Chelima and Zangamarajupalle); (ii) its NE margin (Krishna lamproites); and (iii) various subalkaline and alkaline plutons in the Eastern Ghats Mobile Belt (EGMB) demonstrate that they were generated from ancient (>2 Ga) enriched mantle sources (Chalapathi Rao *et al.* 1998; Chakrabarti *et al.* 2007; Kumar *et al.* 2007). This enrichment is suggested to be the direct consequence of a regional metasomatic event that influenced the subcontinental lithospheric mantle in this domain (Chalapathi Rao *et al.* 1998). However, the average Nd model age (T_{DM2}) of the Racherla apatite comes to 2.6 Ga (Table 4), thereby suggesting that its source enrichment straddled the Archaean–Proterozoic boundary. This study provides strong additional evidence for the existence of an ancient and anomalously enriched mantle beneath the Cuddapah Basin. The existence of such discrete large-scale geodynamic domains in this part of eastern Gondwana at least since the Archaean–Palaeoproterozoic is significant in our understanding of the Gondwana tectonics and in the reconstruction of Proterozoic supercontinents such as Rodinia, Nuna or Columbia (e.g. Rogers & Santosh 2002; Reddy & Evans 2009).

The similarity in ages between the Racherla syenite in the Cuddapah Basin and other syenites within the Eastern Ghats Belt also has important implications in the context of collisional tectonics involving India and Antarctica. Several documented studies, involving structure and metamorphism (e.g. Chetty & Murthy 1994; Ramakrishnan *et al.* 1998; Gupta *et al.* 2000; Dasgupta & Sengupta 2003; Bhadra *et al.* 2004) on the western margin of the EGMB, indicate that its contact with the Archaean cratons of the Bastar and the Eastern Dharwar was a suture related to Indo-Antarctica collision. Some workers (e.g. Dobmeier *et al.* 2006; Biswal *et al.* 2007; Simmat & Raith 2008) opine that the present structural configuration of the Eastern Ghats Belt and its cratonic forelands occurred during the Pan-African (Neoproterozoic–Early Palaeozoic) period when the Grenvillian Eastern Ghats Province–Rayner Complex terrane collided with India. However, the apparent synchroneity in emplacement age of the Racherla syenite (located within the craton *sensu stricto*) and the other alkaline rocks in the EGMB strongly suggest that the basement of at least a part of the Eastern Ghat Belt was already amalgamated to the Eastern Dharwar

Craton by about 1326 Ma. This would necessitate the collision-related tectonic contact between the EGMB and the Indian craton to be of pre-Grenvillian age.

Summary

An *in situ* Sm–Nd isochron age of 1326 ± 73 Ma, determined by LA-MC-ICP-MS on crystalline apatite grains, is reported for the Racherla alkali syenite occurring in the Palaeo-Mesoproterozoic Cuddapah Basin, southern India. The obtained Mesoproterozoic age necessitate the Racherla syenite to be an intrusive into the sedimentary rocks of the Cuddapah Basin contrary to an earlier view that it represents an inlier of the basement pluton. Racherla syenite is coeval, within the error limits of its age, with that of the nearby Chelima lamproite (1354 ± 52 Ma) and other syenite complexes (*c*. 1352 Ma) located elsewhere within the EGMB, southern India, thereby implying a common geotectonic control in all of their generation. Combined incompatible trace element signatures (e.g. high La/Nb, Zr/Nb), $^{87}Sr/^{86}Sr$ (0.70432 ± 10) and $\varepsilon_{Nd}(t)$ (−8.9) of the apatite suggest the derivation of the Racherla syenite parent magma from an enriched mantle source. This study provides strong evidence for the existence of ancient (*c*. 2.6 Ga) metasomatized lithospheric mantle beneath the Cuddapah Basin at the Archaean–Palaeoproterozoic boundary. The apparent synchroneity in ages between the Racherla syenite located within the craton *sensu stricto* and those within the EGMB strongly suggest that the basement of at least a part of the Eastern Ghat Belt was already amalgamated to the Eastern Dharwar Craton by about 1326 Ma. Therefore, the collision-related tectonic contact between the EGMB and the Indian craton is inferred to be of pre-Grenvillian age.

We thank Profs D. Saha and R. Mazumder for their kind invitation to contribute to this special volume. Thought-provoking and highly constructive reviews by S. Gupta and R. Bhutani helped us in improving the final version of this paper. N. V. Chalapathi Rao thanks the Head, Centre of Advanced Study in Geology, Banaras Hindu University for providing the facilities and for his constant encouragement.

References

ANAND, M., GIBSON, S. A., SUBBARAO, K. V., KELLEY, S. P. & DICKIN, A. P. 2003. Early Proterozoic melt generation processes beneath the Intra-cratonic Cuddapah basin, southern India. *Journal of Petrology*, **44**, 2139–2171.

APPAVADHANULU, K. 1966. Minette and riebeckite–kalisyenite dykes in some Upper Cuddapah rocks, Kurnool district, Andhra Pradesh. *Records of the Geological Survey of India*, **94**, 303–304.

BHADRA, S., GUPTA, S. & BANERJEE, M. 2004. Structural evolution across the Eastern Ghats Mobile Belt – Bastar craton boundary, India: hot over cold thrusting in an ancient collision zone. *Journal of Structural Geology*, **26**, 233–245.

BHATTACHARJI, S. & SINGH, R. N. 1984. Thermomechanical structure of the southern part of the Indian shield and its relevance to Precambrian basin evolution. *Tectonophysics*, **105**, 103–120.

BISWAL, T. K., DE WAELE, B. & AHUJA, H. 2007. Timing and dynamics of the juxtaposition of the Eastern Ghats Mobile Belt against the Bhandara craton, India: a structural and zircon U–Pb SHRIMP study of the fold–thrust belt and associated nepheline syenite plutons. *Tectonics*, **26**, TC4006, doi:10.1029/2006TC002005.

CHAKRABARTI, R., BASU, A. R. & PAUL, D. K. 2007. Nd–Hf–Sr–Pb isotopes and trace element geochemistry of Proterozoic lamproites from southern India: subducted komatiite in the source. *Chemical Geology*, **236**, 291–302.

CHALAPATHI RAO, N. V. 2008. Precambrian alkaline potassic–ultrapotassic, mafic–ultramafic magmatism in peninsular India. *Journal of the Geological Society of India*, **72**, 57–84.

CHALAPATHI RAO, N. V., MILLER, J. A., PYLE, D. M. & MADHAVAN, V. 1996. New Proterozoic K–Ar ages for some kimberlites and lamproites from the Cuddapah basin and Dharwar craton, southern India. *Precambrian Research*, **79**, 363–369.

CHALAPATHI RAO, N. V., GIBSON, S. A., PYLE, D. M. & DICKIN, A. P. 1998. Contrasting isotopic mantle sources for Proterozoic lamproites and kimberlites from the Cuddapah basin and Eastern Dharwar craton: implication for Proterozoic mantle heterogeneity beneath southern India. *Journal of the Geological Society of India*, **52**, 683–694.

CHALAPATHI RAO, N. V., MILLER, J. A., GIBSON, S. A., PYLE, D. M. & MADHAVAN, V. 1999. $^{40}Ar/^{39}Ar$ dating of Kotakonda kimberlite and Chelima lamproite, India: implication to the timing of mafic dyke swarm activity in the Eastern Dharwar craton. *Journal of the Geological Society of India*, **53**, 425–432.

CHALAPATHI RAO, N. V., GIBSON, S. A., PYLE, D. M. & DICKIN, A. P. 2004. Petrogenesis of Proterozoic lamproites and kimberlites from the Cuddapah basin and Dharwar craton, southern India. *Journal of Petrology*, **45**, 907–948.

CHALAPATHI RAO, N. V., KAMDE, G., KALE, H. S. & DONGRE, A. 2010. Petrogenesis of the Mesoproterozoic lamproites from the Krishna Valley, Eastern Dharwar craton, southern India. *Precambrian Research*, **177**, 103–130.

CHAUDHURI, A. K., SAHA, D., DEB, G. K., DEB, S. P., MUKHERJEE, M. K. & GHOSH, G. 2002. The Purana basins of southern cratonic province of India – a case for Mesoproterozoic fossil rifts. *Gondwana Research*, **5**, 23–33.

CHETTY, T. R. K. & MURTHY, D. S. N. 1994. Collision tectonics in the Late Precambrian Eastern Ghats mobile belt: mesoscopic to satellite scale structural observations. *Terra Nova*, **6**, 72–81.

Chu, M. F., Wang, K. L., Griffin, W. L., Chung, S. L., O'Reilly, S. L., Pearson, N. J. & Iizuka, Y. 2009. Apatite composition: tracing petrogenetic processes in Transhimalayan granitoids. *Journal of Petrology*, **50**, 1829–1850.

Dasgupta, S. & Sengupta, P. 2003. Indo-Antarctic correlation: a perspective from the Eastern Ghats belt. *In*: Yoshida, M., Windley, B. F. & Dasgupta, S. (eds) *Proterozoic East Gondwana: Supercontinent Assembly and Breakup*. Geological Society, London, Special Publications, **206**, 131–143.

Dobmeier, C., Lutke, S., Hammerschmidt, K. & Mezger, K. 2006. Emplacement and deformation of the Vinukonda meta-granite (Eastern Ghats, India) – implications for the geological evolution of peninsular India and for Rodinia reconstructions. *Precambrian Research*, **146**, 165–178.

Drury, S. A. 1984. A Proterozoic intracratonic basin, dyke swarms and thermal evolution in South India. *Journal of the Geological Society of India*, **25**, 437–444.

Dutt, N. V. B. S. 1964. Stratigraphic relation of the Kurnool system with the Upper Cuddapahs. *Bulletin of the Geological Society of India*, **1**, 31–35.

Evenssen, N. M., Hamilton, P. J. & O'Nions, R. K. 1977. Rare earth abundances in chondritic meteorites. *Geochimica et Cosmochimica Acta*, **42**, 1199–1212.

French, J. E., Heaman, L. M., Chacko, T. & Srivastava, R. K. 2007. 1891–1883 Ma Southern Bastar-Cuddapah mafic igneous events, India: a newly recognised large igneous province. *Precambrian Research*, **160**, 308–322.

Goutham, M. R. & Patil, S. K. 2005. Are the alkali syenite bodies from the Nallamalai fold belt of Cuddapah basin, India related to plume or rift system? *In*: *AOGS (Asia Oceania Geoscience Society) 2nd Annual General Assembly, 20 June, Singapore, 2005*. Asia Oceania Geosciences Society, Singapore, 58-SE-A1512.

Gupta, S., Bhattacharya, A., Raith, M. & Nanda, J. K. 2000. Pressure – temperature – deformational history across a vestigial craton-mobile belt boundary: The western margin of the Eastern Ghats Belt at Deobhog, India. *Journal of Metamorphic Geology*, **18**, 683–697.

King, W. 1872. *The Kadapah and Karnul Formations in the Madras Presidency. Geological Survey of India, Memoir*, **8**.

Krishna Brahmam, N. & Dutt, N. V. B. S. 1992. A meteoritic impact theory for the initiation of the Cuddapah (Proterozoic) basin of India. *Bulletin of the Indian Geologists Association*, **25**, 43–60.

Kumar, A., Gopalan, K., Rao, K. R. P. & Nayak, S. S. 2001. Rb–Sr ages of kimberlites and lamproites from Eastern Dharwar craton, south India. *Journal of the Geological Society of India*, **58**, 135–142.

Kumar, K. V., Frost, C. D., Frost, B. R. & Chamberlain, B. R. 2007. The Chimakurthi, Errakonda, and Uppalapadu plutons, Eastern Ghats belt, India: an unusual association of tholeiitic and alkaline magmatism. *Lithos*, **97**, 30–57.

Madhavan, V., Mallikharjuna Rao, J., Srinivas, M., Natarajan, R. & Sayeed, A. 1994. Petrology and petrogenesis of syenites from the Cuddapah basin, Andhra Pradesh. *Journal of the Geological Society of India*, **43**, 225–237.

Madhavan, V., Mallikharjuna Rao, J., Chalapathi Rao, N. V. & Srinivas, M. 1995. Multi-faceted manifestations of an intrusive province around the intra-cratonic Cuddapah basin, India. *In*: Srivastava, R. K. & Chandra, R. (eds) *Magmatism in Relation to Diverse Tectonic Settings*. Balkema, Rotterdam, 93–105.

Meijerink, A. M. J., Rao, D. P. & Rupke, J. 1984. Stratigraphic and structural development of the Precambrian Cuddapah basin, S.E. India. *Precambrian Research*, **26**, 57–97.

Nagaraja Rao, B. K., Rajurkar, S. T., Ramalingaswamy, G. & Ravindra Babu, B. 1987. Stratigraphy, structure and evolution of the Cuddapah Basin. *Geological Society of India, Memoir*, **6**, 33–86.

Naqvi, S. M. 2005. *Geology and Evolution of the Indian Plate (from Hadean to Holocene 4 Ga to 4 Ka)*. Capital Publishers, New Delhi.

Paul, D. K., Nayak, S. S. & Pant, N. C. 2006. Indian kimberlites and related rocks: petrology and geochemistry. *Journal of the Geological Society of India*, **67**, 328–355.

Ramakrishnan, M., Nanda, J. K. & Augustine, P. F. 1998. Geological evolution of the Proterozoic Eastern Ghats Mobile Belt. *In*: Paul, D. K., Ramakrishnan, M. & Mishra, R. N. (eds) *Proceedings of the Workshop on the Eastern Ghats Mobile Belt*. Geological Survey of India, Special Publications, **44**, 1–21.

Ravikant, V. 2009. Palaeoproterozoic (~1.9 Ga) extension and breakup along the eastern margin of the Eastern Dharwar craton, SE India: New Sm–Nd isochron age constraints from anorogenic magmatism in the Neoarchaean Nellore greenstone belt. *Journal of Asian Earth Sciences*, **37**, 67–81.

Reddy, S. M. & Evans, D. A. D. 2009. Palaeoproterozoic supercontinents and global evolution: correlations from core to atmosphere. *In*: Reddy, S. M., Mazumder, R., Evans, D. A. D. & Collins, A. (eds) *Palaeoproterozoic Supercontinents and Global Evolution*. Geological Society, London, Special Publications, **323**, 1–26.

Rogers, J. J. W. & Santosh, M. 2002. Configuration of Columbia, a Mesoproterozoic supercontinent. *Gondwana Research*, **5**, 5–22.

Sarkar, A. & Paul, D. K. 1998. Geochronology of the Eastern Ghats Precambrian mobile belt – a review. *In*: Reddi, A. G. B., Murty, B. S. R. & Kesavamani, M. (eds) *A Compendium of Four Decades of Geophysical Activity in the Geological Survey of India*. Geological Survey of India, Special Publications, **36**, 985–997.

Sen, S. N. & Narasimha Rao, C. H. 1967. Igneous activity in Cuddapah Basin and adjacent areas and suggestions on the Palaeogeography of the basin. *In*: *Proceedings of the Symposium on Upper Mantle Project, January, NGRI, Hyderabad*. Geophysics Research Board, Hyderabad, 261–285.

Simmat, R. & Raith, M. M. 2008. Th–U–Pb monazite geochronometry of the Eastern Ghats Belt, India: timing an spatial disposition of poly-metamorphism. *Precambrian Research*, **162**, 16–39.

Somasekhar, B., Machigad, B. S. & Naganna, C. 1966. Riebeckite syenite from Idamakallu-Racherla, Kurnool district, Andhra Pradesh. *Proceedings of the Indian Academy of Sciences*, **64**, 267–272.

UPADHYAY, D. 2008. Alkaline magmatism along the south-eastern margin of the Indian shield: implications for regional geodynamics and constraints on craton-Eastern Ghats Belt suturing. *Precambrian Research*, **162**, 59–69.

UPADHYAY, D., RAITH, M. M., MEZGER, K. & HEMMERSCHMIDT, K. 2006. Mesoproterozoic rift-related alkaline magmatism at Elchuru, Prakasam Alkaline Province, SE India. *Lithos*, **89**, 447–477.

VIJAYAM, B. E. 1969. Plutonic rocks in the Cuddapah basin. *Journal of the Indian Geoscience Association*, **11**, 75–78.

WOODHEAD, J., SWEARER, S., HERGT, J. & MAAS, R. 2005. *In situ* Sr isotope analysis of carbonates by LA-MC-ICP-MS: interference corrections, high spatial resolution and an example from otolith studies. *Journal of Analytical and Atomic Spectroscopy*, **20**, 22–27.

WU, F. Y., SUN, D. Y., LI, H., JAHN, B. & WILDE, S. 2002. A-type granites in northeastern China: age and geochemical constraints on their petrogenesis. *Chemical Geology*, **187**, 143–173.

WU, F. Y., YANG, Y. H. ET AL. 2010. *In situ* U–Pb, Sr, Nd and Hf isotopic analysis of eudialite by LA-(MC)-ICP-MS. *Chemical Geology*, **273**, 8–34.

YANG, Y. H., SUN, J. F., XIE, L. W., FAN, H. R. & WU, F. Y. 2008. *In situ* Nd isotopic measurement of natural geological materials by LA-MC-ICP-MS. *Chinese Science Bulletin*, **53**, 1062–1070.

Geology and diamond provenance of the Proterozoic Banganapalle conglomerates, Kurnool Group, India

SOJEN JOY[1]*, H. A. JELSMA[1], R. F. PRESTON[1] & S. KOTA[2]

[1]*De Beers Group Exploration, 77 Crownwood Road, Theta Extension 4, Johannesburg, 2013, South Africa*

[2]*De Beers India Exploration, 36/A, Peenya Industrial Area, IInd Phase, Bangalore 560058, India*

**Corresponding author (e-mail: sojen.joy@debeersgroup.com)*

Abstract: The Banganapalle Quartzite Formation occurs in the Cuddapah Basin in India and is characterized by a basal diamond-bearing conglomerate horizon. The diamonds within this placer are generally thought to have been sourced from the erosion of kimberlites of the Wajrakurur cluster. De Beers' India's exploration efforts have resulted in the discovery of a number of dykes within the basin, with petrographical and geochemical similarities to lamproites. It is proposed that far-field stresses related to the Eastern Ghats Mobile Belt (EGMB) provided extensional sites during the time of lamproite emplacement. The dykes have not been dated. However, zircons recovered from heavy mineral stream samples in the area exhibit a number of age groupings, including one in the range of 1287–1370 Ma. This age is interpreted as the emplacement age of the dykes in this region. Kimberlitic indicator minerals (KIMs), recovered from conglomerate waste dumps, indicate the uniqueness of the garnet population relative to that of the known kimberlite clusters to the west of the basin. We propose that the emplacement of lamproites occurred as dyke–sill complexes at 1.4–1.3 Ga and that the lamproites represent the source of the diamonds in the Banganapalle conglomerates.

Supplementary material: A list of the isotope ratios used in this paper is available at www.geolsoc.org.uk/18508

Diamonds have been known and mined from India since the fourth century BC. During the reign of Ibrahim Qutb Shah (1550–1580 AD), at Golconda, Karwan, located about 5 km to the east of Golconda fort, became a thriving cutting and trading centre for diamonds. Merchants from different parts of the world stayed there, and carried out trading and cutting of diamonds and other semi-precious stones (Satyanarayana 1997). The major source of the diamonds traded at Golconda appears to have been the Krishna/Tungabhadra river gravels and the Banganapalle conglomerate workings. The latter were well known near Banganapalle village in the Kurnool district, Andhra Pradesh, India (Fig. 1).

The Banganapalle Quartzite Formation, with a locally developed basal diamond-bearing conglomerate horizon, occurs over a strike length of 240 km (Kurnool) and 120 km (Palnad) within the Cuddapah Basin. The thickness of the conglomerate horizon varies between 10 and 80 cm). The ancient diamond miners mined the conglomerate horizon using pits up to 5 m deep. The diamond-bearing bed was mined in different directions as the stability of the overlying rock would permit. Recent miners generally limit themselves to the processing of the tailings and waste dumps of abandoned mines in search of smaller stones. The diamond grade of the Banganapalle conglomerate horizon varies from 0 to 35 cpht (carats per hundred tons) with an average grade of only 2–3 cpht. About 75% of the diamonds are reportedly gem quality and the largest stone recorded had a weight of 6 carats (Balasubrahmanyan 2006).

Under the National Diamond Project, the Geological Survey of India, the National Mineral Development Corporation and the Mineral Exploration Corporation Ltd together carried out exploration work on the Banganapalle conglomerate deposit between 1980 and 1984, but did not report any primary source in the area. Few, if any, exhaustive studies have been conducted that aimed to trace the Banganapalle diamonds from sink to source. The detailed discussion of southern African placer deposits by Bluck *et al.* (2005) offers a viable benchmark from which to conduct similar studies in India.

Regional geology

The study area occurs within the central part of the Cuddapah Basin in SE India (Fig. 2). The Cuddapah

From: MAZUMDER, R. & SAHA, D. (eds) 2012. *Palaeoproterozoic of India*. Geological Society, London, Special Publications, **365**, 197–218. http://dx.doi.org/10.1144/SP365.11

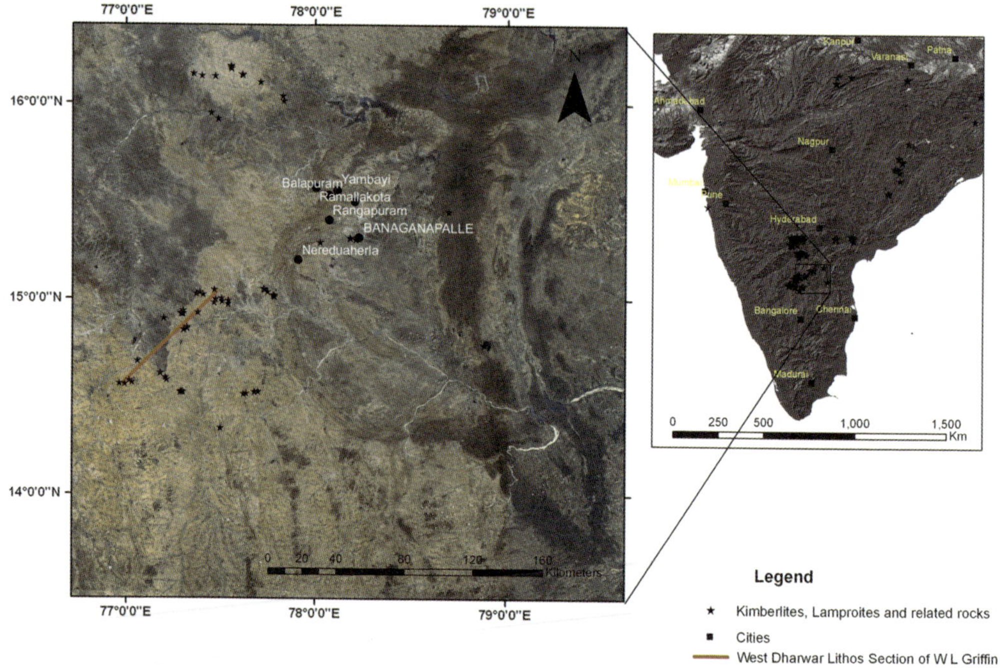

Fig. 1. Location map of the Banganapalle area. The Griffin's line as discussed in the text is also shown. The *Google Earth* imagery of the area is the background for the Banganapalle area map. A hill shaded digital elevation model is the background for the India map.

Basin is the southernmost occurrence of the Mesoproterozoic–Neoproterozoic intracratonic basins, which have been referred to as Purāna basins (Holland 1907). The crescent-shaped basin covers an area of 44 000 km^2. It overlies the granite–greenstone terrain of the Eastern Dharwar Craton of the Indian Shield along its western boundary. The arcuate eastern boundary is marked by a prominent thrust, which is parallel to the Nellore terrane of the Eastern Ghats Mobile Belt (EGMB).

The Cuddapah Basin succession comprises arenaceous, argillaceous and calcareous sediment types that have been assigned to the lower Cuddapah Supergroup and the upper Kurnool Group. The succession has an estimated aggregate thickness of 6–12 km in the east, with the Kurnool Group accounting for approximately 450 m. The Cuddapah Basin sediments are generally weakly metamorphosed and weakly deformed, but show an increase in metamorphic grade and intensity of deformation from west to east.

The sediments were deposited in a shallow-marine continental shelf environment and bear resemblance to sediments encountered in lagoons and tidal flats, carbonate shelf, beach and sand bar, coastal dune, as well as coastal alluvial-plain settings. The early sediments of the basin are intercalated with basaltic volcanics and dolerite sills and dykes. Field relationships suggest that the dolerite dykes were the feeders to the overlying basaltic lavas.

The stratigraphy of the Cuddapah Basin sediments is shown in Figure 3 (after Ramakrishnan & Vaidyanadhan 2008).

Cuddapah Supergroup

The Cuddapah Supergroup succession commences with the Gulcheru Formation conglomerates, grits and arkoses. These sediments unconformably overlie Archaean basement rocks to the west (King 1872; Nagaraja Rao *et al.* 1987). The conglomerates include clast-supported and matrix-supported beds, representing alluvial-fan and braided-stream environments. The clast assemblage is dominated by vein quartz, as well as granite, quartzite, chert, jasper and ironstone. The basal sediments grade into shallow-marine and tidal glauconite-bearing quartzitic sandstones, laminated siltstones and shales, possibly related to subsidence of the depocentre. Subsequent intertidal mudflat conditions during Vempalle deposition is indicated by the presence of dolomitic limestones and the development of stromatolites and algal laminates, which are

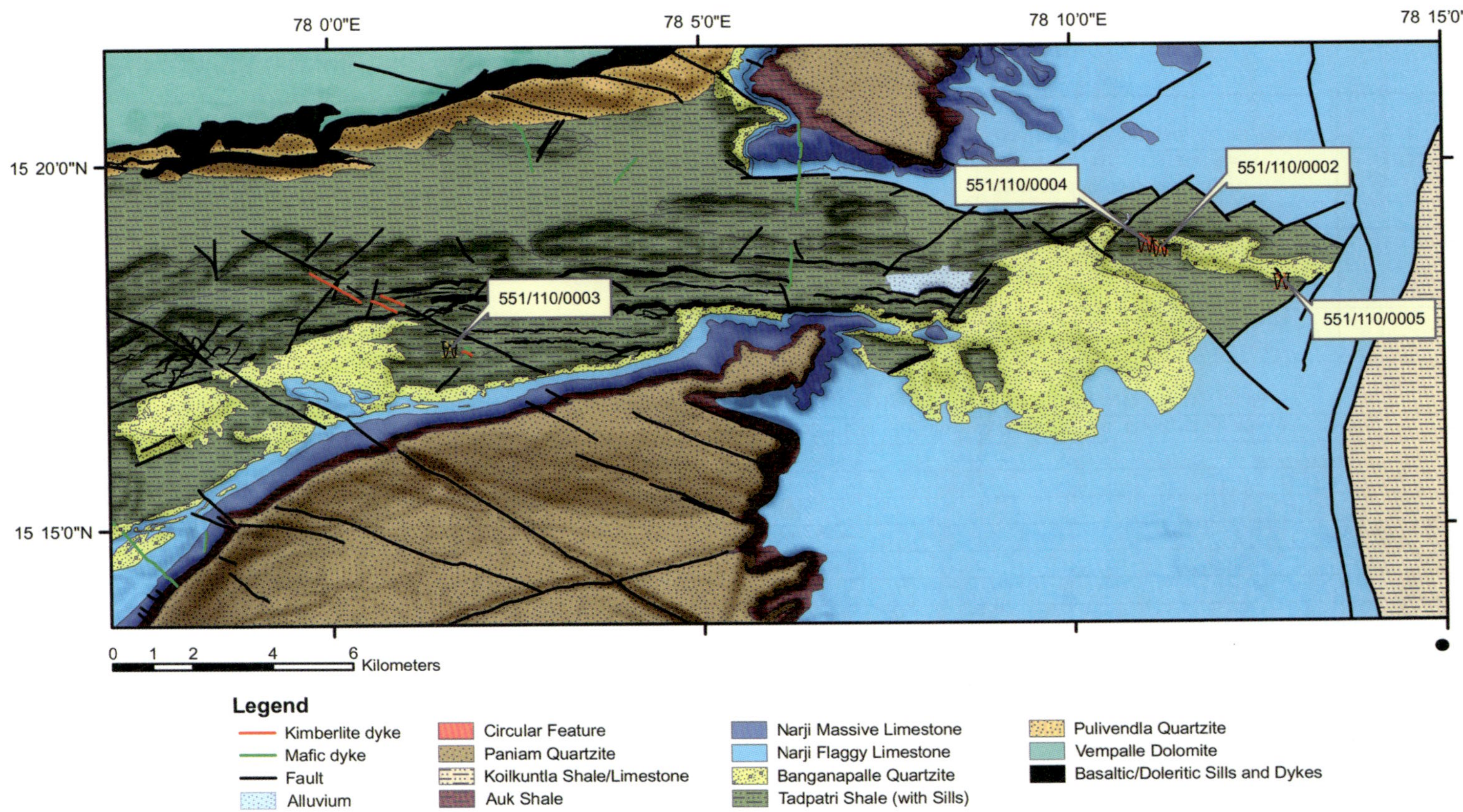

Fig. 2. Geology of the Banganapalle area (with SRTM_VD1 background), with structural elements and the locations of lamproite dykes. The Banganapalle quartzite is the lowermost formation of the Kurnool Group, contains diamonds and unconformably overlies the Cuddapah succession as well as the lamproite dykes. Note the block-faulted horst around Banganapalle Hill within the central-eastern part of the area, as well as the parallellism of lamproite dykes with WNW–ESE-trending faults.

Supergroup	Group	Formation		Thickness (m)
	Kurnool Group	Nandyal Shale, 100m	Shale, in places calcareous	100
		Koilkuntla Limestone, 50m	Siliceous, shaly limestone, quartzite	50
		Paniam Quartzite, 35m	Massive and pinnacled quartzite, siliceous shale, conglomerate	35
		Auk Shale, 15m	Laminated ochrous shale, siltstone	15
		Narji Limestone, 200m	Flaggy and massive limestone	200
		Banganapalle Quartzite, 50m	Oligomict conglomerate, grit, quartzite, shale	50
Kimberlite emplacement level (1.16–1.09 Ga)				
		~~ Unconformity ~~~~		
		Deformation		
Lamproite emplacement level (1.4 Ga)				
Cuddapah Supergroup	Nallamalai Group	Cumbum/Pullampet Shale (2000m)	Slate, phyllite, chert, dolomite	2000
		Bairenkonda/Nagari Quartzite (1500–4000m)	Quartzite, quartz wacke, shale, conglomerate	4000
	Chitravati Group	Gandikota Quartzite (300m)	Shale-quartzite alternations, with glauconite-bearing quartzite beds	300
		Tadpatri Shale (4600m)	Shale, felsic tuff, basic sills and flows, chert, stromatolitic dolomite, quartzite	4600
		Pulivendla Quartzite (10–75m)	Quartzite, conglomerate	10–75
	Papaghni Group	Vempalle Dolomite (1250–2100m)	Stromatolitic dolomite, chert, quartzite, basic sills and flows	1250–2100
		Gulcheru Quartzite (30–210m)	Conglomerate, arkose, quartzite with shale	35–320

Fig. 3. Stratigraphy of the Cuddapah Basin (after Ramakrishnan & Vaidyanadhan 2008).

interbedded with dolomites, calcareous shales, siltstones, sandstones and cherts. Towards the top of the succession basalt and basaltic andesite flows and dolerite–gabbro sills are encountered, with pillow basalts, flows, volcanic glass, ash beds and agglomerates reported (Meijerink *et al.* 1984). These are referred to as the Kuppalapalle Volcanics (Srikantia 1984).

The disconformably overlying Pulivendla Formation (10–75 m) shows shallow-marine basal oligomict conglomerates, clean (quartzitic) arenites and ripple marked siltstones, marking an alluvial–strandline interface (Ramakrishnan & Vaidyanadhan 2008). The overlying Tadpatri Formation reaches a thickness of approximately 4600 m and is made up of monotonous argillaceous sediments, with interbedded thin quartz arenites, as well as stromatolitic dolomites and cherts and felsic pyroclastics. Low-angle bedding-parallel faults caused duplication of the stratigraphy, and an estimation of the true stratigraphic thickness is difficult. Sills and dykes and minor basaltic lava flows are similar to those encountered within the Vempalle Formation (Meijerink *et al.* 1984). The presence of linser–flaser bedding suggests a tidal-flat to shallow-marine environment. The upper Gandikota Formation comprises a 300 m-thick succession of (quartzitic) arenites, with the presence of glauconite and herringbone cross-bedding suggesting an intertidal to shallow-marine environment.

The Nallamalai Group sediments to the east probably form the deformed equivalents of Cuddapah formations to the west. The Nallamalai Group comprises lower, shallow-marine quartzites and pelites (Bairenkonda/Nagari Quartzite) and upper deeper-marine siltstones, shales, slates and phyllites, with interbedded quartzites, cherts and dolomites (Cumbum/Pullampet Shale). The spatial and temporal relationships between the Bairenkonda, Nagari and Gandikota formations and the Cumbum, Pullampet and Tadpatri formations have not been resolved (Ramakrishnan & Vaidyanadhan 2008).

Radiometric dating of volcanics and intrusives within the Vempalle and Tadpatri formations suggests that sedimentation was initiated before 1.9 Ga. The Vempalle lavas and intrusive dolerites have been dated between 1879 and 1899 Ma (Chatterjee & Bhattacharji 2001; Anand *et al.* 2003). An upper age constraint of approximately 1.4 Ga for the Cuddapah deposition is provided by the ages for intrusive lamproite dykes, between 1354 and 1418 Ma (Chalapathi Rao *et al.* 1996, 1999; Kumar *et al.* 2001).

The Cuddapah Supergroup was gently folded and faulted prior to the deposition of the overlying Kurnool Group sediments. The Kurnool Group sediments outcrop in the separate Kurnool and Palnad sub-basins and unconformably overlie the Cuddapah Supergroup succession (Meijerink *et al.* 1984). The attitude of the Kurnool beds is generally horizontal or very gently tilted.

The Banganapalle Formation heralds the onset of deposition of the Kurnool Group. The formation ranges in thickness from 1 to 30 m. It is of importance because of the diamondiferous nature of some of the conglomerate horizons. Diamond placers have been identified in four major belts with historic workings, namely: (1) Banganapalle–Nereducherla; (2) Ramallakota–Yambai; (3) Undutla–Cementnagar; and (4) Balapuram–Rangapuram (Satyanarayana 1997) (Fig. 4). The provenance of the diamonds in the conglomerates is generally linked to the approximately 1.1 Ga kimberlite occurrences to the west of the Cuddapah Basin (Wajrakarur and Raichur kimberlite clusters).

The Banganapalle Formation siliciclastic sediments were deposited in a predominantly littoral-zone environment. In terms of sedimentary facies, the following can be recognized.

- A proximal braided plain setting between the western basin margin and a distal shoreline/littoral setting. This environment is characterized by basin-fining conglomerates and trough cross-bedded quartzose sandstones along the axial tracts of fault-controlled river valleys. The clast assemblages are characterized by a stable, resistant vein quartz – often rounded to well-rounded, suggesting inheritance from the Gulcheru Formation, angular to subrounded (in places oolitic) chert and shale derived from the Vempalle and Tadpatri formations, and rare, but distinct, jasper. In places, coarse-grained, up to boulder-sized, conglomerate beds have been identified.
- A distal, shoreline setting, typical of beach or shore platform deposits, is characterized by basal lenses of oligomict conglomerate, between 10 and 50 cm thick, reaching up to 80 cm. This facies appears to have yielded the bulk of the kimberlitic indicator minerals (KIMs) and the diamonds around Banganapalle. The conglomerates of this facies (Fig. 5) contain 40–70% clasts, which are rounded to subangular in shape, and poorly sorted, with pebble sizes ranging between 1 and 10 cm. The clast assemblage is composed mainly of pebbles of vein quartz, quartzite, chert, shale and jasper in a gritty, sandy or clayey matrix. Rare shale and limestone pebbles were probably derived from the underlying formations. Upgrading in low-energy littoral settings concentrated diamonds and associated indicator grains. The conglomerates are overlain and interbedded with siliceous quartz arenites and arkoses, siltstones and rare shales. The sandstones are finely bedded to cross-bedded with

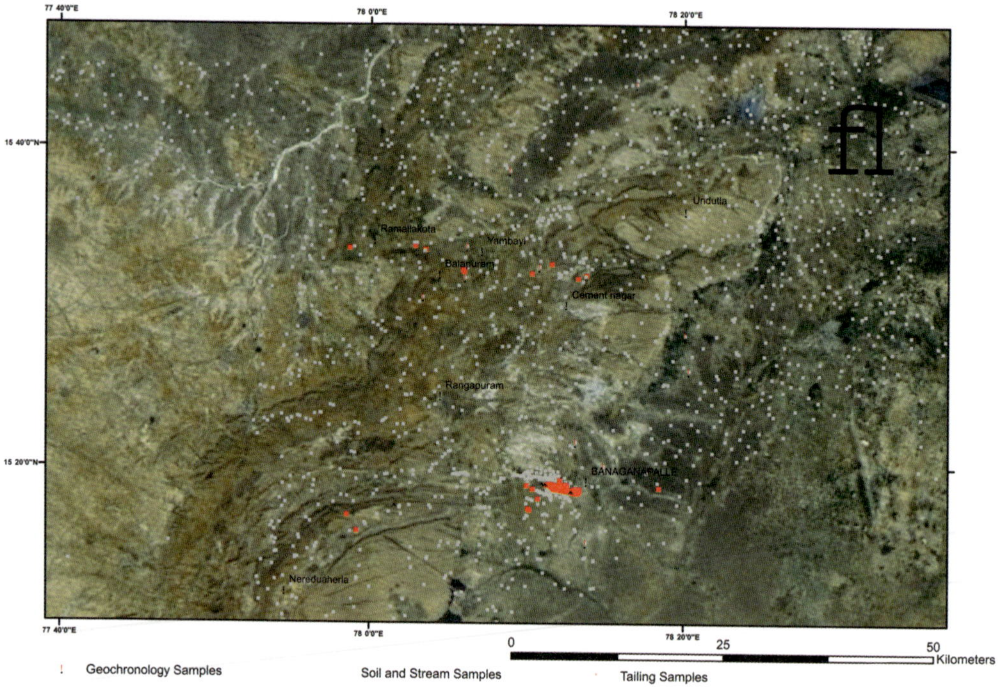

Fig. 4. Banganapalle hill sample locations (with *Google Earth* imagery background), geochronology samples and locations of diamond workings are also shown.

local evidence of tidal action, with hummocky cross-stratification and the presence of interbedded thin glauconite-bearing beds. Heavy minerals in the conglomerate, apart from diamond, are garnet, spinel, ilmenite, magnetite, zircon, tourmaline and rutile.

A sedimentological study of the Banganapalle conglomerates in the Manimadugu–Racherla area by Sivaji & Rao (1989) showed that the palaeocurrent direction from cross-bedding, ripple marks and conglomerate pebble orientations is towards the ENE and SE, indicating a westerly provenance for the sediments. The authors inferred that the palaeoflow direction was slope controlled or perpendicular to the palaeoslope, and probably formed in a fluvio-marine environment, with NE-directed longshore drift. Pebble size was also found to be directly correlated with diamond size, and (higher-energy) ferruginous and sand–matrix conglomerates were found to be richer in diamonds compared to (lower-energy) clay–matrix conglomerates. The association of the proximal facies with fault-controlled river valleys such as the Kalava Fault suggests fault activity during the deposition of the Banganapalle Formation. Lakshminarayana *et al.* (1999) also indicated the provenance of the Banganapalle to be within the Cuddapah Basin.

The Banganapalle Formation is overlain by Narji limestones, Auk shales, Paniam quartzites, Koilkontla limestones and Nandyal shales. The Narji Formation is a thin but extremely persistent horizon that rests conformably on the Banganapalle Formation and consists of flaggy limestones with intercalated glauconitic sandstones, with a massive micritic limestone member. Well-laminated, thinly bedded siliceous or ochrous Auk shale overlies the Narji limestone. The Auk shale is overlain by Paniam Formation quartzites, with a thin basal conglomerate development. The succeeding Koilkuntla Formation consists of grey flaggy limestone with quartzite intercalations. The topmost formation of the Kurnool Group is the Nandyal Formation, which consists of purple calcareous shale and upper shaly limestone, deposited in a shallow-marine reducing environment.

The presence of diamonds in the Banganapalle conglomerates, with approximately 1.1 Ga kimberlites possible source regions, provides a lower age constraint to the deposition of the Kurnool Group, if the diamonds were sourced from these kimberlites. An upper age constraint is not available but could probably be provided by the folding and shearing constrained within the EGMB between 617 Ma (reworked basement) and 517 Ma (youngest syenites) (Biswal *et al.* 2002). Ramakrishnan &

Fig. 5. 78.2282°E, 15.5331°N, 360 m elevation. (**a**)–(**c**) Road cutting through the Banganapalle Formation quartz arenites that are pale purple in colour and display hummocky cross-stratification (b) and soft sediment deformation (a). Interbedded are narrow shale intervals. The basal conglomerates (c) are moderate reddish brown–greyish purple in colour, and comprise subrounded and spherical pebbles. The conglomerates are moderately sorted and clast supported. (**d**) Hilltop with pavement showing the exposed top of the Banganapalle Formation, with cross-bedded and pebbly arenites, with subrounded, spherical pebbles with moderate sorting. A deflation sample recovered abundant garnets.

Vaidyanadhan (2008) reported trace fossils from the Auk shale, Paniam quartzite and Koilkuntla limestone, which indicate a Neoproterozoic age for the Kurnool Group.

Lamproite dykes

Lamproite dykes intrusive within and adjacent to the Cuddapah Basin have been mapped at four different localities: along the Krishna River; near Chelima town; Zangamarajupalle town; and to the west of Banganapalle town.

Krishna River

The Krishna River lamproites intrude over an area of some 300 km^2 to the north of the Krishna River. The lamproites occur as sets of dykes and rare plugs, and are intrusive in granites and gneisses of the Dharwar Craton, adjacent to the Kurnool Group sediments of the Cuddapah Basin (Chalapathi Rao *et al.* 2010*a*, *b*). The dykes have strike lengths of between 10 and 600 m, and widths of between 0.5 and 2 m. The emplacement of the dykes appears to have been controlled by sets of NW–SE-trending faults (Grady 1971; Venkatakrishnan & Dottiwalla 1987). Amongst the 12 dykes reported in Chalapathi Rao *et al.* (2010*a*, *b*), 11 cluster around a NW–SE orientation (310°–130°). Only the Ramannapeta dyke has been dated, at 1384 Ma (Chalapathi Rao *et al.* 1996) and 1224 ± 14 Ma (Kumar *et al.* 2001).

Chelima

The Chelima dykes occur in an area of about 40 km^2 around Chelima village, and intrude slates and phyllites of the Nallamalai Group. The dykes have been alternatively classified as lamprophyres (Appavadhanulu 1962, 1966), kimberlites (Sen & Narasimha Rao 1967, 1970, 1971), minettes (Scott-Smith 1989) and lamproites (Chalapathi Rao & Madhavan 1996; Chalapathi Rao 2007). Old opencast and underground workings in the dykes and proximal alluvial diamond recoveries bear testimony to their diamond potential. The dykes are composite, en echelon bodies, 1–5 m wide and up to 7 km long, with breccias noted along dyke margins (Chalapathi Rao 2007). The dykes were intruded along steeply dipping en echelon fractures trending WNW–ESE (Chalapathi Rao 2007). The Chelima lamproites have been dated at between 1354 and 1418 Ma (Chalapathi Rao *et al.* 1999; Kumar *et al.* 2001).

Zangamarajupalle

The Zangamarajupalle dykes occur some 80 km to the SE of Chelima town. Two ultramafic dykes, 2–3 m wide, were intersected in drill core, and are intrusive within the Cumbum shales (Bhaskar Rao 1976). The dykes are similar in composition to the Chelima dykes. Their emplacement age appears to be younger than that of the Krishna and Chelima dykes, at approximately 1090 Ma, but their source region and petrogenetic evolution is similar to that of the other dykes (Chalapathi Rao 2007).

Banganapalle

As part of this study we report on four dykes that were discovered by De Beers India Exploration. The dykes occur west of Banganapalle town and are referred to as 551/110/0002–551/110/0005.

Dyke 551/110/0004 (78.183°E, 15.315°N) is located in a road metal quarry 4.5 km due west of Banganapalle town. The lamproite dyke intrudes the Tadpatri shales and three 10–20 m-thick dolerite sills (Fig. 6). The dyke can be traced 250 m to the SE where the thickness varies to 0.5–2 m. High-resolution satellite imagery shows the total strike length of the dyke to be in the order of 1 km. In the quarry, the dyke has a width of 0.5 m, and shows refraction and minor offset at the contact between shale and dolerite. It is tilted to the NE at 043°/78° and shows faulted contact surfaces with near-horizontal slickenlines. The adjacent Tadpatri shales and dolerite sills dip to the south at 17°. These field relationships suggest that the dyke has been deformed and tilted together with the adjacent Chitravatri Group sediments, prior to Kurnool deposition. Adjacent dyke 551/110/0002 (78.186°E, 15.313°N) has a strike length of approximately 500 m, and occurs 125 m to the east of, and parallel to, dyke 551/110/0004.

Dyke 551/110/0005 (78.214°E, 15.305°N) occurs 1.7 km to the SW of Banganapalle town.

A set of five en echelon lamproite dykes (551/110/0003) was encountered in a valley, 24 km due west of Banganapalle town (78.001°E, 15.305°N). The dykes are well exposed along a stream bed, intruding the Tadpatri shales, with the outer members some 20 m spaced (Fig. 7). The dykes

Fig. 6. 78.18361°E, 15.31556°N, 280 m elevation. Quarry with Tadpatri shales and two dolerite sills (**a**) and (**b**), cut by lamproite dyke. A third dolerite sill (**c**) is exposed in a lower quarry. From top to bottom the section shows sill A (>5 m thick), followed by shale (*c.* 10 m), sill B (20 m) and further shale with interbedded dolomite. Note the dyke refraction (and possible minor offset) at the contact between the shale and the dolerite.

Fig. 7. 78.00153°E, 15.30519°N, 367 m elevation. Set of five lamproite dykes (551/110/0003) cutting Tadpatri shales. The average thickness of the dykes is 1.2 m (dyke 1), 0.4 m (dyke 2), 1 m (dyke 3), 0.6 m (dyke 4) and 1 m (dyke 5), giving a combined thickness of approximately 4 m.

show a common trend between 120° and 128°, and they have been tilted to the NE at approximately 80°. The dykes bifurcate and pinch and swell. The thickness of the dykes varies between 0.4 and 1.2 m, with a combined thickness of about 4 m. The wall-rock shales are oriented at 187°/28°. The dykes have been traced over a distance of 800 m to the WNW and 6 km to the ESE, where they reportedly combine as a single dyke.

Structure

The Cuddapah Basin is a broad, crescent-shaped and asymmetric north–south-trending synclinorium, in which the lower Cuddapah Supergroup formations have been folded. The arcuate north, south and western boundary of the Cuddapah Basin is an unconformity with basement granites and gneisses, as well as greenstone belts. The basin structure is characterized by a gently tilted, eastward-dipping western limb, and a steeply tilted and overthrust eastern limb. The latter is marked by the Nallamalai and Velikonda ranges of hills. The foot of the Nallamalai Range marks the Rudravaram Line, which shows as: (a) a fault contact juxtaposing flat-lying Kurnool and tilted Nallamalai formations; and (b) a thrust contact between tilted Cuddapah and Nallamalai formations, suggesting prolonged and repeated reactivation. East of this line the Nallamalai formations become progressively folded and thrusted, and transposed into steeply tilted to subvertical orientations. Further east, the Velikonda Range marks the contact with the Nellore Schist Belt and the Eastern Ghats Mobile Belt gneisses and granulites. The western half of the Cuddapah Basin consists of four sub-basins, from north to south, namely: the Palnad, Srisailam, Kurnool and Papaghni sub-basins. In the Kurnool and Palnad sub-basins, the younger Kurnool formations rest with unconformable contact on the tilted Cuddapah formations. The attitude of the Kurnool beds is horizontal or subhorizontal, with a gentle tilt to the SE or south, and with local drape folding along fault contacts. Kurnool deposition was limited to shallow basins, the extent of which was governed by fault-controlled relief. The eastern half of the Cuddapah Basin is occupied by the Nallamalai Fold Belt, consisting of deformed upper Cuddapah sediments that show cleavage development, faulting, folding and thrusting (Saha *et al.* 2010).

Basement-rooted wrench faults have played a role during the deposition of the Kurnool formations and were reactivated after the end of sedimentation.

These are long-lived zones of structural weakness that were active prior to and during Kurnool deposition, and controlled the deposition and subsequent block faulting of the Kurnool and older formations. Examples are the ENE–WSW-trending Kalava and Chenakapalle faults – bounding the Gani-Kalva ('window') anticline – as well as the Nandyal and Vempalle faults and the ESE–WNW-trending Atmakur and Nossam faults. These faults are truncated by the north–south-trending faults of the Rudravaram Line.

On a local scale, the Banganapalle area shown in Figure 2 can be separated into a western and eastern structural domain. The contact between the two domains is a north–south-trending lineament that parallels the Nallamalai Range of hills to the east. The western domain shows a gently southwards-tilted succession comprising, from north to south: Vempalle dolomites, Pulivendla quartzites and Tadpatri shales, with intrusive dolerite sills. In the SW, the succession is unconformably overlain by flat-lying Kurnool formations. The eastern domain shows a central west–east-trending horst structure that exposes tilted Tadpatri shales with dolerite sills and unconformably overlying Banganapalle conglomerates (Banganapalle Hill). Along its northern and eastern sides, the horst juxtaposes Tadpatri shales and the overlying Banganapalle conglomerates against Narji limestones. Along the south side of the horst, Banganapalle conglomerates overlie the Tadpatri shales and are in turn overlain by Narji limestones.

The Banganapalle area is dissected by sets of WSW–ENE- and WNW–ESE- to NW–SE-trending faults, which transect both the Cuddapah and the Kurnool formations. These faults can be observed along the eastern side of the horst structure, but also transecting the western domain. Deformation associated with the Eastern Ghats Mobile Belt may have imparted far-field stresses on the Cuddapah Basin sediments causing (re)activation of faults and affecting even the younger Kurnool Group formations.

The WNW–ESE-trending lamproite dykes around Banganapalle Hill (551/110/0002, 551/110/0004 and 551/110/0005) and further west (551/110/0003), as well as the Chelima, the Krishna and, possibly, Zangamarajupalle dykes, parallel the WNW–ESE faults that dissect the area. The dykes occur as segmented dyke–fracture arrays with anastomosing (branching and bifurcating) and en echelon dyke segments that can be followed as discontinuities or broken linear features striking WNW–ESE on satellite imagery and that disappear beneath the Kurnool sediment cover. The amount of erosion between the emplacement of the Banganapalle lamproites and the Kurnool sedimentation is unknown, and the question remains as to whether the lamproites could have been a source for some of the diamonds.

Amongst contemporary models, the formation of the Cuddapah Basin represents a rifted Palaeoproterozoic continental margin (Chaudhuri *et al.* 2002), which was inverted during a high-grade tectonothermal event within the Eastern Ghats Mobile Belt at approximately 1.6 Ga (Mezger & Cosca 1999; Simmat 2003). Subsequent alkaline magmatism, including the emplacement of lamproites at around 1.3 Ga and monzosyenite and nepheline syenite intrusions at 1.5–1.3 Ga (Upadhyay & Raith 2006; Upadhyay 2008), has been related to rifting along the SE margin of the Indian Shield. It is proposed that lamproite emplacement was related to the regional extensional processes associated with the above rifting episode (Chalapathi Rao *et al.* 2004). Rift basin inversion during the Mesoproterozoic (1.1–1.0 Ga) led to the collision of eastern India with East Antarctica during Rodinia assembly (Upadhyay 2008), the formation of a deep-mantle thermal anomaly, and the emplacement of the Wajrakurur, Naryanpet and Raichur kimberlites.

Exploration by De Beers India Private Limited

De Beers India Private Limited (DBI Pvt Ltd) obtained reconnaissance permits in the Banganapalle area in 2001 to explore for primary sources of diamonds. The exploration was initiated with systematic stream and soil sampling for heavy minerals in the area. The samples were processed in the heavy mineral processing centre of De Beers India, Bangalore. The KIMs were analysed at the University of Melbourne (samples processed in 2002) and at Microbeam Services, Melbourne (for samples processed during 2004). Along with the stream and soil samples, the conglomerate tailings dumps were also sampled. Airborne multispectral, magnetic and electromagnetic surveys were conducted as part of the exploration work.

Over 300 stream and soil samples were collected around the Banganapalle Hill area by De Beers India. About 250 conglomerate tailings samples were treated as part of the study (Fig. 8). The heavy mineral concentrates generated by the processing of conglomerate tailings yielded pyrope garnet, chrome spinel, picro-ilmenite, magnetite, ilmenite, zircon, rutile, tourmaline, pyrite, epidote, amphibole, staurolite and barite. The historic workings located west of Banganapalle town are well known as 'Vajragiri', and all the tailings from the old workings in and around this area were sampled (Fig. 4). Abundant fresh kimberlitic indicator minerals were recovered. The quantity of kimberlitic minerals in

Fig. 8. 78.20521°E, 15.31279°N, 298 m elevation. (**a**)–(**c**) Dumps with Banganapalle Formation ore composed of pebbles of conglomerate, grit, sandstone, siltstone (intraformational?), chert, vein quartz, jasper. (**d**)–(**f**) Mine drives and a 12 m shaft through Banganapalle quartzites with discontinuous conglomerate lenses at the base, with fining-upwards cycles through quartzitic sandstone to siltstones and thin green-coloured shale horizons. The contact with the underlying Tadpatri Formation has an elevation of 280 m at this location.

the Vajragiri dumps is very high and these minerals can be easily identified with the naked eye. Most of the indicators appear to be fresh and unabraded. Some of the garnets have remnant kelyphite material attached, indicating their proximity to the primary source(s).

The exploration efforts resulted in the discovery of a number of lamproite dykes in the area. The dykes have widths of 1 m or less, with the longest dyke segment traced for nearly 7 km. The dykes were sampled for KIMs, but grain yields were too low to offer any meaningful opportunity for comparison with the surrounding sediment samples.

The petrographical sections of the dykes show the rock to be extremely altered with uniform texture. Carbonate pseudomorphs after olivine phenocrysts are common. Carbonatized/silicified olivine phenocrysts and macrocrysts are observed in partly clay-altered micaceous groundmass. The groundmass comprises mainly clay and carbonate with remnants of micaceous minerals. The rock is petrographically classified as magmatic, micaceous lamproite.

Whole-rock geochemistry (X-ray fluorescence (XRF)) corresponds with the highly altered nature of the rock. The contamination index (Clement 1982) is over 2 for all of the samples (Table 1). All the samples indicate a high TiO_2 content, low Na_2O and have low Sc values as normally reported from lamproites (Mitchell & Bergman 1991). La and Ce are the only rare earth elements (REEs) analysed and are highly enriched in the samples. The chondrite-normalized values of La, Ce, Nb, Sr and Ti show enrichments comparable to the kimberlite/lamproite clan of rocks (Table 1). Furthermore, the samples are characterized by low Y and Sc, and high Ba, Sr, Nb and Zr. The Zr/Nb ratios are elevated (Table 1) compared to kimberlites, alkali basalts and lamprophyres (normally 0.4–4), and show a closer affinity to lamproites (Mitchell & Bergman 1995). A detailed classification of the rock requires petrological and geochemical analysis of fresher samples, and has not been attempted as part of the current study.

Terrain U–Pb zircon geochronology

Eleven of the heavy mineral sediment samples collected from the study area have reported zircons in the 0.5–1 mm fraction. The U–Pb geochronology was conducted using the laser ablation-inductively

Table 1. *Whole-rock geochemistry results of the dykes*

	EEH719	EEH724	EEH720	EEH721	EEH722	EEH723	EEH726
Longitude (°)	78.03	78.19	78.03	78.03	78.19	78.18	78.19
Latitude (°)	15.29	15.31	15.29	15.29	15.31	15.31	15.31
Major elements (wt%)							
SiO_2	50.26	31.04	45.9	45.71	44.21	23.11	41.13
TiO_2	4.02	1.53	3.89	3.53	1.17	2.05	4
Al_2O_3	5.81	3.7	6.57	6.26	3.04	6.54	5.24
Fe_2O_3	7.29	8.27	8.26	7.92	5.55	9.71	9.15
MnO	0.13	0.13	0.13	0.1	0.07	0.14	0.14
MgO	10.8	15.06	10.34	12.79	4.24	11.99	7.92
CaO	9.25	20.14	10.12	9.99	21.71	23.25	16.72
Na_2O	0.01	0	0.06	0.43	0.02	0	0.1
K_2O	0.94	0.01	0.07	0.79	0	0.01	0.28
P_2O_5	1.14	0.26	1.2	1.14	0.25	4.66	2.83
Cr_2O_3	0.1	0.22	0.11	0.11	0.14	0.12	0.08
LOI	9.56	18.75	11.53	9.48	17.94	16.82	10.77
Trace elements (ppm)							
S	787	178	802	16	16	16	129
Sc	10	7	5	4	bd	bd	bd
V	105	40	101	101	72	96	88
Cr	744	2025	776	782	1176	838	538
Co	62	76	77	60	74	70	73
Ni	463	891	541	537	757	212	411
Cu	22	bd	21	16	18	125	34
Zn	84	71	109	88	91	85	64
Ga	12	6	14	14	5	12	13
As	4	10	4	11	4	8	12
Rb	30	3	7	10	4	5	6
Sr	1425	144	1401	1581	409	2275	3823
Y	16	8	17	24	26	50	18
Zr	1026	481	976	868	287	291	709
Nb	174	61	146	126	139	212	149
Mo	2	bd	2	Bd	4	bd	bd
Ba	1317	799	1158	549	151	523	1262
La	240	122	205	212	311	743	307
Ce	546	217	456	455	611	1587	689
Th	28	bd	21	20	36	177	19
U	bd	bd	bd	Bd	bd	bd	bd
W	16	6	73	45	60	16	93
Pb	14	bd	8	Bd	24	27	6
Contamination index	4.78	2.31	5.05	3.86	11.15	2.47	5.67
Zr/Nb	5.9	7.89	6.68	6.89	2.06	1.37	4.76
La: chondrite-normalized	729.48	370.82	623.1	644.38	945.29	2258.36	933.13
Ce: chondrite-normalized	631.21	250.87	527.17	526.01	706.36	1834.68	796.53
Nb: chondrite-normalized	497.14	174.29	417.14	360	397.14	605.71	425.71
Sr: chondrite-normalized	120.76	12.2	118.73	133.98	34.66	192.8	323.98
Ti: chondrite-normalized	38.53	14.66	37.29	33.83	11.21	19.65	38.34

bd, below detection.

coupled plasma-mass spectrometry (LA-ICP-MS) located at the Geosciences Facility of De Beers Group Exploration in Johannesburg, South Africa, in 2004. The analyses were completed and interpreted by E. S. Barton and C. M. Appleyard while they were working for De Beers (internal reports of De Beers). The present contribution reassesses the geological significance of the analyses and attempts to relate the ages obtained with the emplacement of the dykes discovered in the area.

The stream samples were collected at various trap sites in the stream course. As only the 0.5–1 mm fraction was analysed for zircons, the obtained ages do not represent the complete spectrum of ages expected for the province.

Methodology

Zircons were mounted in Epoxy resin and polished to expose the central portions of the grains. A Perkin–Elmer ELAN 6000 ICP-MS coupled to a New Wave Research 213 (NWR-213) Nd:YAG UV laser system with an excitation wavelength of 213 nm. The data were reduced using LAMTRACE (by S. Jackson, Macquarie University, Sydney) and was used for all isotope ratio calculations. The ages of the zircons were calculated using the following isotope ratios: $^{207}Pb/^{206}Pb$, $^{207}Pb/^{235}U$ and $^{206}Pb/^{238}U$, and the concordia diagrams were plotted and all statistical analysis were performed utilizing IsoPlot, version 2.49 (Ludwig 2003). The ages quoted are at 2σ (95% confidence) levels. The standard GJ-1 from GEMOC (provided by W. L. Griffin), with a verified age of 608 Ma, was used during the analysis and yielded an average age of 609 ± 11 Ma (23 analyses). The mean standard weighted deviation (MSWD) values quoted are an indication of the probability that the analyses are within error.

The samples demonstrate different clusters of ages. Discrete populations could be classified at 3088 ± 11, 3143 ± 11, 3187 ± 11 and possible older ages (*c.* 3500 Ma), as well as a very prominent component at 2620 ± 5 Ma. These populations (Fig. 9a, b & Table 2) reflect a spectrum of ancient and recent provenance ages, with recycling from Cuddapah sediments a likely major component.

Six zircons from five samples record younger ages in the range of 1287–1370 Ma (Fig. 9a, b & Table 2). These zircons also record a high uranium content consistent with zircons of mantle origin. This age population is interpreted as the age of the dykes in this region, corresponding well with the known ages of other lamproites intrusive within and in proximity of the Cuddapah Basin.

Kimberlitic indicators

Griffin *et al.* (2009) recently presented garnet xenocryst ($n = 1187$) data from 15 kimberlites exposed along an 80 km-traverse crossing the eastern margin of the Closepet Granite (Fig. 1) (Griffin *et al.* 2009, p. 1111, fig. 2). The NE extremity of the traverse line terminates approximately 80 km WSW of the study area, representing the only readily available and comparable kimberlite-borne xenocryst dataset for the East Dharwar Craton. The data presented in Griffin *et al.* (2009) thus provides a useful means by which to compare and contrast the garnet xenocryst data collected from the Banganapalle Formation conglomerate tailings samples. Comparison of these data with published results aims to demonstrate a unique, within the context of the East Dharwar Craton, provenance for the subcontinental lithospheric mantle (SCLM) sample, represented principally by garnets obtained from the sediments under examination in the present study.

Griffin and co-workers, in mapping the composition and structure of the SCLM along this traverse, demonstrated abrupt changes in the characteristics of the SCLM. The samples obtained from the Kalyandurg–Brahamanapalle cluster, defining the SW extremity of the traverse, were characterized by abundant harzburgitic garnets with high median olivine magnesium numbers and yttrium content extending well below 10 ppm, indicating a significant depletion of the source rocks at the time of entrainment. The associated nickel thermometry (utilizing the method of Ryan *et al.* 1996) yields a tightly constrained temperature distribution (Griffin *et al.* 2009, p. 1112, fig. 3) terminating at approximately 1040 °C and, when coupled with the Ryan *et al.* (1996) Cr-in-garnet barometer, indicates a conductive cratonic geotherm of approximately 35 mW m^{-2}. The base of the depleted lithospheric mantle was constrained at 190 km. In contrast, garnet xenocryst data recovered from Uravakonda/WK-7 cluster kimberlites ($n_{garnet} = 172$) is characterized by a much lower abundance of strongly depleted garnets sourced from low-calcium harzburgite – the low-calcium (<2.0 wt% CaO) portion of the harzburgite field (Griffin *et al.* 2009, p. 1113, fig. 4) is sparsely populated in comparison with the Kalyandurg–Brahamanapalle samples. Yttrium-depleted garnets are scarce and absent within the range exhibited for the Kalyandurg–Brahamanapalle bodies (Griffin *et al.* 2009, p. 1112, fig. 3). The available data constrain the geotherm at approximately 40 mW m^{-2}. Griffin *et al.* (2009) kinked the 40 mW m^{-2} geotherm at a temperature of 1150 °C based on the presence of high-temperature Y-depleted garnets with an elevated titanium content, thereby decreasing the maximum depth of depleted SCLM to 160–170 km, albeit with a low degree of confidence.

The NE extremity of the traverse, closest to the present study area contrasts both the Kalyandurg and Uravakonda clusters in the uniqueness of the xenocryst sample. The concentrate data, collected from selected Anumpalle bodies, is more extensive and the low-chrome 'splay' is a characteristic feature of the concentrate sample. Once again, the strongly depleted harzburgitic garnets present in the Kalyandurg samples are not present in the samples from the NE extremity of the traverse. Garnet thermometry yields a bimodal temperature distribution, with the bulk of the more fertile lherzolitic component (with yttrium contents above 10 ppm) confined to temperatures below 1100 °C. The most yttrium-depleted garnets also record the highest temperatures (+1200 °C) but, according to

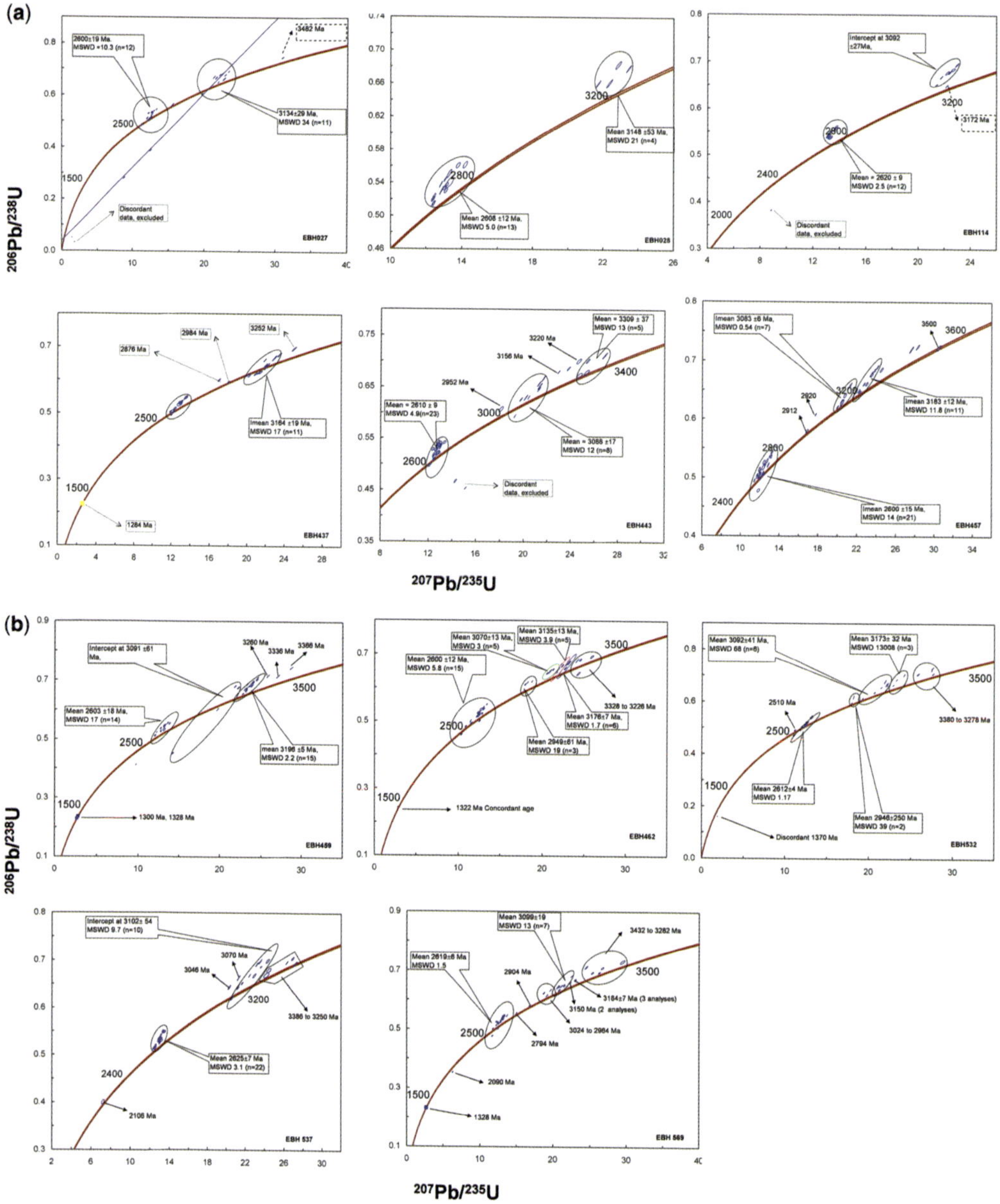

Fig. 9. (**a**) and (**b**) U–Pb isotope concordia plots of the zircons recovered from exploration samples.

Griffin *et al.* (2009), are enriched in titanium and zirconium. Griffin *et al.* (2009) constrained the geotherm at 37.5 mW m^{-2} and, with a kink at 1100–1150 °C, defined the maximum depth of depleted SCLM at approximately 190 km, consistent with the SW end of the traverse.

These observations from selected kimberlites exposed along this traverse form the basis for further discussion and argument in favour of the notion that the garnet xenocrysts recovered from the Banganapalle Formation conglomerates represent clastic input from a discrete and localized source that is characterized by a unique SCLM signature within the context of the point sources of garnet-bearing SCLM samples examined by Griffin and co-workers.

Table 2. *Zircon age groups obtained from the samples*

Sample		Age groups obtained (Ma)					
	Longitude (°)/latitude (°) (WGS 84)	>3200	3200–3000	3000–2900	*c.* 2600	2500–2000	1300–1200
EBH027	78.33444 15.47128	3482	3134 ± 29		2600 ± 19		
EBH028	78.34 15.43		3148 ± 53		2608 ± 12		
EBH114	78.28 15.73		3111 ± 10		2620 ± 9		
EBH437	78.1 15.56		3164 ± 19	2984, 2876	2622 ± 10		1284
EBH443	78.1 15.53	3309 ± 7	3088 ± 17		2610 ± 9		
EBH457	78.21825 15.35978	3500	3183 ± 12, 3083 ± 6		2600 ± 15		
EBH459	78.25 15.33	3400–3300	3196 ± 5, 3091 ± 61		2603 ± 18		13, 281, 300
EBH462	78.23 15.25	3328–3226	3176 ± 7, 3135 ± 13, 3070 ± 13	2949 ± 61	2600 ± 12		1322
EBH532	78.06 15.51	3380–3278	3173 ± 32, 3092 ± 41	2946 ± 250	2612 ± 4		1370
EBH537	78.18 15.54	3386–3250	3102 ± 54		2625 ± 7	2106	
EBH569	78.15 15.64	3432–3282	3184 ± 7, 3150, 3099 ± 19	2904	2619 ± 6	2090	1328

Sampling and methodology

Kimberlitic indicator minerals (KIMs), including garnet, were recovered from almost 350 stream and conglomerate tailings (sourced from abandoned workings) samples collected in close proximity to the town of Banganapalle (Fig. 4).

KIMs were extracted in various size fractions over the history of the sampling campaign (commencing in 2001) – the bulk of the conglomerate tailings samples were reduced to two fractions (0.3–0.5 and 0.5–1.0 mm) and produced the highest yields of heavy minerals. A total of 4874 kimberlitic garnets were extracted for further analysis. The contribution of KIMs from the stream samples was negligible. More than 5064 kimberlitic garnets were analysed for major elements and a subset of approximately 1000 grains were selected for trace element analysis.

The garnet grains (spinel, chrome diopside and picro-ilmenite were also recovered but are not discussed here) were mounted in epoxy resin discs that were ground and polished to expose the mineral surfaces for major and trace element analysis. Electron probe microanalysis (EPMA) for major elements was conducted by Melbourne University and Microbeam Services in 2002 and 2004, respectively (both facilities in Melbourne, Australia), on behalf of De Beers Australia Exploration. Major element data, reported in the form of oxides, were generated using Cameca SX50 electron microprobes with four vertical wavelength dispersive (WDS) spectrometers using a range of natural and synthetic standards. The electron beam parameters were set at 15 kV for the accelerating voltage, with a beam current of 35 or 25 nA. The counting times for all elements were set at 20 s for the peak and 10 s on two backgrounds, either side of the peak. Detection limits were better than 0.05 elemental weight per cent for all elements except zinc (which was 0.09%). Matrix correction calculations were performed using the PAP procedure after Pouchou & Pichoir (1984). Trace elements in garnet, for a subset of 954 grains (Fig. 10a, b), were acquired via laser ablation ICP-MS (LA-ICP-MS) utilizing a Perkin–Elmer ELAN 6000 quadrupole ICP-MS coupled to a CETAC LSX-200 UV Nd:YAG UV laser system with an excitation wavelength of 266 nm. All analyses were generated at the De Beers Group Exploration central analytical facility in Johannesburg. All measurements were carried out on the same Epoxy grain mounts that were prepared for EPMA. NIST SRM-610 (a homogeneous, synthetic doped glass) and a natural garnet megacryst (GHR-1) were used as external standards, and analysed and cross-checked against the certified and accepted compositions throughout the testing work. ^{43}Ca was used as an internal standard in

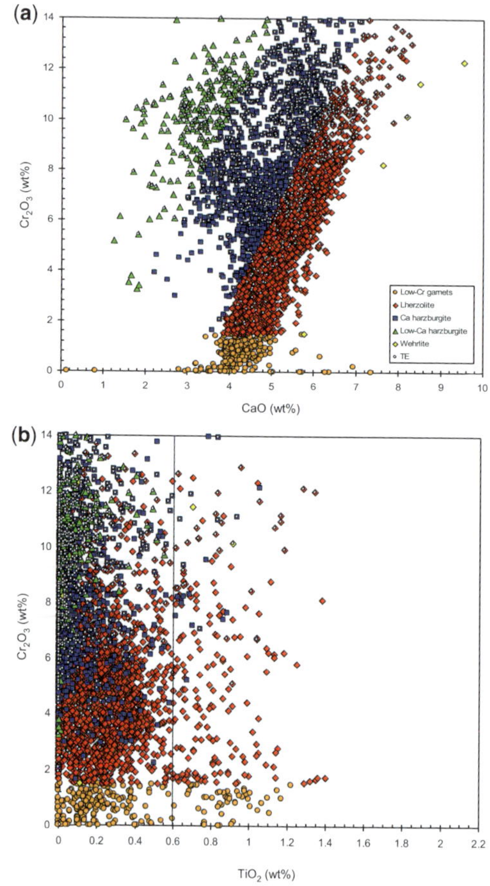

Fig. 10. (**a**) Chrome–calcium discrimination diagram for garnets ($n = 5064$) recovered from the Banganapalle Formation conglomerate tailing samples as part of this study. The classification of the garnets is after Griffin *et al.* (2002), showing the high proportion of harzburgitic compositions. (**b**) Chrome–titanium data. A cut-off of 0.6 wt% TiO_2 is applied (see the annotation) to differentiate high- and low-Ti garnets. Ti-enriched garnets (i.e. $>$0.6 wt% TiO_2) are not abundant. White dots identify specific garnets that were analysed for various trace elements.

order to compensate for ionization and matrix effects (values for CaO were obtained from the electron microprobe analyses for garnet). GLITTER 4.0 (Macquarie University) was used to reduce the time-resolved analysis spectra to calculate the concentrations of the various analytes (in ppm).

For comparison with the available published data (e.g. Griffin *et al.* 2009), the garnets collected from the study area have been classified, on the basis of their major element compositions, using the approach adopted by Griffin *et al.* (2002) to define subpopulations encompassing depleted (low-calcium harzburgite and harzburgite) to fertile compositions (e.g. lherzolite). Similarly, garnet temperatures have been calculated using the nickel-in-garnet thermometer proposed by Ryan *et al.* (1996) in order to compare the temperature distribution (and other chemical parameters) of the kimberlitic garnets collected during the present study directly with the data provided by Griffin *et al.* (2009). The combination of garnet temperature and pressure data allows the geotherm for the portion of lithosphere sampled by the kimberlite (location presently not constrained) to be constrained according to the methodology proposed by Ryan *et al.* (1996).

Garnet major and trace element chemistry

The 5064 kimberlitic garnets recovered from the conglomerates of the Banganapalle Formation, once classified, show significant compositional spread using the conventional binary chrome–calcium systematics (Fig. 10a).

The concentrate data set is indicative of sampling from a range of upper-mantle lithologies dominated by low-titanium lherzolite (*c.* 65%); the chrome–titanium plot in Figure 10b indicates the cut-off for titanium-enriched garnets at more than 0.6 wt%. Garnets from harzburgite (classified as either low-calcium harzburgite or calcium harzburgite) are abundant and account for more than 30% of the concentrate sample, yielding a pattern most similar to the sample of garnets recovered from the Kalyandurg–Brahamanapalle cluster of kimberlites approximately 155 km to the SW (Griffin *et al.* 2009, p. 1113, fig. 4). The vast majority of the harzburgitic garnets are also depleted (<0.6 wt%) with respect to TiO_2 (Fig. 10b). The abundance of harzburgitic garnets recovered from the Banganapalle sediments appears to be unique when considered in the context of the garnet concentrate samples obtained from known occurrences to the SW. As Griffin *et al* (2009) demonstrated, although the garnet concentrate samples from other kimberlite clusters along their traverse are similar in terms of the range of chrome contents, the abundance of harzburgitic garnets present in the samples appears to taper off significantly towards the NE. In addition, garnets from the Anumpalle cluster of kimberlites exhibit a diagnostic 'splay' of low-chrome garnets in chrome–calcium compositional space (Griffin *et al.* 2009, p. 1113, fig. 4) that is clearly not represented in the garnet concentrate sample from the Banganapalle sediments approximately 80 km to the NE. Where low-calcium harzburgitic garnets from the Kalyandurg–Brahmanpalle cluster of kimberlites extend to calcium (CaO) concentrations well below 1 wt% (Griffin *et al.* 2009, p. 1113, fig. 4), grains of the same paragenetic class (i.e. low-calcium harzburgite) recovered from the conglomerate horizons examined as part of the current study terminate at approximately 1.4 wt% Ca (Fig. 10a), highlighting another, albeit more subtle, difference between the kimberlitic garnets recovered in the two regions. This suggests that the source of the KIMs recovered from the Banganapalle Formation is unlikely to be the known kimberlites to the SW.

The uniqueness of the Banganapalle conglomerate concentrate sample is demonstrated more convincingly in the examination and comparison of the trace element data and subsequent construction of the xenocryst geotherm. Trace element data, including nickel, were obtained for a subset of 954 garnets, spanning the entire range garnet compositions represented in the concentrate sample (Fig. 10a, b). Nickel thermometry yields a tight distribution of values, especially with respect to the harzburgitic component – the yttrium–temperature plot (Fig. 11) confirms that the most depleted garnets (i.e. the low-calcium harzburgite) existed/equilibrated over a relatively narrow temperature range, between approximately 1170 and 1330 °C, in comparison to other components. These garnets lie on or close to a 39 mW m^{-2} conductive model (using the models proposed by Pollack & Chapman 1977) and are clearly depleted with respect to yttrium in accordance with the threshold of 10 ppm proposed by Griffin *et al.* (1999, 2002) for depleted garnets. In contrast to the garnet concentrate samples from the kimberlite clusters to the SW (Griffin *et al.* 2009, p. 1113, fig. 3), depleted garnets, with respect to yttrium (Fig. 11) and titanium and zirconium (Fig. 12a), from the Banganapalle conglomerate are abundant to temperatures above 1400 °C, a feature that is unique to the Banganapalle data. Evidence for a high-temperature input is clear in the case of the Anumpalle cluster data, but these garnets are clearly enriched in yttrium with values exceeding 10 ppm (Griffin *et al.* 2009, p. 1113, fig. 3). This is clearly not the case for the Banganapalle data, where the low-calcium harzburgitic, calcium harzburgitic and lherzolitic (to a lesser extent) garnets exhibit very well-defined depletion trends albeit at high temperatures (generally above 1200 °C). The authors present these observations as clear evidence *against* a source region for the kimberlitic indicator minerals recovered from the Banganapalle sediments from the known kimberlites to the SW.

Where Griffin *et al.* (2009), in the case of the Anumpalle cluster, kinked the geotherm at approximately 1200 °C based on the demarcation of the so-called 'yttrium edge', there is no evidence from the Banganapalle data (Fig. 11) of the same thing occurring. In contrast, the low-yttrium trend only commences at approximately 1200 °C, suggesting

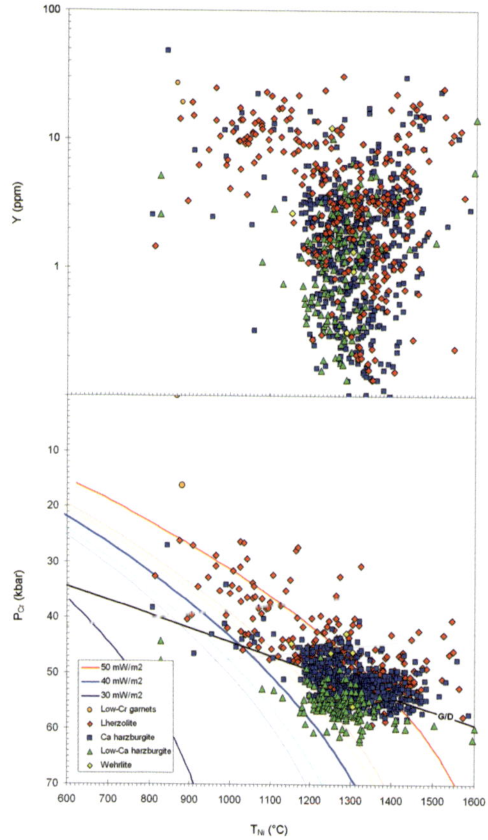

Fig. 11. Garnet pressure–temperature array for a subset of 954 grains selected from the concentrate sample. Pressures and temperatures for single xenocrysts have been calculated following the methodology of Ryan *et al.* (1996). The yttrium–temperature plot highlights the strongly depleted nature of the garnets even at very high temperatures approaching the mantle adiabat.

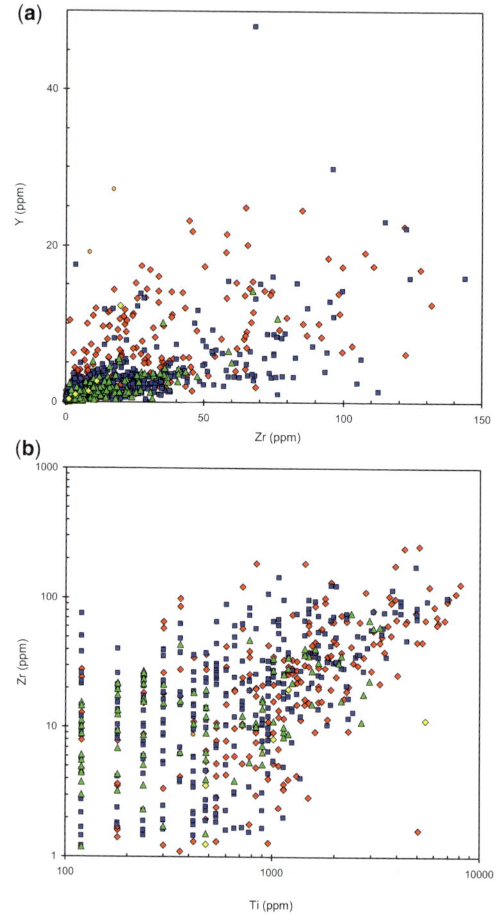

Fig 12. (**a**) Yttrium–zirconium data ($n = 954$). The concentration of data in the bottom left-hand corner of the plot confirms the depleted nature of the bulk garnets analysed, with little evidence of pervasive high-temperature melt-related metasomatism. (**b**) Zirconium–titanium data for the same garnets shown in Figure 10a.

that the portion of the subcontinental lithospheric mantle (SCLM) sampled by the primary source(s)/vector(s) to the KIMs in the Banganapalle Formation sediments is characterized by deep-seated depleted harzburgite at pressures in excess of 60 kbar (Fig. 11). Median yttrium values remain relatively static with depth and to the deepest levels within the SCLM section (profile not shown), with garnet yttrium concentrations rarely rising above 10 ppm. These features of the Banganapalle dataset are clearly not evident in the concentrate samples collected from the kimberlite clusters to the SW. An examination of other trace elements, namely zirconium and titanium (Fig. 12), confirms that the vast majority of the garnets analysed for their trace element contents, and especially the higher-temperature ones, are also depleted with respect to zirconium and titanium (Fig. 12b). Zirconium-rich (i.e. >100 ppm) garnets are rare, with the bulk of the harzburgitic garnets recording values below 40 ppm (Fig. 12a). Thus, evidence for pervasive heating and or modification/refertilization of the deepest portions of the SCLM in the source region is extremely limited (an argument that may likely be used to account for the predominance of high-temperature garnets in the present dataset): for example, the variation in the calculated median X_{Mg} of olivine co-existing with garnet (after Gaul *et al.* 2000) with depth (not shown) is minimal, with median values only dropping as low as 0.92.

Although the ultimate source of the KIMs that were incorporated into the Banganapalle Formation sediments remains unresolved, the present analysis of the garnet major and trace element chemistry confirms that the known kimberlites to the SW, as examined by Griffin *et al.* (2009), do not represent a plausible source. The predominance of high-temperature, depleted garnets is indicative of a deep-seated (+60 kbar) source of depleted harzburgite that is not evidenced in the any of the concentrate samples collected from kimberlites to the SW.

Discussion and conclusion

The WNW–ESE-trending lamproite dykes around Banganapalle Hill, as well as the Chelima, the Krishna and, possibly, the Zangamarajupalle dykes, parallel the WNW–ESE faults that dissect the area. The dykes occur as segmented dyke–fracture arrays with anastomosing (branching and bifurcating) and en echelon dyke segments that can be followed as discontinuities or broken linear features striking WNW–ESE on satellite imagery. The dykes cut through the Tadpatri Shales and the Kuppalapalle Volcanics. The diamondiferous Banganapalle conglomerates were deposited in a shallow-marine tidal setting.

The lamproite dykes are extremely carbonatized and silicified, and as a result are very hard compared to unaltered lamproites. The crushing of large rock samples from 551/110/0004 (close to 1000 kg) released very few indicator minerals and only tens of garnets. Even though chemically these garnets form a subset of the garnet chemistry described above, the data set is too small to make any confident assessment. Destruction of the garnets and other indicators during carbonatization and silicification of the lamproite may account for the poor grain recoveries recorded to date. In addition, the crushed lamproite sample did not yield any datable minerals, including zircon.

Based on the difference in mantle characteristics derived from the garnet xenocrysts extracted from the Banganapalle Formation conglomerate horizon and those from the kimberlites to the SW, the authors believe that the provenance for the Banganapalle diamonds is unlikely to be the 1100 Ma old kimberlites to the SW. Assuming that the diamonds and other kimberlitic indicator minerals recovered from the Banganapalle conglomerate are linked in terms of provenance (i.e. the ultimate source of the diamonds was also the source of the other indicators), a mixed source for the diamonds from a combination of the Banganapalle lamproites and the kimberlites to SW also seems unlikely as the analysis of the extensive garnet xenocryst sample yields no similarities to the concentrate samples presented by Griffin *et al.* (2009) either in terms of major or trace elements. The preferential analysis of more depleted garnets (calcium and low-calcium harzburgitic components) has been highlighted as one possibility that may account for the predominance of high-temperature garnets but, in a xenocryst sample numbering in excess of 5000 individual grains, this is not likely. Secondly, the chrome–calcium systematics confirm the broad compositional range over which garnets were extracted, and the population discrimination also indicates that the more fertile lherzolitic varieties are dominant – approximately 65% (3322 grains) of the concentrate sample comprises lherzolitic garnets. Furthermore, there is no evidence of the presence of the low-temperature depleted garnets present in the Banganapalle samples, a feature that is characteristic of the kimberlites to the SW, especially the Kalyandurg bodies. Another suggestion relates to the preferential concentration or preservation of high temperatures in the surficial environment as a way to explain why the Banganapalle concentrate sample is artificially unique. The authors are aware of no studies that highlight natural processes capable of selectively preserving garnets with higher nickel and chrome contents (that in turn calculate high temperatures) and most often it is these high interest garnets that are more unstable and preferentially destroyed. It should also be noted that the regional geochronology does not provide any evidence of the 1100 Ma age for the kimberlites in the region from the zircons recovered in the area. The presence of kelyphite rims preserved on the garnets also indicates a very short transport distance from the source to the conglomerates. The predominance of high-temperature, depleted garnets is indicative of a deep-seated (+60 kbar) source of depleted harzburgite and, in the absence of any other available data, appears to be a feature unique to the Banganapalle region.

Exploration by De Beers led to the discovery of a number of lamproite dykes in the area, but not to any lamproite pipes. This may be related to the style of

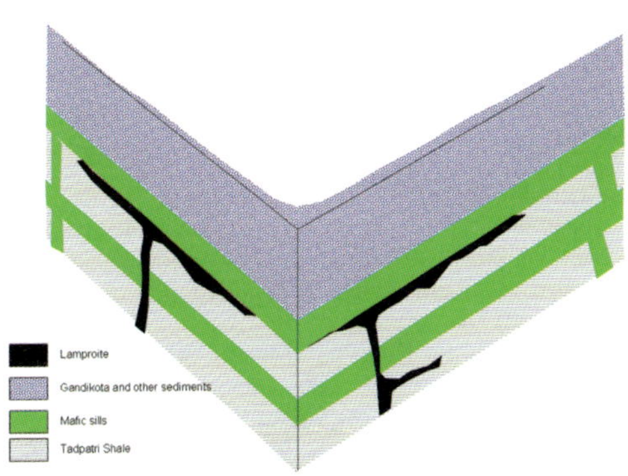

Fig. 13. Cartoon demonstrating the dyke–sill model for the Banganapalle lamproite.

emplacement as well as to the erosion level. The Cuddapah stratigraphy of the Banganapalle area is characterized by widespread dolerite sills that were emplaced within the Tadpatri shale succession at different stratigraphic levels. Along similar lines, we propose that emplacement of the lamproites occurred as dykes and sills intruding as dykes along faults and at upper-crustal levels as sills following the flat-lying contacts of the heterogeneous shale–dolerite stratigraphy (Fig. 13). This is similar to the styles of emplacement observed in other kimberlite–lamproite clusters, such as the Benfontein sill complex in Kimberley, South Africa where the sill is exposed as multiple layers within a stratified dolerite and shale succession (Dawson & Hawthorne 1973). Although, as part of this work, we have not managed to recover any chronometers from the lamproite dykes themselves, we believe the 1300 Ma age obtained from the zircons in exploration samples to indicate the most probable emplacement age of the Banganapalle lamproites.

The lower age constraint marking the beginning of Kurnool sedimentation has to be revised by considering the 1.3 Ga lamproite as the diamond provenance for the Banganapalle conglomerates. This indicates a potentially prolonged depositional episode of the Kurnool Group for about 800 Ma. It should also be noted here that Chalapathi Rao *et al.* (2010*a*, *b*) has proposed a Mesoproterozoic age for Kurnools based on xenoliths in Sidanapalli kimberlite.

The collection of more samples from all of the lamproites trying to recover datable minerals and KIMs is planned for the future to constrain some of the uncertainties discussed in this paper.

We thank Dr D. Saha for inviting us to contribute this research paper in this special publication. This research paper has built on the outputs of the exploration team of De Beers India and we would like to acknowledge this team for their contribution. We thank the management of De Beers Exploration for the permission to publish the data. A. F. Kobussen and W. Griffin provided very insightful comments to the indicator mineral section.

References

Anand, M., Gibson, S. A., Subbarao, K. V., Kelley, S. P. & Dickin, A. P. 2003. Early Proterozoic melt generation processes beneath the intracratonic Cuddapah basin, southern India. *Journal of Petrology*, **44**, 2139–2171.

Appavadhanulu, K. 1962. Minette and riebeckite–kalisyenite dykes in some Upper Cuddapah rocks, Kurnool district. *Indian Minerals*, **16**, 310–311.

Appavadhanulu, K. 1966. Minette and riebeckite–kalisyenite dykes in some Upper Cuddapah rocks, Kurnool district, Andhra Pradesh. *Records of the Geological Survey of India*, **94**, 303–305.

Balasubrhamanyan, M. N. 2006. *Geology and Tectonics of India: An Overview*. International Association for Gondwana Research, Memoir, **9**.

Bhaskar Rao, B. 1976. A note on the micaceous kimberlite in the Cumbum Formation near Zangamarajupalle, Cuddapah district, Andhra Pradesh. *Indian Minerals*, **30**, 55–58.

Biswal, T. K., Biswal, B., Mitra, S. & Moulik, M. R. 2002. Deformation pattern of the NW Terrane Boundary of the Eastern Ghats Mobile Belt, India: a tectonic model and correlation with Antarctica. *Gondwana Research*, **5**, 45–52.

Bluck, B. J., Ward, J. D. & De Wit, M. C. J. 2005. Diamond mega-placers: southern Africa and Kaapvaal craton in global context. *In*: McDonald, I., Boyce, A. J., Butler, I. B., Herrington, R. J. & Polya, D. A. (eds) *Mineral Deposits and Earth Evolution*. Geological Society, London, Special Publications, **248**, 213–245.

Chalapathi Rao, N. V. 2007. Chelima dykes, Cuddapah Basin, southern India: a review of the age, petrology, geochemistry and petrogenesis of world's oldest lamproites. *Journal of the Geological Society of India*, **69**, 523–538.

Chalapathi Rao, N. V. & Madhavan, V. 1996. Some observations on the geochemistry of Ramannapeta–Ustapalle lamproitic body, Krishna district, Andhra Pradesh. *Journal of the Geological Society of India*, **47**, 409–418.

Chalapathi Rao, N. V., Anand, M., Dongre, A. & Osborne, I. 2010*a*. Carbonate xenoliths hosted by the Mesoproterozoic Siddanpalli kimberlite cluster (Eastern Dharwar craton): implications for the geodynamic evolution of southern India and its diamond and uranium metallogenesis. *International Journal of Earth Sciences (Geologische Rundshau)*, **99**, 1791–1804, doi: 10.1007/s00531-009-0484-7.

Chalapathi Rao, N. V., Gibson, S. A., Pyle, D. M. & Dickin, A. P. 2004. Petrogenesis of Proterozoic kimberlites and lamproites from the Cuddapah basin and the Dharwar Craton, Southern India. *Journal of Petrology*, **45**, 907–947.

Chalapathi Rao, N. V., Kamde, G., Kale, H. S. & Dongre, A. 2010*b*. Petrogenesis of the Mesoproterozoic lamproites from the Krishna Valley, Eastern Dharwar Craton, Southern India. *Precambrian Research*, **177**, 103–130, doi:10.1016/j.precamres.2009.11.006.

Chalapathi Rao, N. V., Miller, J. A., Gibson, S. A., Pyle, D. M. & Madhavan, V. 1999. Precise 40Ar/39Ar dating of Kotakonda kimberlite and Chelima lamproite, India: implication to the timing of mafic dyke swarm activity in the Eastern Dhawar craton. *Journal of the Geological Society of India*, **53**, 425–432.

Chalapathi Rao, N. V., Miller, J. A., Pyle, D. M. & Madhavan, V. 1996. New Proterozoic K–Ar age for some kimberlites and lamproites from Cuddapah Basin and Dharwar Craton, South India: evidence for non contemporaneous emplacement. *Precambrian Research*, **79**, 363–369.

Chaudhuri, A. K., Saha, D., Deb, G. K., Deb, S. P., Mukherjee, M. K. & Ghosh, G. 2002. The Purana basins of southern cratonic province of India – a case for Mesoproterozoic fossil rifts. *Gondwana Research*, **5**, 23–33.

Clement, C. R. 1982. *A comparative geological study of some Major Kimberlite pipes in Northern Cape and Orange Free State*. Unpublished PhD thesis, University of Cape Town.

Chatterjee, N. & Bhattacharji, S. 2001. Petrology, geochemistry and tectonic settings of the mafic dykes and sills associated with the evolution of the proterozoic Cuddapah basin of South India. *Proceedings of the Indian Academy of Science (Earth and Planetary Sciences)*, **110**, 433–453.

Dawson, J. B. & Hawthorne, J. B. 1973. Magmatic sedimentation and carbonatitic differentiation in kimberlite sills at Benfontein, South Africa. *Journal of the Geological Society, London*, **129**, 61–85.

Gaul, O. F., Griffin, W. L., O'Reilly, S. Y. & Pearson, N. J. 2000. Mapping olivine composition in the lithospheric mantle. *Earth and Planetary Science Letters*, **182**, 223–235.

Grady, J. C. 1971. Deep main faults in south India. *Journal of the Geological Society of India*, **12**, 56–62.

Griffin, W. L., Fisher, N. I., Friedman, J. H., O'Reilly, S. Y. & Ryan, C. G. 2002. Cr-pyrope garnets in the lithospheric mantle II. Compositional populations and their distribution in time and space. *Geochemistry, Geophysics and Geosystems*, **3**, 1073.

Griffin, W. L., Fisher, N. I., Friedman, J. H., Ryan, C. G. & O'Reilly, S. Y. 1999. Cr-pyrope garnets in the lithospheric mantle I. Compositional systematics and relations to tectonic setting. *Journal of Petrology*, **40**, 679–704.

Griffin, W. L., Kobussen, A. F., Babu, E. V. S. S. K., O'Reilly, S. Y., Norris, R. & Sengupta, P. 2009. A translithospheric suture in the vanished 1-Ga lithospheric root of South India: evidence from contrasting lithosphere sections in the Dharwar Craton. *Lithos*, **112S**, 1109–1119, doi: 10.1016/j.lithos.2009.05.015.

Holland, T. H. 1907. Indian Empire. Geology. *Imperial Gazetteer of India*, **1**, 50–103.

King, W. 1872. Kudapah and Karnul Formations in the Madras Presidency. *Geological Survey of India, Memoir*, **8**, 346 .

Kumar, A., Gopalan, K., Rao, K. R. P. & Nayak, S. S. 2001. Rb–Sr ages of kimberlites and lamproites from Eastern Dharwar craton, south India. *Journal of the Geological Society of India*, **58**, 135–142.

Lakshminarayana, G., Bhattacharjee, S. & Kumar, A. 1999. Palaeocurrents and depositional setiing in the Banganapalle Formation, Kurnool sub-basin, Cuddapah basin, Andhra Pradesh. *Journal of the Geological Society of India*, **53**, 255–260.

Ludwig, K. R. 2003. *Isoplot/EX (version 3.00), A Geochronological Toolkit for Microsoft Excel*. Berkeley Geochronological Centre, Special Publications, **4**.

Meijerink, A. M. J., Rao, D. P. & Rupke, J. 1984. Stratigraphic and structural development of the Precambrian Cuddapah Basin, SE India. *Precambrian Research*, **26**, 57–104.

Mezger, K. & Cosca, M. 1999. The thermal history of the Eastern Ghats Belt (India), as revealed by U–Pb and $^{40}Ar/^{39}Ar$ dating of metamorphic and magmatic minerals: implications for the SWEAT correlation. *Precambrian Research*, **94**, 251–271.

Mitchell, R. H. & Bergman, S. C. 1991. *Petrology of Lamproites*. Plenum Press, New York.

Nagaraja Rao, B. K., Rajurkar, S. T., Ramalingaswamy, G. & Ravindra Babu, B. 1987. Stratigraphy, structure and evolution of the Cuddapah Basin. *Geological Society of India, Memoir*, **6**, 33–86.

Pollack, H. N. & Chapman, D. S. 1977. On the regional variation of heat flow, geotherm and lithospheric thickness. *Tectonophysics*, **38**, 279–296.

Pouchou, J. L. & Pichoir, F. 1984. A new model for quantitative X-ray micro-analysis. I. Application to the analysis of homogeneous samples. *La Recherche Aérospatiale*, **3**, 13–36.

Ramakrishnan, M. & Vaidyanadhan, R. 2008. *Geology of India*, Volume 1. Geological Society of India, Bangalore, 492–509.

Ryan, C. G., Griffin, W. L. & Pearson, N. J. 1996. Garnet geotherms: pressure–temperature data from Cr-pyrope garnet xenocrysts in volcanic rocks. *Journal for Geophysical Research*, **101**, 5611–5625.

Saha, D., Chakraborti, S. & Tripathy, V. 2010. Intracontinental thrusts and inclined transpression along eastern margin of the East Dharwar Craton, India. *Journal of the Geological Society of India*, **75**, 323–337.

Satyanarayana, S. V. 1997. Diamonds in Deccan – an overview. *In*: Gupta, H. K., Sen, A. P. & Balasubramanyam, D. (eds) *Deccan Heritage*. University Press (India) Ltd, 135–156.

Scott-Smith, B. H. 1989. Lamproites and kimberlites in India. *Neues Jahrbuch für Mineralogie – Abhandlungen*, **161**, 193–225.

Sen, S. N. & Narasimha Rao, C. H. 1967. Igneous activity in Cuddapah basin, adjacent areas and suggestions in the palaeogeography of the basin. *In*: *Proceedings of the Symposium on the Upper Mantle Project*. National Geophysical Research Institute, Hyderabad, 261–285.

Sen, S. N. & Narasimha Rao, C. H. 1970. Chelima dykes. *In*: *Proceedings of the II Symposium on the Upper Mantle Project Session 5: Combined Geosurveys in Dharwar and Cuddapah Basin*. National Geophysical Research Institute, Hyderabad, India. 435–439.

Sen, S. N. & Narasimha Rao, C. H. 1971. Chelima dykes – a source for diamonds in Kurnool district, A.P. *Geological Survey of India Miscellaneous Publications*, **19**, 92–94.

Simmat, R. 2003. *Identifizierung hochgradig metamorpher Provinzen des Eastern Ghats Belt in Indien anhand einer EMS-Studie von Monazit-Altersmustern*. Unpublished PhD thesis, University of Bonn.

Sivaji, K. & Rao, K. R. P. 1989. Sedimentological studies of the Banganapalle conglomerates in connection with the assessment of diamond resources. *Proceedings of the Geological Survey of India*, **122**, 51–54.

Srikantia, S. V. 1984. Kuppalapalle volcanics – a distinct Upper Papaghni volcanic activity in the Cuddapah Basin. *Journal of the Geological Society of India*, **25**, 775–779.

Upadhyay, D. 2008. Alkaline magmatism along the southeastern margin of the Indian Shield: implications for regional geodynamics and constraints on craton-Eastern Ghats Belt suturing. *Precambrian Research*, **162**, 59–69.

UPADHYAY, D. & RAITH, M. M. 2006. Intrusion age, geochemistry and metamorphic conditions of a quartz-monzosyenite intrusion at the craton-Eastern Ghats Belt contact near Jojuru, India. *Gondwana Research*, **10**, 267–276.

VENKATAKRISHNAN, R. & DOTTIWALLA, F. E. 1987. The Cuddapah Salient: a tectonic model for the Cuddapah basin, India based on LANDSAT image interpretation. *Tectonophysics*, **136**, 237–253.

Reactivation of basement: example from the Anasagar Granite Gneiss Complex, Rajasthan, western India

N. CHATTOPADHYAY[1]*, D. MUKHOPADHYAY[2] & P. SENGUPTA[1]

[1]*Department of Geological Sciences, Jadavpur University, Kolkata 700032, India*

[2]*Department of Geology, University of Calcutta, 35 Ballygunge Circular Road, Kolkata 700019, India*

**Corresponding author (e-mail: nandinidsg@gmail.com)*

Abstract: Amidst the Meso- to Neoproterozoic South Delhi Fold Belt (SDFB) of Rajasthan, India, a sheet-like body of megacrystic Anasagar Granite Gneiss (AGG) embedded in a supracrustal unit consisting of metapelites, quartzite and calc gneiss is exposed. Detailed analyses of mesoscopic and microscopic structures identify four phases of deformation. Lithological relationships coupled with U–Pb dates of zircon indicate that the protolith of the AGG was emplaced within the supracrustal unit during D_1 folding at approximately 1.85 Ga. This event is significantly older than the age of volcanism in the SDFB (0.99 Ga) but probably synchronous with Aravalli Orogeny. Thrusts associated with the easterly-vergent D_2 folds have a ramp–flat geometry and are refolded by coaxial D_3 folds. The petrology of the metapelites indicates that porphyroblasts of staurolite and/or garnet were formed as a function of bulk-rock composition between D_1 and D_3 folding, at the time of the culmination of metamorphism (5.7 ± 1.5 kbar, 560 ± 50 °C). Combining the petrological and structural attributes, it is proposed that the AGG and its enveloping supracrustals might represent the basement of the Delhi Supergroup, which was folded, thrusted and domed up during the South Delhi Orogeny. The cause of the thermal perturbation that triggered the growth of the porphyroblasts in the metapelites is explored.

It is now well known that rocks of the fold belts that girdle the cratons hold the key to understanding the birth and demise of the ancient supercontinents (reviewed in Moores 1991; Condie 1997, 2002; Rogers 2000; Zhao *et al.* 2002, 2004; Reddy & Evans 2009; Rogers & Santosh 2009). A large volume of published literature has documented that mobile belts evolve in cycles that begins with basin formation and sedimentation that eventually undergo inversion to form fold belts showing spatially and temporally varying metamorphic style and deformation patterns (Bott 1971; Condie 1997, 2002; Mahato *et al.* 2008). Rocks of the mobile belts commonly record multiple deformation and metamorphic episodes that may be the end product of single or multiple orogenic pulses (Jamieson *et al.* 1992; Goscombe & Hand 2000; Regnier *et al.* 2003; Mahato *et al.* 2008). Understanding the deformation style and metamorphic signatures present in the metamorphic rocks of the fold belt is, therefore, crucial to decode the response of these rocks during superimposed orogenic processes.

The NE–SW-trending Aravalli–Delhi Fold Belt (ADFB) in the Aravalli mountain chain of Rajasthan, western India contains two principal fold belts – the Palaeoproterozoic Aravalli Fold Belt in the east and the Meso- to Neoproterozoic Delhi Fold Belt in the west – that are separated by an unit called the Banded Gneissic Complex (BGC), consisting of amphibolite-facies migmatitic gneisses with slivers of granulite-facies rocks (Fig. 1a). The BGC, which shows approximately 1.7 Ga high-grade metamorphism and felsic magmatism (Buick *et al.* 2006; Bhowmik *et al.* 2010), is unconformably overlain by the supracrustal rocks of the Delhi Supergroup (reviewed in Roy & Jakhar 2002). On the basis of the age of intrusive granites, the Delhi Fold Belt was divided into North Delhi Fold Belt (NDFB) and South Delhi Fold Belt (SDFB) (reviewed in Deb *et al.* 2001; Roy & Jakhar 2002). Studies made by a number of workers in different parts of the main Delhi Basin suggests a common pattern of superposed folding involving three main generations of folding (Mukhopadhyay & Dasgupta 1978; Roday 1979; Naha *et al.* 1984, 1988; Naha & Mohanty 1990; Gupta *et al.* 1991).

In the northern part of the SDFB, an oval-shaped body composed of a megacrystic granite gneiss (the Anasagar Granite Gneiss, AGG) is enveloped by a supracrustal unit consisting of metapelites, quartzite and calc gneiss (reviewed in Mukhopadhyay *et al.* 2010). Owing to quaternary cover rocks. the contact between the AGG and its enclosing rocks with the rocks of Delhi fold belts are not visible. The U–Pb date of zircon precisely fixes the time of emplacement of the protolith of the AGG at

From: Mazumder, R. & Saha, D. (eds) 2012. *Palaeoproterozoic of India*. Geological Society, London, Special Publications, **365**, 219–245. http://dx.doi.org/10.1144/SP365.12

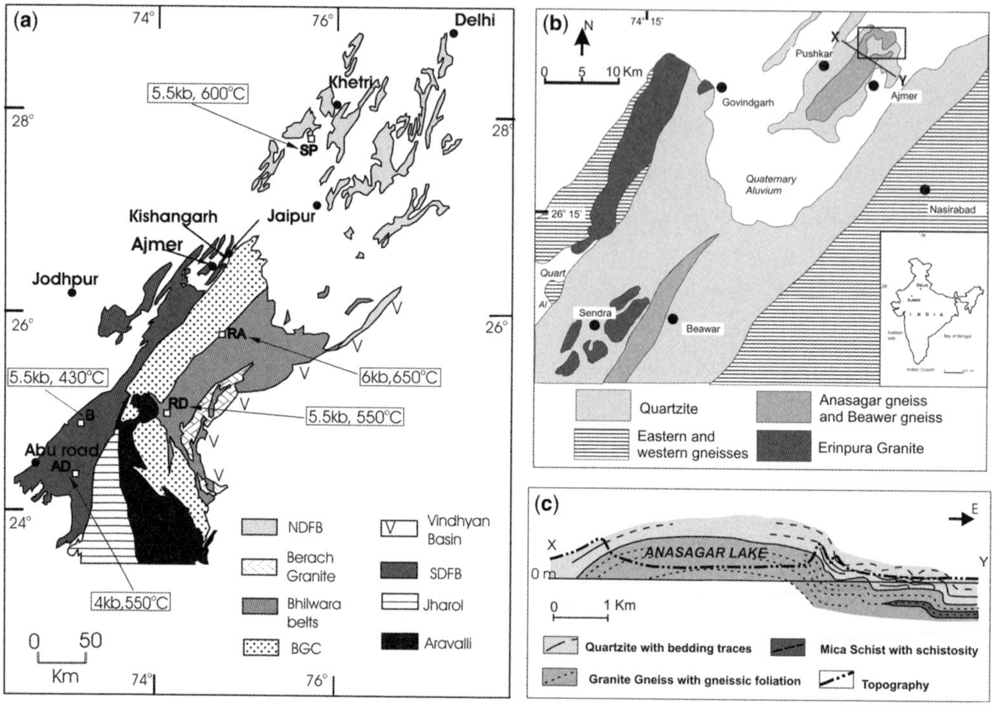

Fig. 1. (**a**) Simplified regional geological map of the Aravalli–Delhi orogenic belt modified from Heron (1953) and Gupta *et al.* (1980), with reported values of *P–T* from different parts of the Delhi Fold Belt. Abbreviations used are: RA, Rampura–Agucha; RD, Rajpura–Dariba; SP, Saladipura; AD, Ambaji Deri; B, Basantgarh. (**b**) Generalized geological map of the northern part of the South Delhi Fold Belt, adapted from Tobisch *et al.* (1994) after minor modifications. The inset shows the location of the study area marked in a rectangular box. (**c**) Schematic section along X–Y marked in (b), across the AGG and its envelope, showing the truncation of the vertical bedding at the horizontal contact surface between the AGG and the enveloping supracrustals.

approximately 1.85 Ga (Mukhopadhyay *et al.* 2000), which is significantly older than the age of volcanism of the rocks of the SDFB (*c.* 1 Ga: Deb *et al.* 2001; Pandit *et al.* 2003; Singh *et al.* 2010). However, the intrusion age of the AGG can be related to the approximately 1.82 Ga intrusion age of the Jasrapur granitoid pluton and the 1.83 Ga age of felsic tuff near the Khetri area of North Delhi Fold Belt (Kaur *et al.* 2009) and the 1.85 Ga age of Hindoli felsic volcanic rocks of SDFB (Deb *et al.* 2002). These age data, therefore, point to a protracted crustal history of rocks exposed in the Delhi Fold Belt. A combined petrological and structural analysis of the AGG and the enclosing supracrustal rocks is expected to provide valuable insight about the Proterozoic crustal evolution within the Delhi Fold Belt and to test the possibility that the AGG and the enclosing rocks represent reworked basement of the SDFB.

In this backdrop, we present detailed structural analysis and metamorphic characterization of a suite of metapelites within the supracrustal envelope of the AGG at the Anasagar Lake area of Ajmer, Rajasthan. Integrating the deformation and metamorphic attributes, it is proposed that the AGG and the enclosing supracrustal rocks possibly represent basement of the rocks of the SDFB, which was again deformed and metamorphosed during the Meso- to Neoproterozoic Delhi Orogeny. This study also explores the source of energy required for metamorphism and demonstrates the control of bulk composition in determining the metamorphic assemblage under a set of physical conditions.

Background geology

The Precambrian fold belt of Rajasthan in NW India, commonly known as the Aravalli–Delhi Fold Belt (ADFB), consists of a suite of gneissic rocks containing enclaves of diverse nature generally known as the Banded Gneissic Complex (BGC), and linear belts of supracrustal rocks belonging to the Aravalli and Delhi supergroups of Palaeo- to Neoproterozoic age (Fig. 1a) (Heron 1953; Roy 1988; Sinha Roy *et al.* 1998; Deb *et al.* 2001; Roy & Jakhar 2002).

The BGC in central Rajasthan that is chiefly composed of amphibolite-facies migmatitic gneisses and schists (Mangalwar Complex, MC) with slivers of medium- to high-pressure granulites and metamorphosed orthopyroxene-bearing granitoids (Sandmata Complex, SMC), is bounded by two metasedimentary sequences: the Aravalli Supergroup to the east and the Delhi Supergroup to the west, with Palaeo- to Mesoproterozoic and Meso- to Neoproterozoic depositional ages, respectively (Buick *et al.* 2006; Saha *et al.* 2008; Bhowmik *et al.* 2010). The Delhi Fold Belt forms the axial zone of the Aravalli mountain chain in Rajasthan (Fig. 1a, b). Extensive field mapping and structural analyses, coupled with limited petrological and geochronological studies by a number of workers, have revealed following information about the BGC and the Aravalli–Delhi fold belts.

- In the earlier literature, the migmatitic gneiss, schists and granulites of the BGC in central Rajasthan was considered to be of Archaean age, on which the supracrustal rocks of the Delhi Supergroup was deposited with a prominent unconformity surface in-between (reviewed in Roy & Jakhar 2002). Refuting the Archaean ancestry of the BGC, the recent geochronological and petrological studies identified the following geological events in the rocks of BGC. (a) Approximately 1.7 Ga old medium-pressure (sillimanite-facies) metamorphism of granulites and accompanying charno-enderbite magmatism in SMC. Protoliths of the migmatitic gneiss of the MC also formed during this time (Buick *et al.* 2006; Bhowmik *et al.* 2009, 2010). Detrital components of metapelitic–psammopelitic rocks of the MC and SMC were deposited after approximately 1.8 Ga (Buick *et al.* 2006). (b) High-pressure amphibolite- to granulite-facies migmatization in the MC and limited reworking of the Palaeoproterozoic granulite-facies rocks of the SMC occurred in the time span of approximately 0.98–0.87 Ga (reviewed in Bhowmik *et al.* 2010). Bhowmik *et al.* (2010) considered this younger high-pressure metamorphism to occur in a collisional setting.
- Quartzite, marble, calc-silicate and metapelites with mafic–ultramafic intrusive and extrusive rocks (now metamorphosed) are the major lithologies of the Aravalli Supergroup. Phosphorite and extensive lead–zinc deposits are the two characteristic lithotypes of the Aravalli Supergroup that distinguish it from the lithologies of adjacent Delhi Fold Belt (reviewed in Deb & Sarkar 1990; Roy & Jakhar 2002). Aravalli Supergroup rocks in the Udaipur region are generally of low- to medium-grade (greenschist- to lower-amphibolite-facies) metamorphism. High-grade supracrustal rocks near Bhilwara belong to the Aravalli Supergroup according to Roy & Jakhar (2002), and to the Archaean Bhilwara Supergroup according to Gupta *et al.* (1980). Existing geochronological data suggest that the Aravalli sedimentation and orogeny took place during Palaeo- to Mesoproterozoic times (Roy & Jakhar 2002). There is considerable variation in structural style in different parts of the belt of the Aravalli Supergroup (summarized in Roy & Jakhar 2002). In the type area near Udaipur, the complex pattern results from three major phases of folding (Roy & Jakhar 2002).
- Kaur *et al.* (2009) documented extensive arc magmatism in a thickened crust at approximately 1.82 Ga in the North Delhi Fold Belt that is distinctly older than the felsic magmatism and high-grade metamorphism in the adjoining BGC. The 'hourglass'-shaped Delhi Fold Belt is divided into an older northern segment (North Delhi Fold Belt, NDFB) with intrusive granitoids of approximately 1.7–1.5 Ga age and a younger southern segment (South Delhi Fold Belt, SDFB) with intrusive granitoids of 0.9–0.8 Ga age (Choudhary *et al.* 1984; Sinha Roy *et al.* 1988; Deb *et al.* 2001; Roy & Jakhar 2002). Geologically, the SDFB is dominated by metamorphosed shelf sediments (sandstone–shale–limestone), acid and basic flows and tuffs, an ophiolitic suite along the western margin (chrome-bearing serpentinites-gabbro, amphibolite and chert; the Phulad ophiolite) and intrusive granitoids (Deb *et al.* 1989, 2001; Gupta *et al.* 1997). In the southern part of the SDFB, the rocks are divided into the Gogunda Group consisting of arenaceous rocks (equivalent to the Alwar Group of the NDFB) and the Kumbhalgarh Group that is dominated by calcareous rocks with minor argillaceous rocks (equivalent to the Ajabgarh Group of the NDFB: Gupta *et al.* 1997). In the northern part of the SDFB, the rocks are classified into the Basantagarh Group, the Barotiya Group and the Sendra Group in the western sub-basin, and into the Devgarh Group, the Rajgarh Group and the Bhim Group in the eastern sub-basin (Gupta *et al.* 1995). Geochemical characteristics of metabasalts (amphibolites) around Phulad have characteristics similar to modern mid-ocean ridge basalts, whereas metabasalts of the Ranakpur (part of the Kumbhalgarh Group) bear the geochemical signals of the oceanic arc (Sugden & Windley 1984; Deb & Sarkar 1990; Deb *et al.* 2001). Synthesizing the available structural information, Mukhopadhyay & Bhattacharya (2000) concluded that the SDFB is separated into two tectonic belts by a thrust wedge of the basement, as suggested by Heron (1953). The dominant

fold sets throughout the SDFB are north- to NE-striking axial planes; the axial plunges being gentle in the eastern segment and more variable, often steep, in the western segment. The earliest isoclinal folds (D_1) are identified in the supracrustal envelope of the thrusted gneiss wedge near Anasagar. The dominant fold set (D_2) in the Delhi supracrustal belt has a consistent eastern sense of vergence and is refolded coaxially by D_3 with subvertical axial planes. Evidence of dextral simple shear is present throughout the SDFB and particularly in the two movement zones along its two boundaries. In contrast to low-grade metamorphism in most parts of the Aravalli Fold Belt, the petrological studies from a few scattered areas show that regional metamorphism of Delhi Fold Belt culminated in the pressure–temperature (P–T) range of 5 ± 1 kbar, 600 ± 50 °C, which was followed by thermal metamorphism (2.5–4 kbar, 550–600 °C) under static conditions in a few places (Sharma 1988; Mukherjee *et al.* 1992; Deb & Sehgal 1997; Sarkar 2000; Pant *et al.* 2008). Emplacement of approximately 967 Ga granite during deformation in the Sendra area gives the timing of the amphibolite-facies metamorphism in the SDFB (Pandit *et al.* 2003). Extant geochronological data support that sedimentation and volcanism in parts of the South Delhi Fold Belt occurred at 1.00 Ga (Choudhary *et al.* 1984; Volpe & Macdougall 1990; Mukhopadhyay *et al.* 2000; Deb *et al.* 2001). Singh *et al.* (2010) documented 0.86–0.80 Ga high-grade metamorphism followed by exhumation of the high-grade rocks at 0.80–0.76 Ga from the Balarampur–Kui–Surpagla–Kengora area lying in the SW part of the SDFB. The findings of Singh *et al.* (2010) that provide the younger metamorphic event in the SDFB is in good agreement with the geochronological study of Tobisch *et al.* (1994) and Volpe & Macdougall (1990). It is thus evident from the existing studies that rocks of the SDFB are polydeformed and polymetamorphosed.

Rock association

Disposition of the different rock types in the studied area is shown in Figure 2a (Mukhopadhyay *et al.* 2010). On the basis of lithological attributes, two major units – namely, the supracrustal unit and the granite gneiss unit (Anasagar Granite Gneiss, hereafter referred to as the AGG) – are recognized. Quartzite, the dominant rock type of the area, along with metasedimentary rocks (metapelite, psammopelite and minor calc-gneisses) and amphibolite constitute the supracrustals unit that overlies the AGG (Fig. 2a, b). Intercalation of quartzite and metapelite is a common feature of the supracrustal packet (Fig. 3a). In most places, quartzite has direct contact with granite gneiss, although locally thin bands of amphibolite separate the former rock from the latter. The granite gneiss typically contains euhedral megacrysts of K-feldspar (up to 30 vol%) that are mostly arranged in parallel, in a particular direction, presumably during magmatic flow (Fig. 3b). K-feldspar megacrysts, which commonly show concentric zones separated by biotite septa, are set in fine-grained aggregates of feldspar, quartz, muscovite and biotite. A prominent foliation defined by the parallel orientation of mica flakes swerves around the K-feldspar megacrysts. In hand specimen, as well as under a microscope, most K-feldspar megacrysts consist of granular aggregates but retain a euhedral outline. This feature is interpreted as the consequence of the recrystallization of strained magmatic K-feldspars. Close to the contact with the overlying quartzite and amphibolite, the K-feldspar megacrysts have a circular to elliptical outline, which in many places grades to a 'streaky gneiss' containing extremely flattened K-feldspar grains parallel to the schistosity (Fig. 3c). The granite gneiss does not send apophyges in the overlying supracrustal unit but contain xenoliths of metapelite (Fig. 3d) and calc-gneiss that are also elongated parallel to the foliation. Parallelism of the direction of the orientation of the K-feldspar megacrysts and the foliation corroborate the view that the original magmatic flow structure was accentuated during deformation. The floor of the granite gneiss is not visible anywhere. However, at Lohagal, which lies on the northern part of the AGG, a pelitic schist horizon divides the gneiss into two sheets that merge together in the western part (Fig. 2a). All of the aforesaid features corroborate the view that the protolith of the granite gneiss was emplaced as a sheet-like body within the supracrustal unit (cf. Mukhopadhyay *et al.* 2010). Muscovite, biotite and rare chlorite define a prominent schistosity in pelitic and psammopelitic rocks in which large porphyroblasts of garnet and staurolite are set in (Fig. 3e). These centimetre-long porphyroblastic staurolite grains occur across the schistosity of the pelitic/psammopelitic schist; often they occur in a random fashion signifying a static condition of growth. Staurolite is restricted to pelitic schist but not all pelitic schists contain staurolite.

Deformational events

The supracrustal rocks and the granitic gneiss have been affected by four phases of deformation (Mukhopadhyay *et al.* 2010). The earliest

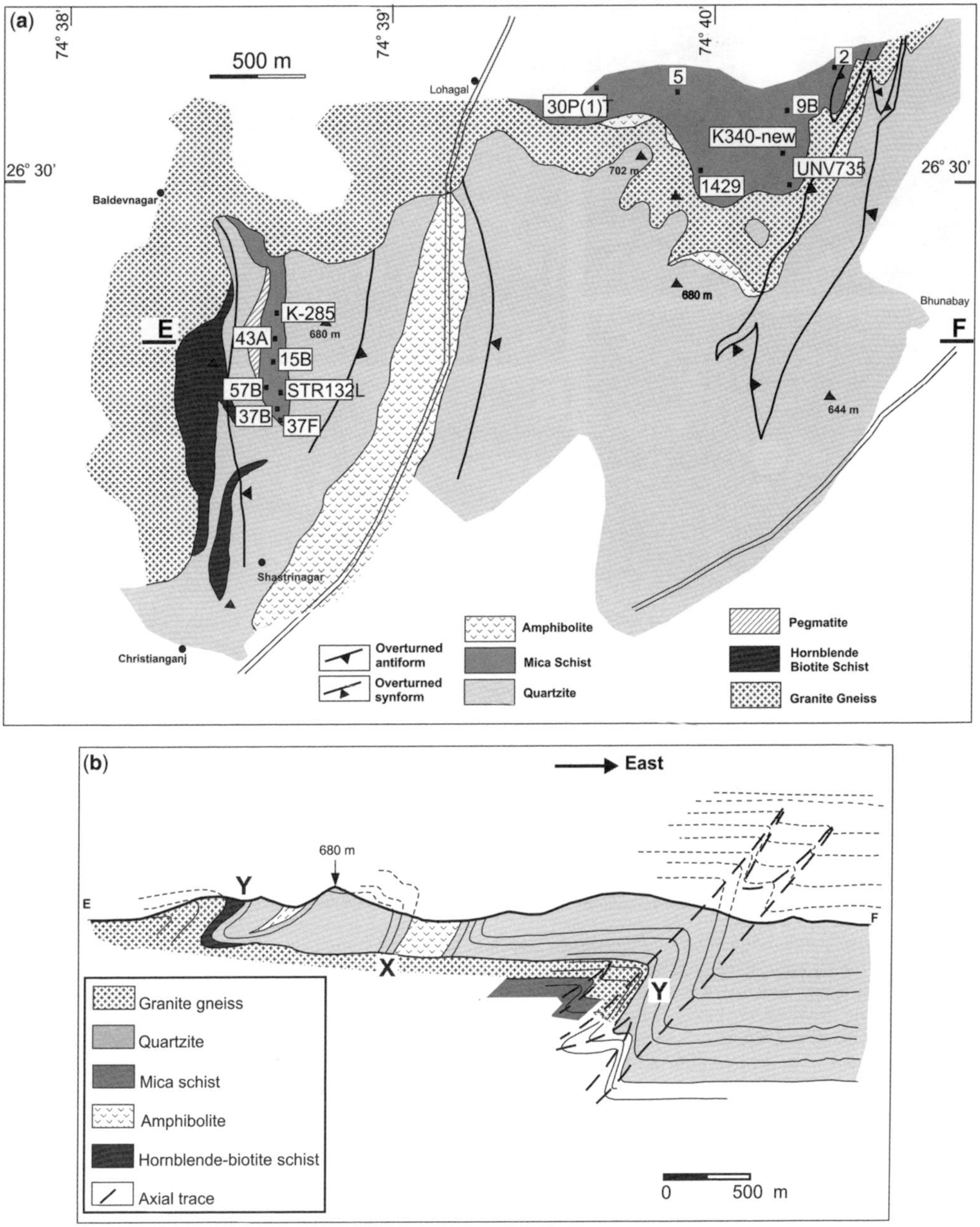

Fig. 2. (**a**) Geological map of the Shastrinagar–Lohagal area after Mukhopadhyay *et al.* (2010). E–F shows the line of section. (**b**) Cross-section across E–F in the map of the Shastrinagar–Lohagal area, after Mukhopadhyay *et al.* (2010). The depth projection is on the basis of the plunge of the fold axis. X is the truncation of vertical bedding at the contact; Y is the bedding parallel to the contact.

deformation produced the mesoscopic isoclinal folds with axial-plane schistosity (S_1) in the supracrustal rocks. The granite was intruded synkinematically with the first deformation, and the gneissic foliation is parallel to the axial-plane schistosity. The second set of folds (D_2) occur on all scales and (Fig. 4a) are asymmetric, and have alternate gentle-dipping and steep-dipping (locally overturned) limbs with subhorizontal to gentle westerly-dipping axial planes. The D_2 folds have coaxially refolded the

Fig. 3. (**a**) Intercalation of quartzite and metapelitic schistose layers in the supracrustal packet. While the quartzite layers define the D_2 folds, the schistose layers show D_2 crenulations. (**b**) Euhedral megacrysts of K-feldspar arranged parallel in a particular direction, presumably during magmatic flow. (**c**) Close to the contact with the overlying quartzite and amphibolite, the K-feldspar megacrysts are stretched, boudinaged and drawn into thin lenses, which in places grade to a 'streaky gneiss' containing extremely flattened K-feldspar grains parallel to the schistosity. (**d**) Xenoliths of fine-grained quartz mica schist within the AGG parallel to the foliation of the gneiss. (**e**) Large porphyroblasts of garnet and staurolite set in a biotite– muscovite–chlorite–quartz matrix. Note the random orientation of the staurolite porphyroblasts.

earlier isoclines and folded the S_1 schistosity giving rise to a crenulation cleavage (Fig. 4b). The gently plunging D_2 folds are generally coaxial with D_1 folds. Consistent easterly vergence of the D_2 folds indicates a simple-shear regime with a top-to-the-east sense of movement. The crenulation cleavage, has at many places, obliterated the S_1 and becomes the dominant planar fabric of the rocks. The contact between the AGG and the enveloping supracrustals rocks is sheared, as manifested by the formation of decimetre- to metre-thick mylonite. The shear zone, that converted the megacrystic AGG to a thinly banded rock (Fig. 4c), is parallel with the gneissosity of the AGG and also parallel to bedding of the horizontal limb of the overlying quartzite, but truncates the steep limb. Shear-sense indicators – namely, asymmetric tails of megacrysts (Fig. 4d), bookshelf sliding and C' shear bands – are consistent with a top-to-the-east sense of shear movement. The shear movement is presumed to be synchronous with D_2 folding because at places the shear zone truncates the steep limbs in the overlying rocks, and at other places it is folded by D_2 folds (Fig. 4c.). This is further corroborated by a top-to-the-east sense of both the vergence of the D_2 folds and the movement sense on the shear zone.

D_3 folds that are coaxial with D_2 are upright and are found on the gentle limb of D_2 folds. Rarely,

Fig. 4. (**a**) Asymmetric D_2 folds with alternate gentle- and steep-dipping (locally overturned) limbs with subhorizontal to gentle westerly-dipping axial planes in the enveloping quartzitic rocks. (**b**) Superimposition of D_2 folds over D_1 isoclinal folds in the quartzite layers of the area near Shastrinagar. (**c**) Megacrystic AGG converted to a thinly banded rock within the shear zone at the contact of the granite gneiss with the supracrustal rocks. The foliation is seen to be folded by the D_2 folds. (**d**) Asymmetric tails of megacrysts within the shear zone at the contact of the granite gneiss with the supracrustal rocks, which were used as shear-sense indicators.

crenulation cleavage develops along the axial plane of the D_3 folds. Bending of D_2 crenulation cleavage by D_3 folds is observed in thin sections. Through detailed structural analyses of the rocks exposed in and around the Anasagar Lake, Mukhopadhyay *et al.* (2010) came to the conclusion that D_2 and D_3 folds represent the early and late stages of the same episode of deformation in a simple-shear regime during which the axial planes were progressively rotated to produce the observed variation of the attitude of axial planes of the D_2 and D_3 folds. D_4 folds with transverse (east–west to WNW–ESE) axial planes bend the D_2–D_3 folds. The axis of the D_4 pucker curves around the hinge area of the D_2 fold.

Petrography of pelitic and psammopelitic rocks

In this section, mutual relationships between different minerals are discussed. Special emphasis is given to elucidate the relationship between the porphyroblastic phases and the matrix minerals. Mineralogy of the studied metapelite and psammopelite is presented in Table 1.

Mineralogically, psammopelites show a higher ratio of biotite/muscovite, have a greater amount of plagioclase and are completely devoid of staurolite compared to the adjoining metapelites. However, staurolite is also absent in a few pelitic rocks.

Both the pelites and the psammopelites show millimetre-thick layers that are alternately rich in quartz + feldspar (Q-domains) and mica (P-domains). The preferred orientation of the mica flakes defines a prominent schistosity, which, on the basis of field features, is identified as an S_1 fabric. Open to tight folding of the S_1 fabric produced D_2 or D_3 crenulation. In Q-domains, quartz and plagioclase (when present) are either flattened parallel to S_1 or show a mosaic of strain-free polygonal grains. In areas where macroscopic D_1–D_3 folds are present, the S_1 schistosity is crenulated by D_2. The axial plane of the D_2 crenulation is bent to produce the D_3 open folds. Commonly,

Table 1. *Mineral assemblages in the studied metapelitic and psammopelitic rocks*

Sample	Rock type	Bt	Ms	Qtz	Pl	Kfs	Grt	St	Chl	Tur_a*	Mag_a	Ilm_a
5	Psammo-pelite	√		√	√		√		√			√
15B	Pelite	√		√			√	√	√		√	√
30P(1)T	Pelite	√	√	√	√					√		
K285-1	Pelite	√		√			√	√	√			√
K340new	Psammo-pelite	√		√	√		√		√			√
STR132L	Pelite	√		√			√	√				√
2D	Pelite	√	√	√	√	√	√					
9B	Pelite	√		√	√		√					√
37B	Pelite	√	√	√			√	√				
1429	Pelite	√	√	√	√		√			√		√
43A	Pelite	√	√	√	√		√					
37F	Pelite	√		√	√		√					
57B	Pelite	√	√	√	√		√	√				√
UNV735	Pelite	√		√	√		√					

*Accessory minerals.

both garnet and staurolite porphyroblasts are riddled with inclusions of quartz and, rarely, plagioclase and ilmenite. Less commonly, porphyroblasts that are almost devoid of any inclusions are also present. Garnet and staurolite porphyroblasts show a wide variation in terms of size, shape, inclusion pattern and orientation with respect to the matrix foliation. Commonly, prismatic staurolite grains are sieved with quartz inclusions that define an internal fabric, which is either straight or curved (S_1, Fig. 5a, b). Staurolite porphyroblasts with an internal sigmoidal fabric of quartz inclusions is common (Fig. 5c). Garnet and staurolite porphyroblasts are seen to grow over or truncate the S_1 schistosity. (Fig. 5d, e). In places, skeletal to euhedral stumpy porphyroblasts of garnet (Fig. 5f) and staurolite (Fig. 5g) develop over the S_1 foliation, with their internal foliation being parallel to the external S_1 schistosity in some places (helicitic texture, Fig. 5h, i). D_2 crenulations deform and develop fracture in the stumpy staurolite grain (Fig. 5g). Textural features (Fig. 5a–c, h, i) indicate that growth of staurolite porphyroblasts started during but outlasted the D_2 folding. Garnet and staurolite porphyroblasts do not show reaction a relationship. However, inclusion of euhedral garnet in euhedral staurolite is noted in several places (Fig. 5j). This texture can best be interpreted as simultaneous growth between the two minerals (Vernon 2004). Chlorite was found in only a few staurolite-bearing rocks. Here, the flakes of chlorite replace biotite and occur across the S_1 fabric (Fig. 5k, l). The contacts of chlorite with both garnet and staurolite are always sharp, which is also consistent with their simultaneous growth. Integrating all of the textural features, it can be presumed that garnet and staurolite grew after D_1 but before the D_3 folding, and was probably broadly synchronous with D_2, which probably outlasted the porphyroblast formation, or earlier than D_2. The relationships between the internal and external fabrics in these minerals are consistent with their growth after the development of the S_1 foliation.

Mineral chemistry

Compositions of minerals were analysed with a CAMECA SX100 microprobe at the Geological Survey of India, Kolkata. Natural and synthetic standards were used and the raw analyses were corrected using ZAF (Zeeman anisotropy fluorescence). Beam diameters were varied from 1 to 3 μm, with an accelerating voltage of 15 kV and a 12 nA beam current. In the following section, some salient compositional features of the minerals observed in the metapelitic rocks are described.

The garnet is rich in almandine (68–85 mol%), with variable proportions of pyrope (3–17 mol%), spessertine (1–18 mol%) and grossular (1–14 mol%) (Table 2). However, no significant compositional variation was noted in individual garnet grains in a particular sample (Table 2). X_{Mg} (Mg/Mg + Fe) and grossular percentage of garnet vary sympathetically with the X_{Mg} of biotite and X_{Ca} (Ca/Ca + Na + K) of plagioclase (Fig. 5m, n). This reflects the dependence of mineral chemistry on local bulk compositions. There exists a marked compositional difference of garnet in staurolite-free (almandine, *c.* 68–80 mol%; grossular + spessertine, 9–28 mol%) and staurolite-bearing rocks (almandine, 79–85 mol%; grossular + spessertine, 4–8 mol%: Table 2). Garnet in staurolite-bearing rocks has a higher X_{Mg}

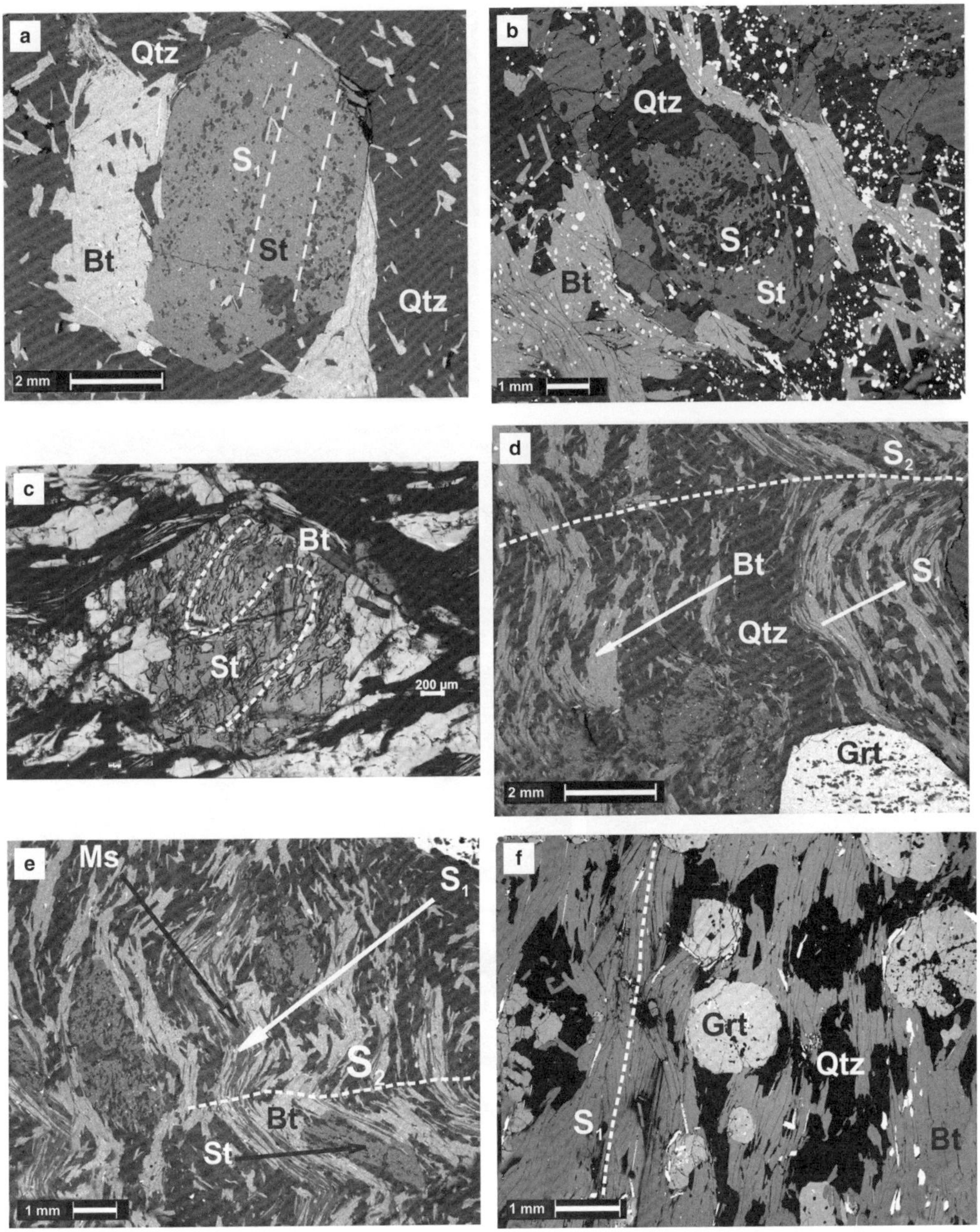

Fig. 5. (**a**) Back-scattered electron (BSE) image showing straight inclusion trails defined by quartz and opaque minerals parallel or oblique to the external S_1 schistosity in staurolite porphyroblast. (**b**) BSE image showing a curved inclusion trail within staurolite porphyroblasts. (**c**) Sigmoidal inclusion trail in staurolite porphyroblasts in metapelites in plain polarized light (PPL). (**d**) BSE image of post-D_1 garnet porphyroblasts with an earlier relict foliation defined by quartz inclusions. Garnet porphyroblasts are seen to grow over or truncate external S_1 schistosity defined by subparallely aligned muscovite and biotite flakes. S_1 schistosity is further seen to be folded by S_2 or S_3 schistosity. (**e**) BSE image of post-D_1 staurolite porphyroblast growing over or truncating the external S_1 schistosity defined by muscovite and biotite flakes. S_1 schistosity gets further crenulated to form D_2 folds, which also folds the staurolite porphyroblasts in places. (**f**) BSE image of circular, post-D_1 garnet porphyroblast with a post-tectonic relationship with the S_1 schistosity. (**g**) BSE image of euhedral stumpy porphyroblasts of post-D_1 staurolite porphyroblast developing over the S_1 foliation. (**h**) & (**i**) BSE image of skeletal porphyroblasts of staurolite developed over the S_1 foliation, with their internal foliation

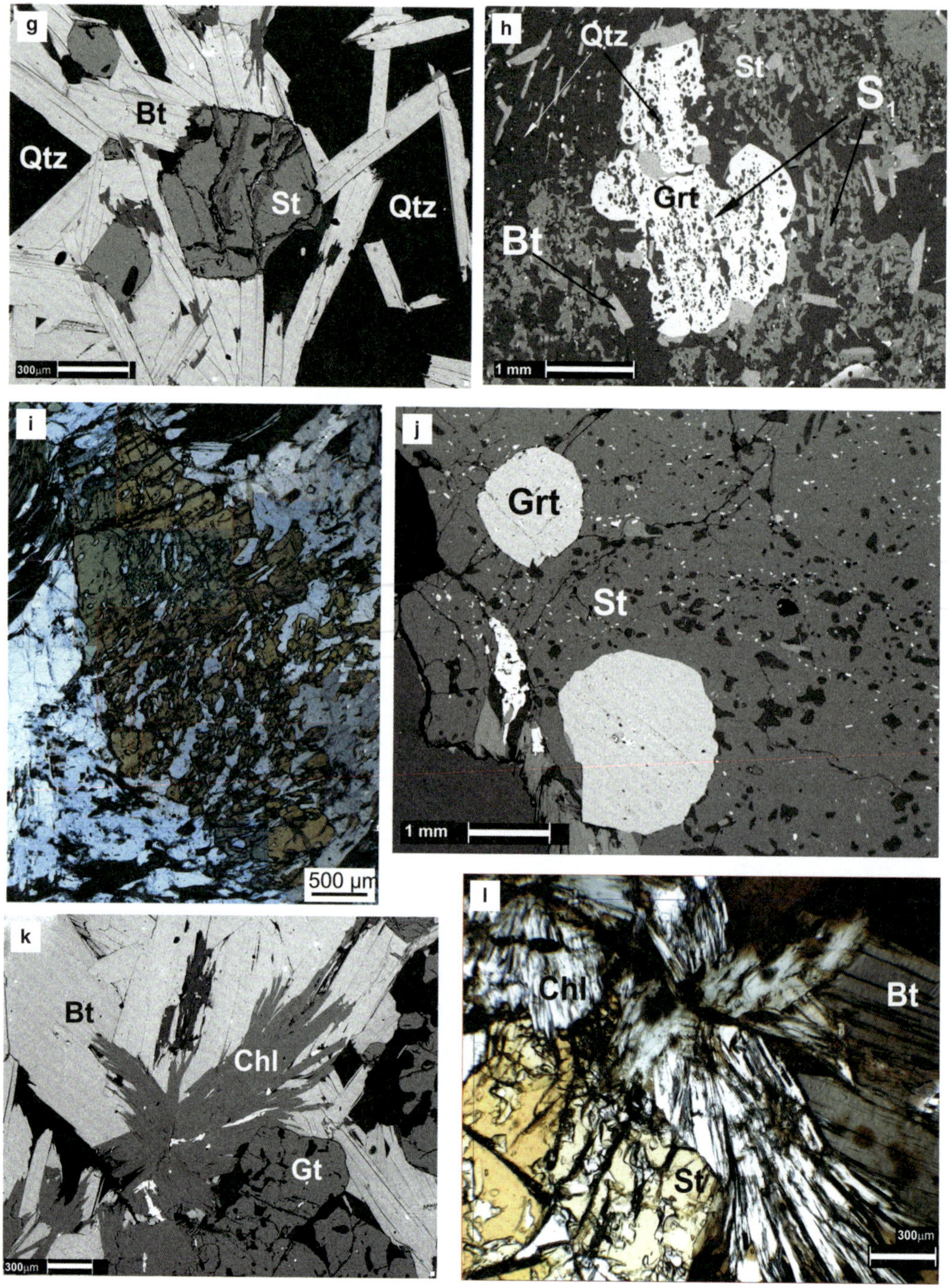

Fig. 5. (*Continued*) parallel to the external S_1 schistosity in some places (helicitic texture) (**j**) BSE image of the inclusion of a euhedral garnet porphyroblast in a euhedral staurolite porphyroblast, depicting the simultaneous growth of the two minerals. (**k**) & (**l**) BSE image of flakes of chlorite replacing biotite at the contact of biotite with garnet, and staurolite porphyroblasts occuring across the S_1 fabric. (**m**) Sympathetic variation of X_{Mg} (Mg/Mg + Fe) in Grt (garnet) with X_{Mg} (Mg/Mg + Fe) in Bt (biotite). (**n**) Variation of X_{Ca} (Ca/Ca + Mg + Mn + Fe) in Grt with X_{Ca} (Ca/Ca + Na + K) in Pl (plagioclase). These variations suggest the control of bulk composition on the composition of minerals of the rock (see the text).

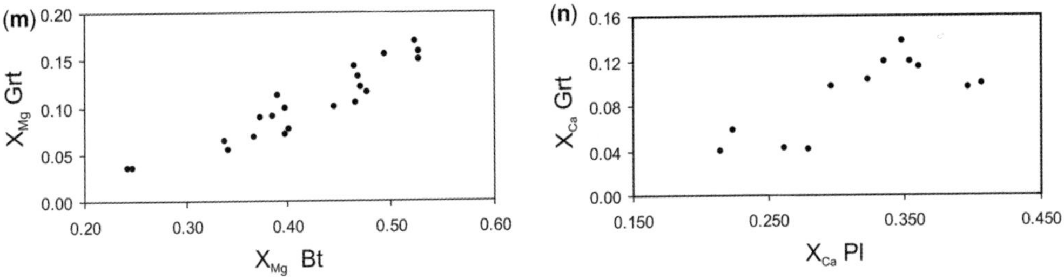

Fig. 5. *Continued.*

(Mg/Mg + Fe *c.* 0.11–0.18) than garnet in staurolite-free rocks ($X_{Mg} = 0.05 - 0.13$).

Individual grains of biotite are compositionally homogeneous but large compositional variations were noted from one sample to the other ($X_{Mg} =$ 0.24–0.61: Table 3). The TiO_2 content of the biotite never exceeds 2.90 wt% (1.73 ± 0.68 wt%: Table 3). A significant proportion of Al occurs in the octahedron sites of biotite (X_{AlVI} 0.57–0.94: Table 3).

Like the biotite and staurolite, individual grains of plagioclase are unzoned but X_{An} (Ca/Ca + Na) of plagioclase show a large variation in different samples ($X_{An} = 0.21–0.42$: Table 4). X_{Ca} of plagioclase shows a systematic increase with an increase of X_{Ca} in garnet.

The staurolite is Zn-poor (0–0.45 wt% ZnO, Table 5), rich in iron ($X_{Mg} = Mg/Mg + Fe^{+2} =$ 0.15–0.24) and does not show any significant compositional zoning (Table 5). However, staurolite is always more magnesian (by 3–8 mol%) compared to coexisting garnets (Table 2).

$KAl_3Si_3O_{10}(OH)$ is the most dominant component in muscovite (60–64%), with significant pyrophyllite (25–27 mol%) and paragonite (5–6 mol%) molecules. The K/(K + Na + Ca) ratio of muscovite varies within 0.87–0.91.

Compositionally homogeneous chlorite shows a large variation in terms of X_{Mg} (0.15–0.59 mol %: Table 5). In one sample (3K340new), chlorite inclusions in very iron-rich garnet ($X_{Mg} = 0.07$: Table 2) is distinctly Fe-rich ($X_{Mg} = 0.15$) compared to chlorite in other rocks (Table 5). Composition of host garnet appears to have controlled the composition of included chlorite. The Al/(Al + Si) ratio of chlorite varies from 0.44 to 0.54.

Physical conditions of metamorphism

Methodology

Physical conditions of the metamorphism of the supracrustal rocks have been constrained by the following methods.

- Quantitative geothermobarometry on the post-D_1 assemblage garnet–staurolite–muscovite–biotite–chlorite–plagioclase. For quantitative geothermobarometry we have used the compositions of minerals that are touching each other or lie within a distance of a few micrometres. In view of the lack of significant compositional zoning of porphyroblasts in most samples, the choice of grains does not affect the *P–T* calculations.
- Interpretation of assemblages in numerically computed phase diagrams for a specific bulk composition (pseudosection approach). For this purpose, the computer program Perple_X-07 (Connolly 2005) is used that incorporates the thermodynamic data of Holland & Powell (1998). Activity–composition relationships are used for the different solid-solution phases used in the calculations: garnet, Holland & Powell (1998); staurolite, Holland & Powell (1998); chlorite, Holland *et al.* (1998); biotite, Powell & Holland (1999); muscovite, Powell & Holland (1999); plagioclase, Newton *et al.* (1980); chloritoid, White *et al.* (2000). Considering the whole-rock compositions, phase diagrams are calculated in the $MnO–Na_2O–CaO–K_2O–FeO–MgO–Al_2O_3–SiO_2–H_2O$ (MnNCKFMASH) system. In view of the fact that TiO_2 is preferentially restricted in ilmenite and that biotite is not very rich in TiO_2 (Table 3), it is expected that discarding TiO_2 from the system component will not have any major impact on the relative phase relationship (Wei & Powell 2004; Jerabek *et al.* 2008). Several studies have demonstrated that *P–T* pseudosections are very sensitive to the 'effective bulk composition' – the bulk compositions that the metamorphic reactions actually see (Evans 2004). The large grain size of the pelitic and psammopelitic rocks makes an accurate determination of bulk compositions difficult. For the purpose of a numerical modelling study, compositions of large chunk of rocks (weighing about 3 kg) with a high and nearly uniform

Table 2. *Composition of garnet*

	Staurolite-free rocks										Staurolite-free rock											
Sample	K285/1		15B/5		STR132L/6		K285-1/7		37B/9		K 340/2		UNV735/3		K340new/4		9B/8		43A/10		37F/11	
Position	c	r	c	r	c	r	c	r	c	r	c	r	c	r	c	R	c	r	c	r	c	r
SiO_2	37.34	36.89	37.07	37.59	37.31	37.13	36.94	37.18	36.74	36.90	37.71	38.09	37.07	36.68	37.60	37.37	37.10	36.92	37.09	37.01	36.62	36.80
TiO_2	0.11	0.11	0.05	0.03	0.00	0.02	0.02	0.01	0.03	0.00	0.11	0.11	0.11	0.11	0.09	0.06	0.00	0.02	0.00	0.00	0.02	0.00
Al_2O_3	20.90	21.51	20.92	21.17	21.28	21.00	20.79	20.96	20.83	20.99	20.75	20.95	20.80	20.71	20.85	21.41	20.55	21.05	20.80	20.84	20.83	20.90
Cr_2O_3	0.06	0.02	0.06	0.03	0.00	0.06	0.00	0.00	0.00	0.03	0.06	0.09	0.02	0.01	0.00	0.05	0.00	0.02	0.07	0.05	0.08	0.00
FeO	36.54	36.21	36.61	36.85	34.79	35.45	35.78	36.58	36.91	36.97	33.02	34.33	31.42	29.24	33.44	34.08	30.88	29.64	34.67	34.68	36.05	36.28
MnO	0.78	0.69	1.01	0.79	1.99	1.83	0.60	0.58	1.40	1.46	1.42	1.96	7.67	7.54	1.13	1.31	7.08	6.98	4.46	5.01	2.08	1.98
MgO	3.91	4.30	3.61	3.03	3.26	2.40	3.97	3.77	2.80	2.43	2.20	1.93	0.90	0.85	2.21	1.76	1.36	1.33	1.73	1.62	2.89	2.61
CaO	0.80	0.79	0.94	0.94	1.14	0.74	1.15	0.84	0.56	0.53	4.75	4.17	3.39	3.45	4.06	3.92	3.44	3.23	2.06	1.40	1.46	1.50
Na_2O	0.12	0.12	0.05	0.01	0.10	0.09	0.03	0.08	0.02	0.02	0.02	0.02	0.12	0.08	0.04	0.03	0.04	0.06	0.02	0.00	0.08	0.05
K_2O	0.01	0.01	0.00	0.00	0.03	0.00	0.05	0.00	0.00	0.02	0.01	0.01	0.01	0.01	0.02	0.00	0.00	0.00	0.00	0.00	0.02	0.01
Total	**100.55**	**100.64**	**100.32**	**100.44**	**99.90**	**98.72**	**99.33**	**100.00**	**99.29**	**99.35**	**100.05**	**101.66**	**101.49**	**98.67**	**99.44**	**99.99**	**100.45**	**99.25**	**100.90**	**100.61**	**100.13**	**100.13**
O_2 basis	12.00	12.00	12.00	12.00	12.00	12.00	12.00	12.00	12.00	12.00	12.00	12.00	12.00	12.00	12.00	12.00	12.00	12.00	12.00	12.00	12.00	12.00
Si	2.990	2.947	2.982	3.013	3.000	3.028	2.987	2.992	2.995	3.005	3.026	3.022	2.986	3.014	3.032	3.005	3.006	3.007	2.995	2.999	2.968	2.981
Ti	0.006	0.006	0.003	0.002	0.000	0.001	0.001	0.001	0.002	0.000	0.007	0.007	0.006	0.007	0.005	0.004	0.000	0.001	0.000	0.000	0.001	0.000
Al	1.973	2.026	1.984	2.000	2.017	2.019	1.982	1.988	2.002	2.016	1.963	1.960	1.975	2.006	1.982	2.030	1.963	2.021	1.980	1.991	1.990	1.996
Cr	0.004	0.001	0.004	0.002	0.000	0.004	0.000	0.000	0.000	0.002	0.004	0.006	0.001	0.001	0.000	0.003	0.000	0.001	0.004	0.003	0.005	0.000
Fe^{3+}	0.034	0.027	0.031	0.000	0.000	0.000	0.031	0.020	0.003	0.000	0.033	0.035	0.037	0.000	0.018	0.000	0.037	0.000	0.021	0.007	0.037	0.023
Fe^{2+}	2.412	2.392	2.432	2.470	2.340	2.418	2.388	2.442	2.513	2.518	2.182	2.243	2.080	2.010	2.237	2.292	2.055	2.019	2.320	2.343	2.406	2.435
Mn	0.053	0.047	0.069	0.054	0.136	0.126	0.041	0.040	0.097	0.101	0.097	0.132	0.524	0.525	0.077	0.089	0.486	0.482	0.305	0.344	0.143	0.136
Mg	0.466	0.511	0.433	0.362	0.391	0.292	0.478	0.452	0.340	0.295	0.263	0.228	0.107	0.104	0.266	0.211	0.164	0.161	0.208	0.196	0.349	0.315
Ca	0.069	0.068	0.081	0.081	0.098	0.065	0.100	0.072	0.049	0.046	0.408	0.355	0.292	0.303	0.351	0.338	0.299	0.282	0.178	0.122	0.127	0.130
Na	0.018	0.018	0.008	0.002	0.016	0.014	0.005	0.012	0.003	0.003	0.003	0.003	0.018	0.013	0.006	0.005	0.006	0.009	0.003	0.000	0.013	0.008
K	0.001	0.001	0.000	0.000	0.003	0.000	0.005	0.000	0.000	0.002	0.001	0.001	0.001	0.001	0.002	0.000	0.000	0.000	0.000	0.000	0.002	0.001
#Cation	8.025	8.043	8.025	7.985	8.000	7.966	8.019	8.019	8.004	7.988	7.987	7.991	8.028	7.983	7.976	7.977	8.016	7.984	8.015	8.004	8.041	8.025
X_{Fe+2}	0.804	0.793	0.807	0.833	0.789	0.834	0.794	0.812	0.838	0.851	0.740	0.758	0.693	0.683	0.763	0.782	0.684	0.686	0.770	0.780	0.795	0.807
X_{Mg}	0.155	0.169	0.144	0.122	0.132	0.101	0.159	0.150	0.113	0.100	0.089	0.077	0.036	0.035	0.091	0.072	0.055	0.055	0.069	0.065	0.115	0.104
X_{Ca}	0.023	0.022	0.027	0.027	0.033	0.022	0.033	0.024	0.016	0.016	0.138	0.120	0.097	0.103	0.120	0.115	0.099	0.096	0.059	0.040	0.042	0.043
X_{Mn}	0.018	0.015	0.023	0.018	0.046	0.044	0.014	0.013	0.032	0.034	0.033	0.045	0.174	0.178	0.026	0.030	0.162	0.164	0.101	0.114	0.047	0.045
$X_{Ca} + X_{Mn}$	0.040	0.038	0.050	0.045	0.079	0.066	0.047	0.037	0.049	0.050	0.171	0.164	0.272	0.282	0.146	0.146	0.261	0.259	0.160	0.155	0.089	0.088
Mg No.	0.162	0.176	0.151	0.128	0.143	0.108	0.167	0.156	0.119	0.105	0.11	0.09	0.05	0.05	0.11	0.08	0.07	0.07	0.08	0.08	0.13	0.11

c, core; r, rim.

Table 3. *Composition of biotite*

	Staurolite-bearing rocks										Staurolite-free rocks													
Sample No.	K285/1		15B/5		STR132L/6		K285-1/7		37B/9		K340/2		UNV735/3		K340new/4		9B/8		43A/10		37F/11		UNV 735/12	K285-1 / 13
Point No.	c	r	c	r	r	c	c	r	c	r	c	r	r	c	r	c	c	r	c	r	c	r	c	r
SiO_2	36.39	36.12	35.50	35.60	35.73	35.70	35.85	36.79	34.61	33.96	35.20	35.54	34.39	35.34	34.91	34.96	34.22	33.99	36.63	34.39	35.20	35.16	34.28	36.68
TiO_2	1.44	1.19	1.79	1.84	1.87	1.79	1.39	1.53	2.19	2.17	0.29	0.37	0.61	2.57	1.77	2.12	2.40	2.42	2.67	2.90	1.74	1.52	1.75	1.19
Al_2O_3	18.65	17.11	18.24	18.54	18.42	18.59	18.38	17.72	18.50	18.72	17.05	16.33	17.39	18.04	16.53	16.84	17.26	17.34	18.11	17.95	18.19	18.86	17.97	18.14
Cr_2O_3	0.06	0.06	0.00	0.02	0.00	0.01	0.03	0.00	0.01	0.00	0.13	0.06	0.06	0.06	0.04	0.00	0.06	0.05	0.08	0.00	0.12	0.07	0.06	0.12
FeO	18.66	18.95	20.23	20.24	20.58	19.28	18.62	17.79	21.62	21.60	24.42	23.94	27.34	24.35	22.92	23.08	23.78	24.01	20.52	23.33	19.70	19.63	27.19	15.64
MnO	0.06	0.06	0.05	0.00	0.00	0.07	0.00	0.05	0.03	0.10	0.02	0.02	0.15	0.24	0.04	0.00	0.24	0.24	0.11	0.06	0.00	0.00	0.16	0.12
MgO	10.21	11.67	9.83	10.08	9.25	9.51	11.63	11.13	7.76	7.99	8.13	8.98	5.00	4.36	8.46	8.08	6.90	6.98	6.66	6.65	10.07	9.57	4.70	13.77
CaO	0.08	0.02	0.00	0.00	0.00	0.04	0.03	0.00	0.00	0.00	0.09	0.09	0.04	0.01	0.04	0.01	0.00	0.06	0.11	0.07	0.00	0.01	0.02	0.03
Na_2O	0.63	0.29	0.41	0.51	0.50	0.74	0.69	0.71	0.44	0.37	0.44	0.23	0.16	0.05	0.42	0.43	0.12	0.20	0.17	0.21	0.41	0.38	0.11	0.74
K_2O	8.49	8.80	8.89	9.01	9.05	8.61	7.89	8.39	9.13	8.52	9.46	9.17	9.61	9.56	8.87	9.07	9.31	9.35	9.03	8.89	9.06	9.05	9.47	8.56
Total	**94.67**	**94.26**	**94.94**	**95.84**	**95.40**	**94.34**	**94.51**	**94.11**	**94.29**	**93.43**	**95.20**	**94.73**	**94.75**	**94.58**	**94.00**	**94.59**	**94.29**	**94.64**	**94.09**	**94.45**	**94.49**	**94.25**	**95.70**	**94.99**
O_2 basis	22.00	22.00	22.00	22.00	22.00	22.00	22.00	22.00	22.00	22.00	22.00	22.00	22.00	22.00	22.00	22.00	22.00	22.00	22.00	22.00	22.00	22.00	22.00	22.00
Si	5.521	5.535	5.438	5.403	5.455	5.470	5.440	5.591	5.394	5.331	5.522	5.578	5.510	5.564	5.498	5.476	5.413	5.370	5.650	5.392	5.416	5.414	5.423	5.479
Ti	0.165	0.138	0.206	0.210	0.215	0.206	0.159	0.175	0.257	0.256	0.034	0.044	0.074	0.304	0.210	0.250	0.286	0.288	0.310	0.342	0.201	0.176	0.208	0.134
Al_{IV}	2.479	2.465	2.562	2.597	2.545	2.530	2.560	2.409	2.606	2.669	2.478	2.422	2.490	2.436	2.502	2.524	2.587	2.630	2.350	2.608	2.584	2.586	2.577	2.521
Al_{VI}	0.858	0.626	0.731	0.720	0.770	0.828	0.729	0.766	0.793	0.796	0.676	0.599	0.795	0.912	0.568	0.586	0.632	0.600	0.944	0.710	0.715	0.838	0.774	0.673
Cr	0.007	0.007	0.000	0.002	0.000	0.001	0.004	0.000	0.001	0.000	0.016	0.007	0.007	0.007	0.005	0.000	0.008	0.006	0.010	0.000	0.015	0.009	0.007	0.014
Fe	2.368	2.428	2.592	2.569	2.628	2.471	2.363	2.261	2.818	2.836	3.204	3.143	3.664	3.206	3.019	3.024	3.146	3.172	2.647	3.059	2.535	2.528	3.597	1.954
Mn	0.008	0.008	0.006	0.000	0.000	0.009	0.000	0.006	0.004	0.013	0.002	0.003	0.020	0.033	0.005	0.000	0.032	0.032	0.014	0.008	0.000	0.000	0.022	0.015
Mg	2.310	2.666	2.244	2.280	2.105	2.172	2.630	2.521	1.802	1.869	1.900	2.101	1.195	1.022	1.986	1.886	1.627	1.643	1.531	1.554	2.309	2.196	1.107	3.065
Ca	0.013	0.003	0.000	0.000	0.000	0.007	0.005	0.000	0.000	0.000	0.015	0.015	0.006	0.002	0.007	0.002	0.000	0.010	0.018	0.012	0.000	0.002	0.004	0.005
Na	0.185	0.085	0.122	0.150	0.148	0.220	0.203	0.209	0.133	0.113	0.133	0.070	0.050	0.016	0.128	0.131	0.037	0.061	0.051	0.064	0.122	0.113	0.033	0.214
K	1.643	1.720	1.737	1.744	1.763	1.683	1.528	1.627	1.815	1.706	1.894	1.836	1.965	1.920	1.782	1.813	1.879	1.885	1.777	1.778	1.778	1.778	1.911	1.631
#Cation	15.556	15.681	15.639	15.675	15.628	15.596	15.620	15.565	15.623	15.590	15.873	15.817	15.778	15.422	15.710	15.691	15.646	15.697	15.302	15.528	15.676	15.639	15.663	15.706
X_{Mg} = Mg/(Fe + Mg)	0.494	0.523	0.464	0.470	0.445	0.468	0.527	0.527	0.390	0.397	0.372	0.401	0.246	0.242	0.397	0.384	0.341	0.341	0.366	0.337	0.477	0.465	0.235	0.611
X_{Ti}	0.029	0.023	0.036	0.036	0.038	0.036	0.027	0.031	0.045	0.044	0.006	0.007	0.013	0.056	0.036	0.043	0.050	0.050	0.057	0.060	0.035	0.031	0.037	0.023
X_{AlVI}	0.151	0.107	0.127	0.125	0.135	0.146	0.124	0.134	0.140	0.138	0.116	0.102	0.139	0.168	0.098	0.102	0.111	0.105	0.174	0.125	0.124	0.146	0.136	0.116

c, core; r, rim.

Table 4. *Composition of plagioclase in staurolite-free rocks*

Sample	K340/2		UNV735/3		K340new/4		9B/8		43A/10		37F/11		9B/14
Position	c	r	r	c	r	c	c	r	c	r	c	r	r
SiO_2	61.27	62.37	61.39	62.01	59.54	59.91	59.50	59.40	63.25	62.89	61.07	61.44	57.59
TiO_2	0.12	0.12	0.11	0.01	0.01	0.06	0.00	0.07	0.00	0.00	0.00	0.04	0.00
Al_2O_3	24.01	24.31	23.78	23.74	25.97	25.69	25.65	25.71	23.75	23.45	24.18	23.95	26.42
Cr_2O_3	0.06	0.06	0.06	0.01	0.04	0.00	0.00	0.01	0.01	0.00	0.00	0.01	0.00
FeO	0.04	0.03	0.03	0.06	0.15	0.08	0.00	0.00	0.14	0.00	0.05	0.16	0.07
MnO	0.06	0.11	0.06	0.06	0.01	0.00	0.01	0.05	0.02	0.01	0.09	0.02	0.00
MgO	0.04	0.04	0.06	0.04	0.01	0.01	0.04	0.02	0.00	0.00	0.00	0.00	0.00
CaO	6.38	6.05	5.86	5.46	7.65	7.19	8.20	8.11	4.75	4.61	5.90	5.50	8.56
Na_2O	6.57	6.06	6.73	7.07	7.50	7.82	6.53	6.80	9.03	9.29	8.40	8.56	6.51
K_2O	0.07	0.10	0.11	0.17	0.03	0.09	0.15	0.04	0.11	0.12	0.06	0.04	0.05
NiO	0.32	0.32	0.32	0.32	0.00	0.06	0.10	0.05	0.00	0.00	0.00	0.00	0.02
Total	**98.95**	**99.56**	**98.52**	**98.95**	**100.91**	**100.91**	**100.18**	**100.26**	**101.06**	**100.37**	**99.75**	**99.72**	**99.22**
O_2 basis	8.00	8.00	8.00	8.00	8.00	8.00	8.00	8.00	8.00	8.00	8.00	8.00	8.00
Si	2.740	2.759	2.753	2.767	2.636	2.650	2.648	2.643	2.772	2.775	2.721	2.736	2.596
Ti	0.004	0.004	0.004	0.000	0.000	0.002	0.000	0.002	0.000	0.000	0.000	0.001	0.000
Al	1.266	1.268	1.257	1.249	1.355	1.340	1.346	1.349	1.227	1.220	1.270	1.257	1.404
Cr	0.002	0.002	0.002	0.000	0.001	0.000	0.000	0.000	0.000	0.000	0.000	0.000	0.000
Fe	0.002	0.001	0.001	0.002	0.006	0.003	0.000	0.000	0.005	0.000	0.002	0.006	0.003
Mn	0.002	0.004	0.002	0.002	0.000	0.000	0.000	0.002	0.001	0.000	0.003	0.001	0.000
Mg	0.003	0.003	0.004	0.003	0.001	0.001	0.003	0.001	0.000	0.000	0.000	0.000	0.000
Ca	0.306	0.287	0.282	0.261	0.363	0.341	0.391	0.387	0.223	0.218	0.282	0.262	0.413
Na	0.569	0.519	0.585	0.612	0.644	0.671	0.564	0.587	0.767	0.795	0.726	0.739	0.569
K	0.004	0.006	0.006	0.009	0.002	0.005	0.009	0.002	0.006	0.007	0.003	0.002	0.003
Ni	0.012	0.011	0.012	0.011	0.000	0.002	0.004	0.002	0.000	0.000	0.000	0.000	0.001
#Cation	4.909	4.864	4.909	4.918	5.008	5.014	4.964	4.975	5.001	5.015	5.008	5.005	4.988
X_{An}	0.348	0.353	0.323	0.296	0.360	0.335	0.406	0.396	0.224	0.214	0.279	0.261	0.420
X_{Ab}	0.647	0.640	0.670	0.694	0.638	0.660	0.585	0.601	0.770	0.780	0.718	0.736	0.577
X_{Or}	0.005	0.007	0.007	0.011	0.002	0.005	0.009	0.002	0.006	0.007	0.003	0.002	0.003

c, core; r, rim.

distribution of porphyroblasts have been chosen. These compositions may slightly differ from the actual compositions of the rocks but will help recover the $P-T$ conditions for the formation of porphyroblasts. Modelling was performed on three representative bulk compositions, two are staurolite-bearing (with 27.48 and 28.74 wt% Al_2O_3) and the other one is staurolite-free (with 15.32 wt% Al_2O_3). Computations were carried out in two steps. In the first step, a $P-T$ section showing the stability limit of different mineral assemblages are computed for a given bulk composition in H_2O-saturated conditions. In view of the published $P-T$ stability limit of the mineral assemblages present in the metasedimentary rocks, pseudosections were computed within the $P-T$ range of 3–7 kbar and 450–650 °C. Studies have shown that pelitic–psammopelitic rocks do not produce any significant melt at temperature below 700 °C (Wei *et al.* 2004). For this reason, melt was excluded in modelling studies. In the second step, the output $P-T$ section was contoured for modal volume and the composition of garnet (X_{Fe} and X_{Ca}). The subprogram WERMI, which is linked to Perplex-X 07, is used for the construction of modal compositional contours. To demonstrate the control of bulk composition on the appearance (or disappearance) of staurolite in the studied rock, several $T-X$ sections have been computed. Temperature in these diagrams was varied along the Barrovian geothermal gradient, as this is the geothermal gradient along which rocks in most of the fold belts evolve. These diagrams are preferred over conventional isobaric $T-X$ (or isothermal $P-X$) diagrams as mineralogy of metamorphic rocks is a product of the interaction of the bulk composition of that rock and the ambient geothermal gradient.

Results

Quantitative geothermobarometry

The result of the quantitative geothermobarometry is presented in Table 6. It is evident from Table 6 that the results of all of the geothermobarometric formulations converge within approximately 6.0 ± 1.6 kbar and 560 ± 60 °C and approximately 5.5 ± 1.4 kbar and 540 ± 45 °C for core and rim positions of garnet, respectively (Table 2). Much tighter $P-T$ brackets are obtained if the formulation of Holdaway (2000) and Wu *et al.* (2004) are used (core 5.8 ± 1.3 kbar, 585 ± 50 °C; rim 5.0 ± 0.9 kbar, 575 ± 15 °C). Barometric formulation of Wu *et al.* (2004), which was calibrated over a larger range of mineral compositions, is compatible with the thermometric formulation of Holdaway (2000). In view of this observation, $P-T$ estimates that are derived from Holdaway (2000) and Wu *et al.* (2004) will be considered. Considering the uncertainties of the calibrations, these geothermobarometric formulations fail to distinguish the $P-T$ clusters determined for core and rim compositions of garnet.

Numerical modelling (pseudosection)

Results of the numerical modelling on the two staurolite-bearing (#37B and #K285) and one staurolite-free (#K340new) samples are presented in Figures 6–8. In all of the pseudosections, mineral assemblages with variances of 4 and 5 dominate. Only a few fields in these diagrams show a trivariant assemblage, with no field showing either a uni- or bivariant assemblage. The quadrivariant peak metamorphic assemblages that are developed in staurolite-bearing (garnet + staurolite + muscovite + biotite + chlorite + quartz + fluid) and staurolite-free (garnet + chlorite + muscovite + biotite + plagioclase + quartz + fluid) rocks are strongly sensitive to temperature but insensitive to pressure (Figs 6–8). The stability fields of these assemblages tightly constrain the temperature within 560–610 and 530–575 °C for staurolite-bearing and staurolite-free rocks (Figs 6–8). Tighter temperature ranges are obtained when isopleths of X_{Fe} and X_{Ca} computed for the observed garnet compositions are superimposed on the $P-T$ section. These isopleths, which pass through the stability fields of the observed quadrivariant assemblages, narrow down the peak temperature of metamorphism within 550–580 °C for sample #37B, 560–600 °C for #K285 and 530–540 °C for #K340new (Figs 6–8). Notwithstanding the uncertainties that are associated with the computation of pseudosections, the inferred temperature values are in excellent agreement with the temperature estimates derived from independent geothermometers (560 ± 50 °C: Table 6) and geobarometers (5.7 ± 1.5 kbar: Table 6).

To explain the mineralogical variation in adjacent metasedimentary rocks, $T-Al_2O_3$, $T-MnO$, $T-CaO$ and $T-X_{MgO}$ diagrams have been constructed using the bulk composition of #37B as the monitor (Fig. 9). Temperature, in these diagrams, is varied along the Barrovian geothermal gradient (P (in bars) $= -2502.5 + 9.167/T$ (in K)). It is evident from the $T-Al_2O_3$ diagram that for constant MnCNKFMS components, staurolite stabilizes only in aluminous bulk compositions ($Al_2O_3 > 19$ wt%) at approximately 550 °C (Fig. 9a). At higher temperatures (>575 °C), staurolite can develop at low aluminous bulk compositions (Fig. 9a). Like Al_2O_3, the Mg# (molar MgO/MgO + FeO) of bulk-rock composition exerts a strong influence, with staurolite

Table 5. *Composition of muscovite, chlorite and staurolite*

	Muscovite				Chlorite				Staurolite							
Sample No.	43A/10		43A/15	37B/16	15B/5*		K340new/4	K 285-1/7*	K285/1		STR132L/6		K285-1/7		37B/9	
Position	c	r	r	c	r	i	i	m	c	r	c	r	c	r	c	r
SiO_2	45.360	45.250	45.760	45.060	25.210	25.030	21.800	29.360	27.207	28.839	27.720	27.320	27.500	27.420	26.940	27.710
TiO_2	0.560	0.300	0.500	0.680	0.080	0.080	0.000	0.520	0.453	0.376	0.580	0.510	0.340	0.520	0.620	0.510
Al_2O_3	35.450	35.870	35.220	34.670	22.660	22.850	21.620	19.230	51.173	51.695	53.010	53.100	52.500	52.280	52.470	52.800
Cr_2O_3	0.050	0.010	0.000	0.030	0.060	0.050	0.000	0.000	0.058	0.058	0.020	0.010	0.030	0.040	0.000	0.000
FeO	1.500	1.180	1.320	1.490	21.810	21.350	38.180	22.060	14.856	14.288	14.910	14.920	14.260	15.160	15.650	14.680
MnO	0.030	0.000	0.000	0.000	0.000	0.060	0.070	0.130	0.060	0.108	0.050	0.050	0.070	0.090	0.000	0.190
MgO	0.650	0.440	0.510	0.760	16.680	17.390	3.850	11.950	2.628	2.082	1.780	1.840	2.230	2.180	1.750	1.480
CaO	0.120	0.000	0.000	0.000	0.010	0.010	0.090	0.470	0.015	0.017	0.020	0.000	0.000	0.080	0.020	0.030
Na_2O	0.900	0.820	0.700	0.900	0.020	0.050	0.290	0.270	0.009	0.018	0.040	0.010	0.060	0.420	0.030	0.020
K_2O	9.500	10.320	10.230	10.070	0.140	0.040	0.090	1.740	0.010	0.010	0.000	0.020	0.040	0.030	0.000	0.010
ZnO	0.000	0.000	0.000	0.000	0.160	0.000	0.000	0.000	0.000	0.000	0.000	0.130	0.240	0.020	0.450	0.440
Total	**94.120**	**94.190**	**94.240**	**93.660**	**86.830**	**86.910**	**85.990**	**85.730**	**96.469**	**97.491**	**98.130**	**97.910**	**97.270**	**98.240**	**97.930**	**97.870**
O_2 basis	11.00	11.00	11.00	11.00	28.00	28.00	28.00	28.00	48.00	48.00	48.00	48.00	48.00	48.00	48.00	48.00
Si	3.049	3.046	3.076	3.057	5.267	5.204	5.085	6.238	8.045	8.378	8.049	7.960	8.053	7.999	7.901	8.082
Ti	0.028	0.015	0.025	0.035	0.013	0.013	0.000	0.083	0.101	0.082	0.127	0.112	0.075	0.114	0.137	0.112
Al	2.810	2.847	2.791	2.773	5.581	5.601	5.946	4.817	17.839	17.704	18.147	18.239	18.125	17.981	18.142	18.155
Cr	0.003	0.001	0.000	0.002	0.010	0.008	0.000	0.000	0.014	0.013	0.005	0.002	0.007	0.009	0.000	0.000
Fe	0.084	0.066	0.074	0.085	3.811	3.713	7.449	3.920	3.674	3.471	3.621	3.636	3.492	3.699	3.839	3.581

Mn	0.002	0.000	0.000	0.000	0.000	0.011	0.014	0.023	0.015	0.027	0.012	0.012	0.017	0.022	0.000	0.047
Mg	0.065	0.044	0.051	0.077	5.194	5.389	1.339	3.784	1.158	0.901	0.770	0.799	0.973	0.948	0.765	0.643
Ca	0.009	0.000	0.000	0.000	0.002	0.002	0.022	0.107	0.005	0.005	0.006	0.000	0.000	0.025	0.006	0.009
Na	0.117	0.107	0.091	0.118	0.008	0.020	0.131	0.111	0.003	0.005	0.011	0.003	0.017	0.119	0.009	0.006
K	0.815	0.886	0.877	0.871	0.037	0.011	0.027	0.472	0.002	0.002	0.000	0.004	0.007	0.006	0.000	0.002
Zn	0.000	0.000	0.000	0.000	0.025	0.000	0.000	0.000	0.000	0.000	0.000	0.028	0.052	0.004	0.097	0.095
#Cation	6.982	7.012	6.987	7.017	19.948	19.971	20.013	19.555	30.855	30.589	30.749	30.794	30.819	30.926	30.896	30.732
Ti- muscovite	0.014	0.008	0.013	0.017												
Phlogopite-annite	0.000	0.000	0.000	0.000												
Celadonite	0.061	0.048	0.050	0.063												
Pyrophyllite	0.267	0.252	0.258	0.253												
Paragonite	0.059	0.054	0.046	0.059												
Margarite	0.004	0.000	0.000	0.000												
Muscovite	0.595	0.640	0.634	0.608												
K/(K + Ca + Na)	0.866	0.892	0.906	0.880												
X_{Mg} = Mg/(Mg + Fe)					0.577	0.592	0.152	0.491	0.240	0.206	0.175	0.180	0.218	0.204	0.166	0.152
Al/(Al + Si)					0.514	0.518	0.539	0.436								
X_{Zn} = Zn/Zn + Fe + Mg + Mn									0.000	0.000	0.000	0.006	0.011	0.001	0.021	0.022

c, core; r, rim; m, matrix grain; i, inclusion within garnet.
*Staurolite-bearing rock.

Table 6. *Pressure–temperature (P–T) estimates from metapelites*

	I T (°C)	II T (°C)	III T (°C)	IV T (°C)	V T (°C)	VI T (°C)	VII T (°C)	VIII T (°C)	IX T (°C)	X P (kbar)	GASP* P (kbar)	GPMB-Mg† P (kbar)	GPMB-Fe‡ P (kbar)	GB-GASP§ P (kbar)
K285 1-core	581		596	612	538	585	523	567						
K285 1-rim	585		593	607	517	581	561	557						
K340 2-core	604	607	650	621	640	591	603	652	759	7.4	8.7	8.8	6.7	
K340 2-rim	544	554	580	524	552	519	526	553	637	6.3	6.8	7.1	5.4	
UNV 735 3-rim	483	562	593	541	589	531	478	610	671	5.8	5.4	5.5	3.5	9.2
UNV 735 3-core	483	563	592	543	592	533	434	607	661	6.0	5.5	6.0	4.0	9.1
K 340new 4-rim	528	542	562	504	539	504	479	525	605	6.0	6.4	6.9	5.3	
K 340new 4-core	588	596	623	600	616	576	554	617	712	7.0	8.0	8.5	6.5	
15B 5-core	591		447	626	565	595	540	373						
15B 5-rim	552		565	554	522	542	481	519						
STR 132L 6-rim	526		552	526	514	520	447	502						
STR 132L 6-core	569		596	602	550	577	512	571						
K285a 7-core	569		581	578	505	560	529	539						
K285a 7-rim	552		566	554	490	541	493	512						
9B 8-core	498	556	585	528	556	522	485	585	631	5.1	4.9	5.3	3.6	7.6
9B 8-rim	497	559	588	533	559	526	486	584	634	5.0	4.8	5.3	3.6	7.7
37B 9-core	591		605	642	613	606	508	600						
37B 9-rim	555		575	578	569	560	463	545						
43A 10-core	508	558	582	532	550	525	422	549	566	5.6	5.3	6.4	4.7	6.8
43A 10-rim	515	570	583	552	570	540	448	563	620	4.5	4.1	5.4	3.6	6.5
37F 11-core	541	559	574	538	512	530	487	524	585	3.7	3.7	5.3	4.0	5.3
37F 11-rim	527	549	566	520	506	516	457	508	574	3.9	3.8	5.5	4.1	5.4

I, Garnet–biotite (Bhattacharya *et al.* 1992).
II, Garnet–biotite (Perchuck & Lavrent'eva 1983).
III, Garnet–biotite (Holdaway 2000).
IV, Garnet–biotite (Ferry & Spear 1978).
V, Garnet–biotite (Ganguly & Saxena 1984).
VI, Garnet–biotite (Thompson 1976).
VII, Garnet–biotite (Dasgupta *et al.* 1991).
VIII, Garnet–biotite (Williams & Grambling 1990).
IX, Garnet–biotite (Ganguly 2006).
X, Garnet–biotite–plagioclase–quartz (Wu *et al.* 2004).
*Garnet–aluminosilicate–plagioclase–quartz (Holland & Powell 1998), input temperature is from Bhattcharya *et al.* (1992).
†Garnet–plagioclase–muscovite–biotite, Mg end member, input temperature is from Bhattcharya *et al.* (1992).
‡Garnet–plagioclase–muscovite–biotite, Fe end member, input temperature is from Bhattcharya *et al.* (1992).
§Garnet–aluminosilicate–plagioclase–biotite (Ganguly 2006).
Note: In samples K340, UNV735, K340new, 9B and 37F, the composition of end member muscovite was used for thermobarometric calculations.

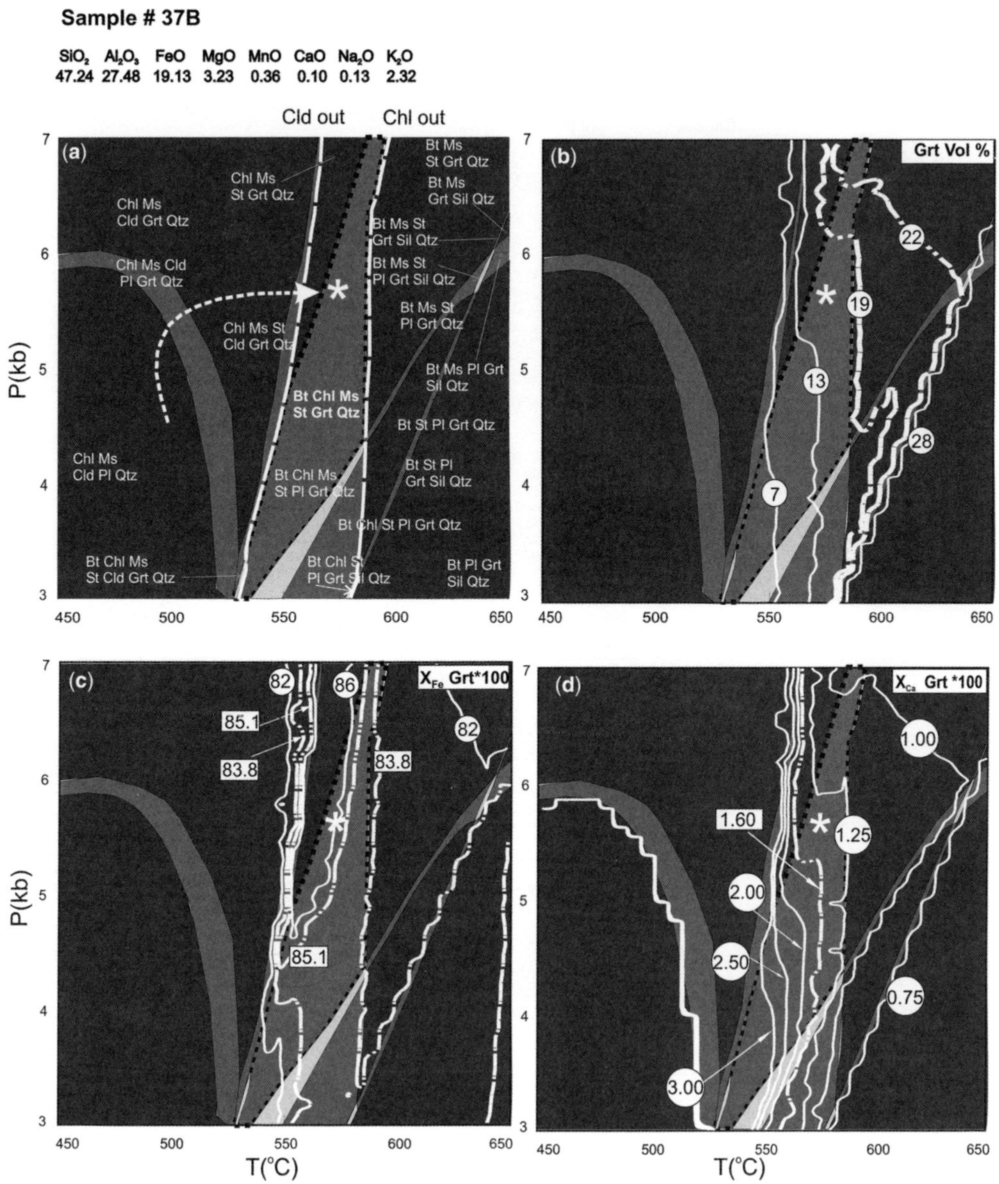

Fig. 6. (**a**) Numerically computed phase diagram (pseudosection) for a representative bulk composition of metapelite (37B). The stability field of the observed assemblage (biotite–chlorite–muscovite–staurolite–garnet–quartz) is bounded by the dashed line. (**b**) Volume proportion of garnet is contoured on this pseudosection. (**c**) & (**d**) The almandine and grossular content of garnet are contoured on this pseudosection, where isopleth values marked in boxes show the composition of this sample, whereas those marked in circles show the isopleth values for a wider range. Assemblages with a higher variance have deeper shades. The star represents the mean *P–T* values derived from geothermobarometry. The dashed line with an arrowhead represents the presumed *P–T* path (see the text).

appearing in rocks with a higher Mg# (Fig. 9d). CaO and MnO, on the other hand, have little influence on the stability of staurolite-bearing assemblages within the chosen range of concentrations of these parameters (Fig. 9b, c). Higher CaO in the bulk, however, stabilizes plagioclase in the rock (Fig. 9b).

In view of the phase relationship in the *T–X* diagrams, it is presumed that the appearance of

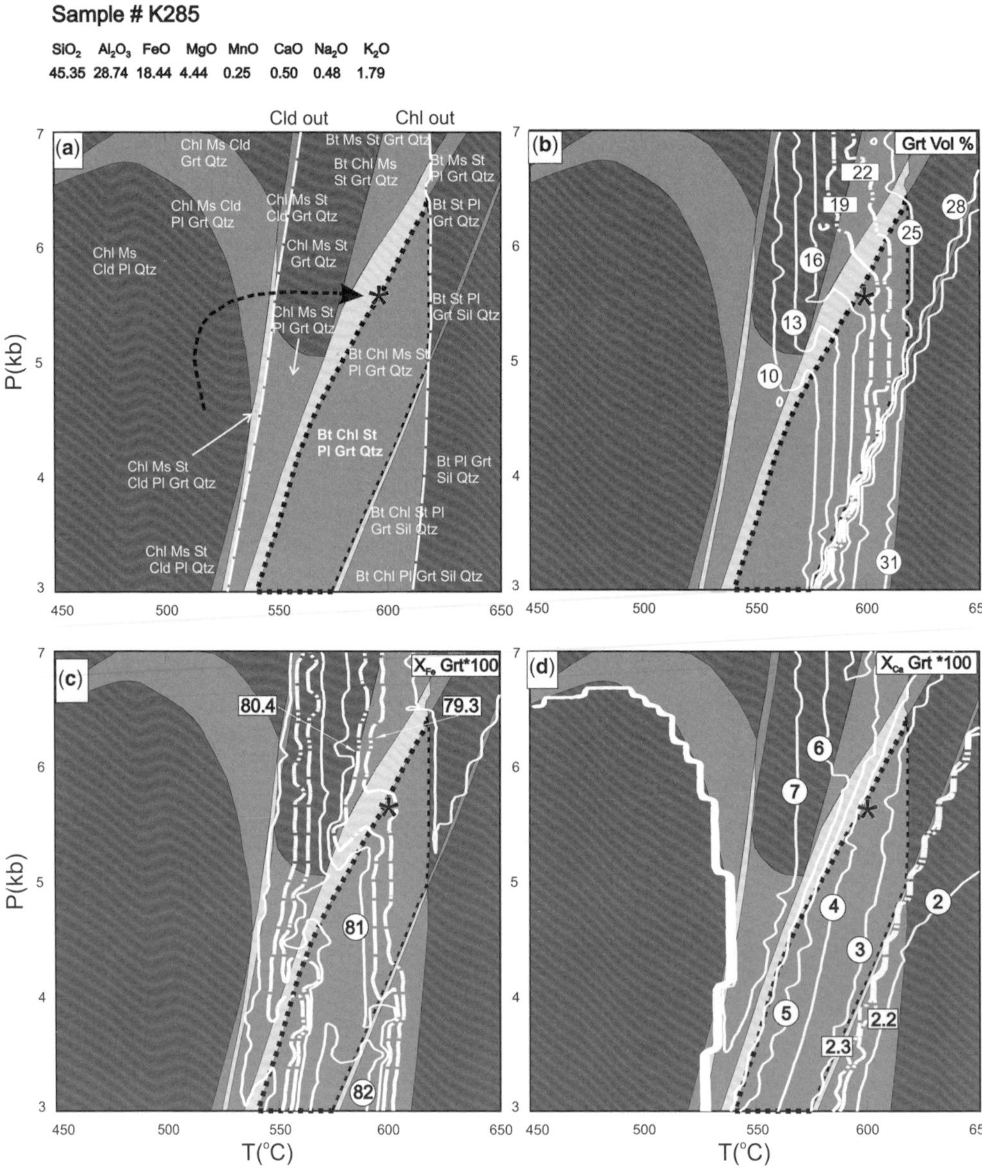

Fig. 7. (**a**) Numerically computed phase diagram (pseudosection) for a representative bulk composition of metapelite (K285). The stability field of the observed assemblage (biotite–chlorite–plagioclase–staurolite–garnet–quartz) is bounded by the dashed line. (**b**) Volume proportion of garnet is contoured on this pseudosection. (**c**) & (**d**) The almandine and grossular content of garnet are contoured on this pseudosection, where isopleth values marked in boxes show the composition of this sample, whereas those marked in circles show the isopleth values for a wider range. Assemblages with a higher variance have deeper shades. The star represents the mean *P–T* values derived from geothermobarometry. The dashed line with an arrowhead represents the presumed *P–T* path (see the text).

staurolite in rocks #37B (Al_2O_3 wt% = 26.89) and #K285 (Al_2O_3 wt% = 28.03), and its absence in rock #K340new (Al_2O_3 = 15.32), are primarily related to Al_2O_3 content in their bulk compositions.

Discussion

It is evident from the foregoing analyses that the supracrustals rocks of the studied area underwent

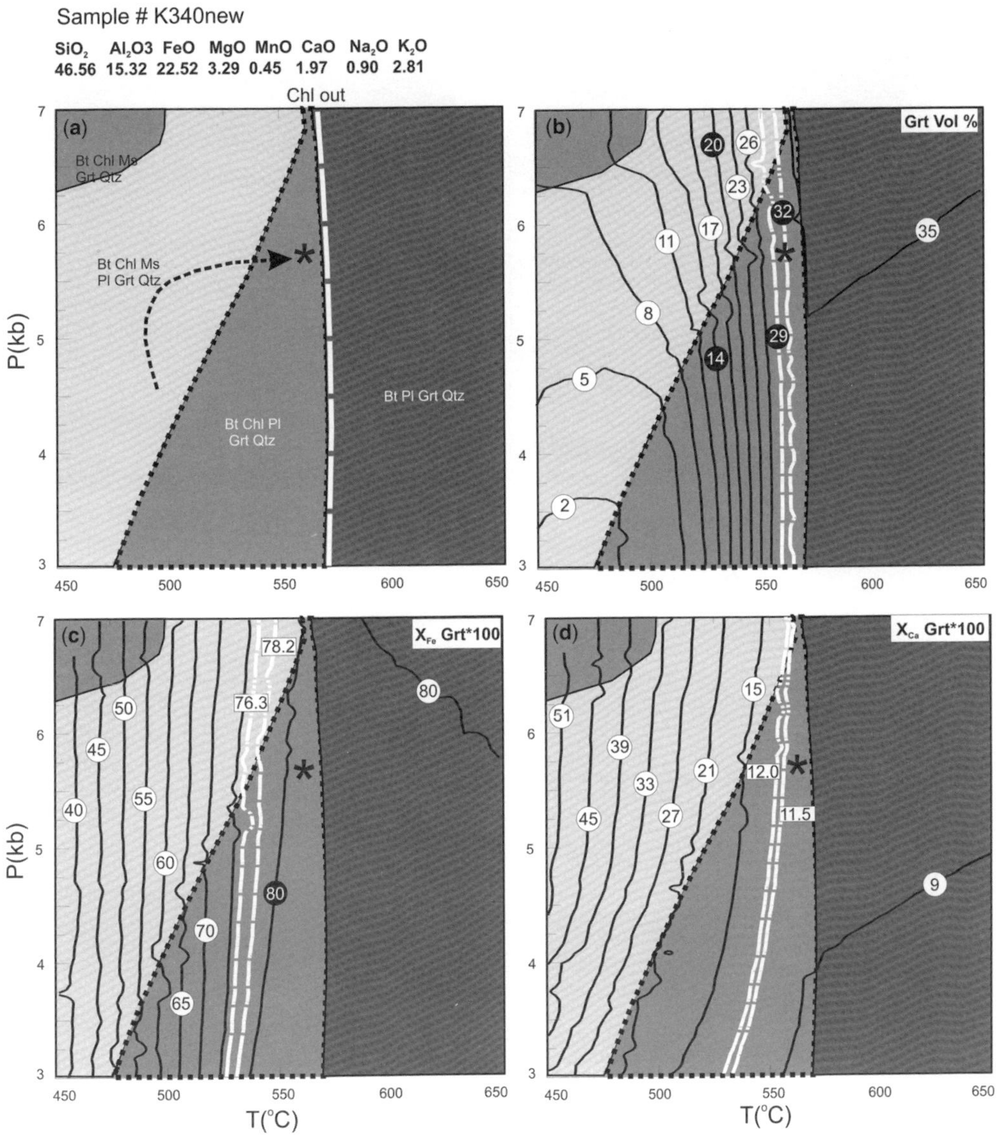

Fig. 8. (**a**) Numerically computed phase diagram (pseudosection) for a representative bulk composition of metapelite (K340new). The stability field of the observed assemblage (biotite–chlorite–plagioclase–garnet–quartz) is bounded by the dashed line. (**b**) The volume proportion of garnet is contoured on this pseudosection. (**c**) & (**d**) The almandine and grossular content of garnet are contoured on this pseudosection, where isopleth values marked in boxes show the composition of this sample, whereas those marked in circles show the isopleth values for a wider range. Assemblages with a higher variance have deeper shades. The star represents the mean P–T values derived from geothermobarometry. The dashed line with an arrowhead represents the presumed P–T path (see the text).

polyphase deformation and metamorphism over a protracted period.

Structural elements preserved in the rocks indicate four phases of deformation (D_1–D_4), with the prololith of the AGG emplaced during D_1 folding. The U–Pb date of zircon precisely constrains the time of the acid magmatism and D_1 folding at 1.849 ± 0.008 Ga (Mukhopadhyay *et al.* 2000). Recently, Kaur *et al.* (2009) demonstrated that the AGG and the Jasrapur granitoids pluton (*c.* 1.82 Ga crystallization age) of the NDFB are geochemically indistinguishable, and they have interpreted them to

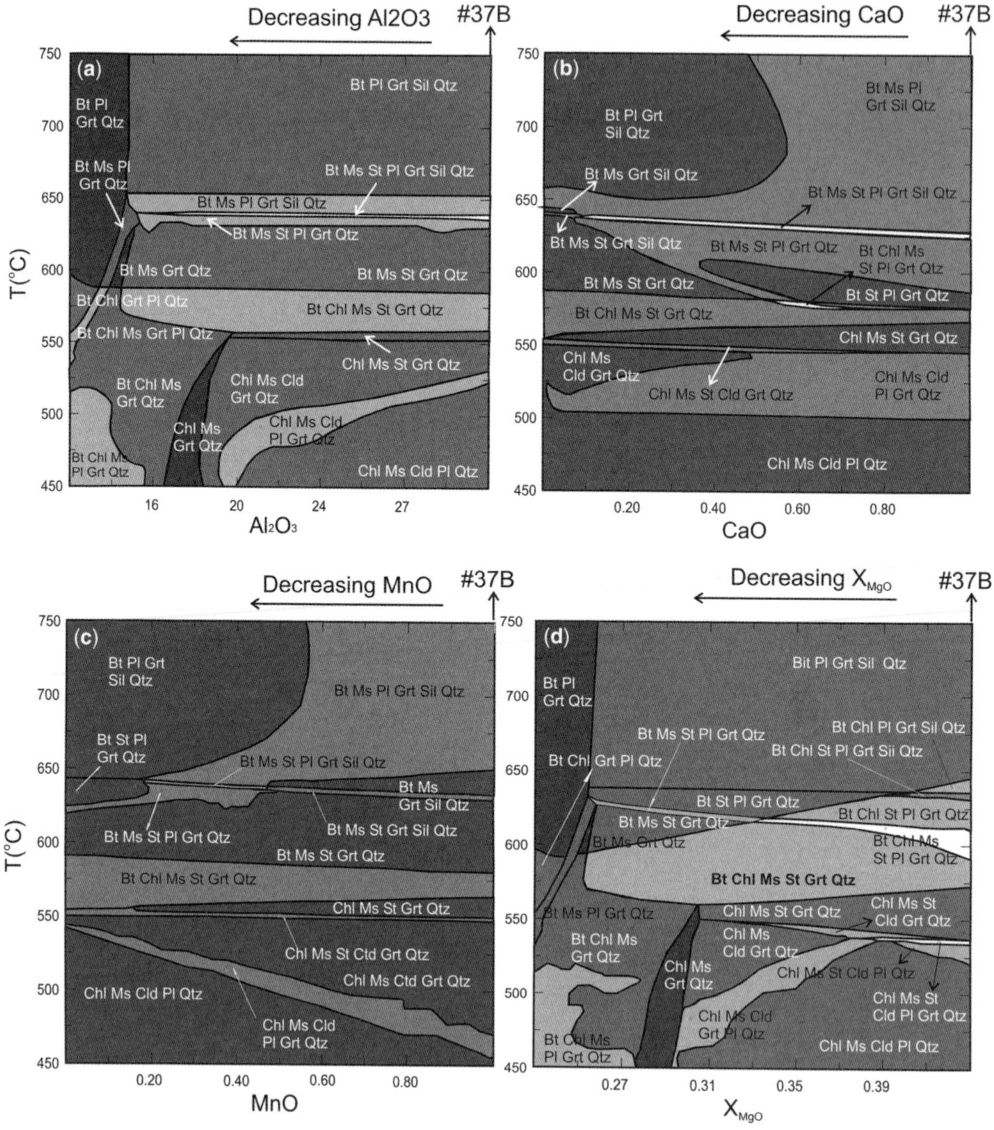

Fig. 9. The figure shows a numerically computed phase diagram (T–Al_2O_3, T–CaO, T–MnO and T–X_{MgO} pseudosection) for a representative bulk composition of metapelite (sample 37B– SiO_2, 47.24; Al_2O_3, 27.48; FeO, 19.13; MgO, 3.23; MnO, 0.36; CaO, 0.10; Na_2O, 0.13; K_2O, 2.32). (**a**) T–Al_2O_3 diagram, marking the stability of staurolite with variation of Al_2O_3 content. (**b**) & (**c**) T–CaO and T–MnO diagram marking the stability of staurolite with variation of CaO and MnO content. (**d**) T–X_{MgO} diagram marking the variation in stability with the variation of Mg number in the rock. Temperature, in these diagrams, is varied along the geothermal gradient, inferred from the P–T estimates (P (bars) $= -2502.5 + 9.167/\mathrm{T}$ (°K)).

be the product of extensive Palaeoproterozoic arc magmatism in a thickened continental crust. This magmatic event is coeval with the geological events observed in the Aravalli Fold Belt (reviewed in Roy & Jakhar 2002). The timing of D_1 folding in, and the accompanying low-grade metamorphism (inferred from the assemblage muscovite + biotite + quartz ± chlorite ± plagioclase) of, the studied metapelitic–psammopelitic rocks are indistinguishable from the timing and grade of metamorphism of the Aravalli Fold Belt (Roy & Jakhar 2002). Subsequently, the AGG and the enveloping supracrustals rocks were involved in three phases of superposed folding (D_2–D_4). Uniform easterly-vergence of the D_2 folds and the sense of drag at the mylonitic zone occurring at the contact between the AGG and the supracrustals rocks are consistent with a top-to-the-east sense of shear movement. The style of the D_2 and D_3 folds, the orientations of their axes, the easterly vergence of the D_2 folds, and the intensity of metamorphism constrained from syn- to post-D_2 porphyroblasts (5.7 ± 1.5 kbar, 560 ± 50 °C) compares well with the style of deformation and the *P–T* condition of metamorphism hitherto published from the enclosing SDFB (Roy & Jakhar 2002).

From a suite of mantle xenoliths, Ganguly *et al.* (1995) calculated the steady-state geothermal gradient for the crust and mantle during Proterozoic and Archaean times (Fig. 10). It is intriguing to note that the estimated *P–T* values constrained from syn- to post-D_2 porphyroblasts, which are presumed to represent the thermobaric conditions of the South Delhi Orogeny, are plotted above the steady-state geotherm of Proterozoic time (Fig. 10). This indicates that the rocks of the studied area were thermally perturbed during the South Delhi Orogeny. Reviewing thc published literature, Harley (1989) identified the following mechanisms that can cause thermal perturbation in regionally metamorphosed rocks:

- advective heat transport through magma or fluid;
- the thinning of crust and mantle with or without magmatism in zones of continental extension;
- heat production through the radioactivity of thickened crust in a continent–continent collision zone.

In view of the lack of syn-D_2 magmas and the preponderance of compressional structures, the first two options do not seem to explain the thermal perturbation. Ganguly *et al.* (1995) calculated transient geothermal gradients that might be expected when a section of crust is overridden a by thrust sheet of 25 and 35 km in thickness (Fig. 10). It is evident that the estimated *P–T* values fall on the transient geotherm calculated for the 25 km thickened crust (Fig. 10). This then follows that the exposed section of the crust now seen in the studied area was buried under approximately 23 km of rock pile (corresponding to the maximum lithostatic pressure of 7.1 kbar) during the evolution of the SDFB in a continental collisional setting. Intense folding (D_2–D_3) during the growth of the garnet and staurolite porphyroblasts is in agreement with the thickening event. The clockwise *P–T* path is considered to be a fingerprint of collisional tectonics (reviewed in Harley 1989). Because of the lack of definitive textural features and significant compositional zoning in the porphyroblastic phases, and the nearly pressure-insensitive stability fields

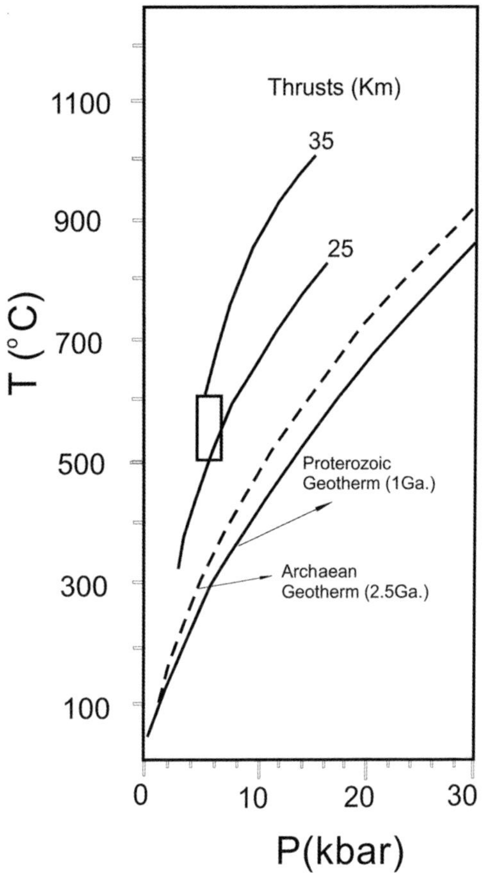

Fig. 10. The estimated *P–T* values constrained from syn- to post-D_2 porphyroblasts, plotted above the steady-state geotherm for crust and mantle of Proterozoic time, after Ganguly *et al.* (1995).

of the metamorphic assemblages in the pseudosections, the sense of the prograde *P–T* path cannot be constrained with certainty. Nevertheless, a clockwise *P–T* path appears to be a plausible mechanism for the growth of porphyroblastic garnet and staurolite during and after the D_2 deformation (Figs 6a & 7a). The inferred clockwise *P–T* path passes through the chloritoid stability fields and cuts the 'chloritoid-out' line (Figs 6a & 7a). This explains the simultaneous growth of garnet and staurolite by the high variant model reaction chloritoid + quartz → garnet + staurolite + H_2O (Spear 1993). The required thermal perturbation of the studied rocks at the time of the South Delhi Orogeny was caused by the radioactive decay of radionuclides in the overburden rocks. The timing of the metamorphism that accompanied D_2–D_3 folding in the Anasagar dome is not known. It is likely that the approximately 0.97 Ga crystallization age of zircon in granitoids that intruded and co-deformed with the supracrustal rocks of the South Delhi Fold Belt from the adjoining Sendra area (Pandit *et al.* 2003) provides the age of D_2 deformation and the accompanying metamorphism of the AGG complex. Our study supports the view that the AGG and its enveloping supracrustals rocks represent the basement onto which sedimentation of the Delhi Supergroup within the SDFB was deposited. The cover sediments along with their Palaeoproterozoic basement were folded, thrusted and domed up during the early Neoproterozoic South Delhi Orogeny. Differential uplift during younger orogeny (or orogenies) led the AGG and its enclosing supracrustal rocks to pop up, thus forming a tectonic window amidst the SDFB (Fig. 1c). The timing of these exhumation events awaits detailed geochronological study.

A recent study by Bhowmik *et al.* (2010) has demonstrated that during the approximately 1.0–0.95 Ga orogenic event (Grenvillian) the roughly 1.72 Ga old Sandmata Granulite Complex of the BGC was reworked in a collisional setting. The authors also demonstrated that the Grenvillian collisional orogenic event affected a vast expanse of the Aravalli–Delhi Mobile Belt (ADMB). The Grenvillian belt of Rajasthan was then correlated with metamorphic belts of similar age in the CITZ (Central Indian Tectonic Zone) and the CGGC (Chotanagpur Granite Gneiss Complex) of central and east India (Bhowmik *et al.* 2010). On the basis of this correlation, Bhowmik *et al.* (2010) supported the view that at least three pre-Grenvillian crustal blocks of Peninsular India were stitched together by Grenvillian mobile belts to form a larger landmass that was part of the supercontinent Rodinia. Reactivation of the approximately 1.85 Ga AGG, presumably during the approximately 0.97 Ga tectonothermal event, is in good agreement with the observations made by Bhowmik *et al.* (2010). However, existing information has demonstrated that the Grenvillian orogenesis in the SDFB affected the Palaeoproterozoic continental basement and the overlying sedimentary cover of the Delhi Supergroup (Pandit *et al.* 2003; Kaur *et al.* 2009; this study). Grenvillian reworking of older continental basement is also reported from the CGGC (Chatterjee *et al.* 2010). More studies are warranted to support the view that the Grenvillian orogenic belts did not result from closure of the intracratonic sedimentary basins but were formed between two or more disparate crustal segments. This information is very important to put a constraint on the configuration of the Indian Shield prior to the Grenvillian Orogeny.

The authors are grateful to Prof. T. Bhattacharya for his useful suggestions and discussions in the field. Funding for the fieldwork was sourced from DST Women Scientist (WOS-A) Project, grant No. SR/WOS-A/ES-02/2006, awarded to Dr N. Chattopadhyay. P. Sengupta acknowledges CAS, Department of Geological Sciences, Jadavpur University for their financial assistance. We thank Prof. S.K. Bhowmik and an anonymous reviewer for their erudite comments, which has substantially improved the clarity of the manuscript. This is a contribution to IGCP 509.

References

Bhattacharya, A., Mohanty, L., Maji, A., Sen, S. K. & Raith, M. 1992. Non-ideal mixing in the phlogopite–annite boundary: Constraints from experimental data on Mg–Fe partitioning and a reformulation of the biotite–garnet geothermometer. *Contributions to Mineralogy and Petrology*, **111**, 87–93.

Bhowmik, S. K., Bernhardt, H. J. & Dasgupta, S. 2010. Grenvillian age high-pressure upper amphibolite–granulite metamorphism in the Aravalli–Delhi Mobile Belt, Northwestern India: New evidence from monazite chemical age and its implication. *Precambrian Research*, **178**, 168–184.

Bhowmik, S. K., Saha, L., Dasgupta, S. & Fukuoka, M. 2009. Metamorphic phase relations in orthopyroxene-bearing granitoids: implication for high-pressure metamorphism and prograde melting in the continental crust. *Journal of Metamorphic Geology*, **27**, 295–315.

Bott, M. H. P. 1971. Evolution of young continental margins and formation of shelf basins. *Tectonophysics*, **11**, 319–327.

Buick, I. S., Allen, C., Pandit, M., Rubatto, D. & Hermann, J. 2006. The Proterozoic magmatic and metamorphic history of the Banded Gneissic Complex, central Rajasthan, India: LA-ICP-MS U-Pb zircon constraints. *Precambrian Research*, **151**, 119–142.

Chatterjee, N., Banerjee, M., Bhattacharya, A. & Maji, A. K. 2010. Monazite chronology, metamorphism-anatexis and tectonic relevance of the mid-Neoproterozoic Eastern Indian Tectonic Zone. *Precambrian Research*, **179**, 99–120.

Choudhary, A. K., Gopalan, K. & Shastry, C. A. 1984. Present status of the geochronology of the Precambrian rocks of Rajasthan. *Tectonophysics*, **105**, 131–140.

Condie, K. C. 1997. *Plate Tectonics and Crustal Evolution*, 4th edn. Butterworth-Heinemann, Oxford.

Condie, K. C. 2002. Breakup of a paleoproterozoic supercontinent. *Gondwana Research*, **5**, 41–43.

Connolly, J. A. D. 2005. Computation of phase equilibria by linear programming: a tool for geodynamic modeling and its application to subduction zone decarbonation. *Earth and Planetary Science Letters*, **236**, 524–541.

Dasgupta, S., Sengupta, P., Guha, D. & Fukuoka, M. 1991. A refined garnet-biotite Fe–Mg exchange geothermometer and its application in amphibolites and granulites. *Contributions to Mineralogy and Petrology*, **109**, 130–137.

Deb, M. & Sarkar, S. C. 1990. Proterozoic tectonic evolution and metallogenesis in the Aravalli–Delhi orogenic complex, north western India. *Precambrian Research*, **46**, 115–137.

DEB, M. & SEHGAL, U. 1997. Petrology and geothermobarometry and C–O–H–S fluid compositions in the environs of Rampura–Agucha Zn (–Pb) ore deposit, Bhilwara district, Rajasthan. *Proceedings of the Indian Academy of Sciences (Earth and Planetary Sciences)*, **106**, 343–356.

DEB, M., THORPE, R. I., CUMMING, G. L. & WAGNER, P. A. 1989. Age, source and stratigraphic implications of Pb isotope data for conformable, sediment-hosted, base metal deposits in the Proterozoic Aravalli–Delhi orogenic belt, northwestern India. *Precambrian Research*, **43**, 1–22.

DEB, M., THORPE, R. I. & KRSTIC, D. 2002. Hindoli group of rocks in the eastern fringe of the Aravalli–Delhi orogenic belt-Archean secondary greenstone belt or Proterozoic supracrustals? *Gondwana Research*, **5**, 879–883.

DEB, M., THORPE, R. I., KRSTIC, D., CORFU, F. & DAVIS, D. W. 2001. Zircon U–Pb and galena Pb isotope evidence for an approximate 1.0 Ga terrane constituting the western margin of the Aravalli –Delhi orogenic belt, northwestern India. *Precambrian Research*, **108**, 195–213.

EVANS, T. P. 2004. A method for calculating effective bulk composition modification due to crystal fractionation in garnet-bearing schist: implications for isopleth thermobarometry. *Journal of Metamorphic Geology*, **22**, 547–557.

FERRY, J. M. & SPEAR, F. S. 1978. Experimental calibration of the partitioning of Fe and Mg between biotite and garnet. *Contributions to Mineralogy and Petrology*, **66**, 113–117.

GANGULY, J. & SAXENA, S. K. 1984. Mixing properties of aluminosilicate garnets: Constraints from natural and experimental data, and applications to geothermobarometry. *American Mineralogist*, **69**, 88–97.

GANGULY, J., SINGH, R. N. & RAMANA, D. V. 1995. Thermal perturbation during charnockitization and granulite facies metamorphism in southern India. *Journal of Metamorphic Geology*, **13**, 419–430.

GANGULY, J., CHENG, W. & TIRONE, M. 1996. Thermodynamics of aluminosilicate garnet solid solution: New experimental data, an optimized model, and thermometric applications. *Contributions to Mineralogy and Petrology*, **123**, 137–151.

GOSCOMBE, B. D. & HAND, M. 2000. Contrasting *P–T* paths in the Eastern Himalaya, Nepal: Inverted isograds in a Paired Metamorphic Mountain Belt. *Journal of Petrology*, **41**, 1673–1719.

GUPTA, P., FAREEDUDDIN, MUKHOPADHYAY, K. & REDDY, M. S. 1991. Structural style in the rocks of the Delhi Supergroup along the eastern part of Aravalli range in central Rajasthan. *Indian Journal of Geology*, **63**, 190–205.

GUPTA, P., FAREEDUDDIN, REDDY, M. S. & MUKHOPADHYAY, K. 1995. Stratigraphy and structure of Delhi Supergroup of rocks in central part of Aravalli range. *Recordings in Geological Survey of India*, **120**, 12–26.

GUPTA, S. N., ARORA, Y. K., MATHUR, R. K., IQBALUDDIN, PRASAD, B., SAHAI, T. N. & SHARMA, S. B. 1980. *Lithostratigraphic Map of the Aravalli Region, Southern Rajasthan and Northeastern Gujarat, Scale 1:1 000 000*. Geological Survey of India, Calcutta.

GUPTA, S. N., ARORA, Y. K., MATHUR, R. K., IQBALUDDIN, PRASAD, B., SAHAI, T. N. & SHARMA, S. B. 1997. *The Precambrian Geology of the Aravalli Region, Southern Rajasthan and Northeastern Gujarat*. Geological Society of India, Memoir, **123**, 262.

HARLEY, S. L. 1989. The origin of granulites: a metamorphic perspective. *Geological Magazine*, **126**, 215–247.

HERON, A. M. 1953. *The Geology of Central Rajputana*. Geological Society of India, Memoir, **79**.

HOLLAND, T. J. B. H. & POWELL, R. 1998. An internally consistent thermodynamic data set for phases of petrological interest. *Journal of Metamorphic Geology*, **16**, 309–343.

HOLLAND, T., BAKER, J. & POWELL, R. 1998. Mixing properties and activity–composition relationships of chlorites in the system $MgO-FeO-Al_2O_3-SiO_2-H_2O$. *European Journal of Mineralogy*, **10**, 395–406.

HOLDAWAY, M. J. 2000. Application of new experimental and garnet Margules data to the garnet–biotite geothermometer. *American Mineralogist*, **85**, 881–892.

JAMIESON, R. A., CULSHAW, N. G., WODICKA, N. & CORRIGAN, D. 1992. Timing and tectonic setting of Grenvillian metamorphism – constraints from a transect along Georgian Bay, Ontario. *Journal of Metamorphic Geology*, **10**, 321–332.

JERABEK, P., JANAK, M. K., FARYAD, S. W., FINGER, F. & KONECNY, P. 2008. Polymetamorphic evolution of pelitic schists and evidence for Permian low-pressure metamorphism in the Vepor Unit, West Carpathians. *Journal of Metamorphic Geology*, **26**, 465–485.

KAUR, P., CHAUDHRI, N., RACZEK, I., KRÖNER, A. & HOFMANN, A. W. 2009. Record of 1.82 Ga Andean-type continental arc magmatism in NE Rajasthan, India: Insights from zircon and Sm–Nd ages, combined with Nd–Sr isotope geochemistry. *Gondwana Research*, **16**, 56–71.

MAHATO, S., GOON, S., BHATTACHARYA, A., MISHRA, B. & BERNHARDT, H.-J. 2008. Thermotectonic evolution of the North Singhbhum Mobile Belt (Eastern India): a view from the western part of the belt. *Precambrian Research*, **162**, 102–127.

MOORES, E. M. 1991. Southwest U.S.-East Antarctica (SWEAT) connection: a hypothesis. *Geology*, **19**, 425–428.

MUKHERJEE, A. D., BHATTACHARYYA, P. K. & BHATTACHARYA, H. N. 1992. Petrotectonics of a southern segment of the polymetalic sulphide-bearing Proterozoic Delhi mobile belt, India, and its implications. *In*: SARKAR, S. C. (ed.) *Metallogeny Related to Tectonics of the Proterozoic Mobile Belts*. Balkema, Rotterdam, 253–270.

MUKHOPADHYAY, D. & BHATTACHARYA, T. 2000. Tectono-statigraphic framework of the South Delhi Fold Belt in the Ajmer – Beawar region, central Rajasthan, India: A critical review. *In*: DEB, M. (ed.) *Crustal Evolution and Metallogeny in the Northwestern Indian Shield (A Festschrift for A. Mookherjee)*. Narosa, New Delhi, 126–137.

MUKHOPADHYAY, D. & DASGUPTA, S. 1978. Delhi–pre Delhi relations near Badnor, Central Rajasthan. *Indian Journal of Earth Sciences*, **5**, 183–190.

Mukhopadhyay, D., Bhattacharya, T., Chattopadhyay, N., Lopez, R. & Tobisch, O. T. 2000. Anasagar gneiss: a folded granitoid pluton in the Proterozoic South Delhi Fold Belt, central Rajasthan. *Proceedings of the Indian Academy of Sciences (Earth and Planetary Science)*, **109**, 21–37.

Mukhopadhyay, D., Chattopadhyay, N. & Bhattacharya, T. 2010. Structural evolution of a gneiss dome in the axial zone of the Proterozoic South Delhi Fold Belt in Central Rajasthan. *Journal of Geological Society of India*, **75**, 18–31.

Naha, K. & Mohanty, S. 1990. Structural studies in the Pre-Vindhyan rocks of Rajasthan: A summary of work of the last three decades. *In*: Naha, K., Ghosh, S. K. & Mukhopadhyay, D. (eds) *Structure and Tectonics: The Indian Scene. Proceedings of the Indian Academy of Sciences (Earth and Planetary Sciences)*, **99**, 279–291.

Naha, K., Mukhopadhyay, D. K. & Mohanty, R. 1988. Structural evolution of the rocks of the Delhi Group around Khetri, northeastern Rajasthan. *In*: Roy, A. B. (ed.) *Precambrian of the Aravalli Mountain, Rajasthan, India*. Geological Society of India, Memoir, **7**, 207–245.

Naha, K., Mukhopadhyay, D. K., Mohanty, R., Milra, S. K. & Biswal, T. K. 1984. Significance of contrast in the early stages of the structural history of the Delhi and pre-Delhi rock groups in the Proterozoic of Rajasthan, western India. *Tectonophysics*, **105**, 193–206.

Newton, R. C., Charlu, T. V. & Kleppa, O. J. 1980. Thermochemistry of the high structural state plagioclases. *Geochimica et Cosmochimica Acta.*, **44**, 933–941.

Pandit, M. K., Carter, L. M., Ashwal, L. D., Tucker, R. D., Torsvik, T. H., Jamtveit, B. & Bhushan, S. K. 2003. Age, petrogenesis and significance of 1 Ga granitoids and related rocks from the Sendra area, Aravalli Craton, NW India. *Journal of Asian Earth Sciences*, **22**, 363–381.

Pant, N. C., Kundu, A. & Joshi, S. 2008. Age of metamorphism of Delhi supergroup rocks—electron microprobe ages from Mahendragarh district, Haryana. *Journal of the Geological Society of India*, **72**, 365–372.

Perchuk, L. L. & Lavrent'eva. 1983. Experimental investigation of exchange equilibria in the system cordierite-garnet-biotite. *In*: Saxena, S. K. (ed.) *Kinetics and Equilibrium in Mineral Reactions. Advances in Physical Geochemistry*, Springer, New York, **3**, 199–239.

Powell, R. & Holland, T. 1999. Relating formulations of the thermodynamics of mineral solid solutions: activity modeling of pyroxenes, amphiboles, and micas. *American Mineralogist*, **84**, 1–14.

Reddy, S. M. & Evans, A. D. 2009. Palaeoproterozoic supercontinents and global evolution: correlations from core to atmosphere. *In*: Reddy, S. M., Mazumder, R., Evans, D. A. D. & Collins, A. S. (eds) *Palaeoproterozoic Supercontinents and Global Evolution*. Geological Society, London, Special Publications, **323**, 1–26.

Regnier, J. L., Ring, U., Passchier, C. W., Gessner, K. & Gungor, T. 2003. Contrasting metamorphic evolution of metasedimentary rocks from the Cine and Selimiye nappes in the Anatolide belt, western Turkey. *Journal of Metamorphic Geology*, **21**, 699–721.

Roday, P. P. 1979. Structural patterns in the Ajabgarh rocks around Anakhar, district Ajmer. *Journal of the Geological Society of India*, **20**, 441–449.

Rogers, J. J. W. 2000. Origin and fragmentation of the possible approximately 1.5-Ga supercontinent Columbia, Abstracts with Programs. *Geological Society of America*, **32**, 455.

Rogers, J. J. W. & Santosh, M. 2009. Tectonics and surface effects of the supercontinent Columbia. *Gondwana Research*, **15**, 373–380.

Roy, A. B. 1988. Stratigraphic and tectonic framework of the Aravalli Mountain Range. *In*: Roy, A. B. (ed.) *Precambrian of the Aravalli Mountain, Rajasthan, India*. Geological Society of India, Memoir, **7**, 3–31.

Roy, A. B. & Jakhar, S. R. 2002. *Geology of Rajasthan (Northwest India) Precambrian to Recent*. Scientific Publishers, Jodhpur.

Saha, L., Bhowmik, S. K., Fukuoka, M. & Dasgupta, S. 2008. Contrasting episodes of regional granulite–facies metamorphism in enclaves and host gneisses from the Aravalli–Delhi Mobile Belt, NW India. *Journal of Petrology*, **49**, 107–128.

Sarkar, S. C. 2000. Geological setting, characteristics, origin and evolution of sediment-hosted sulfide ore deposits of Rajasthan: a critique with comments on their implications for future exploration. *In*: Deb, M. (ed.) *Crustal Evolution and Metallogeny in the Northwestern Indian Shield (A Festschrift for A. Mookherjee)*. Narosa, New Delhi, 240–292.

Sharma, R. S. 1988. Patterns of metamorphism in the Precambrian rocks of the Aravalli mountain belt. *In*: Roy, A. B. (ed.) *Precambrian of the Aravalli Mountain, Rajasthan, India*. Geological Society of India, Memoir, **7**, 33–76.

Singh, Y. K., de Waele, B., Karmakar, S., Sarkar, S. & Biswal, T. K. 2010. Tectonic setting of the Balaram–Kui–Surpagla–Kengora granulites of the South Delhi Terrane of the Aravalli Mobile Belt, NW India and its implication on correlation with the East African Orogen in the Gondwana assembly. *Precambrian Research*, **183**, 669–688.

Sinha Roy, S. 1988. Proterozoic Wilson cycles in Rajasthan. *In*: Roy, A. B. (ed.) *Precambrian of the Aravalli Mountain, Rajasthan, India*. Geological Society of India, Memoir, **7**, 95–107.

Sinha Roy, S., Malhotra, G. & Mohanty, M. 1998. *Geology of Rajasthan*. Geological Society of India, Bangalore.

Spear, F. S. 1993. *Metamorphic Phase Equilibria and Pressure–Temperature–Time Paths*. Mineralogical Society of America, Washington, DC.

Sugden, T. J. & Windley, B. F. 1984. Geotectonic framework of the early–mid Proterozoic Aravalli–Delhi orogenic belt, N. W. India. *Geological Association of Canada, Programs with Abstracts*, **9**, 109.

Thompson, A. B. 1976. Mineral reactions in pelitic rocks: I. Prediction of P-7″-.Y(Fe–Mg) phase relations. *American Journal of Science*, **276**, 401–424.

Tobisch, O. T., Collerson, K. D., Bhattacharya, T. & Mukhopadhyay, D. 1994. Structural relationship and

Sm–Nd isotope systematics of polymetamorphic granitic gneisses and granitic rocks from central Rajasthan, India: implications for the evolution of the Aravalli craton. *Precambrian Research*, **65**, 319–339.

VERNON, R. 2004. *A Practical Guide to Rock Microstructure.* Cambridge University Press, Cambridge.

VOLPE, A. M. & MACDOUGALL, J. D. 1990. Geochemistry and isotope characteristics of mafic (Phulad ophiolite) and related rocks in the Delhi Supergroup, Rajasthan, India: implications for rifting in the Proterozoic. *Precambrian Research*, **48**, 167–191.

WEI, C. J. & POWELL, R. 2004. Calculated phase relations in high-pressure metapelites in the system NKFMASH (Na_2O–K_2O–FeO–MgO–Al_2O_3–SiO_2–H_2O) with application to natural rocks. *Journal of Petrology*, **44**, 183–202.

WHITE, R. W., POWELL, R., HOLLAND, T. J. B. & WORLEY, B. A. 2000. The effect of TiO_2 and Fe_2O_3 on metapelitic assemblages at greenschist and amphibolite facies conditions: mineral equilibria calculations in the system K_2O–FeO–MgO–Al_2O_3–SiO_2–H_2O–TiO_2–Fe_2O_3. *Journal of Metamorphic Geology*, **18**, 497–511.

WILLIAMS, M. L. & GRAMBLING, J. A. 1990. Manganese, ferric iron, and the equilibrium between garnet and biotite. *American Mineralogist*, **75**, 886–908.

WU, C. M., ZHANG, J. & REN, L. D. 2004. Empirical garnet–biotite–plagioclase– quartz (GBPQ) geobarometry in medium- to high-grade metapelites. *Journal of Petrology*, **45**, 1907–1921.

ZHAO, G. C., CAWOOD, P. A., WILDE, S. A. & SUN, M. 2002. Review of global 2.1–1.8 Ga orogens: implications for a pre-Rodinia supercontinent. *Earth-Science Reviews*, **59**, 125–162.

ZHAO, G. C., SUN, M., WILDE, S. A. & LI, S. 2004. A Paleo-Mesoproterozoic supercontinent: assembly, growth and breakup. *Earth-Science Reviews*, **67**, 91–123.

Morphodiversity, complexity and macroevolution: revealed by the megascopic life of the Palaeo-Neoproterozoic Vindhyan Supergroup, India

PURNIMA SRIVASTAVA

Centre of Advanced Study in Geology, Lucknow University, Lucknow, India
(e-mail: purnimasrivastava_51@rediffmail.com)

Abstract: Evidence of past life found in the Vindhyan Supergroup exhibits extensive diversity and taxonomically unresolved morphological complexities indicating varied biological affinities, advanced life and macroevolutionary dynamics of evolving clades. The megascopic life of the supergroup documents possibly two macroevolutionary modes, demarcated as: (i) the Palaeo-Mesoproterozoic transitional period, which is represented by the fossil assemblages of the Semri and Rewa groups, and exhibits the domination of a diversified prokaryotic cyanobacterial community and moderately diversified megascopic life; (ii) the Neoproterozoic (Bhander groups, Uppermost Vindhyans), represented by the diversified and morphologically complex megascopic eukaryotes, and the emergence of multicellularity among plant and animal clades, viz. structures resembling bryophytes, sporophyte and Ediacara fauna. The overall Vindhyan fossils exhibit episodes of adaptive innovations and the emergence of complex morphologies in separate lineages of evolutionary history, resulting in extensive morphodiversity.

The present assemblage can be represented by that of prokaryotes, eukaryotes, possibly a symbiotic association, a *Jacutianema*-like Neoproterozoic form, specimens possibly of vaucheriacean affinity, mesoscopic intermediate forms inferred as transitional forms of microscopic–megascopic evolution, discs with process-like structures, the emergence of animal life (Ediacaran fauna) and a number of other complex morphologies that have no parallels in time and space. Traditionally and conventionally, the age of the Vindhyan Supergroup is considered to be that of Palaeoproterozoic–Neoproterozoic.

Life in the Vindhyan Supergroup is represented by a wide range in size and variable morphologies of fossils, indicative of different biological affinities, the existence of complex life forms and extensive morphodiversity. Available fossil data support the estimated age of Palaeoproterozoic–Neoproterozoic (*c.* 1700–545 Ma). This period has been considered a significant time that witnessed major changes in Earth's atmosphere, like transition to an oxygenated atmosphere, organic carbon burial, glaciations and biotic events in the evolutionary history. Chronologically, the Vindhyan's life can be marked by the domination of the cyanobacterial prokaryotic community (McMenamin *et al.* 1983; Kumar & Srivastava 1992, 1995; Sharma 2006; Prasad 2007; Kumar & Pandey 2008*b*), the emergence of eukaryotes (especially acritarchs) and their subsequent radiation (Prasad *et al.* 2005*a*, *b*; Prasad 2007), the transition from microscopic to megascopic life (Srivastava 2004; Srivastava & Bali 2006; Kumar & Pandey 2008*b*; Sharma & Shukla 2009*b*), the evolution from unicellular to multicellular organization (Kumar 2001; Srivastava 2009*a*) and the appearance of metazoans and metaphytes (Kumar 1995, 2001*a*; Kumar & Srivastava 1997; De 2003, 2006; Sharma 2006; Srivastava 2006, 2009*b*) on the basis of data available up to now. Recently, Bengtson *et al.* (2009) published a paper supporting the authenticity of controversial fossils, as reported by Azmi (1998) and Azmi *et al.* (2007), which have been under debate for some time (shelly fauna, annular tubes, metazoan embryo and calcified tubes of cyanobacteria). At the same time they retained the traditional Palaeo-Mesoproterozoic age for the Lower Vindhyans and suggested the earlier evolution of different clades.

The supergroup is globally acknowledged as one of the best repositories for evidence of Proterozoic life, comprising exceptionally well-preserved fossils in three taphonomic windows. These are: (i) silicified microfossils in chert; (ii) meso-megascopic carbonaceous compressions in shale partings; and (iii) moulds and casts on sandstone rock surfaces (responsible for the preservation of non-biomineralizing organisms in geological records). Based on available data, bacterial, cyanobacterial, algal, fungal, acritarchean, symbiotic association or lichens and metazoan affinities can be assigned to the Vindhyan fossils ranging from the microscopic to the megascopic level.

The present study is based on the evolutionary history and patterns of morphodiversity among Vindhyan megafossils (considered authentic from

From: MAZUMDER, R. & SAHA, D. (eds) 2012. *Palaeoproterozoic of India*. Geological Society, London, Special Publications, **365**, 247–262. http://dx.doi.org/10.1144/SP365.13

time to time by a number of reviewers (viz. Venkatachala *et al.* 1996; Kumar 2001*b*; Kumar & Gupta 2002; Sharma 2003; Ray 2006; Sharma & Shukla 2009*a*; Kumar & Sharma 2010). This study also provides information for a better understanding of the macroevolutionary dynamics of evolving clades and changes in ecosystems; and endeavours to decipher the physical processes occurring during the sedimentation of the supergroup that governed the evolutionary changes.

Geological setting

The Vindhyan Supergroup, about 4000 m thick, is the least metamorphosed sedimentary succession distributed in central India around the Bundelkhand Massif, spreading from Agra through SE Rajasthan to the Son Valley, up to Sasaram in Bihar to Hoshangabad in the south, occupying an area of about 104 000 km^2 (Fig. 1). The supergroup is subdivided into four groups, viz. the Semri Group, the Kaimur Group, the Rewa Group and the Bhander Group (Auden 1933; Prasad 1984; Sastry & Moitra 1984; Soni *et al.* 1987). Each group is further divided into 'Formations and Members (Fig. 2). Traditionally, the Semri Group is considered as the Lower Vindhyans, whereas the other three groups are categorized within the Upper Vindhyans. The overall lithology is represented by the sediments of calcareous, argillaceous and arenaceous facies (Fig. 2). The supergroup unconformably overlies the Bundelkhand Massif and the slightly metamorphosed Bijawar Group, which is approximately 2500 Ma in age (Crawford & Compston 1970; Mandal *et al.* 2002).

Age

The age of the Vindhyan Supergroup is still a matter of debate but conventionally it is considered to be of Palaeo-Neoproterozoic age. Various parameters for defining the age in different geological units have been summarized in Figure 2 and a revised version of the estimated ages based on megafossils is given in Table 1. Records of 1.1 Ga old triploblastic animal traces (Seilacher *et al.* 1998) from the Churhat Sandstone; small shelly fauna, calcareous algae, metazoan embryo (Azmi 1998; Azmi *et al.* 2007) from the Rohtas and Chitrakut formations, respectively; and Ediacara fauna *Spriggina* reported by Kathal *et al.* (2000), all belonging to the Semri Group, Lower Vindhyans, suggest a far younger

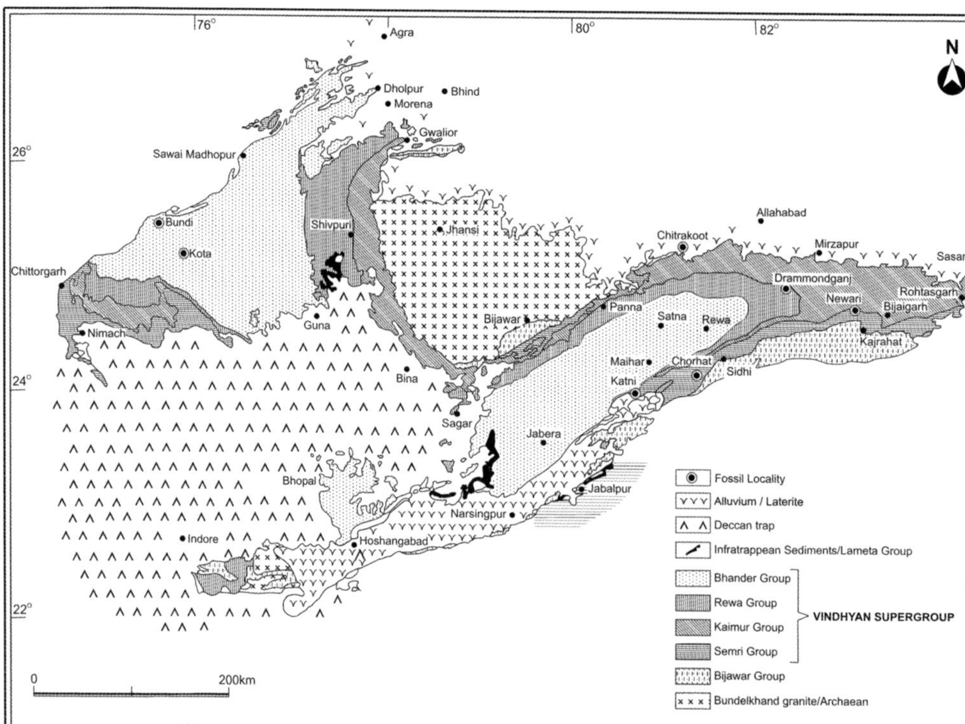

Fig. 1. General geological map of the Vindhyan Supergroup, simplified after Soni *et al.* (1987).

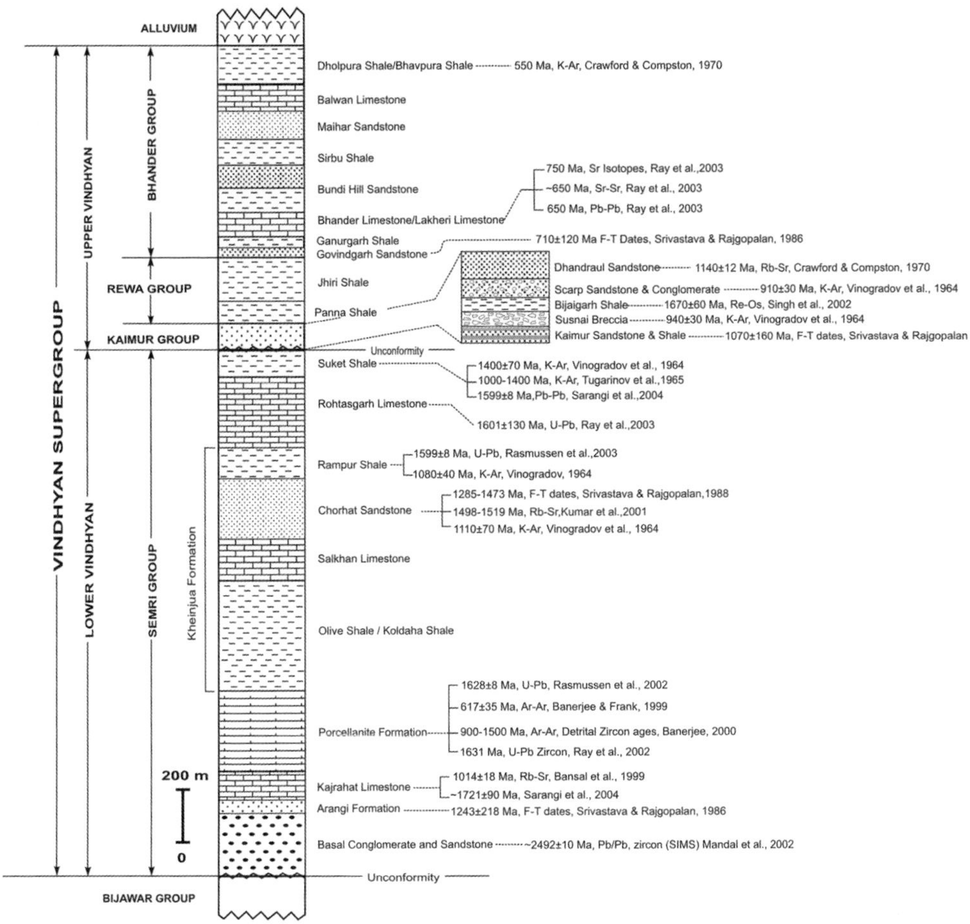

Fig. 2. Generalized lithostratigraphy of the Vindhyan Supergroup.

age than the conventional age of the supergroup. These reports created a hot debatable issue of abiotic origin for the referenced fossils (Conway Morris *et al.* 1998; Brasier 1999; Kerr 2002; Hofmann 2005). The age, biogenicity and reproducibility of these debatable fossils remained inconclusive, and, hence generated interest among geochronologists to resolve the age problem, especially the upper age limit and time of deep metazoan origin in the Vindhyan Supergroup. Recently, a published paper by Bengtson *et al.* (2009) authenticated the controversial fossils of Azmi's shelly fauna, annular tubes, metazoan embryo and calcified tubes of cyanobacteria. At the same time they retained the traditional Palaeo-Mesoproterozoic age for the Lower Vindhyans and suggested the earlier evolution of different clades.

Malone *et al.* (2008) suggested an almost 500 Ma older age for the sedimentation time of the Rewa and Bhander groups on the basis of palaeomagnetic data and detrital zircon dates. The presence of an age-marker acanthomorph *Trachyhystrichosphaera* from the Sirbu Shale (Srivastava 2009*a*) and the Ediacaran fossils from the Bhander Group support the Ediacaran age as an upper age limit for the supergroup. This age assumption is also supported by the absence of any typical Cambrian fossil record from the supergroup.

The megafossils reported by a number of workers from the Vindhyans Supergroup until now and the estimated ages have been summarized in Table 1.

Repository

Specimens, morphologically described in this paper have been deposited in the museum of the Centre of Advanced Study in Geology, Lucknow University, Lucknow, India.

Table 1. *Megafossils reported from the different geological units of the Vindhyan Supergroup, their estimated age, respective workers and year of publication*

Serial no.	Geological unit	Age	Fossils	Reported by:
1	Bhavpura/Dholpura Shale	Vendian	*Chuaria, Tawuia*	Srivastava (2002)
2	Upper Bhander Sandstone/ Maihar Sandstone	Ediacaran	*Skolithos, Planolites*, Medusoids etc., *Arumberia banksi*	Moitra & De (1998), De (2003, 2006), Kumar & Pandey (2008*a*)
3	Sirbu Shale	Late Proterozoic	*Chuaria, Tawuia*	Kumar & Srivastava (1997, 2003)
4	Bundi Hill Sandstone	Ediacaran	Possible Ediacaran metazoans	Srivastava (2006)
5	Bhander Limestone	Late Proterozoic	*Chuaria, Tawuia etc.*, Sponge spicules (?)	Kumar (1999), Kumar & Srivastava (2003)
6	Jhiri Shale	Mesoproterozoic	*Chuaria, Tawuia*, etc.	Rai *et al.* (1997)
7	Panna Shale	Mesoproterozoic	*Chuaria, Tawuia* and mesoscopic carbonaceous films	Srivastava (2004)
8	Suket Shale	Cambrian/Ordovician, Mesoproterozoic	Small shelly fauna, Carbonaceous compressions, calcareous algae, *Orbisiana, Chuaria, Tawuia*	Azmi (1998), Kumar (2001), Azmi *et al.* (2007)
9	Rohtasgarh Limestone	Palaeo-Mesoproterozoic	*Katnia Singhii, Grypania*	Tandon & Kumar (1977), Sharma & Shukla (2009*a*, *b*)
10	Rampur Shale/Chorhat Sandstone	Neoproterozoic, Cambrian/ Ordovician, Palaeoproterozoic	Triploblastic animal traces, horizontal burrow, *Chuaria, Tawuia*, etc., microbial mats	Seilacher *et al.* (1998), Sarkar *et al.* (2006), Srivastava & Bali (2006), Azmi *et al.* (2007)
11	Koldaha/Olive Shale	Palaeoproterozoic	Multicellular macroalga, *Chuaria, Tawuia*, Scolecodonts	Rai & Singh (2006), Sharma (2006), Azmi *et al.* (2007)

Megafossils from the Vindhyan Supergroup

The presence of megascopic life in the Vindhyan Supergroup was established long ago by Jones (1909), who first reported carbonaceous discs from the Suket Shale (Semri Group, Lower Vindhyans). Kumar (1995) reported *Chuaria–Tawuia* from the same stratigraphic unit. Carbonaceous megafossils from the Sirbu Shale (Kumar & Srivastava 1997) was the first report of megafossils from the Bhander Group in the Upper Vindhyans. A number of workers reviewed all of the available records of Vindhyan life (Venkatachala *et al.* 1996; Kumar 2001*b*; Sharma 2006) and concluded that, out of 50 or more reports (at that time), less than a quarter were acceptable as true fossils. Kumar & Gupta (2002) accepted only four reports comprising true fossils and placed the rest among dubiofossils (the author of this paper also supports this observation). Morphologies reported from the Vindhyans as trace fossils and burrows by many workers were later considered as structures produced by inorganic processes (Kumar & Sharma 2010).

Carbonaceous films attributed to multicellular thalloid have been reported from the Koldaha Shale or Olive Shale (Rai & Singh 2006; Sharma 2006*a*). Srivastava & Bali (2006) reported mesoscopic carbonaceous fossils along with *Chuaria–Tawuia*, and considered these mesoscopic fossils as transitional forms and a link between the evolution from micro- to megascopic life. Kumar (1995), Srivastava (2008) and Sharma & Shukla (2009*a*, *b*) reported *Grypania* from the Rohtas Formation, central India. De (2003, 2006) and Srivastava (2006) reported Ediacaran fossils from the Bhander Group, but these fossils are a little dubious. In the absence of any Cambrian fossil, the carbonaceous megafossils of the Vindhyan Supergroup have been bracketed between 1800 and 550 Ma by Sharma & Kumar (2010). Sarkar *et al.* (2006) and Kumar & Pandey (2008*a*) reported microbial mats in the Chorhat Sandstone (Lower Vindhyans) and the Maihar Sandstone (Uppermost litho-unit of Vindhyan's in central India), respectively.

Overall, Vindhyans life can be represented by the presence of bacterial, cyanobacterial, algal, fungal, acritarchean and Ediacaran metazoans ranging in size from the microscopic to the megascopic scale. Like other parts of the world, the Palaeoproterozoic megafossil record of the Vindhyan Supergroup is meager, simple and less diversified in comparison to the younger sequences.

Morphodiversity and complexity

Episodes of morphodiversity, the transition of clades, adaptive innovations and the emergence of complex morphologies in separate lineages can be seen in the evolutionary history of the Vindhyan megafossil record. Evolutionary pathways and the appearance of different clades, observed among megafossils recovered from the different stratigraphic levels of the Vindhyans, ranging in age from the Palaeoproterozoic to the Neoproterozoic are unique and worth describing (Table 1).

The evolution of life from the microscopic to the megascopic level, represented by mesoscopic fossils and interpreted as intermediate forms or a missing link (Srivastava 2004), and the development of the nucleus in cells (eukaryotes) are the landmark evolutionary signals of the Palaeoproterozoic–Mesoproterozoic time.

In general, the Vindhyan fossils exhibit a wide size range, from less than 1 μm to 1 m, and varied biological affinities, such as single-celled simple prokaryotes to complex and advanced multicellular metaphytes and metazoans (Srivastava & Tewari 2011). These fossils have the potential for aiding interpretation of the morphodiversity, complexity and macroevolution in the Vindhyan Supergroup.

The domination of the cyanobacterial mat-building benthic community and the simple morphologies among microfossils is a common feature of most of the Palaeo-Mesoproterozoic microfossil assemblages all over the world. There are few complex fossil forms of this age and they are difficult to compare with any fossil or extant form. Although their number is meagre, their level of preservation is exceptionally good.

Life gradually evolved and morphologies became more complex during the Neoproterozoic, which is well demonstarated by the microscopic as well as the megascopic fossil record of the Bhander Group – the youngest group of the Vindhyan Supergroup. The Neoproterozoic fossil record reflects a picture of a more diverse and dynamic ecosystem than that of the proceeding era.

Palaeobiological evidence and inferences, based on the study of megafossils from the Semri and the Rewa groups of Palaeo-Mesoproterzoic age (as no reliable megafossil record has yet come from the Kaimur Group) of the Vindhyan Supergroup are as follows.

- Carbonaceous megafossils, such as *Chuaria* and *Tawuia*, are commonly occurring forms (although *Tawuia* are very few in number in comparison to *Chuaria*) exhibiting simple morphologies. These are reported from the Suket Shale (Kumar 2001*a*) and the Churhat Sandstone (Srivastava & Bali 2006) of the Semri Group, Lower Vindyans.
- Carbonaceous compressions inferred as multicellular macro alga from the Olive Shale (1800–1600 Ma old), Semri Group, Lower Vindhyans have also been reported (Rai &

Gautam 1998; Rai & Singh 2006; Sharma 2006). Preservation of these algal fossils is moderate to poor; hence, conclusions based on these fossils are unclear.

- A spirally coiled megascopic filamentous form *Grypania* (Fig. 3g), considered to be of cyanobacterial or algal affinity (Knoll *et al.* 2006), has been reported from the Rohtas Formation (*c.* 1700 Ma old), Katni area, Semri Group, Lower Vindhyans (Tandon & Kumar 1977; Kumar 1995; Srivastava 2008; Sharma & Shukla 2009*b*). In other parts of the world, *Grypania* is recorded from approximately 1800 Ma old rocks located in Michigan, USA, where it was compared with a giant, unicellular photosynthetic green algal organism, *Acetabularia* (Han & Runnegar 1992).
- Small-sized semi-transparent discs preserved on the surface of micritic limestone, comprising organic matter, appear as appendages of insects or chromosome-like structures (Fig. 4a). These fossils are unique and remind us of the fossils preserved in amber. It is difficult to find any comparable fossil or extant form of the plant or animal kingdom.
- An impression of a megascopic dichotomously branched structure and associated *Grypania*-like object has also been reported from the Rohtas Formation (Fig. 3i), which is a rare fossil association of this age.

The complex morphologies observed among the Neoproterozoic Vindhyan megafossils are given below.

- Variable morphologies exhibited by carbonaceous discs or *Chuaria* are: excystement, considered to be a characteristic feature of eukaryotic affinity (Fig. 4c); the presence of a cluster of very small spheroids within it (Fig. 5d); the presence of enveloping sheaths (Fig. 5h); a well-preserved intracellular mass (Figs 4g & 5h); process-like structures, as in acanthomorphs (Figs 4l & 5k); *Tappania*-like budding in *Chuaria*-like discs (Figs 4k & 5j); uniquely associated *Chuaria*- and *Tawuia*-like vesicles (Fig. 5a, c, f); and *Chuaria* clustered within irregular fungal hyphae-like structures (Fig. 4i) are the features observed only in Bhander Group, Neoproterozoic Vindhyan fossil assemblages.
- A sporophytic plant-like body with sporangia-like structures (Fig. 4h); a bryophyte-like structure (Fig. 4e); vesicles attached on a branched filaments (Fig. 5o), to some extent resembling vaucheriacean alga; a pre-Ediacaran metazoan or an algal carbonaceous fossil (Fig. 4b); *Jacutianema*-like terminally attached *Chuaria* with *Tawuia* (Fig. 5a, c) considered a vaucheriacean affinity (Butterfield 2004); a carbonaceous spheroidal body attached to an umbrella-like structure (Fig. 5m) resembling the microscopic problematic form *Kakabekia umbellata* from the Gunflint Chert (Barghoorn & Tyler 1965); a dinoflagellate-like structure (Fig. 5e, j); and scar marks on *Tawuia*-like vesicles, suggestive of predation (Fig. 4f, m) are exclusively reported from the present assemblage belonging to the Bhander Group, Neoproterozoic Vindhyan sequences.
- Very small-sized globular carbonaceous structures, scattered within a gel or mucilage-like organic mass (Fig. 5n).
- Well-preserved ellipsoidal body with an arched anterior and an marginal notch-like structure may be a head preserved in siliciclastic rocks, favouring an Ediacaran affinity or a dubiofossil (Fig. 3a). A cup-like three-dimensional structure slightly tapered at one end (without any ornamentation), associated with microbial mats (Fig. 3e), and an *Aristophycus*-like structure indicative of diagenetically originated sediment fluidization along the sand–clay interface are categorized among pseudofossils (Fig. 3h)

Ediacaran fossils (Fig. 3a–f), mostly medusoids, have already been reported from the Maihar Sandstone and Bundi Hill Sandstone Formation of the Bhander Group, exposed in the Son Valley and Chambal Valley respectively (De 2003, 2006; Srivastava 2006). Most of these fossils are considered to be dubious.

The presence of microbial mats and *Arumberia* have been reported from the siliciclastic rocks of the Bundi Hill Sandstone, Chambal Valley, Rajasthan and the Maihar Sandstone Formation of central India (Bhander Group). The association of *Beltanelliformis* with microbial mats has been reported by Kumar & Pandey (2008). *Arumberia* (considered as an organo-sedimentary structure of Ediacaran age) is an additional supporting parameter, assigning the Ediacaran age to the Bhander Group.

Evidence indicating well-diversified megascopic life, predation and symbiotic association supports the existence of an advanced ecosystem, especially during the Neoproterozoic Vindhyan sedimentation.

Description of complex morphologies

The complex morphologies observed among megafossils of the Palaeoproterozoic–Mesoproterozoic Vindhyan Supergroup are listed in this section

Form A

Transparent–translucent discs, 0.5–1.5 mm in diameter, preserved on the bedding surface of

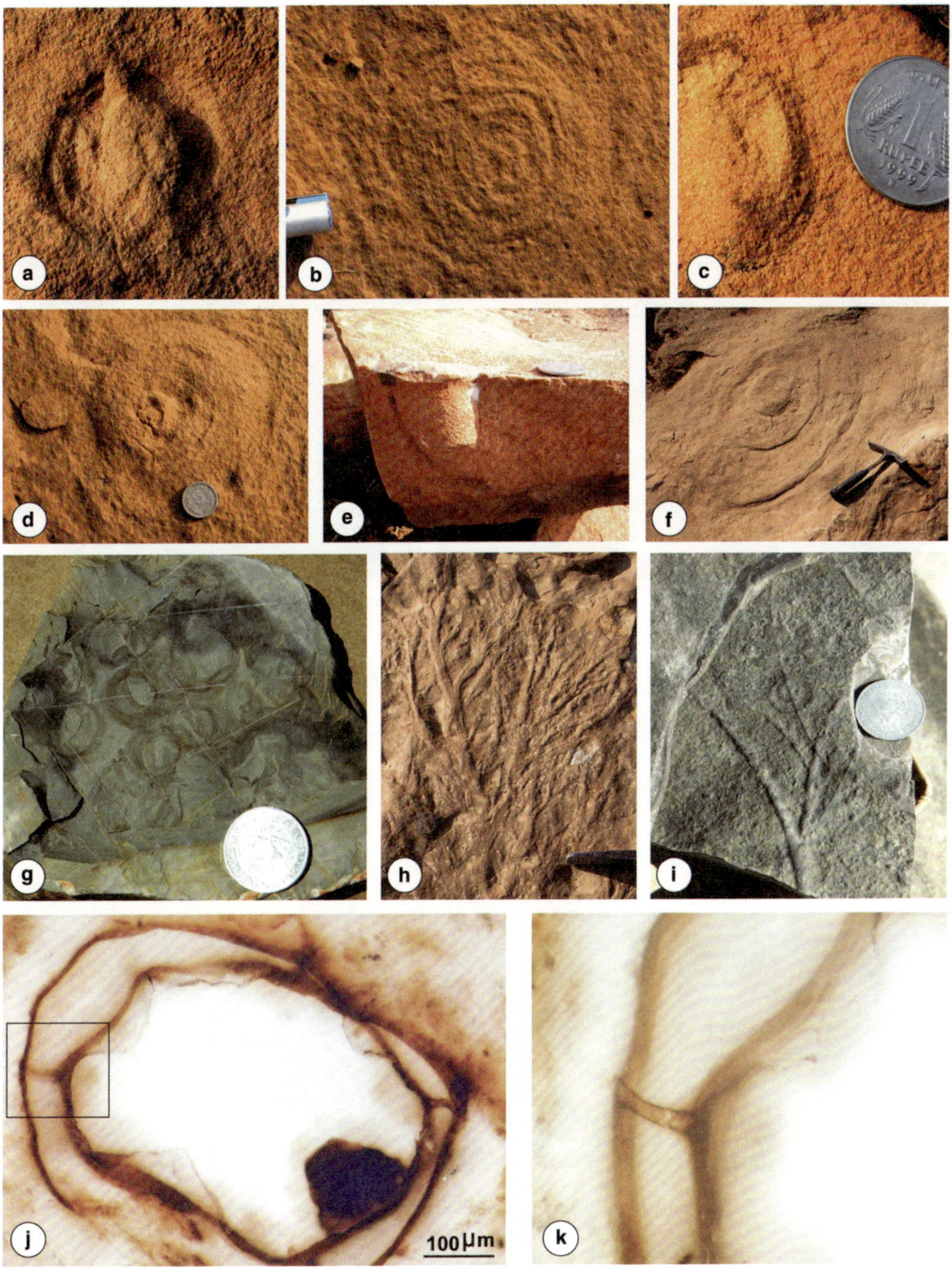

Fig. 3. (**a**)–(**f**) Ediacaran fossils from the Bundi Hill Sandstone, Bhander Group, Rajasthan, preserved in Sandstone rocks. (a) Dubiofossil: a double-walled spheroidal body with an arched anterior and terminal notch. (b) Concentric rings of some medusoid-like structure, cf. *Tirasiana*. (c) Dubiofossil: a frond-like structure, comparable with *Charniodiscus*. (d) Dubiofossil: *Tribrachidium*/medusoid-like body. (e) Dubiofossil: cup-like three-dimensional body of an unidentified fossil. (f) Dubiofossil: *Ediacaria*-like large sized elliptical concentric rings. (**g**) Cluster of *Grypania*, Rohtas Formation, Semri Group, Lower Vindhyans. (**h**) Pseudofossil: Aristophycus from the Bundi Hill Sandstone, Bhander Group, Rajasthan, preserved in Sandstone. (**i**) Megascopic branching and associated *Grypania*, Rohtas Formation, Semri Group, Lower Vindhyans. (**j**) & (**k**) *Trachyhystrichosphaera* – an age-marker acanthomorph from the Sirbu Shale, Rajasthan.

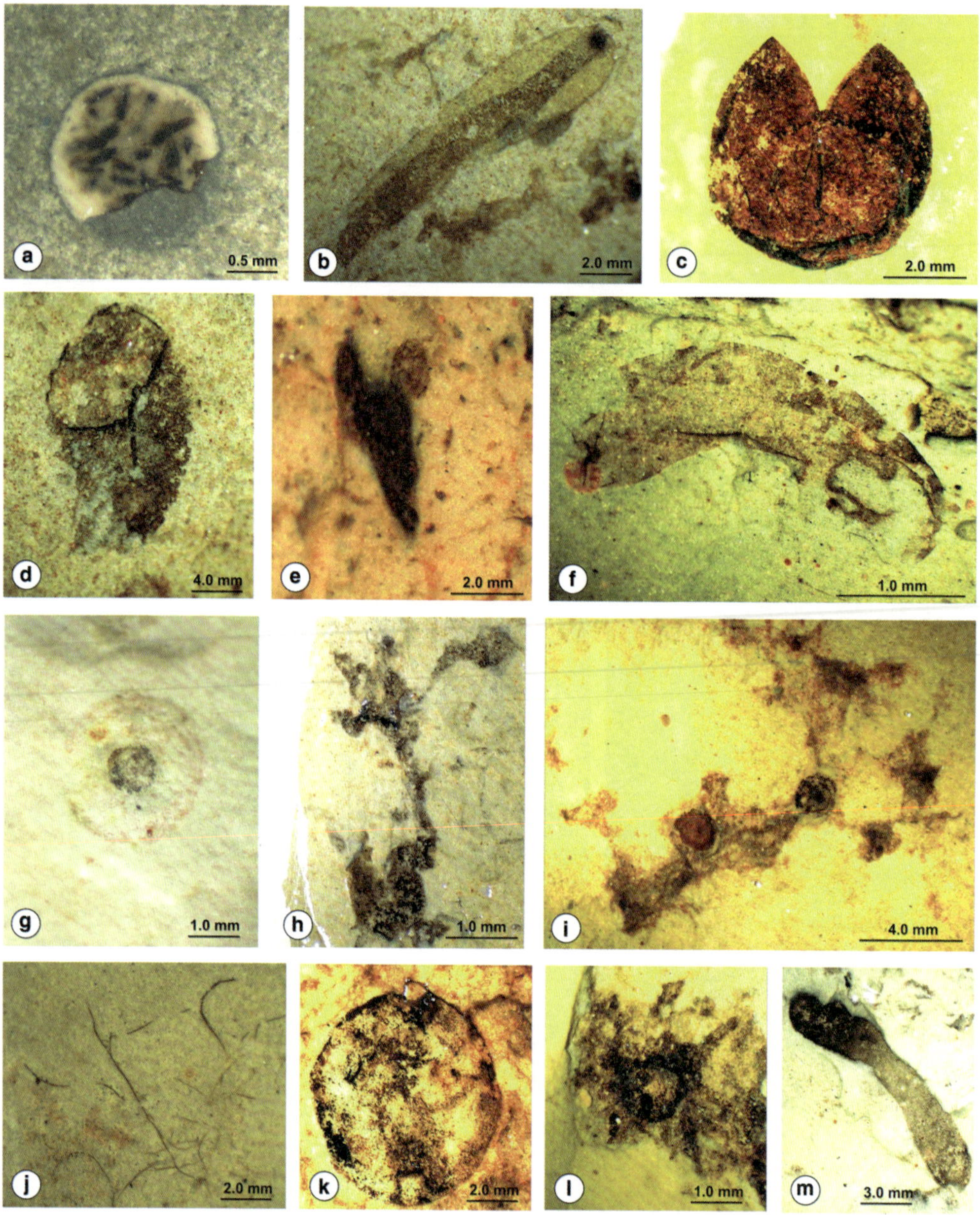

Fig. 4. (**a**) Compressed disc encompassing numerous chromosome- or appendage-like mesoscopic structures, from the Rohtas Formation, Semri Group, Lower Vindhyans. (**b**) Carbonaceous fossil interpreted as a pre-Ediacaran metazoan or a vaucherian alga, Samaria Shale, Bhander Group. (**c**) Excystement in *Chuaria*: a feature characteristic of eukaryotic algae, Sirbu Shale, Bhander Group. (**d**) Carbonaceous fossil interpreted as a pre-Ediacaran metazoan, Samaria Shale, Bhander Group. (**e**) A bryophyte-like organisms with two buds or a feather-like structure, Samaria Shale, Bhander Group. (**f**) *Tawuia*-like elongated body with scar marks possibly an indication of predation, Bhander Group. (**g**) Carbonaceous disc with a perfect inner body/nucleus, Sirbu Shale, Bhander Group. (**h**) A sporophyte-like carbonaceous plant body, with sporangia-like structures, Sirbu Shale, Bhander Group. (**i**) *Chuaria* associated with fungal hyphae-like structures, Sirbu Shale, Bhander Group. (**j**) Very thin thread-like filaments from the Rohtas Formation, Semri Group, Lower Vindhyans. (**k**) Carbonaceous disc with a bud-like structure, to some extent comparable with *Tappania*, Bhander Group. (**l**) Process-/spine-bearing carbonaceous disc, Bhander Group. (**m**) *Tawuia*-like elongated vesicle with scar marks, Bhander Group.

micritic limestone from the Palaeoproterozoic Rohtas Formation. This disc is comprised of organic matter resembling appendages of very small-sized (mesoscopic) unidentified insects or to some extent comparable with chromosome-like structures. Three specimens have been traced with different levels of preservation (Fig. 4a). It is worth mentioning here that during the participation in the World Summit at UCLA, Los Angeles, USA, in August 2008, the specimens were displayed before the internationally renowned palaeobiologists, and no one was able to suggest the nearest analogue for it.

Form B

The impression of a dichotomously branched mega plant fossil, in association with a *Grypania*-like object, preserved on the bedding plane of a micritic limestone sample, from the Rohtas Formation, Lower Vindhyans. The maximum diameter of the branch measured is 1.0 cm, the minimum diameter is 1.5 mm. A single specimen was recovered in both hypo- and epirelief. The maximum length of the branch is 9 cm. The width of the spire in the *Grypania*-like structure is 1.8 cm (Fig. 3i).

Beautifully preserved branching at the megascopic scale has not yet been recorded. *Konglingiphyton* – a megascopically branched algal thallus considered as the first erect plant, which broke the dominance of flat-lying microbial mats – exhibits comparable morphology, except for the hold fast structure reported from China (Menge *et al.* 1995). Such fossils are of ecological and environmental significance as they could raise the photosynthetic efficiency, and the free oxygen level in the atmosphere and hydrosphere. They may also accelerate the metazoan evolution by means of making a submarine grassland, which was a new comfortable niche for animals and improved the food constitution by providing a shelter from storms or predators. Macroscopic algae are, therefore, a treasure for elucidating the earlier evolution of life (Menge *et al.* 1995).

Form C

A cluster of well-preserved mesoscopic thin thread-like filamentous structures preserved on the bedding surface of the siliciclastic limestone. These are smooth, straight or curved, and branched in places (Figs 4j & 5l). The diameter of the filaments ranges from 0.2 to 0.3 mm, and their length varies between 3 and 12 mm. They are hardly visible with the naked eye and are rarely recorded fossils of the Rohtas Formation, Lower Vindhyans. The thickness of the filaments is almost uniform throughout their lengths.

Chambalia minor, reported by Kumar & Srivastava (2003), from the Bhander Group, central India is larger in size, and are visible with the naked eye. Moreover, branching has not been reported in any specimen of *Chambalia minor*; however, it is very prominent in the present form. *Vendotaenia* also exhibits larger dimensions in comparison to present forms. It is therefore categorized among complex morphologies. The mesoscopic size of the filaments also opens the possibility that the fossils may be an intermediate form, and probably representing a link between the evolution from the microscopic to the megascopic filamentous population. Intermediate carbonaceous mesoscopic coccoid forms, inferred as a missing link, have already been reported from the Panna Shale, Rewa Group, Upper Vindhyans (Srivastava 2004).

Megascopic complex morphologies from the Neoproterozoic sequences of the Vindhyan Supergroup are listed below.

Form D

Chuaria-like morphologies, 1.5–2.0 mm in diameter, with well-preserved folds and concentric markings, entrapped within irregularly branched fungal hyphae-like mesh.

The modern marine fungi (mostly ascomycetes) also exhibit a wide range of interactions with cyanobacteria, chlorophytes, phaeophytes and rhodophytes. These structure can be a lichen-like association of phycobionts with macroscopic algae or an obligate lichen-like association (Yuan *et al.* 2005). Present specimens of the *Chuaria*-like objects, embedded within a spongy irregular mesh-like structure (Fig. 4i), may also represent similar association.

Form E

An elongated, compressed and slightly curved elongated *Tawuia*-like carbonaceous vesicle (length measured at 11.2 mm, width of 1.4 mm, single specimen) comprising well-marked impressions of *Chuaria*-like discs (folds and wrinkle markings are very distinct in *Chuaria*) arranged in a single row within the vesicle (Fig. 5f). Adjacent to this structure, a number of spheroids (11 in number) assignable to *Chuaria* are scattered. The elongated vesicle is tapered at one end with an attachment or a hold fast-like structure (Fig. 5g) made of a thin, slightly undulating plate (length measured at 2.2 mm and a thickness of 0.4 mm) at an angle of 80°.

There is a possibility that the specimen belongs to the *Vaucherian* affinity. The presence of a hold fast-like structure may indicate some inclination towards a rhodophycean affinity (Knoll *et al.* 2006).

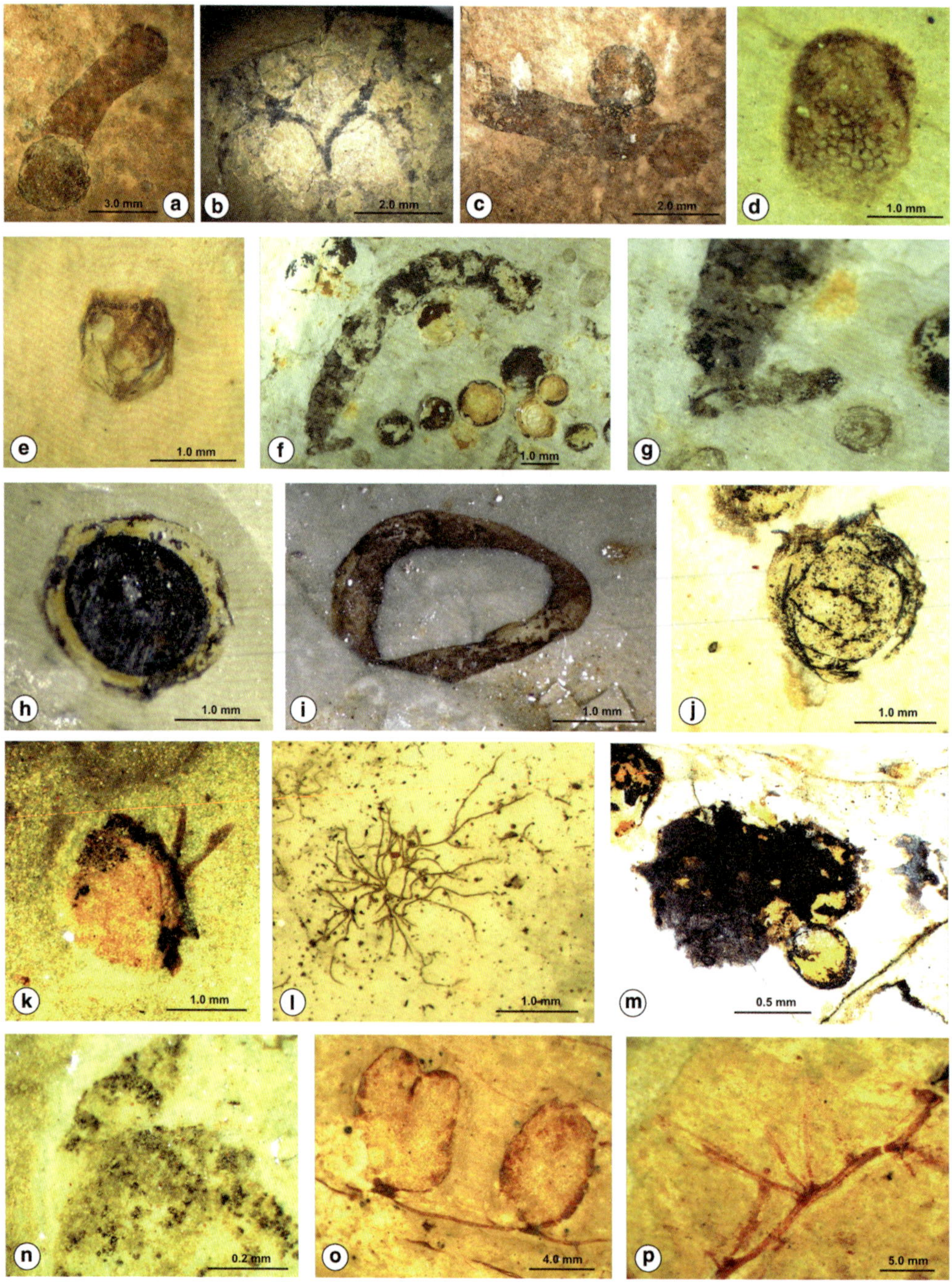

Fig. 5. Megafossils from the Vindhyan Supergroup. (**a**) Association of *Chuaria* and *Tawuia* comparable with *Jacutianema* from the Rewa Group. (**b**) Dichotomous branching from the Samaria Shale, Bhander Group. (**c**) Association of *Chuaria* and *Tawuia* comparable with *Jacutianema* from the Rewa Group. (**d**) Carbonaceous disc with a cluster of small spheroids or a scale of some metazoan, Dholpura Shale, Bhander Group. (**e**) Carbonaceous vesicle with dinoflagellate-like features and two notches, Samaria Shale, Bhander Group. (**f**) *Chuaria*-like carbonaceous discs arranged within a *Tawuia*-like elongated vesicle, Sirbu Shale, Bhander Group. (**g**) Close view of the hold fast-like structure of (f). (**h**) *Chuaria* with prominent and well-preserved inner body/nucleus and outer ring, Sirbu Shale, Bhander Group. (**i**) Carbonaceous ring, a part of the hold fast, according to the model proposed by Kumar (2001) for

Form F

Well-preserved carbonaceous filaments, 0.3–0.5 mm in diameter, exhibiting a repeated lateral branching (Fig. 5p) with two vesicles laterally attached to these filaments, where one vesicle exhibits excystement (Fig. 5o). The attachment structure is very clear in one vesicle. The diameter of vesicles varies between 1 and 2 mm (four vesicles). The preservation of these samples is exceptional, branching is very clear and excystement in the vesicle is undoubted. The morphological features suggest a Vaucheriacean affinity for these fossils; however, no such fossil has so far been reported from any Proterozoic sequence.

Form G

An erect plant-like megascopic carbonaceous fossil exhibiting branching and sporangia-likes features (Fig. 4h). The maximum measured length of the whole plant is 7 mm, the maximum diameter of the sporangia-like structure is 1.5 mm.

The morphological features of the present form suggest a sporophytic affinity to some extent. The presence of structures resembling sporangia, along with rhizoid-like structures or hold fast, are the supporting features. Differentiation of the thalli into hold fast and vegetative parts, like sporangia and setae, resembles the *Konglingiphyton*-like fossil form found in China (Menge *et al.* 1995). However, the evidence is not conclusive as the complete morphology is represented by only a single specimen.

Form H

Very small-sized carbonaceous subspherical vesicles, 2.0 mm in diameter, encompassing a cluster of about 30–40 very small closely packed spheroids (two specimens). The diameter of these small cell-like units ranges between 0.1 and 0.2 mm (Fig. 5d).

The morphology of the present form resembles to some extent the scale fossil *Chilodictyon* (Chrysophycean affinity), a microfossil 15–18 μm in diameter, reported from the 700–600 Ma old Tindir Group, Canada, and interpreted as a member of the *Chromalveolate* clade. Allison & Hilgert (1986) reported scale fossils from the Early Cambrian of Canada. There is also a possibility that the specimen may represent a sheath encompassing numerous cell-like units of a multicellular tissue of a metaphyte. However, the presence of cell tissue in Precambrian carbonaceous compressions is extremely rare, it has only been reported from phosphorites and siliceous rocks, like the 900 Ma old *Bangiophyte* colonies from Somerset Island in northern Canada and the Doushantou Formation of South China (Allison & Hilgert 1986).

Form I

An elongated body, 4 mm in length and 1.5 mm in width, tapered at one end with an opening-like structure seen at the broader end (Fig. 4b). A median gut-like structure, runs along its whole length (single specimen from the Bhander Group, Upper Vindhyan).

As far as affinity of this form is concerned, it is very difficult to decide whether it belongs to the plant or animal kingdom. There is also a possibility that the specimen is a Vaucheriacean alga or a benthic plant body with hold fast-like structure. There is a remote possibility that the form represents a metazoan, in which an opening-like structure is its mouth and the median dark structure running along its length is its gut. Another specimen with a conical hollow body with a well-preserved rim-like structure and a loop-like string may also represent an animal fossil form of unknown affinity (Fig. 4d).

Form J

An elongated, more or less triangular or cuneiform carbonaceous film or thalli with two small, spherical, bud-like structures at the broader end (Fig. 4e). The maximum length of the triangular structure is 6–8.5 mm, the maximum width measured is 2–3 mm (two specimens from the Bhander Group, uppermost Vindhyans).

In general morphology, the present form shows some resemblance to the multicellular thallus *Eopalmaria prinstina* Yan (1995) or *Changchengia stipitata* (Sharma 2006). The difference in the present form is the shape of the bud-like structures with reticulate ornamentation. There is a possibility that the present form belongs to some other unknown biological affinity. The morphology also

Fig. 5. (*Continued*) a multicellular plant from the Sirbu Shale, Bhander Group. (**j**) *Chuaria* with spines/notches/budding, Samaria Shale, Bhander Group. (**k**) Carbonaceous disc with two spines/processes, Samria Shale, Bhander Group. (**l**) Very thin filaments exhibiting branching, Rohtas Formation, Semri Group, Lower Vindhyans. (**m**) Small carbonaceous disc, with an umbrella-like protrusion (Sirbu Shale), comparable with the problematic microfossil *Kakabekia* from the Gunflint Chert (Barghoorn & Tyler 1965). (**n**) Very small-sized carbonaceous globules scattered haphazardly in an organic gel-like matrix, Sirbu Shale, Bhander Group. (**o**) Carbonaceous vesicles attached on a branched filament, Dholpura Shale, Bhander Group. (**p**) Branched filaments of the Dholpura Shale, Bhander Group.

suggests a bryophyte affinity to some extent for this particular form.

Form K

The presence of large-sized acanthomorphs identified as *Trachyhystrichosphaera* from the Sirbu Shale, Upper Vindhyans (Srivastava 2009*a*) is represented by spheroidal vesicles, 480–615 μm in diameter (visible with the naked eye in the petrographical thin section of chert), with tubular hollow processes that extend outwards from the inner wall. The intracellular mass is prominent and about 100 μm in diameter (Fig. 3j, k).

Morphologically, *Trachyhystrichosphaera* is one of the most complex acritarchs of eukaryotic affinity known from the Precambrian deposits (Knoll 1984). According to Corsetti *et al.* (2003), *Trachyhystrichosphaera* and *Tappania* are possibly of fungal affinity. The presence of open-ended, unbranched, tubular processes, the absence of excystement structures, and the conspicuously large and variable central vesicle supports their affinity towards that of fungi. It is considered globally as an age-marker form of the Neoproterozoic (Cryogenian). The large size acquired by the Vindhyan form also supports a Ediacaran age as an upper age limit for the Vindhyans, as in Cambrian period the size of acanthomorphs reduced markedly at the global level (Srivastava 2009*a*).

Macroevolution

Morphological changes occurring above the level of species are considered as 'macroevolution'. Diversified megafossils of the supergroup document possibly two macroevolutionary modes, which can be demarcated at the transitional period of Palaeoproterozoic–Mesoproterozoic and later on in the Neoproterozoic. The former transitional period of the Vindhyans represented broadly by the Semri Group is dominated by the diversified microscopic benthic cyanobacterial community and the moderately diversified megascopic life. The population of eukaryotes was comparatively low.

A number of plausible initial and intermediate fossil forms have been observed while studying the metaphytes from the Panna Shale, Rewa Group, Upper Vindhyans (Srivastava 2004). They exhibit the overlapping size range of micro- and megafossils. These forms have been interpreted as the intermediate forms/meso-forms or a missing link between the evolutions from microscopic to megascopic life (Srivastava 2004).

Neoproterozoic fossil assemblages also exhibit diversified megascopic life, the radiation of eukaryotes, multicelluarity and the emergence of animal clade, represented by Ediacaran fauna (Fig. 3a–f). The indication of a symbiotic relationship (Fig. 5f) and discs embedded within the mesh of irregularly branched fungal hyphae (Fig. 4i) may represent an example of an entirely new realm and life style at the megascopic level. The presence of dividing animal embryo-like structures in thin sections of chert (Sirbu Shale Formation) are possibly microzooplanktons (Srivastava 2008, 2009*b*) and are indicative of the emergence of animal life at the microscopic level, which may suggests an episode of macroevolution.

Discussion and conclusion

The megafossil assemblage comprised of complex morphologies from the Vindhyan Supergroup suggests the following.

- Life during Vindhyan sedimentation was complex, advanced and highly diversified. The overall fossil profile of the Vindhyan Supergroup can be represented by cyanobacterial (dominating community), bacterial, algal, fungal acritarchean, multicellular plant, animal and lichen-like affinities.
- In general, Palaeoproterozoic–Mesoproterozoic acritarchs are simple spheroids (sphaeromorphs) with occasional median split, enveloping sheaths or surface ornamentation; whereas, during the Neoproterozoic, acritarchs are larger in size and more complicated in morphology, such as *Trachyhystrichosphaera* (acanthomorph), at the microscopic level (Srivastava 2009*a*). At the megascopic level, the presence of spines and scale-like features among carbonaceous discs are speculative. Vindhyan acritarchs of Cryogenian age (850–630 Ma) indicate morphological disparity and suggest that the upper age limit of the Vindhyans should not exceed Ediacaran, as in the Cambrian the size of acanthomorphs again reduced globally (Srivastava 2009*a*). It is believed that towards the end of the Neoproterozoic, the phytoplanktons acquired increasing amounts of armour and ornamentation in defence against the grazing pressure from the mesozooplanktons (Janhunen *et al.* 2007).
- At the global level, large process-bearing acritarchs and multicellular organisms without skeletons were found in the Partataka Formation, Spitsbergen, in the Vindhyan Supergroup, India and in the Doushantou Formation, China. The affinity for these complex and peculiar acritarchs is suggested to be green algae, dinoflagellates, and other protists and metazoans.
- The presence of mesoscopic carbonaceous fossils interpreted as intermediate forms or a link between the evolution from microscopic to

megascopic life that opens a new window in Proterozoic palaeobiology.

- *Grypania* is broadly distributed during approximately 1400 Ma and appears to be unaccompanied by other multicellular macroscopic fossils (Butterfield 2000, 2007). In contrast, the Vindhyan Supergroup exhibits the presence of a *Grypania*-like structure (in *c.* 1700 Ma old rocks) accompanied by a megascopic branched alga-like object (Fig. 3i).
- The morphology of *Chuaria–Tawuia* during the Mesoproterozoic Vindhyans is simple. In the Neproterozoic they exhibit complex morphologies, spines or processes, evidence of predation (scar marks) and a close association of both these genera, probably a symbiotic association.
- Multicellularity, sexuality and animal affinity at both the microscopic and megascopic levels might have been established during the sedimentation of the Neoproterozoic Bhander Group. The presence of fossils inclining towards Ediacaran metazoan affinities indicate that such animal clades had already evolved during the Vindhyans.
- After carrying out a detailed study of micro- and megafossils of the Upper Vindhyans, it was noticed that no typical Cambrian fossil has so far been reported. Therefore, a Precambrian age is strongly suggested for the Vindhyan Supergroup.
- The Vindhyan Supergroup documents two macroevolutionary modes, separated by Palaeoproterozoic–Mesoproterozoic and Neoproterozoic, especially during the Ediacaran period (630–545 Ma). The Palaeoproterozoic–Mesoproterozoic biosphere was dominated by microscopic cyanobacterial communities. Evidence of undisputed Pre-Ediacaran micro- or megascopic metazoans has yet to be reported; however, the presence of microfossils comparable to some extent with dividing animal embryo and vase-shaped microfossils opens the probability of pre-Ediacaran animal life at the microscopic level (Yin *et al.* 2004; Srivastava 2008).

The assemblage has an astrobiological significance, as more attention should be paid to the uncommon complex morphologies of primitive life while considering the possibilities of complex extraterrestrial life. The rate of evolution during the Proterozoic was probably tied to solar brightening and cooling of the planetary interior; hence, signals for ice and snow are very significant as they can provide an estimation for the physical conditions that have the potential to provoke the development of complex life out of a pre-existing simple microbial background (Janhunen *et al.* 2007).

In Vindhyans, there is still lot of potential and scope for further detailed studies that may resolve a number of problems related to evolution, and may place this supergroup as one of the most diversified, well-preserved and advanced Proterozoic basins on a global level.

Author is grateful to the Head of the Centre of Advanced Study in Geology, Lucknow University for providing the basic facilities to carry out the present study. Special thanks are due to Dr R. Bali for his immense co-operation throughout the study. Financial assistance from the Department of Science and Technology, New Delhi under WOS scheme (SR/WOS-A//ES-20/2008) is gratefully acknowledged.

References

ALLISON, C. W. & HILGERT, J. W. 1986. Scale microfossils from the Early Cambrian of northwest Canada. *Journal of Palaeontology*, **60**, 973–1015.

AUDEN, J. B. 1933. Vindhyan sedimentation in Son Valley, Mirzapur district. *Geological Society of India, Memoir*, **62**, 141–250.

AZMI, R. J. 1998. Discovery of lower Cambrian small shelly fossils and Brachiopods from the Lr. Vindhyan of Son Valley, Central India. *Journal of the Geological Society of India*, **52**, 381–389.

AZMI, R. J., JOSHI, D., TEWARI, B. N., JOSHI, M. N., MOHAN, K. & SRIVASTAVA, S. S. 2007. Age of the Vindhyan Supergroup of Central India: an exposition of biochronology v. radiochronology. *In*: SINHA, D. K. (ed.) *Micropaleontology: Application in Stratigraphy and Palaeooceanography*. Narosa, New Delhi, 29–62.

BANERJEE, D. M. & FRANK, W. 1999. Preliminary Ar/Ar dates of porcellanite and detrital mica from the Vindhyan of Central India. *In*: KUMAR, S. (ed.) *Abstracts: Workshop on Vindhyan Stratigraphy and Palaeobiology, Lucknow, India*. Lucknow University Department of Geology, Lucknow, 5.

BANSAL, M., VIJAN, A. R., MAHAR, Y. S., GHOSH, N., KRISHNA, P. S. & PRABHY, B. N. 1999. Rb–Sr dating of Lower Vindhyan sediments from Son Valley, M. P., India. *Indian Mineralogist*, **33**, 1–9.

BARGHOORN, E. S. & TYLER, S. A. 1965. Microorganisms from the Gunflint chert. *Science*, **147**, 563–577.

BENGTSON, S., BELIVANOVA, V., RASMUSSEN, B. & WHITEHOUSE, M. 2009. The controversial 'Cambrian fossils' of the Vindhyan are real but more than a billion years old. *Proceedings of the National Academy of Sciences of the United States of America*, **106**, 7729–7734.

BRASIER, M. 1999. Discovery of lower Cambrian small shelly fossils and Brachiopods from the Lr. Vindhyan of Son Valley, Central India. Discussion. *Journal of the Geological Society of India*, **53**, 727–730.

BUTTERFIELD, N. J. 2000. *Bangiomorpha pubescens* n. gen., n. sp.: implications for the evolution of sex, multicellularity and the Mesoproterozoic/Neoproterozoic radiation of eukaryotes. *Palaeobiology*, **26**, 386–404.

BUTTERFIELD, N. J. 2004. A vaucheriacean alga from the middle Neoproterozoic of Spitsbergen: implication for the evolution of Proterozoic eukaryotes and the Cambrian explosion. *Palaeobiology*, **30**, 231–252.

Butterfield, N. J. 2007. Macroevolution and macroecology through deep time. *Palaeontology*, **50**, 41–55.

Conway Morris, S., Jensen, S. & Butterfield, N. J. 1998. Fossil discoveries in India: Continued. *Science*, **282**, 1265.

Corsetti, F. A., Awramik, S. M. & Pierce, D. 2003. A complex microbiota from snowball Earth times: microfossils from the Neoproterozic Kingston Peak Formation, Death Valley, USA. *Proceedings of the National Academy of Sciences of the United States of America*, **100**, 4399–4404.

Crawford, A. R. & Compston, W. 1970. The age of the Vindhyan system of peninsular India. *Journal of the Geological Society, London*, **125**, 351–371.

De, C. 2003. Possible organisms similar to Ediacaran forms from the Bhander Group, Vindhyan Supergroup, Late Neoproterozoic of India. *Journal of Asian Earth Sciences*, **21**, 387–395.

De, C. 2006. Ediacara fossil assemblage in the upper Vindhyans of Central India and its significance. *Journal of Asian Earth Sciences*, **27**, 660–683.

Han, T. M. & Runnegar, B. 1992. Megascopic eukaryotic algae from the 2.1 billion-year-old Negaunee Iron-Formation, Michigan. *Science*, **257**, 232–235.

Hofmann, H. J. 2005. Palaeoproterozoic dubiofossils from India. Revisited-Vindhyan triploblastic animal burrows or pseudofossils? *Journal of the Palaeontological Society of India*, **50**, 113–120.

Janhunen, P., Kaartokallio, H., Oksanen, I., Lehto, K. & Lehto, H. 2007. Biological feedbacks as cause and demise of Neoproterozoic icehouse: astrobiological irospects for faster evolution and importance of cold conditions. *PLoS ONE*, **2**, (2), e214, doi: 10.1371/journal.pone.0000214.

Jones, H. C. 1909. General report. *Records of the Geological Survey of India*, **3**, 66.

Kathal, P. K., Patel, D. R. & Alexander, P. O. 2000. An ediacaran fossil *Spriggina* (?) from the Semri Group and its implication on the age of the proterozoic Vindhyan Basin, Central India. *Neues Jahrbuch für Geologie und Palaeontologie Monatshefte.*, **6**, 321–332.

Kerr, R. A. 2002. Earliest animal trace or just mud cracks. *Science*, **295**, 1209–1210.

Knoll, A. H . 1984. Microbiotas of the Late Precambrian Hunnberg Formation Nordoustlandet, Svalbard. *Journal of Palaeontology*, **58**, 131–162.

Knoll, A. H., Javaux, E. J., Hewitt, D. & Cohen, P. 2006. Eukaryotic organisms in Proterozoic Oceans. *Philosophical Transactions of the Royal Society of Biosciences*, **36**, 1023–1038.

Kumar, S. 1995. Megafossils from the Mesoproterozoic Rohtas Formation (The Vindhyan Supergroup) Katni area, Central India. *Precambrian Research*, **72**, 171–184.

Kumar, S. 1999. Siliceous sponge spicule-like forms from Neoproterozoic Bhander Limestone, Maihar area, Madhya Pradesh. *Journal of the Palaeontological Society of India*, **44**, 141–148.

Kumar, S. 2001*a*. Mesoproterozoic megafossl *Chuaria–Tawuia* association may represent part of a multicellular plant, Vindhyan Supergroup, Central India. *Precambrian Research*, **106**, 187–211.

Kumar, S. 2001*b*. Controversy concerning Cambrian fossils from the Vindhyan sediments: a reassessment. *Journal of the Palaeontological Society of India*, **54**, 115–117.

Kumar, S. & Gupta, S. 2002. *Field Guide Book 'International Field Workshop on the Vindhyan Basin. Central India*. Palaeontological Society of India, Lucknow, 72.

Kumar, S. & Pandey, S. K. 2008*a*. Arumberia and associated fossils from the Neoproterozoic Maihar Sandstone, Vindhyan Supergroup, Central India. *Journal of the Palaeontological Society of India*, **53**, 83–97.

Kumar, S. & Pandey, S. K. 2008*b*. Discovery of organic-walled microbiota from the black bedded chert, Balwan Limestone, the Bhander Group, Lakheri area, Rajasthan. *Current Science*, **94**, 797–800.

Kumar, S. & Sharma, M. 2010. *Field Guide: Vindhyan Basin, Son Valley area, Central India*. Palaeontological Society of India, Lucknow.

Kumar, S. & Srivastava, P. 1992. Discovery of microfossils from the nonstromatolitic middle Proterozoic Vindhyan chert, Chitrakut are, U. P. *Journal of the Geological Society of India*, **38**, 511–515.

Kumar, S. & Srivastava, P. 1995. Microfossils from the Kheinjua Formation, Mesoproterozoic Semri Group, Newari area, Central India. *Precambrian Research*, **74**, 91–117.

Kumar, S. & Srivastava, P. 1997. A note on the carbonaceous megafossils from the Neoproterozoic Bhander Group, Maihar area, M. P. *Journal of the Palaeontological Society of India*, **42**, 141–146.

Kumar, S. & Srivastava, P. 2003. Carbonaceous megafossils from the Neoproterozoic Bhander Group, Central India. *Journal of the Palaeontological Society of India*, **48**, 139–154.

Kumar, A., Gopalan, K. & Rajagopalan, G. 2001. Age of the Lower Vindhyan sediments, Central India. *Current Science*, **81**, 806–809.

Malone, S. J., Meert, J. G. *et al.* 2008. Palaeomagnetism and detrital zircon geochronology of the Upper Vindhyan sequence, Son Valley and Rajasthan, India: a ca 1000 Ma closure age for the Purana Basins? *Precambrian Research*, **164**, 132–159.

Mandal, M. E. A., Goswami, J. N., Deomurari, M. P. & Sharma, K. K. 2002. Ion microprobe Pb/Pb ages of zircons from the Bundelkhand Massif, northern India: implications for crustal evolution of the Bundelkhand–Aravalli Supercontinent. *Precambrian Research*, **117**, 85–100.

McMenamin, D. S., Kuma, S. & Awramik, S. M. 1983. Microbial fossils from the Kheinjua Formation, middle proterozoic, Semri Group, (Lr. Vindhyans) Son Valley, Central India. *Precambrian Research*, **21**, 247–271.

Menge, C., Zongzheng, X. & Xunlai, Y. 1995. A great diversification of Macroscopic algae in Neoproterozoic. *Scientia Geologica Sinica*, **3**, 295–308.

Moitra, A. K. & De, C. 1998. Primitive trace fossils from the Vindhyans. *Geological Survey of India News*, **29**, 47–48.

Prasad, B. 1984. Geology, sedimentation and palaeogeography of the Vindhayn Supergroup, southeastern Rajasthan. *Geological Society of India, Memoir*, **116**, 1–107.

PRASAD, B. 2007. Obruchevella and other Terminal Proterozoic (Vendian) organic-walled microfossils from the Bhander Group, (Vindhyan Supergroup), Madhya Pradesh. *Journal of the Geological Society of India*, **69**, 295–310.

PRASAD, B., UNIYAL, S. N. & ASHER, R. 2005*a*. Organic-walled microfossils from the Proterozoic Vindhyan Supergroup of Son Valley, Madhya Pradesh, India. *Palaeobotanist*, **54**, 13–60.

PRASAD, B., UNIYAL, S. N. & ASHER, R. 2005*b*. Meso-Neoproterozoic organic-walled microfossils. *In*: SINHA, D. K. (ed.) *Micropalaeontology: Application in Stratigraphy and Palaeooceanography*. Narosa, New Delhi, 1–27.

RAI, V. & GAUTAM, R. 1998. New occurrence of carbonaceous megafossls from the Meso-Neoproterozoic horizons, the Vindhyan Supergroup, Kaimur–Katni area, Madhya Pradesh, India. *Geophytology*, **26**, 13–25.

RAI, V. & SINGH, V. K. 2006. Discovery of megascopic multicellularity in deep time: new evidence from the 1.63 billion years old Lower Vindhyan succession, Vindhyan Supergroup, Uttar Pradesh, India. *Journal of Applied Biosciences*, **32**, 196–203.

RAI, V., SHUKLA, M. & GAUTAM, R. 1997. Discovery of carbonaceous megafossils (*Chuaria–Tawuia* assemblage) from the Neoproterozoic Vindhyan succession (Rewa Group), Allahabad–Rewa area, India. *Current Science*, **73**, 783–788.

RASMUSSEN, B., BOSE, P. K., BANERJEE, S., FLETCHER, I. R. & MC NAUGHTON, N. J. 2002. 1.6 Ga U–Pb zircon age for the Chorhat Sandstone, Lower Vindhyan, India: possible implications for early evolution of animals. *Geology*, **30**, 103–106.

RAY, J. S. 2006. Age of the Vindhyan Supergroup: a review of recent findings. *Journal of Earth System Sciences*, **115**, 149–160.

RAY, J. S., MARTIN, M. W., VEIZER, J. & BOWRING, S. A. 2002. U–Pb zircon dating and Sr isotope systematics of the Vindhyan Supergroup, India. *Geology*, **30**, 131–134.

RAY, J. S., VEIZER, J. & DAVIS, W. J. 2003. C, O, Sr and Pb isotope systematics of carbonate sequences of the Vindhyan Supergroup, India: age, digenesis, correlation and implication for global events. *Precambrian Research*, **121**, 103–140.

SARANGI, S., GOPALAN, K. & KUMAR, S. 2004. Pb–Pb age of the earliest megascopic eukaryotic alga bearing Rohtas formation, Vindhyan Supergroup, India: implication for Precambrian atmospheric oxygen evolution. *Precambrian Research*, **132**, 107–121.

SARKAR, S., BANERJEE, S., SAMANT, P. & JEEVANKUMAR, S. 2006. Microbial-mat induced sedimentary structure in siliciclastic sediments; examples from the 1.6 Ga. Chorhat sandstone, Vindhyan Supergroup, M. P., India. *Journal of Earth System Science*, **115**, 49–60.

SASTRY, M. V. A. & MOITRA, A. K. 1984. Vindhyan stratigraphy: a review. *Geological Society of India, Memoir*, **116**, 109–148.

SEILACHER, A., BOSE, P. K. & PFLUGER, F. 1998. Triploblastic animals more than 1 billion years ago: trace fossil evidence from India. *Science*, **282**, 80–83.

SHARMA, M. 2003. Age of Vindhyans-Palaeobiological evidence: a paradigm shift (?). *Journal of the Palaeontological Society of India*, **48**, 191–214.

SHARMA, M. 2006*a*. Late Palaeoproterozoic (Statherian) Carbonaceous films from the Olive Shale (Koldaha Shale), Semri Group, Vindhyan Supergroup, India. *Journal of the Palaeontological Society of India*, **51**, 27–35.

SHARMA, M. 2006*b*. Palaeobiology of Mesoproterozoic Salkhan Limestone, Semri Group, Rohtas, Bihar, India: systematics and significance. *Journal of Earth System Science*, **115**, 67–98.

SHARMA, M. & SHUKLA, Y. 2009*a*. The evolution and distribution of life in the Precambrian eon-global perspective and the Indian record. *Journal of Biosciences*, **34**, 1–12.

SHARMA, M. & SHUKLA, Y. 2009*b*. Mesoproterozoic coiled megascopic fossil Grypania spiralis from the Rohtas Formation, Semri Group, Bihar, India. *Current Science*, **96**, 1636–1640.

SINGH, S. K., RAI, S., SANTOSH, K. & TRIVEDI, J. K. 2002. Re–Os systematics of black shales from the Vindhyans: implications to their geochronology. *In*: *Tenth ISMAS Workshop (10th ISMAS-WS 2002) on 'Advances in Mass Spectrometry' at Toshali Sands, Puri, 25 February–1 March*. Indian Society for Mass Spectrometry, Mumbai, 242–244.

SONI, M. K., CHAKRABORTY, S. & JAIN, V. K. 1987. Vindhyan Supergroup – a review. In Purana Basins of India. *Geological Society of India, Memoir*, **6**, 87–138.

SRIVASTAVA, A. P. & RAJAGOPALAN, G. 1986. F.-T dating of Precambrian deposits of Vindhyan Group at Chitrakut, Banda district (U.P.–M.P.). *In*: SHARMA, K. K. (ed.) *Nuclear Tracks: Application to Earth Sciences*. Space Physics and Nuclear Physics, Dehradun, 44–52.

SRIVASTAVA, A. P. & RAJAGOPALAN, G. 1988. F-T ages of Vindhyan Gauconitic SSt beds exposed around Rawatbbhata area, Rajasthan. *Journal of the Geological Society of India*, **32**, 527–529.

SRIVASTAVA, P. 2002. Carbonaceous megafossils from the Dholpura Shale, Uppermost Vindhyan Supergroup, Rajasthan: an age implication. *Journal of the Palaeontological Society of India*, **47**, 97–105.

SRIVASTAVA, P. 2004. Carbonaceous fossils from the Panna Shale, Rewa Group (Upper Vindhyans), Central India: a possible link between evolutions from Micro-megascopic life. *Current Science*, **86**, 644–646.

SRIVASTAVA, P. 2006. Possible Ediacaran fossils from the Bundi Hill Sandstone, Upper Vindhyans, Rajasthan. (Abstract.) *In*: *Diamond Jubilee International Conference on Changing Scenario in Palaeobotany and Allied Subjects*. BSIP, Lucknow, 188.

SRIVASTAVA, P. 2008. Some bizarre forms of the Meso-Neoproterozoic Vindhyan Supergroup, India. *In*: *World Summit on Ancient Microscopic fossils*, University of California, Los Angeles, USA, 27 July–2 August, 39.

SRIVASTAVA, P. 2009*a*. An age-marker acanthomorph Trachyhystrichosphaera: from the upper Vindhyans, Rajasthan. *Journal of Earth System Sciences*, **118**, 575–582.

SRIVASTAVA, P. 2009*b*. Unidentified complex morphologies among the fossils of the Vindhyan Supergroup, indicating an advanced Ecosystem. (Abstract.) *In*: *International Conference on Palaeoproterozoic Supercontinents and Global Evolution, IGCP-509, October 2009*. Indian Statistical Institute, Kolkata, 8–9.

SRIVASTAVA, P. & BALI, R. 2006. Proterozoic Carbonaceous remains from the Chorhat Sandstone: oldest fossils of the Vindhyan Supergroup. *Geobios*, **39**, 873–878.

SRIVASTAVA, P. & TEWARI, V. C. 2011. Morphological changes in microscopic–megascopic life and stromatolites, recorded during Late Palaeoproterozoic–Neoproterozoic transition: the Vindhyan Supergroup. *In*: TEWARI, V. C. & SECKBACH, J. (eds) *Stromatolites: Interaction of Microbes with Sediments*. Cellular Origin, Life in Extreme Habitats and Astrobiology, **18**. Springer, Dordrecht, 87–114.

TANDON, K. K. & KUMAR, S. 1977. Discovery of annelid and arthropod remains from Lower Vindhyan rocks (Precamb) of Central India. *Geophytology*, **7**, 126–129.

VENKATACHALA, B. S., SHARMA, M. & SHUKLA, M. 1996. Age and life of the Vindhyans- facts and conjectures. *Geological Society of India, Memoir*, **36**, 137–165.

VINOGRADOV, A. P., TUGARINOV, A. I., ZHYRKOV, C. I., STANIKOVA, N., BIBIKOVA, E. V. & KHORRE, K. 1964. *Geochronology of the Indian Precambrian.* XXII International Geological Congress, Delhi, part 10, 553–567.

YAN, Y.-Z. 1995. Discovery and preliminary study of megascopic algae (1700) from Tuanshanzi Formation in Jixian, China. *Acta Micropalaeontologica Sinica*, **12**, 107–126.

YIN, C., BENGTSON, S. & ZHAO, Y. 2004. Silicified and phosphatised *Tianzhushania* spheroidal microfossils of possible animal origin from the Neoproterozoic of south China. *Acta Palaeontologica Sinica*, **49**, 1–12.

YUAN, X., XIAO, S. & TAYLOR, T. N. 2005. Lichen-like symbiosis 600 million years ago. *Science*, **308**, 1017–1020.

India's Palaeoproterozoic legacy

M. SANTOSH

Division of Interdisciplinary Science, Faculty of Science, Kochi University, Kochi 780-8520, Japan (e-mail: santosh@kochi-u.ac.jp)

Abstract: The Indian lithospheric plate assembles the history of geological events covering almost the entire history of our planet. The proto-Indian continent was an integral part of the Palaeoproterozoic supercontinent Columbia. Global mantle warming beneath this supercontinent triggered rising plumes that led to rifting and the emplacement of mafic dyke swarms and large igneous provinces in many of the major cratons of Peninsular India. Continued rifting and separation of crustal blocks led to the formation of thick passive margin successions. The opening of ocean basins and their eventual closure are recorded in the travelogue of oceanic plates from mid-ocean ridge to the trench in the form of 'ocean plate stratigraphy' in Proterozoic belts, including the association of dismembered ophiolites, pelagic sediments and continental margin sequences that were imbricated into accretionary belts such as those in the Aravalli–Delhi domain (Western India Suture), the Central India Tectonic Zone (Central India Suture) and the eastern margin of the East Dharwar Craton (Eastern India Suture). Recent U–Pb zircon chronology from ophiolitic rocks in the eastern margin of the Dharwar Craton shows ages spanning from 1.85 to 1.33 Ga suggesting a prolonged Wilson cycle of subduction–accretion processes before the final collisional event and the extrusion of high-grade metamorphic orogens. The extruded high pressure–temperature (P–T) metamorphic belts are often associated with mafic/ultramafic units with an abyssal signature. The available geophysical information from the Indian Palaeoproterozoic belts, particularly deep seismic reflection data, provides important clues on the architecture of these orogens and the subduction polarity. Whereas the Palaeoproterozoic convergent margins in the NW and SE sectors of the Indian peninsula were characterized by the westward subduction of oceanic plates, ocean closure along the Central India Tectonic Zone probably involved a double-sided subduction. The Wilson cycle traces a continuum from Palaeoproterozoic through Mesoproterozoic to Neoproterozoic in some of these belts, with a prolonged subduction–accretion history similar to the ongoing convergent margin processes in the western Pacific region. Peninsular India thus preserves a complete record from Pacific-type accretionary tectonics along the margins of the Columbia supercontinent to Himalayan-style collisional assembly within the Neoproterozoic Rodinia supercontinent.

Precambrian terrains offer windows into the lower crust and provide opportunities to study the deeper roots of orogens. The development of major orogens during Earth history related to the assembly of continental fragments within major supercontinents involved, in many cases, the consumption of the intervening large oceans (Collins *et al.* 2007; Santosh *et al.* 2009*b*; Santosh 2010*a*). Continental aggregation was not a simple collision process but witnessed a prolonged cycle of construction and destruction of oceanic lithosphere and continental crust, where Pacific-type subduction–accretion was fundamental in building the major components and in defining the architecture of orogenic belts, prior to the final Himalayan-style collision (Santosh *et al.* 2009*a*; Santosh 2010*a*).

The Indian lithospheric plate records the history of geological events covering almost the entire history of our planet, from more than 4.0 Ga up to 4 ka (Naqvi 2005). In a recent study, Pradhan *et al.* (2010) synthesized the geochronological constraints and palaeomagnetic poles from the major cratons in India and showed that the Indian subcontinent was involved in various Precambrian supercontinental assemblies. The major Palaeoproterozoic terranes in Peninsular India are represented by the Aravalli, Delhi, Singhbhum and Bhandara domains, as well as the Cuddapah–Eastern Ghats Belt at the eastern margin of the Dharwar Craton (Fig. 1). Further south, in the crustal blocks constituting the mosaic of the Southern Granulite Terrain, Late Neoproterozoic–Cambrian events dominate with only vestiges of suspect Palaeoproterozoic crust formation events. Although there is robust evidence that the metasedimentary units in some of the blocks were sourced from Palaeoproterozoic provenance (Collins *et al.* 2007; Santosh *et al.* 2009*a*). The major Palaeoproterozoic orogenic belts in India thus extend from the NW through central to the eastern part of the peninsula, with some suspect terranes further south (Fig. 1, inset). These belts, often referred to as mobile belts surrounding Archaean cratons, preserve a broadly similar tectonic history with regard to magmatism, metamorphism and sedimentation (reviewed in Naqvi 2005; Valdiya 2010). The rock records in many of

From: Mazumder, R. & Saha, D. (eds) 2012. *Palaeoproterozoic of India*. Geological Society, London, Special Publications, **365**, 263–288. http://dx.doi.org/10.1144/SP365.14

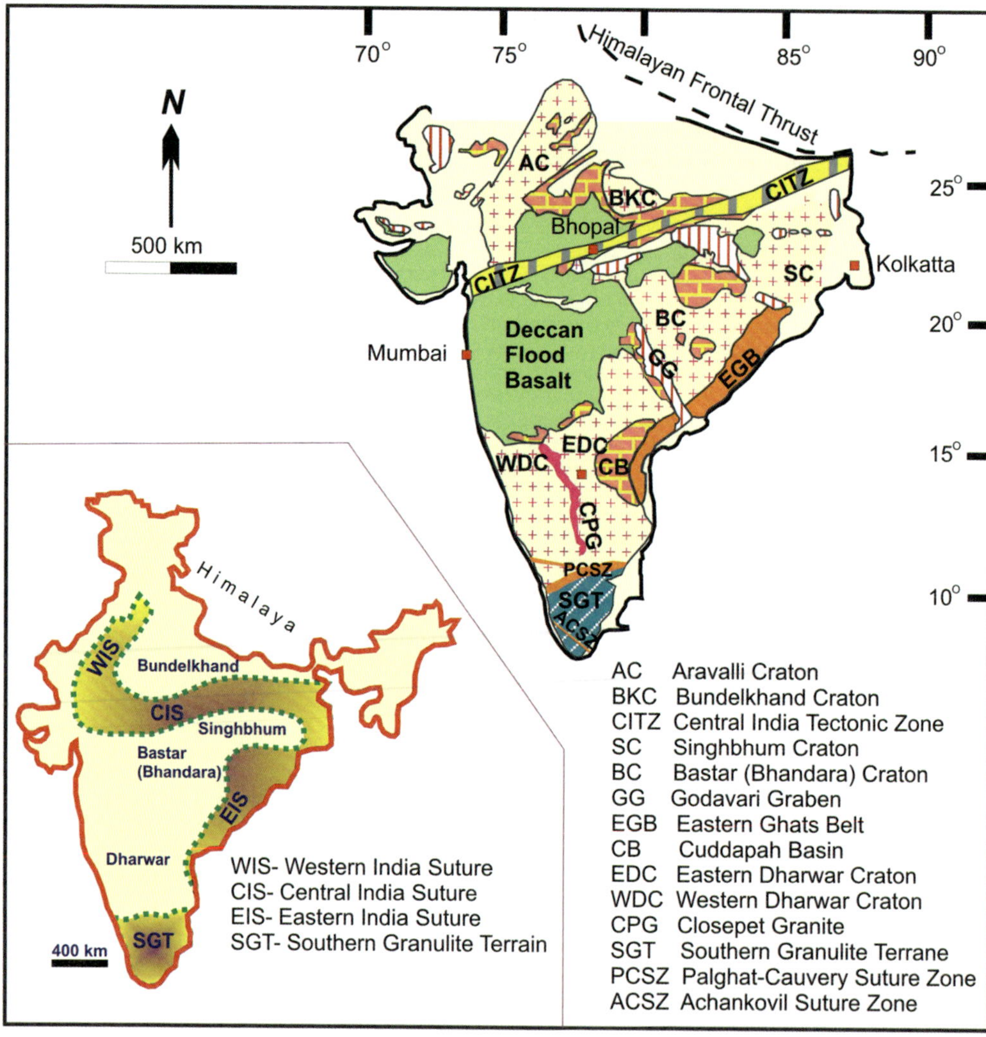

Fig. 1. Geological framework of Peninsular India showing the major Archaean cratonic blocks, Proterozoic belts and Deccan flood basalt cover. The inset shows the major Palaeoproterozoic sutures in Peninsular India classified here as the Eastern India, Central India and Western India Suture. The Southern Granulite Terrain was mostly assembled during late Neoproterozoic–Cambrian during the final assembly of the Gondwana supercontinent, although suspect Palaeoproterozoic terrains and basement rocks occur.

these domains suggest initial intracontinental rifting probably triggered by rising plumes, the emplacement of mafic dyke swarms, the formation of large igneous provinces, and the development of elongate basins with abundant sedimentation and volcanism. In several cases, continued rifting resulted in continental fragmentation, the development of passive margins, the opening of oceans and a prolonged Wilson cycle characterized by a subduction–accretion history before culminating in a final collision. Therefore, here, I broadly define the major Palaeoproterozoic belts as the Western India Suture (WIS), the Central India Suture (CIS) and the Eastern India Suture (EIS).

This paper summarizes the various configurations proposed for the Palaeoproterozoic supercontinent Columbia, where the proto-Indian continent has been identified as an integral component. This is followed by a synthesis of the information from some of the recent geological and geophysical studies from the major Palaeoproterozoic belts of Peninsular India to identify the various stages of the Wilson cycle and to compile the evidence for ocean closure through subduction–accretion–collision tectonics. The tectonic history of the Southern Granulite Terrain is not considered in this contribution, and readers are referred to a recent work (Santosh *et al.* 2009*a*) for further details and a plate tectonic model.

India and the Palaeoproterozoic world

The creation of supercontinents

The early geological configuration of the surface of our planet was dominated by island arcs in an oceanic realm, and the parallel collision of arcs led to the formation of embryonic continents that were subsequently welded together to create continents (see Santosh *et al.* 2009*b* and references therein). Whereas supercontinent cycle describes the periodic assembly and dispersal of continental fragments (e.g. Nance *et al.* 1998; Rogers & Santosh 2004), the term Wilson cycle has been used to explain the periodic opening and closing of ocean basins following the original proposal of Wilson (1966). In general, the supercontinent cycle and the Wilson cycle are considered complementary because the fragmentation of supercontinents opens new ocean basins, and the amalgamation of supercontinents leads to the closure of oceans. However, supercontinent cycles involve more complex processes during the assembly of multiple continental fragments of different ages, including ancient cratons together with accreted terranes welded together in various fashions, as against the rather simple components in the Wilson cycle involving the production and destruction of ocean floor.

The history of growth, evolution and dispersion of supercontinents on the globe through time has received considerable attention in recent years (Santosh & Zhao 2009 and references therein). A recent synthesis of the various conceptual models suggest that supercontinent tectonics in relation to mantle dynamics provides a key to evaluate the history of evolution and destruction of the continental crust, to understand the evolution and extinction of life, and to trace the major surface environmental changes of our planet (Santosh 2010*a*).

Continental rafts impose their own wavelength on mantle convection by impeding downwelling below them (Gurnis 1988). The assembly of supercontinents would force larger length scales and drive the underlying mantle towards higher temperatures. Some models consider that the extensive subduction associated with the assembly of supercontinents involves the transport of large volumes of oceanic, as well as continental, materials into the deep mantle, and that some of these recycled materials act as fuel to generate mantle plumes that rise up and eventually disintegrate the supercontinents (e.g. Maruyama *et al.* 2007). Although models on the assembly and disruption of supercontinents are diverse, most of the studies emphasize the implications of supercontinent tectonics on the rheological properties of the mantle, which in turn control the major geological processes in our planet (Yoshida & Santosh 2011).

Based on the topology of Y-shaped triple junctions in major supercontinental assemblies, Santosh *et al.* (2009*b*) recognized two distinct categories of subduction zones on the globe: the circum-Pacific subduction zone; and the Tethyan subduction zone. In the randomly distributed subduction zones on the surface of the Earth, the low temperature and dense subducted material sinks down to the bottom of the mantle (e.g. Maruyama *et al.* 2007; Fukao *et al.* 2009). In the scenario where the subduction is double-sided, the triangular regions with a Y-shaped topology would selectively refrigerate the underlying mantle and turn down the temperature in these domains compared to the surrounding regions. The Y-shaped domains also accelerate the refrigeration through larger amounts of subduction and, thus, promote stronger downwelling compared to other regions of the mantle. Once this process is initiated, a runaway growth of cold downwelling will start to finally develop into a large zone of super-downwelling to pull together the continental material on the surface into a tight assembly. The 'maximum close packing' of continental fragments within supercontinents suggested by Rogers & Santosh (2004) can be explained through this process. Multiple subduction zones promote the rapid amalgamation of continental fragments into supercontinents, and also act as major zones of material flux into the deep mantle transporting substantial volumes of trench sediments and arc crust through sediment subduction and tectonic erosion (e.g. Santosh *et al.* 2009*b*; Yamamoto *et al.* 2009; Santosh 2010*a*). Owing to buoyancy, the subducted TTG (tonalite–trondhjemite–granodiorite) material is stacked in the mid-mantle region and may not sink down to deeper levels.

The first coherent supercontinent

A synopsis of the past supercontinents and semi-supercontinents back to 3.0 Ga shows that continents were assembled into large landmasses during various periods in Earth history (Rogers & Santosh 2004; Santosh *et al.* 2009*b*). From oldest to the youngest, the supercontinental assemblies include: Ur (3.0 Ga), Kenorland (2.7–2.5 Ga), Columbia (1.9–1.8 Ga), Rodinia (1.1 Ga), Gondwana (0.54 Ga) and Pangaea (0.25 Ga) (e.g. Dalziel 1991; Hoffman 1991; Meert 2001; Rogers & Santosh 2002, 2004, 2009; Collins & Pisarevsky 2005; Li *et al.* 2008; Meert & Lieberman 2008; Eriksson *et al.* 2009).

The first coherent supercontinent in Earth history is considered to be the Palaeoproterozoic assembly termed as Columbia, proposed by Rogers & Santosh (2002) and recently revised by them (Rogers & Santosh 2009) (Fig. 2). The assembly and dispersal of Columbia demonstrates the concept of 'maximum packing' of supercontinents that undergo accretion and fragmentation in different areas at the same

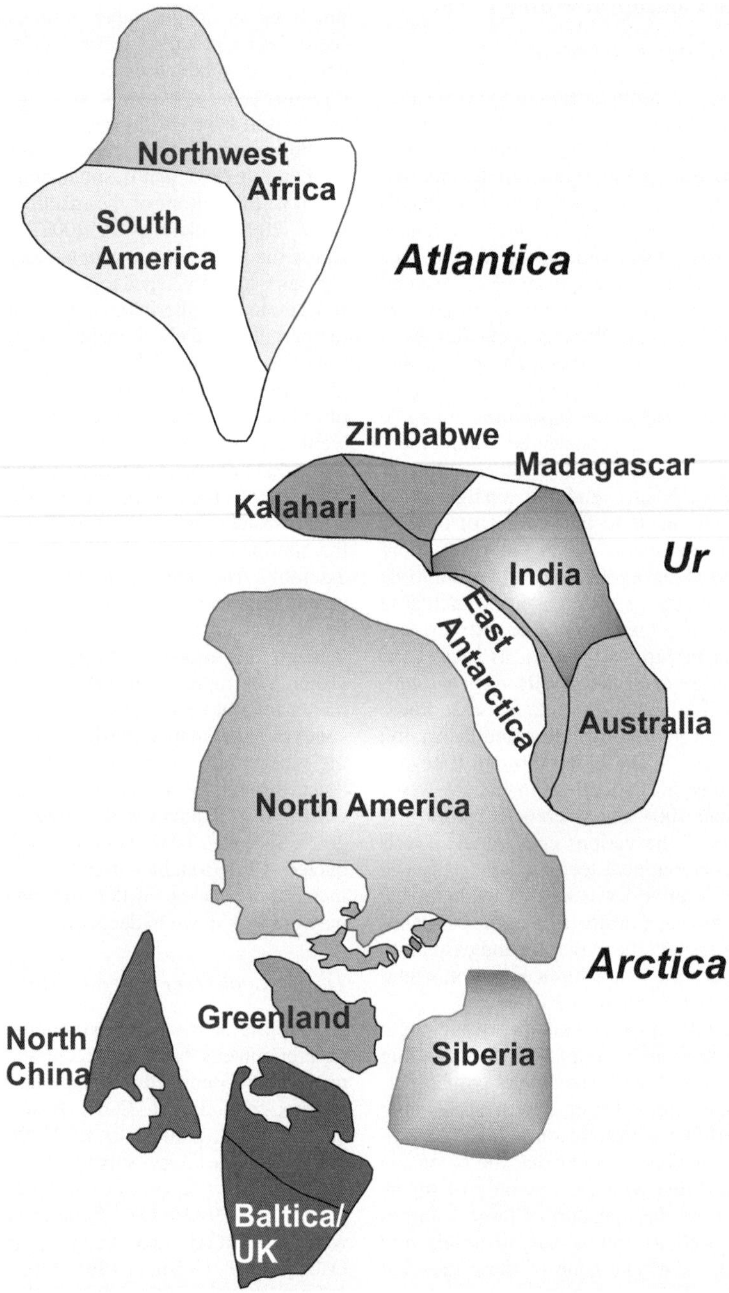

Fig. 2. Configuration of the Palaeoproterozoic supercontinent Columbia as proposed by Rogers & Santosh (2002, 2009).

time. Zhao *et al.* (2002) also proposed that a Palaeoproterozoic supercontinent existed in the period 2.1–1.8 Ga based on coeval orogeny and anorogenic magmatism, and, in their model, the configuration of Columbia largely resembled the Neoproterozoic supercontinent Rodinia (Fig. 3). The existence of a Palaeo-Mesoproterozoic supercontinent was speculated by Piper (1976) based largely on palaeomagnetic data, and in a recent study (Piper 2010) he proposed a revised configuration of this assembly termed as Palaeopangaea (Fig. 4).

Rogers & Santosh (2002) and Zhao *et al.* (2002) proposed the global assembly of Palaeoproterozoic continents. Rogers & Santosh (2002) originally suggested the time of maximum packing of Columbia as being 1.6 Ga, and Zhao *et al.* (2002), in a slightly different model, suggested maximum packing at 1.8 Ga. Condie (2002) and Zhao *et al.* (2002) also called attention to the worldwide distribution of orogenic belts with ages of about 1.8 Ga. In a recent work, Rogers & Santosh (2009) revised their earlier suggestion on the time of maximum close packing of Columbia based on a synthesis of recent petrological and geochronological data, and proposed that the amalgamation of Columbia occurred between 1.85 and 1.90 Ga. The assembly of this supercontinent coincided with several

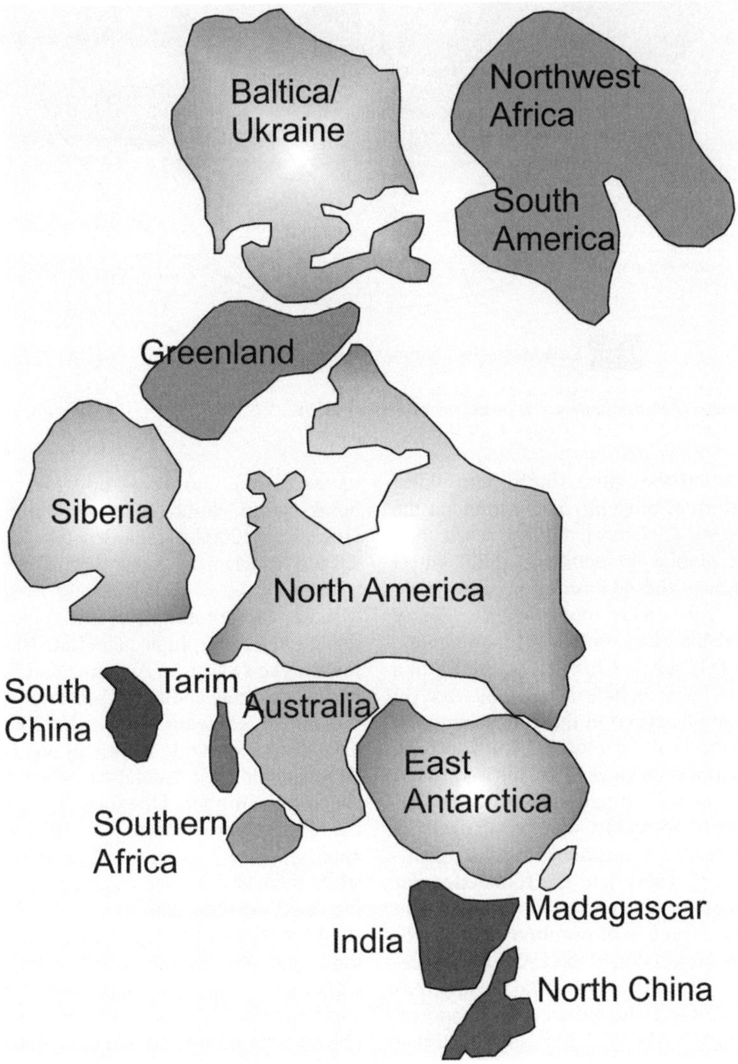

Fig. 3. Configuration of the Palaeoproterozoic supercontinent assembly as proposed by Zhao *et al.* (2002).

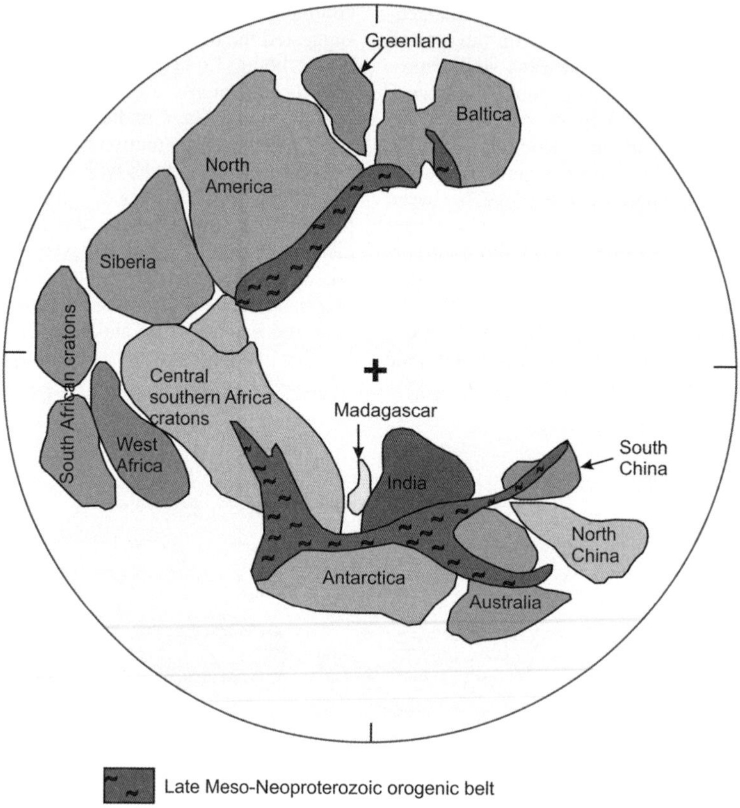

Fig. 4. Configuration of Palaeopangaea as proposed by Piper (2010).

events that affected the entire Earth. The oldest worldwide network of orogenic belts formed at the same time. Rogers & Santosh (2009) noted that, although some granite–granodiorite (GG) suites had formed earlier, the GG suites became more common in the 1.9–1.8 Ga orogenic belts. These suites succeeded the older tonalite–trondhjemite–granodiorite (TTG) suites. Changes on the Earth's surface at 1.9–1.8 Ga include a rapid increase in the concentration of oxygen in the atmosphere and oceans, and probably the evolution of eukaryotes. All of these surface changes may have resulted from the development of a broad shallow-water platform on the newly accreted Columbia. Rogers & Santosh (2009) therefore considered that the assembly of Columbia probably laid the foundation for drastic changes of the Earth's surface environment as well as for the evolution of primitive life forms.

In another reconstruction of the Columbia supercontinent, Hou *et al.* (2008*a*), based on the approximately 1.85–1.75 Ga giant radiating dyke swarm and Large Igneous Provinces (LIPs), suggested that the North China Craton, the Indian Craton and Laurentia were united together within Columbia before its extension and break-up (Fig. 5). The final break-up of Columbia occurred during 1.3–1.2 Ga. Hou *et al.* (2008*a*) regarded the 1.3–1.2 Ga fan-shaped Mackenzie dyke swarm and other dyke swarms in the Canadian Shield as sub-swarms of a Late Mesoproterozoic giant radiating swarm related to a superplume. The Late Mesoproterozoic mafic dyke swarms in Australia and East Antarctica might constitute additional sub-swarms of this giant radiating dyke swarm. Hou *et al.* (2008*a*) placed the late Mesoproterozoic mantle plume between North America and East Antarctica–West Australia. This superplume triggered the final disruption of Columbia, which occurred during a continuous extension at approximately 1.3–1.2 Ga. The Hou *et al.* (2008*a*) study indicates that the fragmentation of Columbia is related to plume activity.

A synthesis of the geological, petrological, tectonic and geochronological history of the North China Craton and its connection to the Columbia supercontinent was proposed by Kusky & Santosh (2009), who postulated that the eastern and western blocks of the North China Craton collided as a single entity with the Columbia supercontinent at

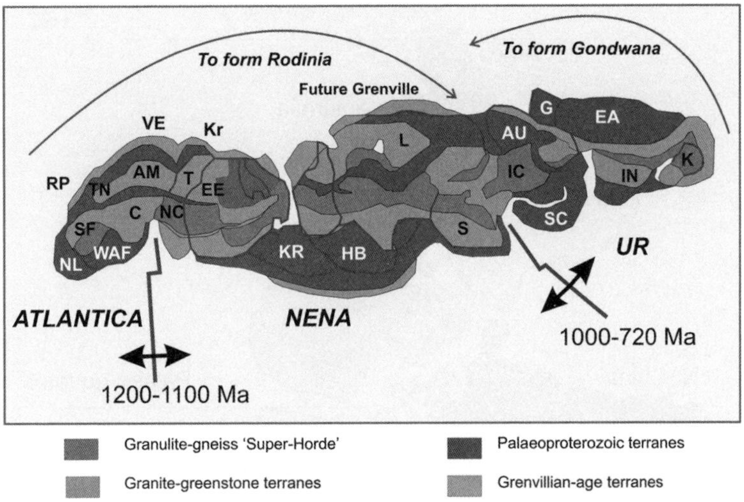

Fig. 5. Configuration of the Palaeoproterozoic supercontinent Columbia as proposed by Hou *et al.* (2008*a*).

1.93 Ga. According to them, the North Hebei orogen in the northern margin of the North China Craton correlates with the Transamazonian–Eburnian belts of Africa and South America, and the Svecofennian of Baltica. The North China Craton joined the Columbia supercontinent amalgam along the North Hebei orogen during the Palaeoproterozoic. Kusky & Santosh (2009) also endorsed that evidence for probable plume-related mafic magmatism is provided by the extensive mafic dyke swarms cutting the region and elsewhere within the North China Craton with geochemical characters testifying to emplacement within rifts that opened up.

In another reconstruction, Yakubchuk (2010) attempted to restore Columbia in a different style (Fig. 6). He used Palaeoproterozoic collisional (internal) and accretionary (external) orogens with additional constraints from the correlations between Archaean granite–greenstone and granulite–gneiss terranes. The Columbia reconstruction proposed by Yakubchuk (2010) is a 30 000 km-long assembly that formed at about 1.85 Ga. In his model, there is no evidence for the break-up of Columbia during the Mesoproterozoic, but instead continued its accretionary growth until 1.25 Ga when the Atlantica group of cratons rifted away. The rest of the assembly remained intact and was incorporated as such into the Rodinia amalgam at 1.0 Ga.

India's Columbia connection

As seen from most of the configurations proposed for Columbia discussed in the previous section, the proto-Indian continent is recognized as an integral component of this supercontinent assembly. The principal evidence supporting the configuration of Columbia as identified in the first proposal of this supercontinent by Rogers & Santosh (2002) is the correlation between eastern India with the western side of North America based on the existence of rifts with similar ages in the two continents. Hou *et al.* (2008*a*) also focused on the Palaeoproterozoic rifting events in India to emphasize the Columbia connection. In their slightly different models, Piper's (2010) Palaeopangaea based on palaeomagnetic evidence, and Yakubchuk's (2010) Palaeoproterozoic restoration pivoted on major orogenic belts, also identify a Columbia lineage for the Indian continent.

Zhao *et al.* (2003) attempted a direct Columbia connection for India and the North China Craton (NCC) based on the correlation between the Trans-North China Orogen (TNCO) along which the eastern and western blocks were sutured in the NCC, and the Central India Tectonic Zone (CITZ), which defines the suture between the South and North India blocks. They proposed that the Eastern Block of the NCC and the South India Block share many similarities in magmatic and sedimentary features. The early Archaean supracrustal sequences in both regions include 3.6–3.4 Ga quartzites, pelitic gneisses, calc-silicates, marbles, banded magnetic quartzites and amphibolites. The middle Archaean assemblages in both of the blocks can be divided into an early (3.4–3.3 Ga) phase and a late (3.1–2.9 Ga) phase. Both the Eastern Block in the NCC and the South India Block are characterized by the near-contemporaneity of late Archaean granitoid intrusive and metamorphic events. The Palaeoproterozoic volcano-sedimentary units in both blocks comprise lower clastic-rich, middle volcanic-rich,

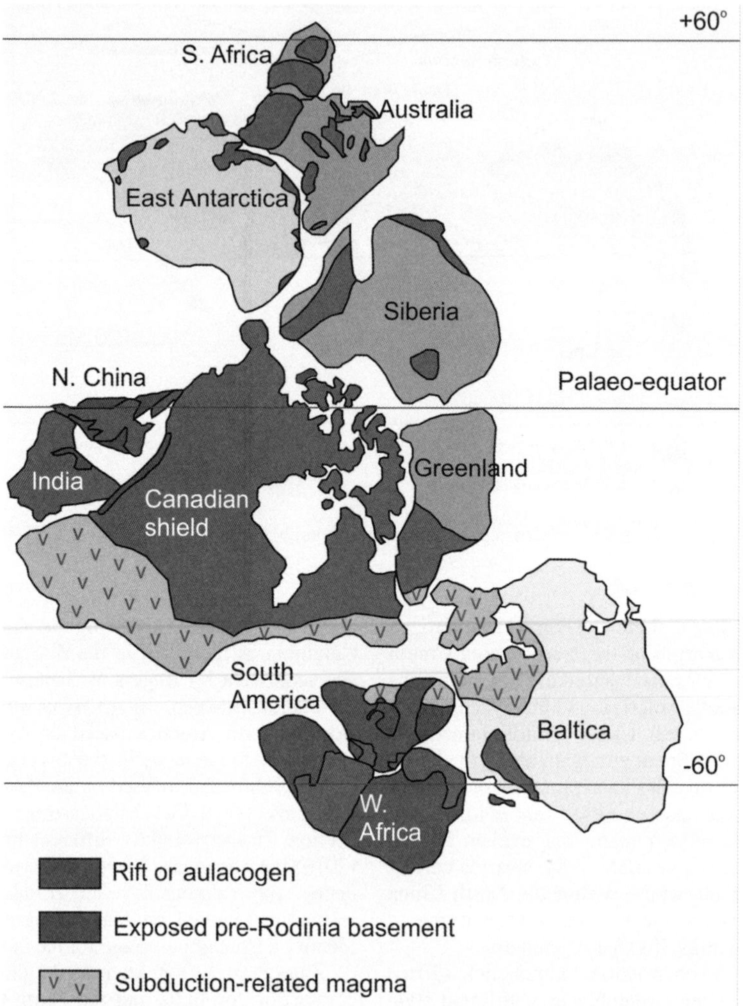

Fig. 6. Restoration of the Palaeoproterozoic supercontinent Columbia after Yakubchuk (2010).

and upper clastic and carbonate sequences. The strong similarity in the Archaean–Palaeoproterozoic magmatic and sedimentary records in both blocks led them to suggest that these blocks represent the dispersed remnants of a single continent. Zhao *et al.* (2003) proposed several possible fits for the reconstruction of the Eastern Block of the NCC and the South India Block (Fig. 7a–d), where their preferred model is the fit of the northern margin of the Eastern Block with the western margin of the South India Block. In their model, the Trans-North China Orogen and the Western Block in the NCC represent the continuations of the CITZ and the North India Block, respectively. Their proposal implies that the Trans-North China Orogen and the Central India Tectonic Zone formed a continuous suture developed during a Palaeoproterozoic collisional event. Figure 7e shows an alternate possible fit on which I speculate herein, based on new information from the eastern margin of the Dharwar Craton and the Eastern Ghats Belt where Palaeo- to Mesoproterozoic ophiolites were recently identified (Vijaya Kumar *et al.* 2010; Dharma Rao *et al.* 2011). The discovery of sapphirine-bearing Palaeoproterozoic ultrahigh-temperature granulites from the North China Craton (Santosh *et al.* 2007) also offers a potential correlation with several occurrences of sapphirine-bearing ultrahigh-temperatures granulites in the Eastern Ghats Belt. However, the postulated intimate relationship between the Eastern Block in the NCC and South India Block did not continue long, and the two blocks

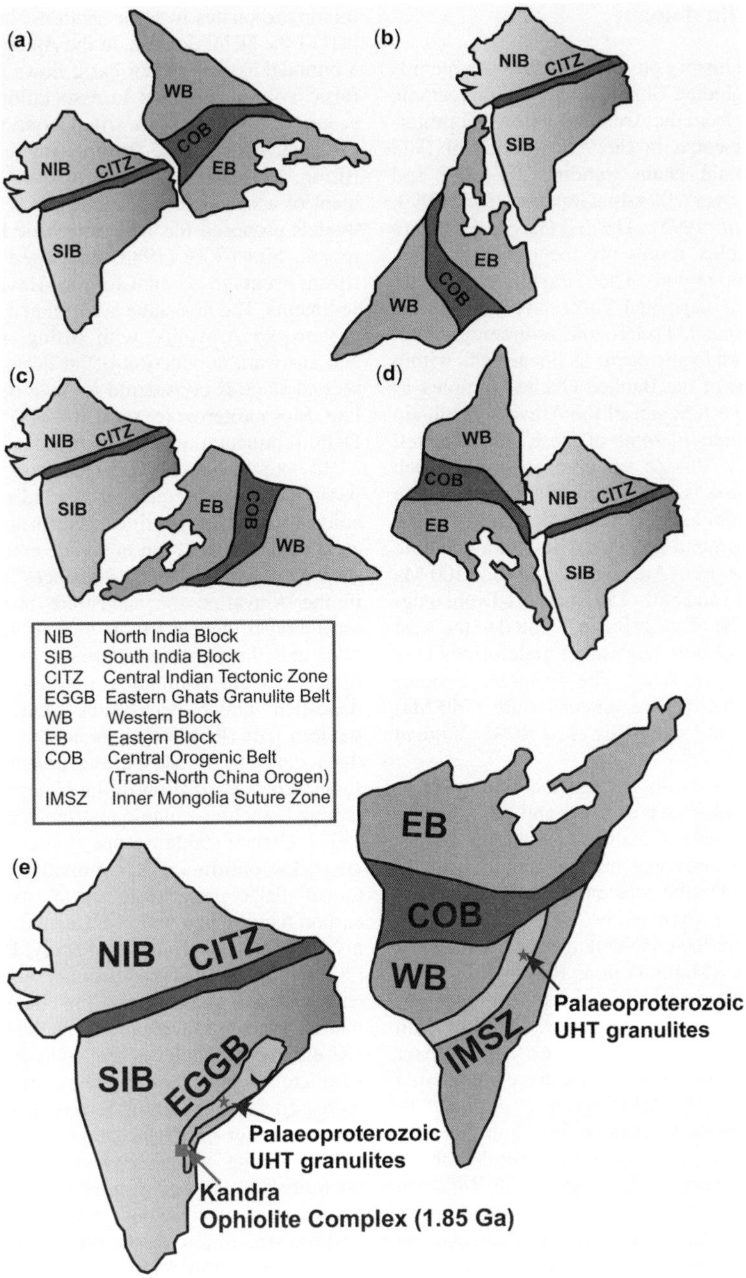

Fig. 7. Possible configurations for the reconstruction of the North China Craton and the Indian Shield: (**a**) (**b**) (**c**) & (**d**) after Zhao *et al.* 2003; (**e**) this study. EB, Eastern Block; WB, Western Block; COB, Central Orogenic Belt; IMSZ, Inner Mongolia Suture Zone; NIB, North India Block; SIB, South India Block; CITZ, Central India Tectonic Zone; EGGB, Eastern Ghats Granulite Belt. See the text for discussion.

were separated possibly by the beginning of Mesoproterozoic with a different tectonic history since then. The SE margin of the Indian plate underwent major convergent tectonics in the late Mesoproterozoic–Neoproterozoic with ocean closure and continental amalgamation associated with the Rodinia supercontinent assembly, which will be discussed in a later section.

Aravalli–Delhi domain

The geological history preserved in the rock records from the NW Indian Shield traces major tectonic events ranging from the Archaean to the Neoproterozoic as represented in the Aravalli–Delhi Fold Belt, a mountain chain trending NE–SW and extending for over 700 km (Gopalan *et al.* 1990; Sinha-Roy *et al.* 1995). The 3.3 Ga old Bhilwara Gneissic Complex represents the oldest cratonic nucleus of the region. The Aravalli and Delhi sediments were deposited successively over this Archaean basement. Proterozoic sedimentary sequences of uncertain age occur as linear belts within the eastern part of the Banded Gneiss Complex as well as in the western part of the Aravalli mountain range. The outliers of some of these belts are well known for their Pb–Zn mineralization (e.g. Deb 1992). The region is considered to have undergone at least four major tectono-magmatic and metamorphic events at around 3000 Ma (Bhilwara Gneissic Complex), 1800 Ma (Aravalli Orogeny), 1100 Ma (Delhi Orogeny) and 850–750 Ma (post-Delhi magmatic event). The Marwar Basin, located to the west of the Delhi Fold Belt, consists of undeformed clay and evaporate sequences. The Erinpura granites (850 Ma) and the Malani igneous suite (740 Ma) form its basement (Sinha-Roy *et al.* 1995; Valdiya 2010) (Fig. 8).

The basal Aravalli succession incorporates extensive volcanics such as lavas and pyroclastics, including tuffs. Several mafic dyke swarms associated with the extensional magmatism in Aravalli have been traced in the adjacent domain of Bundelkhand. Dolerites, pyroxenites and basaltic suites from this domain have yielded ages in the range of 2000–2115 Ma (Mallikharjuna Rao *et al.* 1995) suggesting a continental rifting event. Assemblages characteristic of a shelf sequence, together with basic metavolcanics belonging to the Jhazpaur Group, occur in the Aravallis and have been dated as approximately 1900 Ma (Choudhary *et al.* 1984). A younger granite–greenstone belt belonging to the Hindoli Group is also identified. The Vindhyan Basin to the SE consists of approximately 1400 Ma continental shelf sequences. In the Delhi domain, mafic volcanic rocks occur in all of the successions: at the top of the Riaolo, at the base of the Alwar and at the bottom of the Ajabgarh Groups (Bhattacharjee 1988; Srivastava 1988; Raza *et al.* 2001). In Alwar, at least 18 flows of vesicular basalt intercalated with tuffs have been identified. The lava flows in Ajabgarh show a pillow structure and associated pyroclastic material. The lavas bear a close similarity to low-temperature continental flood basalts derived from an enriched mantle source without any significant crustal contamination, and are interlayered with arkosic sediments, conglomerate and minor carbonates in some sections. In the southern part of the Delhi domain, in the Ajabgarh sequence, a bimodal association of basic flows and tuffs with felsic volcanics occurs in association with clastic sediments typical of a rifted continental margin package. All of these features suggest continental rifting, passive margin formation and enlargement of a wide ocean. Among the several tectonic models proposed for the Proterozoic history of this region, Sinha-Roy (1988) envisaged two separate rifting events to account for the Aravalli and Delhi sediments. The first stage is correlated to the Palaeoproterozoic Aravallis, with rifting, sedimentation and eastward subduction of an oceanic plate. The second stage is considered to have occurred in the late Mesoproterozoic (*c.* 1300 Ma) with rifting, Delhi sedimentation and westward subduction.

In both the Aravalli and Delhi successions, thick passive margin sequences including quartzites, siliceous marbles, phyllites, schists and conglomerates indicate deposition in a wide continental shelf. Two contrasting facies sequences are identified in the Aravallis: the sandstone–shale–carbonate assemblages deposited on a shelf in the eastern part; and the flysch assemblage characterized by lithic- and quartz-arenites interbedded with shale deposited under deep-water conditions in the western part (Roy 2001). A notable feature is the thick deposits of phosphorite associated with the Jhamarkotra Formation in the Udaipur region containing branching columnar stromatolites (Banerjee 1971). Carbon stable isotope studies of the carbonate rocks confirm a $\delta^{13}C$ excursion correlating to a global Palaeoproterozoic cycle and an organic carbon fixation in a wide continental shelf (e.g. Srinivas *et al.* 2001). In a recent study, Papineau *et al.* (2009) identified high primary productivity in carbonaceous shales from the Jhamarkotra Formation, consistent with stratigraphically underlying ^{13}C-enriched carbonates and with their associated stromatolitic phosphorites. They suggested that the increased fluxes of phosphorous in Palaeoproterozoic sea water may have played an important role in stimulating increased levels of primary productivity and the rise of atmospheric oxygen. The phosphorous and other nutrient-enrichment in shallow seas following the break-up of continental assemblies is considered to have played a critical role in the various stages of biological evolution through Earth history (e.g. Maruyama & Santosh 2008).

The wide ocean developed through Palaeoproterozoic rifting in the Aravallis soon started closing down with the initiation of subduction of the oceanic lithosphere. A recent study provides robust evidence for the development of an Andean-type continental arc margin in the NW segment of the Palaeoproterozoic Aravalli domain (Kaur *et al.*

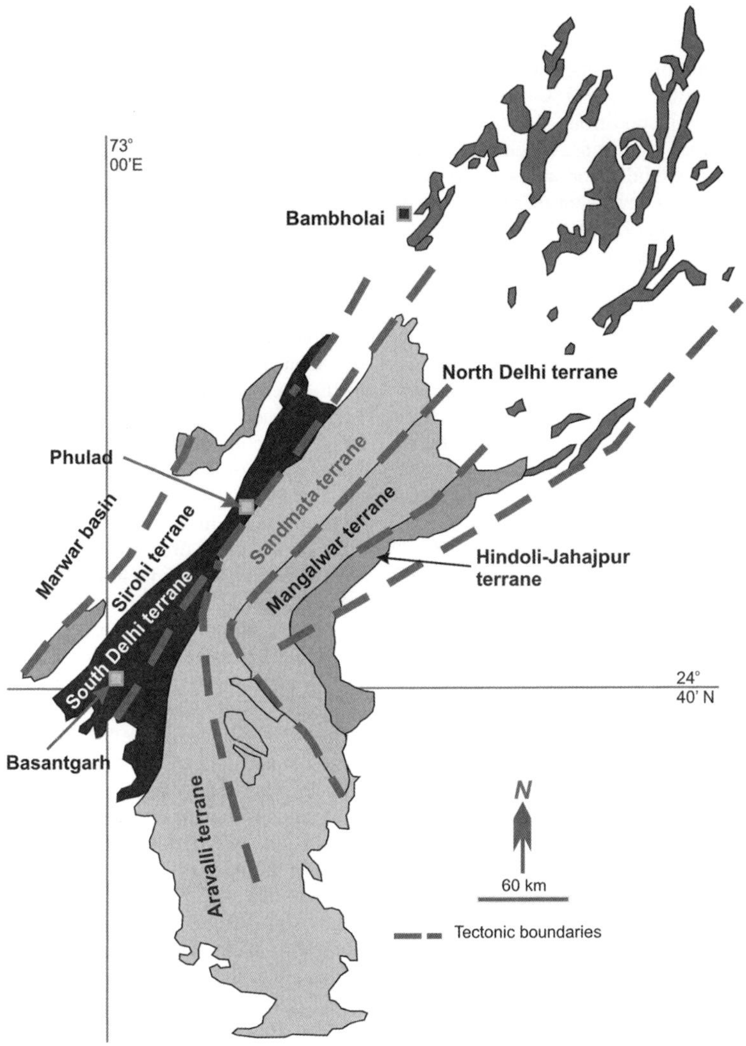

Fig. 8. Generalized tectonic framework of the Aravalli–Delhi domains showing the major geological units, defined as belts (modified from a compilation by Singh *et al.* 2010).

2009). This work described the Jasrapur pluton dominantly comprising monzogranites with a subordinate amount of granodiorite and quartz monzodiorite. The rocks possess typical continental arc geochemical signatures, including high-K calc-alkaline, mildly peraluminous, I-type and magnesian features, as well as high Rb and low Nb, Y and Rb/Zr ratios, decoupling between LILE (large-ion lithophile elements) and HFSE (high field strength elements), and negative Nb–Ta and Ti anomalies. Geochronological studies by Kaur *et al.* (2009) yielded Sm–Nd ages of 1800 ± 59 Ma for the emplacement of the magmas and a more precise zircon Pb/Pb evaporation age of 1821.7 ± 0.4 Ma. The high Ba-enrichment trend (400–1600 ppm), together with mantle-like initial Sr ratios ($Sr_i = 0.7060 \pm 0.0007$) of these rocks, led the authors to suggest that the primary magma of the Jasrapur Complex was generated from partial melting of a depleted mantle source enriched by slab-induced LILE fluids. Negative initial Nd ($\varepsilon_{Nd(t)} = -5.01 \pm 0.13$) and a 2.6 Ga mean crustal residence age suggest the involvement of an Archaean component, with the interpretation that the magma underwent crustal contamination while traversing through a relatively thick continental crust. A correlation with the magmatic suites further south in central Rajasthan indicates extensive continental arc magmatism during 1850–1822 Ma, which the authors linked with the amalgamation of the Palaeoproterozoic supercontinent Columbia. Geothermometric studies suggest high-temperature (*c.* 800 °C) and medium-pressure (≥600 MPa) conditions for the crystallization of the Jasrapur granitoids, which would suggest a probable ridge subduction and slab window setting along an active convergent margin (Santosh & Kusky 2010; Zhang *et al.* 2010).

The final collision between the Aravalli and Delhi domains defining ocean closure and continental amalgamation is critical in understanding the culmination of the Proterozoic Wilson cycle in NW India. In a recent study, Bhowmik *et al.* (2010) reported monazite chemical ages in two granulite suites from the Sandmata and Mangalwar complexes of central Rajasthan. The Sandmata granulites record an earlier medium-pressure sillimanite-facies metamorphic event. The granulites from both complexes also show a subsequent high-pressure upper-amphibolite-facies ($P = 10.3 \pm 0.8$ kbar, $T = 675 \pm 60$ °C) to granulite-facies ($P = 12–14$ kbar, $T = 815$ °C) peak metamorphism. Metamorphic monazites from the Sandmata Complex show distinct age populations of approximately 1700–1600 and 880 Ma. The Late Palaeoproterozoic age populations correlate with the first high-grade metamorphic event in Sandmata, and the Early Neoproterozoic ages represent a subsequent event. However, the monazites from metapelites of the Mangalwar Complex yield characteristic Late Mesoproterozoic–Neoproterozoic ages. Based on these age data, Bhowmik *et al.* (2010) assigned a Grenvillian age for the collisional orogeny between the Aravalli and Delhi domains. Importantly, the late Palaeoproterozoic metamorphic ages correlate with the final stages of collisional orogeny associated with the subduction–accretion tectonics in the Aravalli domain.

Geophysical data, in particular deep seismic reflection data, provide useful information in evaluating crustal architecture and mantle dynamics, and have recently been employed in many studies in conjunction with the surface geological features to understand the subduction–accretion–collision history in Precambrian terranes (e.g. Naganjaneyulu & Santosh 2010, 2011; Santosh 2010*b*). Rajendra Prasad *et al.* (1998) reported deep seismic reflection data along a 400 km-long profile across the Aravalli–Delhi Fold Belt. The data bring out two sets of reflection bands with opposite dips, which the authors correlated to tectonics associated with the Aravalli and Delhi orogenies. Figure 9a, b shows a compilation and reinterpretation of the seismic reflection data and two-dimensional (2D) crustal density model of Rajendra Prasad *et al.* (1998). The prominent west-dipping reflections observed in the central domain, towards the western margin of the Mangalwar Complex, could be related to the Palaeoproterozoic subduction of an oceanic lithosphere, and suggests a clear northwestward subduction polarity, as against the eastward suduction proposed by Sinha-Roy (1988). The high-density layer probably defines the hanging wall of the subducted slab, as I have attempted to trace out from the 2D crustal density model shown in Figure 9b. Resting on the top of this slab are seen still higher density domains which I interpret as eclogitized oceanic crust forming the root of the exhumed metamorphic orogen. The Manglarwar Complex incorporates major Pb–Zn mineralization in Sawar, Banera and Dariba localities, which have been dated as about 1800 Ma (Deb *et al.* 1989). The sediment-hosted mineralization can be correlated to convergent margin processes in the Palaeoproterozoic.

The lithological association and their nature of distribution are in accordance with a Palaeoproterozoic Pacific-type orogeny in the Aravalli region, with a westward subduction of the Archaean cratonic margin and development of a wide accretionary belt, imbricated ocean plate stratigraphy including ophiolites (see below) and the extrusion of a high-grade regional metamorphic belt at the orogenic core following the final collision (Fig. 9c). The seismic reflection data towards the western region (Fig. 9a) show less prominent and east-dipping reflectors, which probably correlate with a subsequent tectonic regime (Rajendra Prasad *et al.* 1998), possibly associated with the collision of the Aravalli and Delhi domains. In order to evaluate this aspect further, we will briefly examine below some of the recent concepts on subduction–accretion–collision tectonics, particularly with reference to the identification of suture zones (see also Santosh *et al.* 2009*a*).

The zone along which two continental plates have joined following the subduction of oceanic lithosphere represents a suture. It is often difficult to trace suture zones in deep structural levels of old deeply eroded orogenic systems (Dewey 1977). Ophiolites and blueschist-facies rocks have been considered to be the hallmarks of subduction tectonics (Condie & Kröner 2008). Sutures have been traditionally defined by the presence of ophiolites as they represent the on-land remnant of fossil oceanic lithosphere, and ophiolites form in a wide range of plate tectonic settings (e.g. Kusky 2004). The three major features considered as diagnostic of plate tectonics are generally rare or absent in the Precambrian geological history: ophiolites, Franciscan-type mélange and blueschists. All of them characterize divergent or consuming plate boundary processes on the modern Earth. However, Kusky (2004) assembled robust evidence for the occurrence of 2.5 Ga ophiolites and related

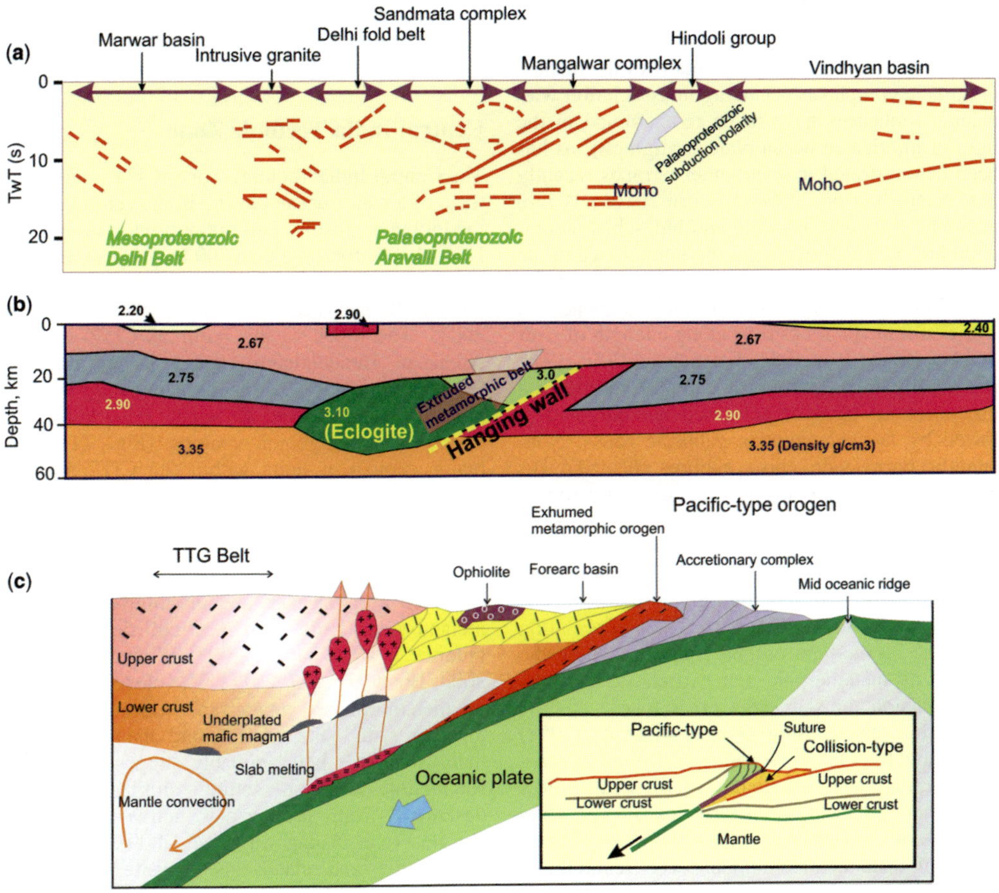

Fig. 9. (**a**) Seismic section from the Aravalli–Delhi domains in NW India (after Rajendra Prasad *et al.* 1998). Note the prominent west-dipping reflections in the central region that are correlated with Palaeoproterozoic subduction. The minor east-dipping reflections below the Marwar Basin are correlated to a younger event. (**b**) True-scale 2D crustal density model computed from seismic reflection data (after Rajendra Prasad *et al.* 1998). In the presented study, the moderate- to high-density layer (density 2.90) is correlated with the 'frozen'-in structure developed from Palaeoproterozoic subduction, and the hanging wall of the subducted lithosphere is also postulated. The 3.0–3.10 density layers over this layer is interpreted as eclogitized oceanic crust, forming the root of the extruded metamorphic orogen. (**c**) A schematic illustration of Pacific-type orogen (as compiled in Santosh 2010*a*) showing the distribution of the various lithounits and extruded metamorphic orogen. The lithological association, nature of distribution and the geophysical data are all consistent with a Pacific-type orogeny for the formation of the Palaeoproterozoic Aravalli Belt.

rocks in North China, Canada and other regions that suggest that the features considered as diagnostic for plate tectonics are present in the older Earth history also. Komiya *et al.* (2002) reported evidence of an accretionary complex in the early Archaean Isua supracrustal belt. The discovery of an accretionary complex, comprising 'oceanic plate stratigraphy', a travelogue of the oceanic plate from mid-oceanic ridge to the trench (Matsuda & Isozaki 1991), and duplex structures, lends strong support for the operation of plate tectonics, including rigid plates, significant lateral movement of the plates, the presence of multiple plates surrounded by plate boundaries and the linearity of plate boundaries. Kusky & Polat (1999) suggested that Archaean granite–greenstone terranes represent juvenile continental crust formed in a variety of plate tectonic settings, and most of them acquired their first-order structural and metamorphic characteristics at convergent plate margins, where large accretionary wedges grew through offscraping and accretion of oceanic plateaus, oceanic crustal fragments, juvenile island arcs, rifted continental margins, and pelagic and terrigenous sediments similar to the Phanerozoic examples (e.g. Hara *et al.* 2009; Safonova *et al.* 2009; Isozaki *et al.* 2010).

In the Palaeoproterozoic terranes of Peninsular India, typical ophiolites are generally rare (with exceptions, see later), although many of the mafic–ultramafic complexes occurring in association with pelagic sequences have to be re-examined in the light of imbricated ocean plate stratigraphy, as has been demonstrated in some other terranes recently (e.g. Santosh *et al.* 2009*a*; Santosh 2010*b*). A rare example of typical ophiolite and blueschist association has been reported from the Phulad area in the Palaeoproterozoic Aravalli mountain range of the South Delhi sector, where a thick sequence of mafic–ultramafic rocks runs along a length of over 300 km (Sinha-Roy 1984; Khan *et al.* 2005). The structurally lowest part of the complex comprises a discontinuous band of deformed harzburgite followed by layered cumulus gabbroic rocks. Non-cumulus hornblende schists, gabbros, sheeted dykes and pillowed basalts occupy the structural top of the layered gabbros. The rocks belonging to the Phulad suite are juxtaposed with accreted fore-arc basin sediments to the east, and are sandwiched between these and the high-grade metamorphic suite of the Sandmata Complex. Thus representing a typical Pacific-type orogen (Fig. 10) where the high-grade metamorphic belt extruded during the final collision and slab break-off in a west-dipping subduction regime. Sinha-Roy & Mohanty (1988) considered the Phulad sequence of layered gabbros and serpentinized ultramafic rocks with pillow basalt and hyaloclastic material as an ophiolitic mélange of contrasting ensialic rift and arc–trench marginal basin tectonic domains. The trench metagreywacke contains relict blueschist assemblages that have been overprinted by regional greenschist-facies metamorphism. Rhyolites, andesites and basalts intercalated with chert and marble overlie this sequence. This succession represents a typical ocean plate stratigraphy, and the presence of ophiolite and blueschist clearly mark subduction tectonics.

Central India Tectonic Zone

The Central India Tectonic Zone (CITZ) marks the trace of a major ocean closure in the central part of the Indian peninsula. This 400 km-wide suture runs ENE–WSW for more than 800 km, and dissects the Indian peninsula into northern and southern crustal blocks (Radhakrishna & Naqvi 1986; Acharyya 2003). The block to the north of CITZ has the Archaean Bundelkhand Craton as its nucleus, whereas the composite block to the south is composed of the Dharwar, Bhandara (also known as Bastar) and Singbhum cratons (Fig. 11). The major geological units within the CITZ include metamorphosed supracrustal belts, metamorphosed mafic and ultramafic bodies, extensive metacarbonates, iron and manganese formations, TTG gneisses, charnockites and related arc magmatic suites, exhumed high-pressure and ultrahigh-temperature metamorphic belts, and post-collisional K-rich granites (e.g. Acharyya 2003; Bhowmik & Roy 2003; Bhandari *et al.* 2011). The close association of metamorphosed mafic and ultramafic rocks, together with belts of iron and manganese formations in this belt, probably suggests a typical subduction–accretion setting similar to those described from present-day active continental margins (e.g. Nakagawa *et al.* 2009). The TTG, charnockites and related igneous suites were probably generated from subduction-related magmas, whereas the potassic granites represent plutons emplaced along suture zones after the collisional assembly, similar to those in other collisional orogenic sutures (e.g. Santosh *et al.* 2009*a*).

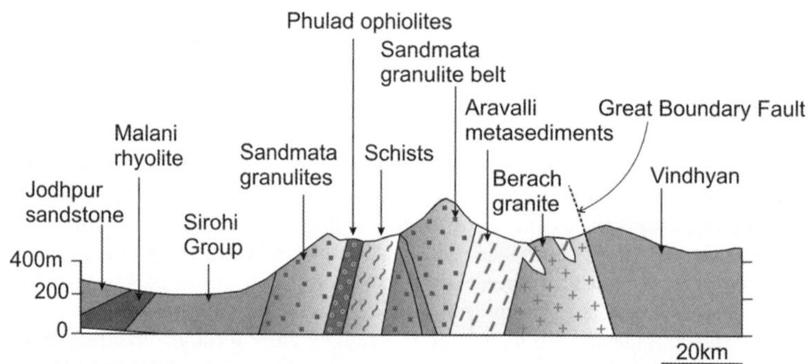

Fig. 10. Geological cross-section of the Aravalli Belt (after Sharma 1995). Note the high-grade metamorphic orogen of the Sandmata Complex, with westerly dips incorporating the Phulad ophiolites, correlating with a westward subduction, and the subsequent extrusion of the metamorphic belt along the Benioff plane following the final collision and slab breakoff.

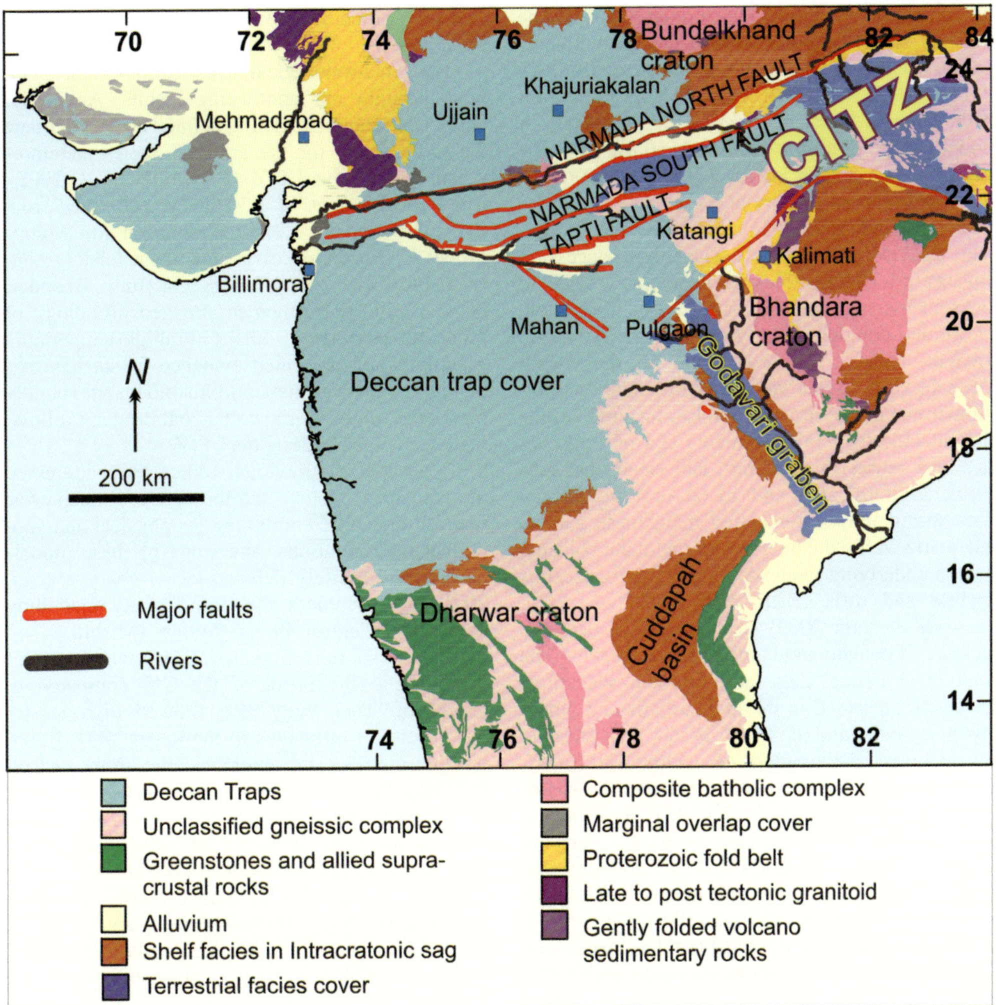

Fig. 11. Geological map of the central part of India showing the Central India Tectonic Zone (CITZ) and the surrounding cratons (modified from a compilation by Naganjaneyulu & Santosh 2010).

The Singhbhum Basin in eastern India exposes a series of Proterozoic volcano-sedimentary units comprising the Dhanjori, Chaibasa, Dhalbhum, Dalma and Chandil formations defining successively younger ages (Mazumder & Sarkar 2004). The Dhanjori volcano-sedimentary succession displays evidence of having passed through passive, as well as active, phases of continental rifting with an increasingly important influence of volcanism on sedimentation through time. The lavas in the succession are considered to be part of a LIP related to mantle plume upwelling in an intracontinental rift setting (Mazumder & Sarkar 2004; Mazumder 2005). The lavas (>1.6 Ga) in the Singhbhum Basin suggest that the plume magmatism was possibly part of the Late Palaeoproterozoic global tectonothermal event (Acharyya 2003). The Dalma Volcanics, which have been subjected to greenschist-facies metamorphism, comprise both mafic and ultramafic volcanic rocks of both eruptive and explosive phases (Bose 1994). The upper unit consists of pyroclastic and mafic lavas that are often vesicular and pillow-shaped, carrying lenses of tuffaceous material. Both a tholeiitic nature, with a plume-related origin in a rift setting (Roy *et al.* 2003), and an andesitic nature, with formation in an island-arc setting close to continental margin (Alvi & Raza 1992), have been proposed for the volcanics. Conglomerate, coarse sandstone and shale representing a fluvial-fan complex occur in association with the volcani-clastic material (Mazumder 2005). The Kuilapal granite exposed within the Chandil Formation yielded a Rb–Sr whole-rock isochron age of 1638 ± 38 Ma, which fixes the upper age limit of the Chandil volcanic rocks (Acharyya 2003). A model Pb age of approximately 1660 Ma from the ore deposits has been interpreted as the age of the Dalma Formation (Acharyya 2003).

The Bhandara domain comprises the Sausar, Sakoli and Dongargarh units. The Sakoli succession consists of a 6000 m-thick sequence of metapelitic rocks in association with banded iron formation, arkose and conglomerate lying over the Archaean Amagon gneiss. The lower part of the Sakoli sequence comprises a volcano-sedimentary association dominated by rhyolites, felsic tuff and basalts intercalated with pelitic sediments. Garnet amphibolites with lenses of metachert and basic tuffs represent an imbricated and metamorphosed ocean plate stratigraphy. Intercalated manganese garnets and quartz within the interface of the mafic volcanic and the underlying sequence of chert, carbonaceous phyllite and tuffs (Bandopadhyay *et al.* 1995; Chawade & Naik 2003) also suggest an accreted package of pelagic and continental sequences. The continental rifting stage and bimodal volcanism are best represented in the Dongargarh unit where rhyolites, tuffs and lavas of basalt and andesite are documented (Neogi *et al.* 1996). The Sausar unit comprises metamorphosed sandstones, shales and carbonate rocks of a stable shelf sequence. A suite of felsic and aluminous granulites carrying sapphirine and spinel in Mg–Al-rich domains, and intruded by gabbro, norite and pyroxenite, records ultrahigh-temperature metamorphism at 950 °C and 9 kbar (Bhowmik *et al.* 2005). The spatial and temporal association of ultrahigh-temperature felsic granulites with metamorphosed mafic and ultramafic bodies in this region is similar to the occurrence reported from some of the major collisional orogens, such as the Cambrian collisional suture of Gondwana in southern India, suggesting intercalation of subducted sediments with mantle-wedge material in an accretionary belt (Santosh *et al.* 2009*a*). Recent studies have also indicated a regional high-pressure metamorphic event at the northern margin of the Sausar Belt (Bhowmik & Roy 2003). The high-pressure–ultrahigh-temperature and high-pressure metamorphic belts appear to define a Mesoproterozoic paired metamorphic belt. Such paired metamorphic belts, that contain ultramafic rocks, linear belts of iron formations, carbonates and accreted continental margin sequences, are products of metamorphism within a subduction zone, and are exposed at the final stages of collision – which is characteristic of an initial Pacific-type orogeny followed by a final Himalayan-style collision when the metamorphic belts exhume through a wedge extrusion process following slab break-off (Santosh *et al.* 2009*a*).

The geochronological investigations in the CITZ have brought out contrasting results. An earlier study from monazite dating defined an approximate 2.1–2.0 Ga age for the granulite-facies metamorphism within the CITZ (Bhowmik *et al.* 2005). More precise ages of 1.59 Ga were derived from SHRIMP (Sensitive High Resolution Ion Microprobe) U–Pb dating of zircons from felsic granulites (Bhowmik 2006). In a recent study, Bhandari *et al.* (2011) performed *in situ* geochronology of monazites associated with the ultrahigh-temperature granulites and presented evidence for an approximately 1.6 Ga peak metamorphism. This age roughly marks the timing of the collisional orogeny following ocean closure along the CITZ.

A number of previous studies have addressed the collisional history and the nature of deep crust beneath the CITZ combining geophysical and geological interpretations, and some of these models have been intensely debated (see Acharyya *et al.* 2001 and references therein). Gravity modelling across the Central Indian Suture (Mishra *et al.* 2000) shows a thick crust (*c.* 50 km) with a greater thickness to the north of the CIS compared to the south. This study also showed high-density (2.81 g cm^{-2}) intrusives in the upper crust. Based on the high-density anomaly, the dome-shaped seismic reflector pattern and the surface exposures of granulite-facies rocks, Mishra *et al.* (2000) inferred an upthrust block from the lower crust. Mall *et al.* (2008) re-evaluated deep seismic reflection data along the CIS, and proposed a crustal thickness of about 50 km in the SE and 40 km in the NE. They interpreted a Mesoproterozoic collisional suturing of two micro-continents. The nature of the subduction–accretion–collision tectonics, including the polarity of subduction in the CITZ, was recently addressed in a recent study by Naganjaneyulu & Santosh (2010). They synthesized seismic, gravity, heat flow and magnetotelluric signatures from the CITZ, and correlated these with the salient geological features to evaluate the tectonics of this major suture. The modelled high-velocity, high-density and high-conductivity layers observed in the upper crust shows a similarity with those of the lower crust, suggesting a stratified nature of the crust resulting from accretionary tectonics. Based on gravity data, subducted slabs and extruded metamorphic orogens were inferred with the high-density anomaly, the dome-shaped seismic reflector pattern and the surface exposures of granulite-facies rocks corresponding to the extrusion of the regional metamorphic orogen and doming up following subduction and slab break-off (Fig. 11). Magnetotelluric profiles from both the eastern and western parts of the CITZ show low-resistivity regions at shallow upper-crustal regions, all suggesting a Pacific-type subduction–accretion process. The lithological association of limestones, banded iron formations and the manganese-bearing horizons, in association with quartzites and metacarbonates within the Mesoproterozoic Sausar Group occurring along the southern domain of the CITZ, probably represents a mixture of accreted oceanic and continental margin sediments. Manganese-bearing chert horizons have been widely reported from the Jurassic accretionary complex in SW Japan (e.g. Nakagawa *et al.* 2009), and are thought to constitute key beds in identifying ocean plate stratigraphy and their imbrication within accretionary belts. The high-pressure granulites, and associated mafic and ultramafic sequences, suggest subducted ocean-floor as well as mantle-wedge components. The protoliths of the orthogneisses, charnockites and related rocks widely distributed in this zone correspond to TTG and adakitic magmas derived by slab melting during the subduction process (e.g. Santosh *et al.* 2009*a*). A number of late K-rich granites occurring in the area mark post-collisional suture plutons. Thus, the geological characteristics and lithological sequences provide robust evidence for the sutures within the CITZ to be zones of ocean closure where subduction–accretion–collision tectonics played a critical role.

Some of the previous studies have also correlated the Sausar Group of rocks belonging to the Satpura Mobile Belt in the southern domain of the CITZ with a Palaeoproterozoic collisional orogeny (e.g. Yedekar *et al.* 1990). Sarbadhikari & Bhowmik (2008), based on detailed studies of a metamorphic orogen at the southern domain of the Sausar Belt, proposed multistage amalgamation processes within the CITZ. They envisaged two possibilities: (a) an Archaean collisional event in the South Indian Block; (b) a Palaeoproterozoic collisional event, in a possible island-arc setting, leading to the development of an early Mesoproterozoic hot orogen. Their model infers a tectonic switching from early lithospheric extension to contraction, resulting from successive stages of slab retreat and flat subduction that led to the suturing of arcs, back arcs and continents in the CITZ. The recent age data on the timing of the ultrahigh-temperature metamorphism (e.g. Bhandari *et al.* 2011) and the paired nature of the high-pressure–ultrahigh-temperature belts suggest a major Mesoproterozoic tectonothermal event. The association of the high-pressure–ultrahigh-temperature paired orogens with accretionary sequences developed through subduction–collision tectonics, the lack of a distinct signature for the associated rock units of back-arc origin and the general hydrous nature of modern back-arcs that inhibit the preservation of dry ultrahigh-temperature assemblages suggest that the paired metamorphic belts within the CITZ were probably generated through subduction-zone metamorphism (e.g. Santosh & Kusky 2010) associated with the final suturing of the South Indian and North Indian continental blocks. A major Mesoproterozoic collisional suturing of micro-continents along the CITZ is also envisaged in the work of Mall *et al.* (2008), a study that combines geological and geophysical data.

The seismic reflection data compiled from various transects by Naganjaneyulu & Santosh (2010) provide some clues on the subduction polarity in the CITZ (Fig. 12). The seismic reflectors in a traverse across the CITZ towards the western domain clearly show a general northward dip. The seismic reflectors across the Central Indian Suture (CIS) in the eastern domain display a dip to the NW. Thus, a north to NW subduction of the South India Block beneath the North India Block was proposed. The seismic data interpretation also shows that the northward dip is terminated towards the northern boundary where the reflectors tend to show an opposite dip, to the south. The Moho also warps up in this domain. One possible explanation for this would be a subduction component from the north. The opposite dipping reflectors probably suggest the case for a double-sided subduction history. Santosh *et al.* (2009*b*) has evaluated the process of double-sided subduction and its role in the rapid amalgamation of continental fragments during the formation of supercontinents. The amalgamation of the major crustal blocks within the North China Craton during the Palaeoproterozoic is proposed to have involved double-sided subduction (Santosh 2010*b*). The western Pacific region illustrates the case of a double-sided subduction zone (Maruyama *et al.* 2007). Another modern analogue for double-sided subduction is illustrated by the Banda Sea region characterized by active subduction, transfer of material at the subduction zone, the collision of oceanic plate buoyant features, arcs and continents, and abundant magmatism (e.g. Hall 2009).

Eastern Ghats domain

Recent studies have brought to light increasing evidence for Palaeoproterozoic rifting and continental fragmentation along the SE margin of the Indian plate. In a recent study, French *et al.* (2008) reported evidence for a Palaeoproterozoic LIP from the southern Bastar Craton and from the adjacent Cuddapah Basin from the U–Pb dating of baddeleyite in mafic dyke swarms. Their results suggest an approximately 1.9 Ga mafic magmatism that spans an area of around 90 000 km^2 in the southern Indian Shield associated with concomitant intracontinental rifting and basin development. Their results are consistent with a major episode of

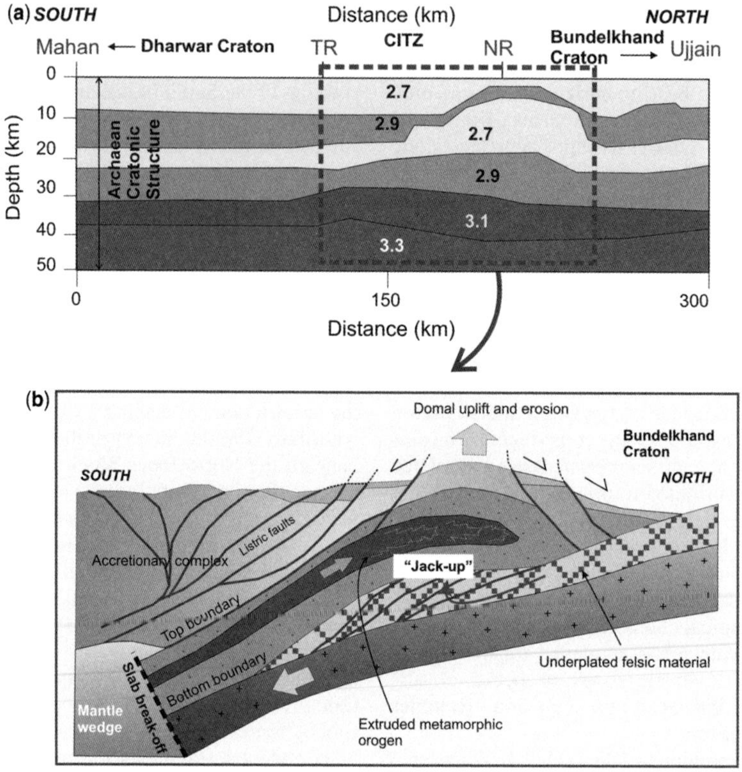

Fig. 12. (**a**) Interpretation of gravity data from the Mahan–Ujjain transect within the Central India Tectonic Zone (after Naganjaneyulu & Santosh 2010). (**b**) Cartoon illustrating the domal structure resulting through the 'jack-up' by underplated sediments during subduction and the exhumation of the orogenic core to explain the sandwiched nature of the crustal layers with a high-density layer trapped between two low-density layers. This model is also consistent with the tectonic extrusion of regional metamorphic belts at the orogenic core and their subsequent exposure through extensive erosion.

1.89–1.87 Ga intraplate mafic magmatism and rifting, spanning a large tract of the Dharwar protocontinent. The southern Bastar–Cuddapah mafic magmatic province also coincides with a synchronous 1.89–1.87 Ga LIP preserved in various other crustal fragments, including the Superior and Kalahari cratons as well as the North China Craton. French *et al.* (2008) correlated the widespread continental LIP to a period of enhanced mantle plume activity and/or a period of focused mantle upwelling.

Previous geochronological studies from various sectors of the Dharwar Craton, the southern Bastar Craton and the Cuddapah region have also recorded possible plume-related mafic magmatism in the Palaeoproterozoic. For example, from the Dharwar Craton, Pandey *et al.* (1997) reported a Sm–Nd age of 2173 $\pm$ 64 Ma from gabbro and dolerite dykes around Mahbubnagar, and French *et al.* (2008) reported a U–Pb baddelyite age of 2180 Ma from dolerite dykes in Shiomoga. The approximately 250 km-long and 100 km-wide Cuddapah Basin offers a generally well-preserved Palaeo-Neoproterozoic volcano-sedimentary sequence, which is estimated to be up to 12 km thick, termed as the Cuddapah Supergroup (Kaila *et al.* 1985). The mafic dyke swarms in the Bastar Craton were emplaced during 1.89–1.88 Ga, earlier than the mafic dyke swarm around the Cuddapah Basin and, in the latter case, most of the dykes were emplaced in the Late Palaeoproterozoic (French *et al.* 2008). A coeval radial mafic dyke swarm that developed around the Cuddapah Basin, which covers approximately 1.25 Mkm^2 ($Mkm^2 = 10^6\ km^2$), has yielded ages in the range of 1.86–1.65 Ga based on Ar/Ar and K–Ar studies and palaeomagnetic correlations (Radhakrishna *et al.* 1999 and references therein). The NW-trending Dharmapuri dykes to the SW of the basin have yielded $^{40}Ar/^{39}Ar$ ages of 1855 $\pm$ 9 Ma. Mafic lava flows in the Vempalle Formation of the Papaghni Group in the Cuddapah Basin have yielded K–Ar ages of 1841 ± 71 Ma (Murthy 1987). Bhaskar Rao *et al.* (1995) reported a Rb/Sr isochron age of 1817 ± 24 Ma for the Tadpatri sill. The K–Ar and Rb–Sr systems are likely to have been disturbed giving rise to younger ages. However, the Rb–Sr (Bhaskar Rao *et al.* 1995) and $^{40}Ar/^{39}Ar$ isotopic work (Anand *et al.* 2003) place an 1.9–1.8 Ga age on the lava flows intruding the Tadpatri shale, the youngest formation of the lower Cuddapah sequence. It is likely that the magmatism is possibly coeval with the dyke magmatism around the Cuddapah Basin (Radhakrishna *et al.* 1999) (Fig. 4). Anand *et al.* (2003) provided an $^{40}Ar/^{39}Ar$ age of 1899 ± 20 Ma for the mafic Tadpatri sill inside the basin. High-precision U–Pb dates of 1885.4 ± 3.1 Ma (baddeleyite) were obtained for the mafic Pulivendla sill from the Cuddapah Basin (French *et al.* 2008). Palaeoproterozoic ages reported from this region by other workers include Chatterjee & Bhattacharji (2001: 1879 ± 5 M Ar–Ar age from mafic dyke SW of the Cuddapah Basin) and Halls *et al.* (2007: 1894 Ma; U–Pb baddeleyite from a mafic dyke west of the Cuddapah Basin). Several late Mesoproterozoic ages have also been recorded from these regions from mafic dykes and related intrusives. These ages correlate well with the prolonged rifting and break-up of the Columbia supercontinent from the Palaeoproterozoic to the end of the Mesoproterozoic (e.g. Hou *et al.* 2008*b*; Rogers & Santosh 2009).

Several mafic–ultramafic complexes and probable Precambrian dismembered ophiolite occurrences decorate the eastern boundary of the Dharwar Craton (Fig. 13). Some of these, including the recently studied Kandra ophiolite complex (Sesha Sai 2009), were previously designated as rift-related magmas (e.g. Ravikant 2010). However, recent results have clearly brought out their clear ophiolite affinity (Dharma Rao & Reddy 2009) with zircon SHRIMP ages of 1.85–1.33 Ga (Vijaya Kumar *et al.* 2010; Dharma Rao *et al.* 2011), which suggest the presence of an active convergent margin from the late Palaeoproterozoic–early Mesoproterozoic. Dharma Rao *et al.* (2011) examined the Kanigiri ophiolite mélange within the Proterozoic Nellore–Khammam Schist Belt, where the ophiolite units were accreted along a NE-trending suture that juxtaposes the Proterozoic Eastern Ghats Granulite Belt (EGGB) against the Archaean Nellore Schist Belt of the Dharwar Craton. The ophiolite components in the Kanigiri mélange are cut by plagiogranites and gabbros that show mutually intrusive relationships indicating their coeval nature. Dharma Rao *et al.* (2011) reported LA-ICP-MS (laser ablation-inductively coupled plasma-mass spectrometry) age data and

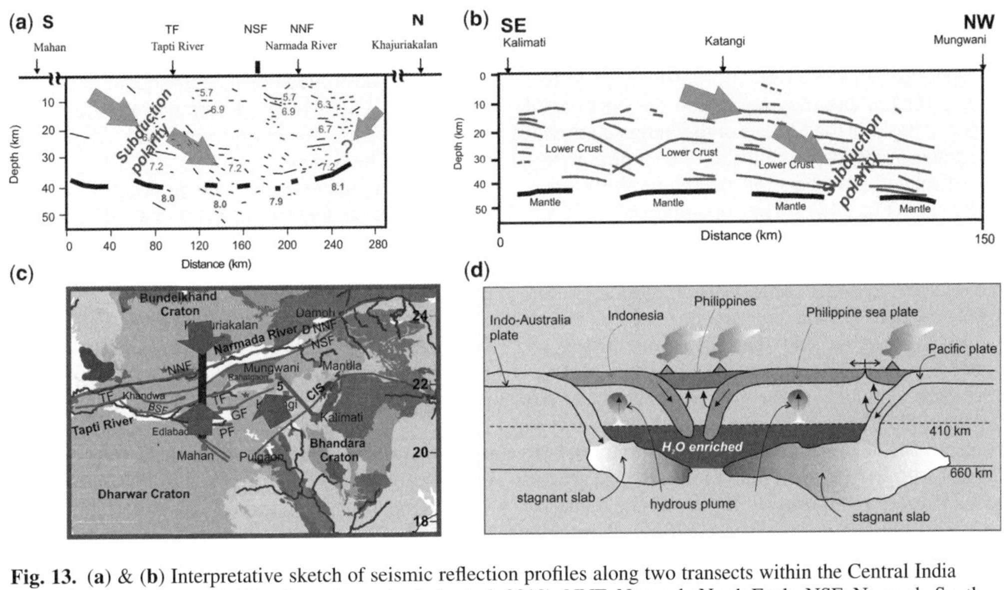

Fig. 13. (**a**) & (**b**) Interpretative sketch of seismic reflection profiles along two transects within the Central India Tectonic Zone (as compiled by Naganjaneyulu & Santosh 2010). NNF, Narmada North Fault; NSF, Narmada South Fault; TF, Tapti Fault. Velocity values are given in km s^{-1}. The inferred subduction polarity is also shown. (**c**) Geological map of the Central India Tectonic Zone with proposed subduction directions suggesting a double-sided subduction process for the assembly of continental blocks along the CITZ. (**d**) Cartoon sketch illustrating the process of ongoing double-sided subduction in the western Pacific region (after Maruyama *et al.* 2007).

REE (rare earth element) geochemistry of zircons from gabbro and granite from Kanigiri. The zircons from both gabbro and granite show high a REE content, prominent HREE (heavy REE) enrichment and a conspicuous negative Eu anomaly, suggesting a common melt source. U–Pb dating of the zircons reveals prominent Mesoproterozoic ages for the plagiogranite, with the approximately 1.33 Ga age of the Kanigiri ophiolite mélange offering important clues for arc–continent collision during the final stages of the amalgamation of the Columbia-derived fragments within the Neoproterozoic supercontinent assembly. In general, ophiolites are emplaced over continental margins within 5–10 Ma of their age of formation (Dewey 2003; Dilek *et al.* 2007). It is possible that the subduction was initiated as early as 1.8 Ga, as indicated by the Kandra ophiolite, and continued for over 500 Ma. The ongoing subduction tectonics in SW Japan provide an excellent analogue where subduction has been continuing for over 500 Ma with repeated accretion of the oceanic plate stratigraphy (Santosh *et al.* 2009*a*; Isozaki *et al.* 2010). The wide range in ages of the dated ophiolite suites from Kandra and Kanigiri would, therefore, indicate accretion along the craton margin during a prolonged history of convergence. The presence of a Proterozoic ophiolite mélange in Kanigiri (Dharma Rao & Reddy 2009) and arc-related magmatism of the Chimalpahad anorthosites and basaltic amphibolites (Dharma Rao *et al.* 2011) in the Nellore–Khammam Schist Belt within the marginal zone between the mobile belt and craton confirms that the marginal zone represents a Mesoproterozoic subduction complex.

Dharma Rao *et al.* (2011) proposed a plate tectonic model where they envisaged the opening of an ocean basin in the SE part of the Indian continental plate associated with the Palaeoproterozoic plume-related break-up of the Columbia supercontinent (Fig. 14). The opening of a NE–SW-trending Mesoproterozoic rift along the present cratonic margin has also been suggested in previous studies (e.g. Kale 1991; Chaudhuri *et al.* 1999), which led to the opening of an ocean towards the south and the formation of several craton margin basins, such as the Chattisgarh, Indravati and Pranhita–Godavari. Sediments of rift milieu were deposited on a stable Atlantic-type margin. The alkaline complex of Khariar in NW Orissa shows interlayered tholeiitic magmatic rocks that are interpreted to represent fragments of an ocean floor (Upadhyay *et al.* 2006). Thus, the early part of the orogenic cycle during the Palaeoproterozoic witnessed a continental rifting stage and the formation of a wide basin. The available geochronological data from LIPs and mafic dykes discussed in a previous section testify to this. Several alkaline magmatic plutons characterizing the rift setting occur along the craton margin. These suites testify to the rifting events during the break-up of the Columbia supercontinent. A typical rift setting is also suggested by the association of passive margin sequences. Some of these units were partly destroyed during subsequent deep subduction or extensively deformed and modified by high-grade metamorphism and retrograde hydration. The wide ocean formed by rifting would have enlarged extensively, and prolonged subduction of the ocean floor in such a case would generate an extensive batholith belt comprising arc and associated felsic magmatic suites. Extensive Mesoproterozoic anorthositic magmatism has been recognized in the Eastern Ghats Belt, and recent field, petrological and geochemical studies clearly establish their arc-related character (Dharma Rao *et al.* 2011). The subduction associated with the closure of a wide ocean would also create a series of accretionary complexes and, if subduction of a mid-oceanic ridge was involved, high-temperature and high-pressure regional metamorphic belts would also be generated (e.g. Santosh & Kusky 2010). The TTG-type arc magmas in association with pelagic sequences and ultrahigh-temperature metamorphic orogens in the Eastern Ghats Belt and adjoining suture therefore represent a typical Pacific-type orogeny that culminated in a Himalayan-style collision.

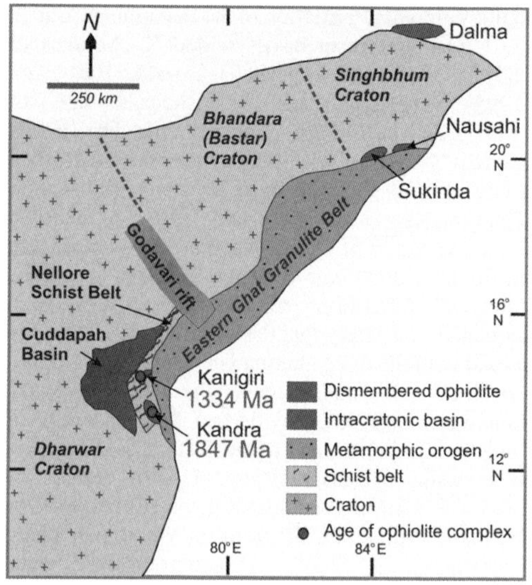

Fig. 14. Geological sketch map of the SE part of the Indian Shield showing the occurrence of a series of ophoilitic mélanges along the margin of the eastern Dharwar Craton. The U–Pb zircon ages for two of the ophiolite complexes (Kandra and Kangiri) are from Vijaya Kumar *et al.* (2010) and Dharma Rao *et al.* (2011).

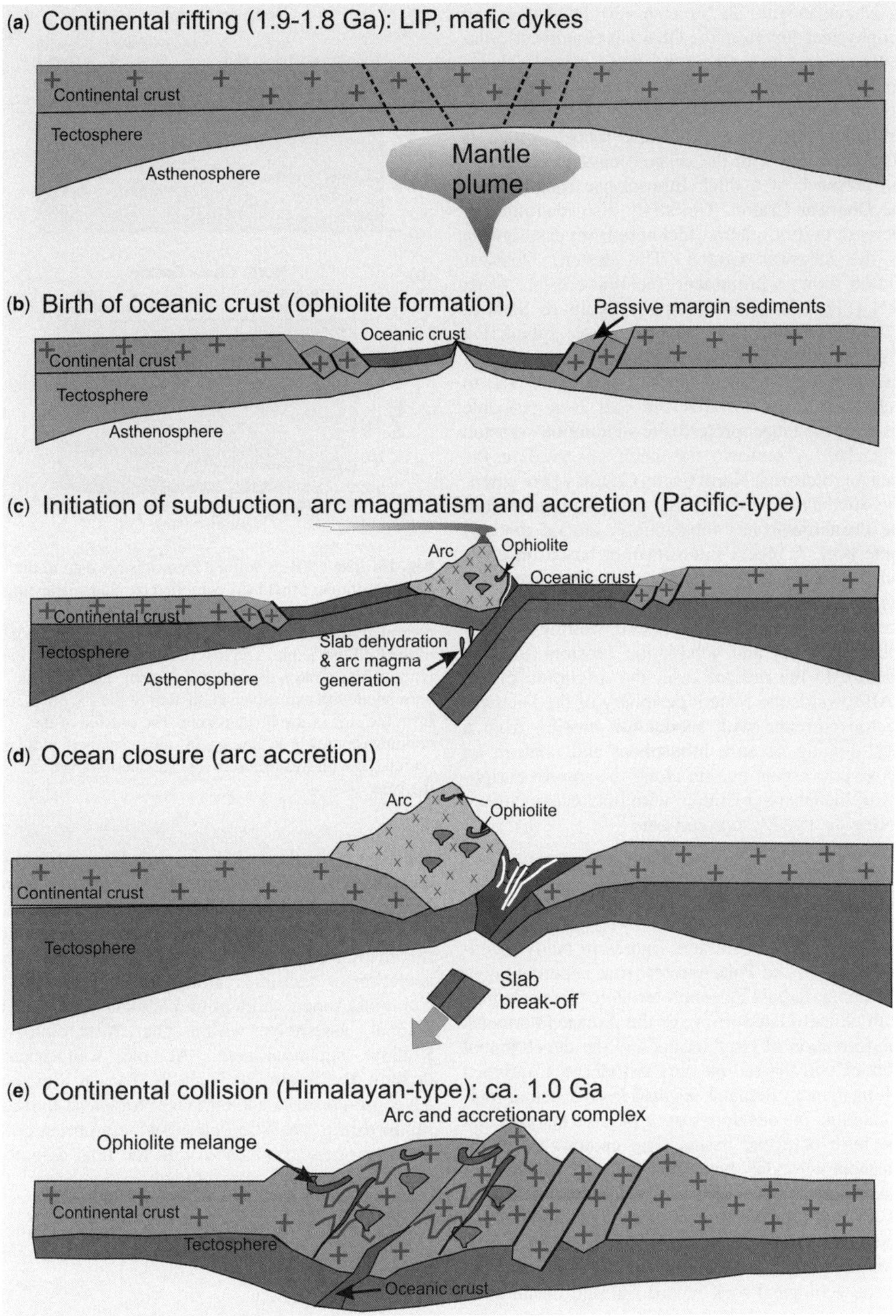

Fig. 15. Plate tectonic model proposed by Dharma Rao *et al.* (2011) for the evolution of the Pacific-type orogens in the SE segment of Peninsular India during the Palaeoproterozoic–Neoproterozoic involving continental rifting and the opening of a wide ocean followed by subduction–accretion–collision tectonics. Dismembered fragments of the destructed ocean floor now occur as ophiolite mélanges imbricated within the accretionary units.

Naganjaneyulu & Santosh (2011) synthesized geophysical data from the Dharwar Craton and adjacent regions, including heat flow, seismic tomography and magnetotellurics. They identified a lithospheric thickness of 230 km in the Dharwar Craton using a magnetotelluric model, which is also consistent with the seismic data and confirms the presence of a thick lithospheric root beneath the Dharwar Craton. The study also identified an increase in lithospheric thickness from east to west in the Dharwar Craton. The eastern Dharwar Craton shows a prominent 'tectonic erosion' of its keel (Fig. 15), which Naganjaneyulu & Santosh (2011) correlated to Palaeoproterozoic subduction from the east. They proposed that the migration of the lithosphere–asthenosphere boundary (LAB) to shallower levels towards the east is a probable result of the Palaeoproterozoic subduction–erosion (Fig. 16). A scenario has been observed in the Palaeoproterozoic North China Craton where extensive subduction of the Pacific plate from the east in the Phanerozoic has substantially eroded the cratonic keel. A recent interpretation based on combined geological and geophysical data from this region suggests that this 'erosion' of the continental keel probably resulted from both upwelling asthenosphere (plume) and subduction erosion (Santosh 2010*b*). In the present case, the migration of the LAB towards the eastern periphery of the Dharwar Craton correlate with subduction erosion from a west-dipping oceanic lithosphere and confirm an active convergent margin along the eastern periphery of the Dharwar Craton, with final ocean closure during the late Mesoproterozoic.

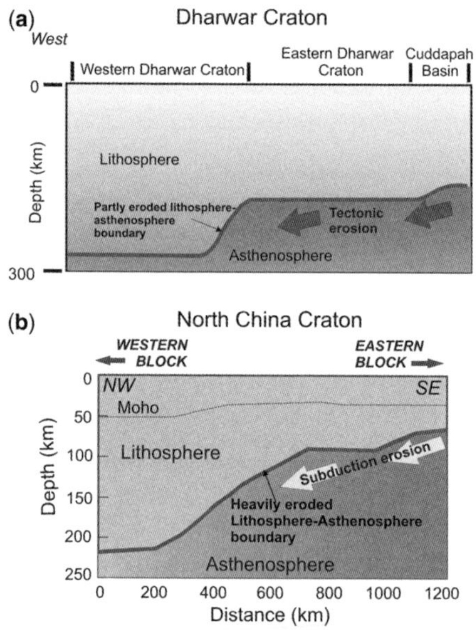

Fig. 16. The LAB as defined from seismic data in the southern Indian Shield (as compiled by Naganjaneyulu & Santosh 2011) and the North China Craton (as compiled by Santosh 2010*b*). In both cases, an 'erosion' of the LAB is visible. The loss of the cratonic root of the North China Craton towards the eastern side has been correlated with extensive subduction of the Pacific plate from the east in the Phanerozoic. The erosion of the cratonic root towards the eastern side in southern India is correlated with Mesoproterozoic subduction from the east.

Summary

The proto-Indian continent figures in many reconstructions of the Palaeoproterozoic supercontinent Columbia. India's Palaeoproterozoic history begins with plume-related rifting of the Archaean cratons, the formation of large basins and the development of thick volcano-sedimentary sequences. Continued rifting in many domains resulted in continental fragmentation, the development of passive margins and the birth of ocean basins. The creation of major Palaeoproterozoic belts in India involved the closure of the intervening oceans and the destruction of oceanic lithosphere in a prolonged subduction–accretion history culminating in final continent–continent collision. Evidence for which is preserved in the geological rock record through ocean plate stratigraphy imbricated within accretionary belts, extruded high pressure and temperature metamorphic orogens, as well as 'frozen-in' subduction architecture imaged from geophysical techniques. The Palaeoproterozoic world in India traces a continuum through Meso- and Neoproterozoic, and has to be evaluated within the context of a prolonged Wilson cycle and supercontinent cycle. Peninsular India thus preserves an unbroken record of a plate tectonic cycle from Pacific-type accretionary tectonics along the margins of the Columbia supercontinent to a Himalayan-style collisional assembly within the Neoproterozoic Rodinia supercontinent. At the southernmost domain, the Southern Granulite Terrain illustrates a Neoproteorozoic accretionary orogenesis that culminated in a late Neoproterozoic–Cambrian collisional orogeny associated with the final assembly of the Gondwana supercontinent.

I thank Dr R. Mazumder for inviting me to contribute to this volume and two anonymous referees for their helpful comments.

References

Acharyya, S. K. 2003. The nature of Mesoproterozoic Central Indian Tectonic Zone with exhumed and reworked older granulites. *Gondwana Research*, **6**, 197–214.

ACHARYYA, S. K., BANDYOPADHYAY, B. K. & ROY, A. 2001. Comment on 'Two cases of continental collisions and related tectonics during the Proterozoic period in India insights from gravity modelling constrained by seismic and magnetotelluric studies' by Mishra, D. C., Singh, B., Tiwari, V. M., Gupta, S. B., Rao, M. B. S. V., 2000 – *Precambrian Research*, 99, 149–169. *Precambrian Research*, **108**, 335–338.

ALVI, S. H. & RAZA, M. 1992. Nature and magma type of Jagannathpur Volcanics, Singhbum, Eastern India. *Journal of the Geological Society of India*, **38**, 524–531.

ANAND, M., GIBSON, S. A., SUBBARAO, K. V., KELLEY, S. P. & DICKIN, A. P. 2003. Early Proterozoic melt generation processes beneath the intracratonic Cuddapah basin, southern India. *Journal of Petrology*, **44**, 2139–2171.

BANDOPADHYAY, B. K., ROY, A. & HUIN, A. K. 1995. Structure and tectonics of a part of the central Indian shield. *Geological Society of India, Memoir*, **31**, 433–468.

BANERJEE, D. M. 1971. Precambrian stromatilitic phosphorite of Udaipur, Rajasthan, India. *Geological Society of America Bulletin*, **82**, 2319–2330.

BHANDARI, A., CHANDRA PANT, N., BHOWMIK, S. K. & GOSWAMI, S. 2011. ~1.6 Ga Ultrahigh-temperature granulite metamorphism in the Central Indian Tectonic Zone: insights from metamorphic reaction history, geothermobarometry and monazite chemical ages. *Geological Journal*, **46**, 198–216, doi: 10.1002/gj.1221.

BHASKAR RAO, Y. J., PANTULU, G. V. C., DAMODARA REDDY, V. & GOPALAN, K. 1995. Time of early sedimentation and volcanism in the Proterozoic Cuddapah basin, south India: evidence from the Rb–Sr age of the Pulivendla mafic sill. *Geological Society of India, Memoir*, **33**, 329–338.

BHATTACHARJEE, P. K. 1988. Petrology of jacobsite-bearing assemblages from Sausar Group, India. *Neues Jahrbuch Mineralogie Abhandlungen*, **159**, 101–111.

BHOWMIK, S. K. 2006. Ultra high temperature-metamorphism and its significance in the Central Indian Tectonic Zone. *In*: SAJEEV, K. & SANTOSH, M. (eds) *Extreme Crustal Metamorphism and Related Crust–Mantle Processes. Lithos*, **92**, 484–505.

BHOWMIK, S. K. & ROY, A. 2003. Garnetiferous metabasites from the Sausar Mobile Belt: petrology, *P–T* path and implications for the tectonothermal evolution of the Central Indian Tectonic Zone. *Journal of Petrology*, **44**, 387–420.

BHOWMIK, S. K., BASU-SARBADHIKARI, A., SPIERING, B. & RAITH, M. M. 2005. Mesoproterozoic reworking of Paleoproterozoic ultrahigh-temperature granulites in the central Indian Tectonic Zone and its implications. *Journal of Petrology*, **46**, 1085–1119.

BHOWMIK, S. K., BERNHARDT, H.-J. & DASGUPTA, S. 2010. Grenville age high-pressure upper amphibolite–granulite metamorphism in the Aravalli–Delhi Mobile Belt, Northwest India: new evidence from monazite chemical age and its implication. *Precambrian Research*, **178**, 168–184.

BOSE, M. K. 1994. Sedimentation pattern and tectonic evolution of the Proterozoic Singhbhum basin in the eastern Indian shield. *Tectonophysics*, **231**, 325–346.

CHATTERJEE, N. & BHATTACHARJI, S. 2001. Petrology, geochemistry and tectonic settings of the mafic dikes and sills associated with the evolution of the Proterozoic Cuddapah Basin of South India. *Proceedings of the Indian Academy of Sciences (Earth and Planetary Sciences)*, **110**, 433–453.

CHAUDHURI, A. K., MUKHOPADHYAY, J., DEB, S. P. & CHANDA, S. K. 1999. The Neoproterozoic cratonic successions of Peninsular India. *Gondwana Research*, **2**, 213–225.

CHAWADE, M. P. & NAIK, K. K. 2003. A reappraisal of tectnostratigraphy of Sakoli fold belt and its implications on mineral occurrences, Central India. *Gondwana Magazine*, **7**, 153–168.

CHOUDHARY, A. K., GOPALAN, K. & SHASTRY, C. A. 1984. Present status of the geochronology of the Precambrian rocks of Rajasthan. *Tectonophysics*, **105**, 131–140.

COLLINS, A. S. & PISAREVSKY, S. A. 2005. Amalgamating eastern Gondwana: the evolution of the Circum-Indian orogens. *Earth-Science Reviews*, **71**, 229–270.

COLLINS, A. S., CLARK, C., SAJEEV, K., SANTOSH, M., KELSEY, D. E. & HAND, M. 2007. Passage through India: the Mozambique Ocean suture, high pressure granulites and the Palghat–Cauvery shear system. *Terra Nova*, **19**, 141–147.

CONDIE, K. C. 2002. Breakup of a Paleoproterozoic supercontinent. *Gondwana Research*, **5**, 41–43.

CONDIE, K. C. & KRÖNER, A. 2008. When did plate tectonics begin? Evidence from the geological record. *Geological Society of America Bulletin*, **440**, 281–294.

DALZIEL, I. W. D. 1991. Pacific margins of Laurentia and East Antarctica–Australia as a conjugate rift pair: evidence and implications for an Eocambrian supercontinent. *Geology*, **19**, 598–601.

DEB, M. 1992. Lithogeochemistry of rocks around Rampura–Agucha massive zinc sulphide ore-body, NW India – implications for evolution of Proterozoic 'Aulacogen'. *In*: SARKAR, S. C. (ed.) *Metallogeny Related to Tectonics of the Proterozoic Mobile Belts*. Oxford & IBH, New Delhi, 1–36.

DEB, M., THORPE, R. I., CUMMING, G. L. & WAGNER, P. A. 1989. Age, source and stratigraphic implications of Pb isotope date for comformable, sediment-hosted base metal deposits in the Proterozoic Aravalli–Delhi orogenic belt, northwestern India. *Precambrian Research*, **108**, 195–213.

DEWEY, J. 2003. Ophiolites and lost oceans: rifts, ridges, arcs, and/or scrapings? *In*: DILEK, Y. & NEWCOMB, S. (eds) *Ophiolite Concept and the Evolution of Geological Thought*. Geological Society of America, Special Paper, **373**, 153–158.

DEWEY, J. F. 1977. Suture zone complexities: a review. *Tectonophysics*, **40**, 53–67.

DHARMA RAO, C. V. & REDDY, U. V. B. 2009. Petrological and geochemical characterization of Ophiolite mélange, Nellore–Khammam schist belt, SE India. *Journal of Asian Earth Sciences*, **65**, 261–276.

DHARMA RAO, C. V., SANTOSH, M. & WU, Y. 2011. Mesoproterozoic ophiolite mélange from the SE periphery of Indian plate: U–Pb zircon ages and tectonic implications. *Gondwana Research*, **19**, 384–401.

Dilek, Y., Furnes, H. & Shallo, M. 2007. Suprasubduction zone ophiolite formation alongthe Mesozoic periphery of Gondwana. *Gondwana Research*, **11**, 435–475.

Eriksson, P. G., Banerjee, S. *et al.* 2009. A Kaapval craton debate: nucleus of an early small supercontinent or affected by an enhanced accretion event? *Gondwana Research*, **15**, 354–372.

French, J. E., Heaman, L. M., Chacko, T. & Srivastava, R. K. 2008. 1891–1883 Ma Southern Bastar–Cuddapah mafic igneous events, India: a newly recognized large igneous province. *Precambrian Research*, **160**, 308–322.

Fukao, Y., Obayashi, M., Nakakuki, T. & DEEP SLAB PROJECT GROUP. 2009. Stagnant slab: a review. *Annual Reviews of Earth and Planetary Sciences*, **37**, 19–46.

Gopalan, K., McDougall, J. D., Roy, A. B. & Murali, A. V. 1990. Sm–Nd evidence for 3.3 Ga old rocks in Rajasthan, northwestern India. *Precambrian Research*, **48**, 287–297.

Gurnis, M. 1988. Large-scale mantle convection and the aggregation and dispersal of supercontinents. *Nature*, **332**, 695–699.

Hall, R. 2009. The Eurasian SE Asian margin as a modern example of an accretionary orogen. *In*: Cawood, P. A. & Kroner, A. (eds) *Earth Accretionary Systems in Space and Time*. The Geological Society, London, Special Publications, **318**, 351–372.

Halls, H. C., Kumar, A., Srinivasan, R. & Hamilton, M. A. 2007. Palaeomagnetic and U–Pb geochronology of easterly trending dykes in the Dharwar Craton, India: feldspar clouding, radiating dyke swarms and the position of India at 2.37 Ga. *Precambrian Research*, **155**, 47–68.

Hara, T., Wakita, K. *et al.* 2009. Nature of accretion related to Paleo-Tethys subduction recorded in northern Thailand: constraints from mélange kinematics and illite crystallinity. *Gondwana Research*, **16**, 310–320.

Hoffman, P. F. 1991. Did the breakout of Laurentia turn Gondwanaland inside out? *Science*, **252**, 1409–1412.

Hou, G., Santosh, M., Qian, X., Lister, G. S. & Li, J. 2008*a*. Configuration of the Late Paleoproterozoic supercontinent Columbia: insights from radiating mafic dyke swarms. *Gondwana Research*, **14**, 395–409.

Hou, G., Santosh, M., Qian, X., Lister, G. S. & Li, J. 2008*b*. Tectonic constraints on 1.3–1.2 Ga final breakup of Columbia supercontinent from a giant radiating mafic dyke swarm. *Gondwana Research*, **14**, 561–566.

Isozaki, Y., Aoki, K., Nakama, T. & Yanai, S. 2010. New insight into a subduction-related orogen: a reappraisal of the geotectonic framework and evolution of the Japanese Islands. *Gondwana Research*, **18**, 82–105.

Kaila, K. L., Reddy, P. R., Dixit, M. M. & Koteswar Rao, P. 1985. Crustal structure across the Narmada–Son lineament, Central India from deep seismic soundings. *Journal of the Geological Society of India*, **26**, 465–480.

Kale, V. S. 1991. Constrains on the evolution of the Purana basins of Peninsular India. *Journal of the Geological Society of India*, **38**, 231–252.

Kaur, P., Chaudhri, N., Racsek, I., Kroner, A. & Hofmann, A. W. 2009. Record of 1.82 Ga Andean-type continental arc magmatism in NE Rajasthan, India: insights from zircon and Sm–Nd ages, combined with Nd–Sr isotope geochemistry. *Gondwana Research*, **16**, 56–71.

Khan, M. S., Smith, T. E., Raza, M. & Huang, J. 2005. Geology, geochemistry and tectonic significance of mafic–ultramafic rocks of Mesoproterozoic Phulad ophiolite suture of South Deli Fold Belt, NW Indian Shield. *Gondwana Research*, **8**, 553–566.

Komiya, T., Hayashi, M., Maruyama, S. & Yurimoto, H. 2002. Intermediate P/T type Archean metamorphism of the Isua supracrustal belt: implications for secular change of geothermal gradients at subduction zones and for Archean plate tectonics. *American Journal of Science*, **302**, 806–826.

Kusky, T. M. 2004. Introduction. *In*: Kusky, T. M. (ed.) *Precambrian Ophiolites and Related Rocks*. Elsevier, Amsterdam, 1–34.

Kusky, T. M. & Polat, A. 1999. Growth of granite–greenstone terranes at convergent margins, and stabilization of Archean cratons. *Tectonophysics*, **305**, 43–73.

Kusky, T. M. & Santosh, M. 2009. The Columbia connection in North China. *In*: Reddy, S. M., Mazumder, R., Evans, D. A. D. & Collins, A. S. (eds) *Palaeoproterozoic Supercontinents and Global Evolution*. Geological Society, London, Special Publications, **323**, 49–71.

Li, Z. X., Bogdanova, S. V. *et al.* 2008. Assembly, configuration and break-up history of Rodinia: a synthesis. *Precambrian Research*, **160**, 179–210.

Mall, D. M., Reddy, P. R. & Mooney, W. D. 2008. Collision tectonics of the Central Indian Suture zone as inferred from a deep seismic sounding study. *Tectonophysics*, **460**, 116–123.

Mallikarjuna Rao, J., Bhattacharji, S., Rao, M. N. & Hermes, O. D. 1995. ^{40}Ar–^{39}Ar ages and geochemical characteristics of dolerite dykes around the Proterozoic Cuddapah Basin, South India. *In*: Devaraju, T. C. (ed.) *Mafic Dyke Swarms of Peninsular India. Geological Society of India, Memoir*, **33**, 307–328.

Maruyama, S. & Santosh, M. 2008. Models on snowball earth and Cambrian explosion. A synopsis. *Gondwana Research*, **14**, 22–32.

Maruyama, S., Santosh, M. & Zhao, D. 2007. Superplume, supercontinent, and post-perovskite: mantle dynamics and anti–plate tectonics on the core-mantle boundary. *Gondwana Research*, **11**, 7–37.

Matsuda, T. & Isozaki, Y. 1991. Well-documented travel history of Mesozoic pelagic chert in Japan: from remote ocean to subduction zone. *Tectonics*, **10**, 475–499.

Mazumder, R. 2005. Proterozoic sedimentation and volcanism in the Singhbhum crustal province, India, and their implications. *Sedimentary Geology*, **176**, 167–193.

Mazumder, R. & Sarkar, S. 2004. Sedimentation history of the Palaeoproterozoic Dhanjori Formation, Singhbhum, eastern India. *Precambrian Research*, **130**, 267–287.

Meert, J. G. 2001. Growing Gondwana and rethinking Rodinia: a paleomagnetic perspective. *Gondwana Research*, **4**, 279–288.

MEERT, J. G. & LIEBERMAN, B. S. 2008. The Neoproterozoic assembly of Gondwana and its relationship to the Ediacaran–Cambrian radiation. *Gondwana Research*, **14**, 5–21.

MISHRA, D. C., SINGH, B., TIWARI, V. M., GUPTA, S. B. & RAO, M. V. S. V. 2000. Two cases of continental collision and related tectonics during the Proterozoic period in India insights from gravity modeling constrained by seismic and magnetotelluric studies. *Precambrian Research*, **99**, 149–169.

MURTHY, N. G. K. 1987. Mafic dyke swarms of the Indian Shield. *In*: HALLS, H. C. & FAHRIG, W. F. (eds) *Mafic Dyke Swarms*. Geological Association of Canada, Special Paper, **34**, 393–400.

NAGANJANEYULU, K. & SANTOSH, M. 2010. The Central India Tectonic Zone: a geophysical perspective on continental amalgamation along a Mesoproterozoic suture. *Gondwana Research*, **18**, 547–564, doi: 10.1016/j.gr.2010.02.017.

NAGANJANEYULU, K. & SANTOSH, M. 2011. The nature and thickness of the lithosphere beneath the Dharwar Craton, southern India: A magnetotelluric model. *Journal of Asian Earth Sciences*, http://dx.doi.org/10.1016/j-jseases.2011.07.002.

NAKAGAWA, M., SANTOSH, M. & MARUYAMA, S. 2009. Distribution and mineral assemblages of bedded manganese deposits in Shikoku, Southwest Japan: implications for accretion tectonics. *Gondwana Research*, **16**, 609–621.

NANCE, R. D., WORSLEY, T. R. & MOODEY, J. B. 1998. The supercontinent cycle. *Scientific American*, **259**, 72–79.

NAQVI, S. M. 2005. *Geology and Evolution of the Indian Plate: From Hadean to Holocene- 4 Ga to 4 ka*. Capital Publishing Company, New Delhi.

NEOGI, S., MIURA, H. & HARIYA, Y. 1996. Geochemistry of the Dongargarh volcanic rocks, central India: implications for the Precambrian mantle. *Precambrian Research*, **76**, 77–91.

PANDEY, B. K., GUPTA, J. N., SARMA, K. J. & SASTRY, C. A. 1997. Sm–Nd, Pb–Pb and Rb–Sr geochronology and petrogenesis of mafic dyke swarm of Mahbubnagar, South India: implications for Paleoproterozoic crustal evolution of the Eastern Dharwar Craton. *Precambrian Research*, **84**, 181–196.

PAPINEAU, D., PUROHIT, R. ET AL. 2009. High primary productivity and nitrogen cycling after the Paleoproterozoic phosphogenic event in the Aravalli Supergroup, India. *Precambrian Research*, **717**, 37–56.

PIPER, J. D. A. 1976. Definition of pre-2000 m.y. apparent polar movements. *Earth and Planetary Science Letters*, **28**, 470–478.

PIPER, J. D. A. 2010. Protopangaea: palaeomagnetic definition of Earth's oldest (mid-Archaean-Palaeoproterozoic) supercontinent. *Journal of Geodynamics*, **50**, 154–165.

PRADHAN, V. R., MEERT, J. G., PANDIT, M. K., KAMENOV, G., GREGORY, L. & MALONE, S. J. 2010. India's changing place in global Proterozoic reconstructions: a review of geochronologic constraints and paleomagnetic poles from the Dharwar, Bundelkhand and Marwar Cratons. *Journal of Geodynamics*, **50**, 224–242.

RADHAKRISHNA, B. P. & NAQVI, S. M. 1986. Precambrian continental crust of India and its evolution. *Journal of Geology*, **94**, 145–166.

RADHAKRISHNA, T., MALUSKI, H., MITCHELL, J. G. & JOSEPH, M. 1999. $^{40}Ar/^{39}Ar$ and K/Ar geochronology of the dykes from the south Indian granulite terrain. *Tectonophysics*, **304**, 109–129.

RAJENDRA PRASAD, B., TEWARI, H. C., VIJAYA RAO, V. V. & DIXIT, M. M. 1998. Structure and tectonics of the Proterozoic Aravalli–Delhi Fold Belt in northwestern India from deep seismic reflection studies. *Tectonophysics*, **288**, 31–33.

RAVIKANT, V. 2010. Palaeoproterozoic (1.9 Ga) extension and breakup along the eastern margin of the Eastern Dharwar Craton, SE India: new Sm–Nd isochron age-constraints from anorogenic mafic magmatism in the Neoarchean Nellore greenstone belt. *Journal of Asian Earth Sciences*, **37**, 67–81.

RAZA, M., CASSHYAP, S. M. & KHAN, A. 2001. Accretionary lapilli from the basal Vindhyan volcanic sequence, south of Chittaurgarh, Rajasthan, and their implication. *Journal of the Geological Society of India*, **57**, 77–82.

ROGERS, J. J. W. & SANTOSH, M. 2002. Configuration of Columbia, a Mesoproterozoic supercontinent. *Gondwana Research*, **5**, 5–22.

ROGERS, J. J. W. & SANTOSH, M. 2004. *Continents and Supercontinents*. Oxford University Press, New York.

ROGERS, J. J. W. & SANTOSH, M. 2009. Tectonics and surface effects of the supercontinent Columbia. *Gondwana Research*, **15**, 373–380.

ROY, A., SARKAR, A., JEYAKUMAR, S., AGGARWAL, S. K. & EBIHARA, M. 2003. Mid-Proterozoic plume-related thermal event in the Eastern Indian Craton: evidence from trace elements, REE geochemistry and Sr–Nd isotope systematic of basic–ultrabasic intrusive from Dalma Volcanic Belt. *Gondwana Research*, **5**, 133–146.

ROY, A. B. 2001. Neoproterozoic crustal evolution of the northwestern Indian Shield: implications on breakup and assembly of supercontinents. *Gondwana Research*, **4**, 289–306.

SAFONOVA, I. Y., UTSUNOMIA, A., KOJIMA, S., NAKAE, S., TOMURFOGOO, O., FILIPPOV, A. N. & KOIZUMI, K. 2009. Pacific superplume-related oceanic basalts hosted by accretionary complex of Central Asia, Russian Far East and Japan. *Gondwana Research*, **16**, 587–608.

SANTOSH, M. 2010*a*. A synopsis of recent conceptual models on supercontinent tectonics in relation to mantle dynamics, life evolution and surface environment. *Journal of Geodynamics*, **50**, 116–133.

SANTOSH, M. 2010*b*. Assembling North China Craton within the Columbia supercontinent: the role of double-sided subduction. *Precambrian Research*, **178**, 149–167.

SANTOSH, M. & KUSKY, T. 2010. Origin of paired high pressure–ultrahigh-temperature orogens: a ridge subduction and slab window model. *Terra Nova*, **22**, 35–42.

SANTOSH, M. & ZHAO, G. 2009. Supercontinent dynamics. *Gondwana Research*, **15**, 225–227.

SANTOSH, M., MARUYAMA, S. & SATO, K. 2009*a*. Anatomy of a Cambrian suture in Gondwana: Pacific-type orogeny in southern India? *Gondwana Research*, **16**, 321–341.

SANTOSH, M., MARUYAMA, S. & YAMAMOTO, S. 2009*b*. The making and breaking of supercontinents: some speculations based on superplumes, superdownwelling and the role of tectosphere. *Gondwana Research*, **15**, 324–341.

SANTOSH, M., TSUNOGAE, T., LI, J. H. & LIU, S. J. 2007. Discovery of sapphirine-bearing Mg–Al granulites in the North China Craton: implications for Paleoproterozoic ultrahigh-temperature metamorphism. *Gondwana Research*, **11**, 263–285.

SARBADHIKARI, A. B. & BHOWMIK, S. K. 2008. Constraining the metamorphic evolution of a cryptic hot Mesoproterozoic orogen in the Central Indian Tectonic Zone, using P–T pseudosection modelling of mafic intrusions and host reworked granulites. *Precambrian Research*, **162**, 128–149.

SESHA SAI, V. V. 2009. Sheeted dykes in Kandra ophiolite complex, Nellore Schist Belt, Andhra Pradesh – vestiges of Precambrian oceanic crust. *Journal of the Geological Society of India*, **74**, 509–514.

SHARMA, R. S. 1995. An evolutionary model for Precambrian crust of Rajasthan: some petrological and geochronological considerations. *Geological Society of India, Memoir*, **34**, 91–115.

SINGH, Y. K., DE WALEE, B., KARMAKAR, S., SARKAR, S. & BISWAL, T. K. 2010. Tectonic setting of the Balaram–KuiSurpagla–Kengora granulites of the South Delhi Terrane of the Aravalli Mobile Belt, NW India and its implication on correlation with the East African Orogen in the Gondwana assembly. *Precambrian Research*, **183**, 669–688.

SINHA-ROY, S. 1984. Precambrian crustal interaction in Rajasthan, NW India. *Indian Journal of Earth Sciences*, **25**, 84–91.

SINHA-ROY, S. 1988. Paleozoic Wilson Cycles in Rajasthan. *Geological Society of India, Memoir*, **7**, 95–108.

SINHA-ROY, S. & MOHANTY, M. 1988. Blueschist facies metamorphism in the ophiolitic mélange of thte Late Proterozoic Delhi Fold Belt, Rajasthan. *Precambrian Research*, **42**, 97–105.

SINHA-ROY, S., MALHOTRA, G. & GUHA, D. B. 1995. A transect across Rajasthan Precambrian terrain in relation to geology, tectonics and crustal evolution of south-central Rajasthan. *Memoir Geological Society of India*, **31**, 63–90.

SRINIVAS, B., SHARMA, D. S., KUMAR, B., PATIL, D. J., ROY, A. B. & SRINIVASAN, R. 2001. Positive $\delta^{13}C$ excursion in carbonate and organic fractions from the Paleoproterozoic Aravalli Supergroup, Northwestern India. *Precambrian Research*, **106**, 277–290.

SRIVASTAVA, R. K. 1988. Magmatism in the Aravalli Mountain Range and its environs. *Memoir Geological Society of India*, **7**, 77–93.

UPADHYAY, D., RAITHH, M. M., MEZGER, K. & HAMMERSCHMIDT, K. 2006. Mesoproterozoic rift-related alkaline magmatism at Elchuru, Prakasam Alkaline Province, SE India. *Lithos*, **89**, 447–477.

VALDIYA, K. S. 2010. *The Making of India: Geodynamic Evolution*. Macmillian, Delhi.

VIJAYA KUMAR, K., ERNST, W. G., LEELANANDAM, C., WOODEN, J. L. & GROVE, M. J. 2010. First Paleoproterozoic ophiolite from Gondwana: Geochronologic–geochemical documentation of ancient oceanic crust from Kandra, SE India. *Tectonophysics*, **487**, 22–32.

WILSON, J. T. 1966. Did the Atlantic close and then re-open? *Nature*, **211**, 676–681.

YAKUBCHUK, A. 2010. Restoring the supercontinent Columbia and tracing its fragments after the breakup: a new configuration and a Super-Horde hypothesis. *Journal of Geodynamics*, **50**, 166–175.

YAMAMOTO, S., SENSHU, H., RINO, S., OMORI, S., MARUYAMA, S. & ZHAO, D. 2009. Granite subduction: arc subduction, tectonic erosion and sediment subduction. *Gondwana Research*, **15**, 433–442.

YEDEKAR, D. B., JAIN, S. C., NAIR, K. K. K. & DUTTA, K. K.1990. The central Indian collision suture. *In*: *Precambrian of Central India*. Geological Survey of India, Special Publications, **28**, 1–43.

YOSHIDA, M. & SANTOSH, M. 2011. Supercontinents, mantle dynamics and plate tectonics: a perspective based on conceptual v. numerical models. *Earth-Science Reviews*, **105**, 1–24.

ZHANG, Z. M., ZHAO, G. C., SANTOSH, M., WANG, J. L., DONG, X. & SHEN, K. 2010. Late Cretaceous charnockite with adakitic affinities from the Gangdese batholith, southeastern Tibet: evidence for Neo-Tethyan mid-ocean ridge subduction? *Gondwana Research*, **17**, 615–631.

ZHAO, G., CAWOOD, P. A., WILDE, S. A. & SUN, M. 2002. Review of global 2.1–1.8 Ga collisional orogens and accreted cratons: a pre-Rodinia supercontinent? *Earth-Science Reviews*, **59**, 125–162.

ZHAO, G. C., SUN, M. & WILDE, S. A. 2003. Correlations between the Eastern Block of the North China Craton and the South Indian Block of the Indian Shield: an Archaean to Palaeoproterozoic link. *Precambrian Research*, **122**, 201–233.

Index

Page numbers in *italics* refer to Figures. Page numbers in **bold** refer to Tables.

Abujhmar Basin *6*
Achankovil Shear Zone *18*
Achankovil Suture Zone *264*
Ajabgarh Group 7, 272
Alwar Group 7, 272
Amet Granite 7
Ampani Basin *6*
Anasagar Granite 7
Anasagar Granite Gneiss Complex 2, **10**, 219, 222, 224
  deformation events 222–225
  metamorphic conditions
    analytical methods 229–233
    results **230**, **231**, **232**, 233, **234**, **235**, **236**, 237–238, *239*, *240*
    results discussed 238–242
  mineral chemistry 226–229
  petrography 225–226
apatite
  chemistry in Racherla alkali syenite
    methods of analysis 187, 189
    results **188**, 189, **190**, 191–192
    summary 192–193
Arangi Formation *249*
Aravalli Craton *264*
Aravalli Fold Belt 219, 222, 240
Aravalli Orogeny 2, 272
Aravalli Supergroup 7, *8*, *9*, **10**, **21**, 221
Aravalli–Bundelkhand Craton 1, 5, *6*, 7, **21**, 22, *58*
  Bundelkhand sector *8*, 9–11
  Rajasthan sector 7–9
  Vindhyan basin *8*, 11
Aravalli–Delhi domain, seismic section *275*
Aravalli–Delhi Fold Belt/ Mobile Belt 7, *118*, 219, 220, 272–276
  see also Aravalli Fold Belt; Delhi Fold Belt
Arumberia banksi **250**, 252
asbestos 185
atmosphere, Palaeoproterozoic 5, 268, 272
Auk Shale *200*, 202, 203

Bababudan Group 17, **19**
Bagalkot Group **19**
Bailadila Group 11
Bairenkonda Quartzite *164*, 168, *169*, *170*, 172, 177, *200*, 201
Balwan Limestone *249*
Banded Gneissic Complex 7, *9*, **10**, 219, 220, 221, 242, 272
banded ironstone formations 1, 11, 34–35
Bangalore dyke swarm **19**
Banganapalle Formation 2, *164*, 170, *200*, 202
  diamond working 206–207, 209–210
  kimberlitic indicators 209–210
    methods of analysis 212
    results 213–215
    results discussed 215–216
Bangur Gabbro 40
Barakhamba Granite 22
barite 185
Barotiya Group 221
Basantagarh Group 221
Bastar (Bhandara) Craton 1, 5, *6*, 11–14, **21**, 22, *58*, *118*, *162*, *264*, 276, 279, 280
Bengal Anorthosite **17**
Berach Granite 7, *9*, **10**, 22, *220*, *276*
Betul belt *12*, **13**
Bhandara Craton *see* Bastar Craton
Bhander Group *9*, **10**, 11, 248, *249*, *253*, *254*, 255, *256*
Bhander Limestone *249*, **250**
Bhavpura Shale *249*, **250**
Bhilwara Gneissic Complex 272
Bhilwara Supergroup 221
Bhim Group 221
Bhima Basin *6*
Bhopalpatnam granulite 5, 14, 20
Bijaigarh Shale *249*
Bijawar Basin *8*, 9
Bijawar Group 9, **10**, 11, 248
Bijli Rhyolites 13
biotite chemistry
  Anasagar Granite Gneiss Complex 229, **231**
  chloritoid–biotite schist 97, **99**, 101
Birtola Formation 35–37
Bisrampur Formation *36*
Black Reef Formation *54*, 55, *56*
blueschists 274
Bomdila Group 20
Bonai Granite *32*, 40
Bouguer anomaly 178, *179*
Bundelkhand Craton *264*
Bundelkhand Granite 7, *9*, **10**, 11, **13**, 22, *162*
Bundelkhand Massif *118*, 248
Bundi Hill Sandstone *249*, **250**, *253*
Bushveld Complex *54*, 57

$\delta^{13}C$ excursion 272
Central Indian Shear Zone *12*, *118*
Central Indian Suture *264*, 278
Central Indian Tectonic Zone *6*, *12*, **21**, 22, *118*, *162*, *264*, 269, 270, *271*, 276–279, *280*
Central Rand Group *54*
Chaibasa Formation 2, 16, *32*, 40, 41, 42, 59, *63*, *64*, 67, 277
  description 77–79
  native iron study
    methods 79
    results
      magnetic susceptibility 79
      microprobe analysis 81–83, **84**, **85**
      SEM 81, 83
      thin sections 79
    results discussed 85–89
Chakradharpur Granite *15*, **17**, *32*, *60*
Chakradharpur Granite Gneiss 40, *92*
Chandil Formation 16, *32*, 42, 62, *78*, 277, 278
Chattisgarh Basin *6*, *118*, 282
Chattsigarh Supergroup *12*

Chelima lamproite 180, 192
Chikmagalur Granite **19**
Chitradurga Boundary Fault *18*, *179*
Chitradurga Granite **19**
Chitradurga Group 17, **19**
Chitravati Group **19**, 21, *163*, *164*, 166–167, 169, *172*, 174
chlorite chemistry
  Anasagar Granite Gneiss Complex 229, **234**, **235**
  chloritoid–biotite schist 97, **101**
chloritoid chemistry
  in chloritoid–biotite schist 93, 94, *95*, 97, **98**, 101
Chorhat Sandstone **10**, 11, *249*, **250**
Chotanagpur Granite Gneiss Complex 2, *15*, **17**, *32*, *60*, 61, *92*, *141*
  classification 119
  metamorphic evolution 138–140
  petrology and geochronology 120–121
    sector 1 121, **122**, 123
    sector 2 **122**, 123–124
    sector 3 **122**, 124
    sector 4 **122**, 124
    sector 5 **122**, 124–128
    sector 6 **122**, 128–132
    sector 7 **122**, 132–135
    sector 8 **123**, 135–136
    sector 9 **123**, 136
    sector 10 **123**, 137
  regional structural trend 119–120
  role in supercontinent cycle 140–142
  setting 117–119
Chotanagpur Orogeny 119
Chuaria **250**, 251, 252, *254*, 255, *256*
Chuniespoort Group *54*, 55, *56*, 57
Closepet Granite 17, 18, **19**, 22, *163*, *179*, *264*
Columbia supercontinent 140, 177, 265, *266*, 267, 268, 269, *270*, 274, 281, 282
Cr-in-garnet barometer 209
Cuddapah Basin 2, 17, *18*, 21, *264*, 279, 280, 281
  diamond exploration 206–207
  lamproites 192
  map *186*
  origin 185
  potential thrusts 172–173
  setting 197–198
  significance of crustal structure 177–178
  stratigraphy 162–165, *200*
    Chitravati Group *164*, 166–167
    Cuddapah Supergroup *164*, 198, *200*, 201–203
    Nallamalai Group *164*, 167–169
    Papaghni Group *164*, 165–166
  structure 205–206
  summary 180–181
  tectonic controls 173–174
    basin opening 174
    large igneous province 174
    palaeocurrent data 174–175
  terrain geochronology 207–208
    method of measurement 209
    results *210*, **211**
  unconformities 169–172
  see also Racherla alkali syenite
Cuddapah Intrusive Province 185
Cuddapah Supergroup *164*, 173, 280
Cumbum Formation *164*, 168–169, *171*, 177, *200*, 201
Cumbum phyllites 186, 192

Dalma Formation *32*, 59, 61, *63*, 65, *66*, 67, 277, 278
Dalma volcanics *15*, 16, 40, 42, *60*, *78*, *92*, *118*
Damodar–Gondwana Basin *118*
Dargaon sill 11
Darjing Group 35
Darwal Granite 7, *8*, **10**
Daspoort Formation *56*
Deccan basalts/trap *6*, 17, *18*, *58*, *118*, *162*
deep seismic sounding 178
Delhi Fold Belt 22, 219, 220, 221, 222
  see also North Delhi Fold Belt; South Delhi Fold Belt
Delhi Orogeny 272
Delhi Supergroup 2, 7, *9*, **10**, 219, 221, 242
Delwara Formation 7
Deoland Formation 11
Devarh Group 221
Dhalbhum Formation 16, *32*, 40, 42, 59, *63*, *65*, 67, *78*, 277
Dhandraul Sandstone *249*
Dhanjori Formation *32*, *36*, 40, 41, 42, 59, *63*, *64*, 67, 68, *78*, 277
Dhanjori Group 7, 14, **17**, **21**, 42, *92*
Dharmapuri dykes 280
Dharwar (Karnataka) Craton 1, 2, 5, *6*, 17–20, **21**, 22, *58*, *162*, 177–178, *179*, *186*, *264*, 276, 280, 281, 284
Dharwar Supergroup **19**
Dholpura Shale *249*, **250**
diamonds 2, 185, 197
  current exploration 206–207
Dongargarh belt *12*
Dongargarh granite *12*, **13**, 22
Dongargarh Supergroup *12*, **13**, **21**, 278
Duitschland Formation 55, *56*, 57
dyke swarms 174

East Indian Tectonic Zone *141*, 142
Eastern Ghats Belt 17, *18*, *141*, *148*, 161, *264*, *271*
Eastern Ghats domain 279–284
Eastern Ghats Granulite Belt 281
Eastern Ghats Mobile Belt *118*, *186*, 192, 193
Eastern Ghats Province khondalites 2
  composition 147
  iron enrichment 149–150
  isotopic analysis 156
  metamorphic history 149
  origin 147
  setting 147–148
  summary of study 156–158
  trace element chemistry
    methods of analysis 150
    results 150–153
    results discussed
      HFSE and Y 153
      REE 154–156
      transition metals 153–154
Eastern Indian Suture *264*
Ediacaran fauna **250**, 252, *253*
Erinpura granite *9*, **10**, 272

fission track age dating, Chotanagpur Granite Gneiss Complex, sector 10 **124**, 137
fossils *see* Vindhyan Supergroup

Gandikota Quartzite Formation *163*, *164*, 167, 170, 171, 172, 174, *200*, 201
Gani–Kalva Fault 174, **175**, 180
Ganurgarh Shale *249*
garnet chemistry
  Anasagar Granite Gneiss Complex 226, **230**
  chloritoid–biotite schist 97, **100**, 101
geochronology *see* K–Ar; Rb–Sr; U–Pb
Ghaap Group *54*, 55
Giddalur alkali syenite 185, 186
glaciation 37, 43, *52*, 53
Godavari graben 11, 22, *162*, *264*
Gogunda Group 7, 221
Gondwana supercontinent 161, 180, 284
Govindgarh Sandstone *249*
gravity data 178, 278
Great Boundary Fault *276*
Great Oxidation Event 43, 51, *52*
Grenvillian orogenic event 242
Grypania **250**, *253*, 255
Gulcheru Quartzite *163*, *164*, 165–166, 170, 174–175, *176*, 198, *200*
Gwalior Basin *8*, 9
Gwalior Group **10**, 11

Harrohalli dyke swarm **19**
high field-strength elements (HFSE), khondalites 153
Hindoli Group 7, *8*, **10**, 272
Hindoli volcanics 220

Idamakallu alkali syenite 185, 186
impact ejecta 77
impact event 2
Indravati Basin *6*, 21, 282
Inner Mongolia Suture Zone *271*
iron
  chemistry in khondalites 149–150, 158
  marine deposits 5
  native 1–2
    Chaibasa Formation
      methods of analysis 79
      results
        magnetic susceptibility 79
        microprobe analysis 81–83, **84**, **85**
        SEM 81, 83
        thin sections 79
      results discussed 85–89
Iron Ore Group 14, *16*, **17**, *32*, 33–35, *36*, 65, *92*
isotopes *see* $^{145}Nd/^{146}Nd$; $^{89}Sr/^{88}Sr$

Jagannathpur volcano-sedimentary succession 43, *63*, 65, 67, 68
Jagannathpur–Malangtoli–Ongarbira volcanics 14, *15*
Jasrapura granite *9*, **10**, 220, 239, 273, 274
Jhamrkotra Formation 7, 272
Jhazpaur Group 272
Jhiri Shale *249*, **250**
Jodhpur sandstone *276*
Jublatola Shear Zone 41

K–Ar geochronology
  Chotanagpur Granite Gneiss Complex
    sector 3 **122**, 124
    sector 7 **123**
    sector 9 **123**
Kaapvaal Craton 53–58
  compared with Singhbhum Craton 67–69
Kaimur Group *9*, **10**, 11, 248, *249*
Kaimur Sandstone and Shale *249*
Kajrahat Limestone **10**, *249*
Kaladgi Basin *6*
Kaladgi Group **19**
Kandra ophiolite 281, 282
Kaptipada Granite *32*
Karimnagar granulite 5, 22
Karnataka Craton *see* Dharwar Craton
Karur–Kambam–Painavu–Trissur Shear Zone *18*
Katnia **250**
Kenorland supercontinent 265
Keonjhargarh palaeosol 68
KFMASH system, chloritoid–biotite schist 105–107, 109–110
Khairagarh Group *12*, 13
Khariar Basin 21
Kheinjua Formation 11
khondalites of Eastern Ghats 2
  composition 147
  iron enrichment 149–150
  isotopic analysis 156
  metamorphic history 149
  origin 147
  setting 147–148
  summary of study 156–158
  trace element chemistry
    methods of analysis 150
    results 150–153
    results discussed
      HFSE and Y 153
      REE 154–156
      transition metals 153–154
kimberlite indicator minerals
  introduction 209–210
  methods of analysis 212
  results 213–215
  results discussed 215–216
Kodarma Group 119
Koegas Subgroup 55
Koilkuntala Limestone *164*, *200*, 202, 203
Kolar–Kadiri Schist Belt **19**
Koldaha Shale 11, *249*, **250**
Kolhan Group 14, *15*, 43, *60*, *63*, 65, 66, 67
Kona Fault 174, **175**, 180
Kondagaon granulite 14, 22
Krishna lamproite field *186*
Kuilapal Granite 16, *32*, *92*, 278
Kumbhalgarh Group 7, 221
Kuppalapalle Volcanics 201
Kurat lava 11
Kurnool Group 2, **19**, 21, *163*, *164*, *172*, 180, 198, *200*, 201
Kurnool sub-basin 162, *163*, 174

Lakheri Limestone *249*
lamproite dykes 2, 192, 203–205, 207
  geochemistry **208**
large igneous provinces (LIP) 22, 161, 170, 174, 268, 277, 279, 280

Magaliesberg Formation *56*
magmatic shutdown (lull) 43, 51, 69
magnetic susceptibility 77, 79
magnetotelluric profiles 278, 284
Mahakoshal belt *12*, **13**
Mahakoshal Group 14, **21**, 119
Mahanadi graben 11, *162*
Maidukuru Thrust (Rudravaram Line) 162, 172, 178, 180
Maihar Sandstone *249*, **250**
Majhgawan kimberlite **10**
Malangtoli volcanics *63*, 65, 67
Malani igneous suite 272
Malani Rhyolites 9, **10**, *276*
Malanjkhand granite *12*, 13, 22
Malmani Subgroup *56*
Mangalwar Complex 221, 274
Mangalwar terrane *273*
Marwar Basin 272, *273*
Marwar Supergroup 9, **10**
Mayurbhanj Gabbro *15*, *60*
Mayurbhanj Granite *32*, 40, *92*
megafossils *see* Vindhyan Supergroup
Mewar Gneissic Complex 7
Moho depth 178, *179*
Moyar–Bhavani Shear Zone *18*
Mozaan Group *54*
Munger Orogeny 119
muscovite chemistry
  Anasagar Granite Gneiss Complex 229, **234**, **235**
  chloritoid–biotite schist 97, 101, **102**

Nagari Quartzite *164*, 168, 169, 172–173, 174–175, *177*, *200*, 201
Nallamalai Fold Belt 2, *6*, 17, *18*, 162, 167, 168, 172, 175, 178–180
Nallamalai Group *163*, *164*, 167–169, 172–173, 175, 176, 177, 186, *200*, 201
Nandgaon Group 13
Nandyal Shale *164*, *200*, 202
Narayanpet kimberlite field *186*
Narji Limestone *164*, *200*, 202
Nayadih Formation *36*, 37
$^{145}Nd/^{146}Nd$
  khondalites 156
  Racherla alkali syenite 189, **190**, 191–192
Nellore Schist Belt 17, *18*, 161, 171, 172, 173, 178, 281
Newer Dolerite dyke swarm 37, 39–40
nickel thermometry 209
Nilgiri Granite *15*, **17**, *60*
North China Craton 177, 268, 269, *271*
North Delhi Fold Belt 7, 219, *220*, 221
North Delhi terrane *273*
North Hebei Orogen 269
North India Block *271*, 279
North Singhbhum Fold Belt/Mobile Belt *32*, *92*, 93, 119, *141*
North Singhbhum (Tamar–Porapahar) Shear Zone 14, *15*, **17**, *60*
Northern Shear Zone 119
numerical modelling, Anasagar Granite Gneiss Complex phase diagrams 229, 233, *237*, *238*, *239*, *240*

Older Metamorphic Group *32*, 33, *60*
Older Metatonalite Gneiss *32*, 33, *60*, *92*
Olive Shale *249*, **250**
Ongarbira volcano-sedimentary succession 43, *63*, 65, 67
ophiolites 274–275
Orbisiana **250**
Owk Shale *164*, *200*, 202, 203
  see also Auk Shale

palaeocurrent studies 174–175, *176*
palaeontology *see* Vindhyan Supergroup
Palaeopangaea *268*
palaeostress regimes **175**
Palghat–Cauvery Shear Zone *18*
Palghat–Cauvery Suture Zone *264*
Palnad Nappe 180
Palnad sub-basin 162, *163*
Pandyan Mobile Belt (Southern Granulite Terrane) *6*, 17, *18*, 263, *264*, 284
Pangaea supercontinent 265
Paniam Quartzite *164*, *200*, 202, 203
Panna Shale *249*, **250**
Papaghni Group **19**, 21, *163*, *164*, 165–166, 169, 174, 175, 281
Papghni sub-basin 162, *163*, 174, 175
Penge Formation *56*
Peninsular Gneiss **19**, 162, *163*, *164*, 170, 172
petrogenetic grid, chloritoid–biotite schist 105–107, 109–110
phosphorite 221, 272
Phulad ophiolite 221, *276*
Phulad Shear Zone 8, *9*
plagioclase chemistry, Anasagar Granite Gneiss Complex 229, **232**
Planolites **250**
Pongola Supergroup 55
Porcellanite Formation *249*
Postmasburg Group 55
Pranhita–Godavari basin 282
Pranhita–Godavari Graben/Rift Valley *6*, 17, 20
Pretoria Group *54*, 55, *56*, 57, 68
Pulivendla Quartzite *163*, *164*, 166, 169, 170, *200*, 201
Pulivendla sill 281
Pullampet Shale *164*, 168, 173, *200*, 201

Racherla alkali syenite 2
  apatite chemical study
    methods 187, 189
    results **188**, 189, **190**, 191–192
    summary 192–193
  geochemistry 187
  petrology 187
  setting 186–187
Raichur kimberlite field *186*
Rajgarh Group 221
Rajgir Group 119
Rajkharsawan conglomerate *38*
Rajmahal Basalt *118*, 119
Ramadugu lamproite field *186*
Ramgiri Schist Belt **19**

Rampur Shale **10**, *249*, **250**
Ranakpur Diorite 8, *9*, **10**
rare earth elements (REE)
khondalites 154–156, 158
Racherla alkali syenite 187, 189, 191–192
Rayanhalla Group 7
Rayner Complex (Antarctica) 117, *141*, 142
Rb–Sr geochronology
Chotanagpur Granite Gneiss Complex
sector 1 121
sector 2 **122**, 124
sector 5 **122**
sector 7 **123**, 134
sector 9 **123**, 136
sector 10 **123** 137
Rewa Group *9*, **10**, 11, 248, *249*, *256*
Riaolo Group 272
Rodinia supercontinent 141, 242, 265, 267, 269, 284
Rohinibera Shear Zone 41
Rohtas Formation *254*
Rohtas Limestone **10**, 11
Rohtasgarh Limestone *249*, **250**
Rooihoogte Formation 57
Rudravaram Line *see* Maidukuru Thrust

Sakoli belt *12*, 22
Sakoli Group *12*, **13**, 278
Salkhan Limestone *249*
Samaria Shale *254*
Sandmata Complex 7, *8*, **10**, 22, 221, 242
Sandmata granulite 274, *276*
Sandmata terrane *273*
Sandur Schist belt **19**
Sargur Group 17, **19**
Satpura mobile belt 11, *12*, *15*, *60*, 279
Satpura Orogeny 119
Sausar belt *12*, **13**, 22
Sausar Group *12*, **13**, 278, 279
Scarp Sandstone and Conglomerate *249*
Segwagwa Group 55, 57
Semri Group *9*, **10**, 11, 21, 248, *249*, *254*
Sendra Granite 8, *9*, **10**
Sendra Group 221
Shillong Group 20
Shillong plateau 20
Shillong–Meghalaya Gneiss Complex *141*
silicon, native 1–2, 81–83
Simlipal volcanics 14, *15*, 43, *60*, *63*, 65, 67
Singhbhum Basin 277
Singhbhum Craton 1, 5, *6*, 14–17, **21**, 22, 58–59, *92*, *118*, *141*, *264*, 276
comparisons
Kaapvaal 67–69
north 59–62
north *v.* south 66–67
south 62–66
Singhbhum Granite 14, *15*, *16*, **17**, *32*, 33, 34, *36*, 59, *60*, *92*
Singhbhum Group 16, **17**, 21, *92*
Singhbhum Mobile Belt 40
Singhbhum Shear Zone *39*, 40–41, *60*, 67, *92*
chloritoid–biotite schist 92–93
fabric 93
mineral assemblages 93–97
mineral chemistry 97–98, **99**, **100**, **101**
mineral reactions 101
mineral stability fields 1–3–107, 109–110, 110–111
P–T path 111–113
thermobarometry 98, 101, **104**
Singhii **250**
Sirbu Shale **10**, *249*, **250**, *254*, *256*
Sirohi Group *276*
Sirohi terrane *273*
Skolithos **250**
snowball earth hypothesis 53
Soda Granite **17**, *32*, 40, 41, *92*
Somanpalli Group 20
Son–Narmada Lineament Zone *12*, *118*
South Delhi Fold Belt 8, 219, *220*, 221
rock associations 222
see also Anasagar Granite Gneiss Complex
South Delhi Orogeny 241, 242
South Delhi terrane *273*
South India Block *271*, 279
South Purulia Shear Zone 119
Southern Granulite Terrane (Pandyan Mobile Belt) *6*, 17, *18*, 263, *264*, 284
Soutpansberg Group *54*
$^{89}Sr/^{88}Sr$
khondalites 156
Racherla alkali syenite 187, **189**, 191
Srisailam Formation *164*
Srisailam sub-basin 162, *163*
staurolite chemistry, Anasagar Granite Gneiss Complex 229, **234**, **235**
steatite 185
stromatolites 34
subduction zones 265
Suket Shale *249*, **250**
Sukinda Thrust 14
Sukma Basin *6*, 21
supercontinent cycle 140–142, 265
supercontinents
Columbia 140, 177, 265, *266*, 267, 268, 269, *270*, 274, 281, 282
Gondwana 161, 180, 284
Kenorland 265
Pangaea 265
Rodinia 141, 242, 265, 267, 269, 284
superplume events 51, 53
Susnai Breccia *249*
suture zones 264, 274

Tadpatri Formation *163*, *164*, 166–167, 170, 171, 172, 173, 174, 178, *200*, 201, 281
talc 185
Tamakhan Granite 22
Tamar–Porapahar (North Singhbhum) Shear Zone 14, *15*, **17**, *60*
Tamar–Porapahar–Khatra Fault Zone 117
Tamparkola Granite 35, 36
Tan Shear Zone *12*
Tapti Fault *12*
Taupone Group 55
Tawuia **250**, 251, 252, *254*, *256*
tempestites 21
thermobarometry
Anasagar Granite Gneiss Complex
methods 229

thermobarometry (*Continued*)
results 233
chloritoid–biotite schist 98, 101
Chotanagpur Granite Gneiss Complex
sector 1 121
sector 5 126, 128, **131**
sector 6 131, 132
sector 7 133–134
kimberlites 209
Tirodi Gneiss 13
trace element chemistry
khondalites
methods of analysis 150
results 150–153
results discussed
HFSE and Y 153
REE 154–156
transition metals 153–154
Racherla alkali syenite 187, **188**, 189, 191–192
Trachyhistrichosphaera 258
Trans-North China Orogen 270, *271*
transition metal chemistry, khondalites 153–154
Transvaal Supergroup 55, *56*, 68
triple junctions 265

U–Pb geochronology
Anasagar Granite Gneiss Complex 219–220
Banganapalle
introduction 207–208
method 209
results *210*, **211**
Central India Tectonic Zone 278
Chotanagpur Granite Gneiss Complex
sector 1 121, **122**
sector 2 **122**
sector 5 128
sector 6 **122**
sector 7 **122**
sector 10 137
Udaigiri domain 173, 174

Vellaturu Granite 180
Vempalle Formation *163*, *164*, 166, 169, 170, 173, 174, 178, *200*, 281
Ventersdorp Supergroup *54*, 55
Vindhyan Basin 11, 21, *118*, *220*, 272
Vindhyan Supergroup 2, *118*
age 248–249
megafossils **250**, 251
macroevolution 258
morphodiversity 251–252
morphology
Form A 252, 255
Form B 255
Form C 255
Form D 255
Form E 255
Form F 257
Form G 257
Form H 257
Form I 257
Form J 257–258
Form K 258
summary 258–259
setting 248
stratigraphy *249*
Vinjamuru domain 173, 174
Vinukonda granite 180
Vryburg Formation 55

Wajrakarur kimberlite field *186*
Waterberg Group *54*
West Rand Group *54*
Western Indian Suture *264*
Wilson cycles 22, 265, 274
Witwatersrand Supergroup 53
Witwatersrand–Pongola Basin 53

Y-shaped domains 265
yttrium (Y), khondalites 153